GEOMETRIC TOPOLOGY

Academic Press Rapid Manuscript Reproduction

Proceedings of the 1977 Georgia Topology Conference Held in Athens, Georgia, August 1–12, 1977.

GEOMETRIC TOPOLOGY

Edited by

JAMES C. CANTRELL
Department of Mathematics
The University of Georgia
Athens, Georgia

ACADEMIC PRESS New York San Francisco London 1979
A Subsidiary of Harcourt Brace Jovanovich, Publishers

ACADEMIC PRESS, INC.
111 Fifth Avenue, New York, New York 10003

United Kingdom Edition published by
ACADEMIC PRESS, INC. (LONDON) LTD.
24/28 Oval Road, London NW1 7DX

Library of Congress Cataloging in Publication Data

Georgia Topology Conference, University of Georgia, 1977.
Geometric topology.

1. Topology—Congresses. 2. Manifolds (Mathematics)–Congresses. I. Cantrell, James Cecil, 1931- II. Title.
QA611.A1G46 Date 514'.2 78-31631
ISBN 0-12-158860-2

PRINTED IN THE UNITED STATES OF AMERICA
79 80 81 82 9 8 7 6 5 4 3 2 1

DEDICATION

This Proceedings of the 1977 Georgia Topology Conference is dedicated to three persons who have been instrumental in the success of the Georgia Conferences of 1961, 1969, and 1977. They are Professors R. H. Bing, O. G. Harrold, and Deane Montgomery. This dedication is in recognition of their contributions to the development of scores of todays topologists.

Professor Bing has been a pioneer and leading figure in many of the most important developments in topology of the past thirty years. He has been especially encouraging and influential with young topologists. Professor Harrold made many contributions to basic taming and embedding problems in manifold theory. He has also been instrumental in the progress of mathematics in the south over the past three decades. Professor Montgomery is one of the outstanding contributors to the theory of transformation groups. As Professor of the School of Mathematics, the Institute for Advanced Study, he has most effectively provided wise counsel and leadership to the mathematical community. He has assisted the professional development of more of the world's mathematicians than any other person in the history of the science. The dedication of these proceedings is but a small expression of gratitude for the many contributions of these three men.

CONTENTS

PART II TOPOLOGY OF MANIFOLDS

PART III SHAPE THEORY AND INFINITE DIMENSIONAL TOPOLOGY

PART IV MISCELLANEOUS PROBLEMS

CONTRIBUTORS

Numbers in parentheses indicate the pages on which the authors' contributions begin.

R. H. Bing (3), University of Texas, Austin, Texas 78712
Joan S. Birman (23), Columbia University, New York, New York 10027
J. L. Bryant (261), Florida State University, Tallahassee, Florida 32306
J. W. Cannon (261), University of Wisconsin, Madison, Wisconsin 53706
Sylvanin Cappell (301), New York University, New York, New York 10012
T. A. Chapman (581), University of Kentucky, Lexington, Kentucky 40506
Robert Connelly (675), Cornell University, Ithaca, New York 14853
Jerome Dancis (305), University of Maryland, College Park, Maryland 20742
Karl H. Dovermann (323), Rutgers University, New Brunswick, New Jersey 08903
Allan L. Edmonds (337), Cornell University, Ithaca, New York 14853
David A. Edwards, (597), University of Georgia, Athens, Georgia 30602
Dan Everett (53), University of Georgia, Athens, Georgia 30602
D. Galewski (345), University of Georgia, Athens, Georgia 30602
Gene G. Garza (73), University of Georgia, Athens, Georgia 30602
Ross Geoghegan (603), State University of New York, Binghamton, New York 13901
O. G. Harrold (79, 87), Florida State University, Tallahassee, Florida 32306
Harold M. Hastings (597), Hastings, Hofstra University, Hempstead, New York 11550
Richard E. Heisey (609), Vanderbilt University, Nashville, Tennessee 37235
John G. Hollingsworth (685), University of Georgia, Athens, Georgia 30602
W.-C. Hsiang (351), Princeton University, Princeton, New Jersey 08540
William Jaco (91), Rice University, Houston, Texas 77001
Klaus Johannson (101), University of Bielefeld, West Germany
Lowell Jones (367), State University of New York, Stony Brook, New York 11790

R. C. Kirby (113), University of California, Berkeley, California 94720
R. C. Lacher (261), Florida State University, Tallahassee, Florida 32306
Vo T. Liem (621), Louisiana State University, Baton Rouge, Louisiana 70813
Richard Mandelbaum (147), University of Rochester, Rochester, New York 14627
Yukio Matsumoto (393), Institute for Advanced Study, Princeton, New Jersey 08540
Dusa McDuff (429), University of Warwick, Coventry, England
José M. Montesinos (219), Institute for Advanced Study, Princeton, New Jersey 08540
Jerome Powell (23), Columbia University, New York, New York 10027
Richard Randell (445), Institute for Advanced Study, Princeton, New Jersey 08540
Jean E. Roberts (693), Oakland University, Rochester, Michigan 48063
T. B. Rushing (631, 649), University of Utah, Salt Lake City, Utah 84112
Martin Scharlemann (113, 475), University of California, Santa Barbara, California 93106
Peter B. Shalen (91), Rice University, Houston, Texas 77001
Julius L. Shaneson (301), Rutgers University, New Brunswick, New Jersey 08903
L. Siebenmann (503), Universite Paris-Sud, Orsay, France (91405)
Michael Starbird (3, 239), University of Texas, Austin, Texas 78712
Ronald Stern (345), University of Utah, Salt Lake City, Utah 84112
Neal Stoltzfus (527), Louisiana State University, Baton Rouge, Louisiana 70803
Dennis Sullivan (503, 543), Institut des Hautes Etudes Scientifiques, Bures/Yvette, France
Gerard Venema (649), Institute for Advanced Study, Princeton, New Jersey 08540
J. B. Wagoner (557), University of California, Berkeley, California 94720
J. E. West (655), Cornell University, Ithaca, New York 14853
Edythe P. Woodruff (239, 253), Trenton State College, Trenton, New Jersey 08625
R.Y.-T. Wong (655), University of California, Santa Barbara, California 93106

PREFACE

The Georgia Topology Conference of 1977 followed the tradition of the earlier conferences of 1961 and 1969. The aim was to bring together in as relaxed and informal way as possible leading contributors to the field of geometric topology. Special attention was given to having invited addresses given by those at the forefront of the area and whose work are likely to influence the directions of geometric topology for the next few years.

The organizing committee consisted of R. H. Bing, J. C. Cantrell, J. G. Hollingsworth, R. C. Kirby, and Deane Montgomery. Financial support for the conference was obtained from the National Science Foundation, the Franklin College of Arts and Sciences, University of Georgia, and the office of the Vice President for Research, University of Georgia.

The conference was held at the University of Georgia from August 1–12, 1977 with approximately 125 participants. Invited addresses were given by Steve Armentrout, J. L. Bryant, J. W. Cannon, T. A. Chapman, D. A. Edwards, D. E. Galewski, W.-C. Hsiang, Klaus Johannson, R. C. Kirby, Richard Mandelbaum, John Mather, J. W. Milnor, R. C. Randell, Dennis Sullivan, W. P. Thurston, and J. E. West.

The daily schedule consisted of two hour-long lectures by the invited speakers and three or more seminar talks by other participants. A problem session, devoted to discussion of a few of the most important open problems in Geometric Topology, was held on Wednesday evening of the last week.

This volume, the proceedings of the 1977 Georgia Topology Conference, contains accounts of most of the invited lectures and talks. Mrs. Dianne Byrd was the typist for the proceedings. For her skill and patient assistance the editor is extremely grateful.

LOW DIMENSIONAL MANIFOLDS

A DECOMPOSITION OF S^3 WITH A NULL SEQUENCE OF CELLULAR ARCS

R. H. Bing[1]
Michael Starbird[1]

Department of Mathematics
The University of Texas
Austin, Texas

An example is described showing the following:

THEOREM. There exists a null sequence of disjoint cellular arcs $\{A_i\}$ in S^3 such that the decomposition space $S^3/\{A_i\}$ is topologically different from S^3.

David Gillman and Joe Martin announced a sequence of arcs implying the above theorem many years ago but their example was never published. The example given here is related to that of Gillman and Martin, but differs in details.

The example is modeled after Bing's example of a null sequence of cellular continua $\{G_i\}$ in S^3 where $S^3/\{G_i\}$ is not a manifold. Each of the G_i's was a planar indecomposable continuum which is the intersection of a decreasing sequence of solid tori. Each of the A_i's in the present example is the intersection of a decreasing sequence of handle bodies each of which is a long tube with a handle on each end. Instead of having big oscillations, the tubes have small ones which hook like knitting.

[1]*Research supported in part by NSF Grant MCS76-07242A01.*

ISBN 0-12-158860-2

This example is of current interest in view of recent results discussed elsewhere in these Proceedings, showing that under many decompositions G of S^3, S^3/G is topologically S^3. The present example establishes a bound on what can be proved.

I. INTRODUCTION

We give an example called the Main Example, which proves the following theorem.

MAIN THEOREM. There exists a null sequence of disjoint cellular arcs $\{A_i\}$ in S^3 such that the decomposition space $S^3/\{A_i\}$ is topologically different from S^3.

A sequence of sets is *null* if for each $\varepsilon > 0$, at most a finite number of them have diameters greater than ε. A set X in S^3 is *cellular* if $X = \bigcap_{i \in \omega_0} B_i$ where for each i, B_i is a 3-cell in S^3 and $B_{i+1} \subset \text{Int } B_i$. We use $S^3/\{A_i\}$ to denote the decomposition space whose only non-degenerate elements are the A_i's.

D. Gillman and J. Martin announced the existence of a sequence of arcs implying the Main Theorem many years ago [8], but their example, which made use of the eyebolt construction employed in [5], was never published. The example given here, which was constructed by Bing with Starbird's aid, is similar to the Gillman-Martin example but differs in details. The decomposition described here has a defining sequence each stage of which is a disjoint collection of solid double tori.

This example is of current interest because of the flurry of recent results [7,9,10,11,12] showing that S^3/G is topologically equivalent to S^3 under various sets of restrictions on G. In particular, in these Proceedings there appears a result [7,10] stating that $S^3/\{A_i\}_{i \in \omega}$ is topologically equivalent to S^3 if each

A_i is a tamely embedded cellular polyhedron. The example in this paper gives a bound on what more can be proved.

The dogbone space [2] gives a decomposition of S^3 into points and tame arcs such that the decomposition space differs from S^3. However, the decomposition in it has a Cantor set of tame arcs rather than a countable number. Bing showed [3, Theorem 3] that if G is an upper semicontinuous decomposition of S^3 with only a countable number of nondegenerate elements $\{G_i\}$ and each G_i is a tame arc, then S^3/G is topologically S^3.

In [6, §2-5], Bing constructed an example of null sequence of disjoint, cellular continua $\{G_i\}$ so that $S^3/\{G_i\}$ is topologically different from S^3. In that example each G_i is an indecomposable continuum. The example here is modeled after that example, but modifications are made to ensure that each non-degenerate element of the present decomposition is a cellular arc rather than an indecomposable continuum.

One feature which the present example shares with Bing's previous example in [6] is the method by which the decomposition space is shown to differ from S^3. That method employs the concept of the two disk property which is defined in the next section. The third section contains theorems which can be used in describing decompositions G of S^3 so that S^3/G is topologically different from S^3. These theorems are used in Section 4 to construct the Main Example. They can also be used in proving the following more general theorem which will appear in a future paper.

THEOREM. Let $\{X_i\}_{i\in\omega_0}$ *be a countable collection of nondegenerate continua each of which admits a cellular embedding in* S^3. *Then there exist embeddings* $h_i : X_i \to S^3$ *so that for each* i, $h_i(X_i)$ *is a cellular embedding,* $\{h_i(X_i)\}$ *is a null sequence of disjoint sets, and* $S^3/\{h_i(X_i)\}$ *is topologically different from* S^3.

Throughout this paper we use PL sets if that is convenient. For example, if we define a cellular set as the intersection of 3-cells, it is assumed that the 3-cells are PL. If D is a meridional disk of a solid torus T, it is assumed that D and T are PL. If D intersects a handlebody H it is understood that H is PL and D, H are in general positon. Of course, the cellular arcs which we describe in the Main Example as an intersection of handlebodies are not PL.

II. THE TWO DISK PROPERTY

To show that a cellular upper semicontinuous decomposition G of S^3 is not shrinkable, it is frequently convenient to construct a defining sequence for G each stage of which is a collection with the two disk property defined below.

Definition. A finite collection M of disjoint closed subsets in the interior of a solid torus T has the *two disk property* if and only if for every pair of disjoint meridional disks D_0, D_1 of T, there is an element of M which intersects both D_0 and D_1. A disk D in T is *meridional* if $D \cap Bd\ T = Bd\ D$ and $Bd\ D$ does not bound a disk in $Bd\ T$.

Example 2.1. The collection M which contains only one element, a core of T, has the two disk property.

Example 2.2. The collection M consisting of the two eyeglasses J_1 and J_2 illustrated in Figure 2.1 has the two disk property. (See Lemma 3.3 for a proof of this fact.)

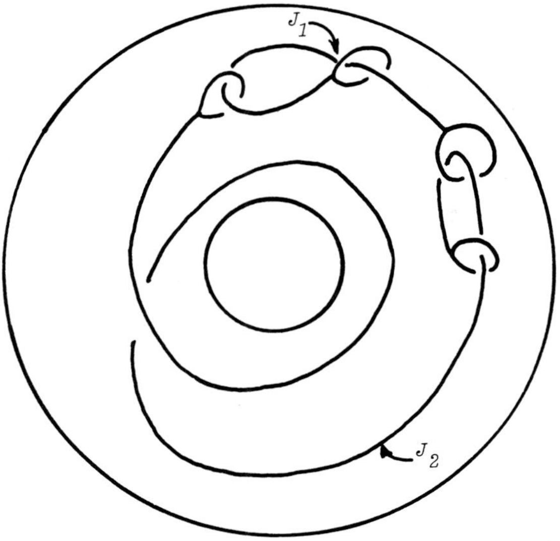

Fig. 2.1.

Definition. A decomposition G of S^3 *is defined by a sequence* $\{M_i\}_{i\in\omega}$ if and only if for each i, M_i is a finite collection of disjoint 3-manifolds with boundary in S^3 so that each element of M_{i+1} is contained in the interior of an element of M_i, and each element of G is either a component of $\bigcap_{i\in\omega_0} (\bigcup M_i)$ or a point of $S^3 - \bigcap_{i\in\omega_0} (\bigcup M_i)$.

THEOREM 2.1. *Let T be a solid torus and G be a cellular decomposition of S^3 defined by the sequence $\{M_i\}_{i\in\omega}$ where each M_i has the two disk property in T. Then S^3/G is not homeomorphic to S^3.*

The properties mentioned in Theorem 2.1 are a favorite way of showing that a decomposition space of S^3 is topologically different from S^3. See for example [2, Lemma for Theorem 12], [6, Section 5] and [1, Theorem 3]. We do not repeat the proof of Theorem 2.1.

III. PRESERVING THE TWO DISK PROPERTY

One method for constructing a defining sequence for an upper semicontinuous decomposition G is to specify how an element at stage i contains the element or elements of stage $i + 1$. In constructing examples of decompositions G for which S^3/G is not homeomorphic to S^3, it is useful to construct a defining sequence for G each stage of which has the two disk property. This section contains theorems which say that if a given collection M has the two disk property then certain modifications of M will yield a new collection which also has the two disk property.

We make an easy start.

LEMMA 3.1. Let T be a solid torus and L be an element of M, a collection with the two disk property. If L^+ is a closed set in T containing L and disjoint from the other sets in M, then the collection M' obtained from M by replacing L by L^+ also has the two disk property.

LEMMA 3.2. Let W be a solid torus and X be a subset of W with n components each of which is a pair of eyeglasses, i.e., two simple closed curves joined by an arc, and so that the chain of eyeglasses is embedded in W as shown in Figure 3.1. Then each meridional disk of W intersects X.

Proof. Let $\{A_i\}_{i=1}^{2n}$ be the planar disks bounded by the simple closed curves in X so that for each i, $X \cap \text{Int } A_i$ is a single point. (See Figure 3.1.) Suppose the lemma is not true. Then there exists a PL, meridional disk D of W so that $D \cap X = \emptyset$, D is in general position with respect to $E = \bigcup_{i=1}^{2n} A_i$ and each component of $D \cap E$ is essential in $E - X$. There is a subdisk

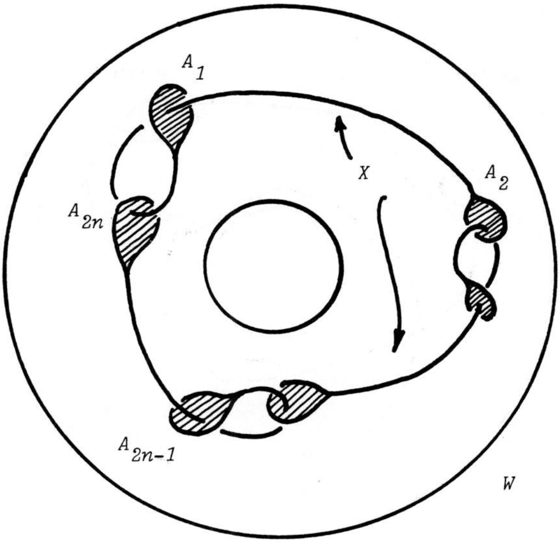

Fig. 3.1.

D' of D such that Int $D' \cap (X \cup E) = \emptyset$ and $Bd\ D' \subset E$. However, this supposed disk D' cannot exist since $Bd\ D' \subset$ Int A_j for some j and is essential in Int $A_j - X$. But each essential curve in Int $A_j - X$ links a simple closed curve in $(E -$ Int $A_j) \cup X$. Therefore D' cannot exist and the lemma is proved.

Figure 3.2 represents a solid torus W in which two disjoint sets Y and Z are embedded. The set Z is one pair of small eyeglasses while Y is the union of several eyeglasses. The chain of eyeglasses $Y \cup Z$ goes around W twice.

LEMMA 3.3. Let M be a collection of subsets of a torus T so that M has the two disk property. Suppose the solid torus W is an element of M. Then the collection M', obtained from M by replacing W by the two sets Y and Z as described above, still has the two disk property.

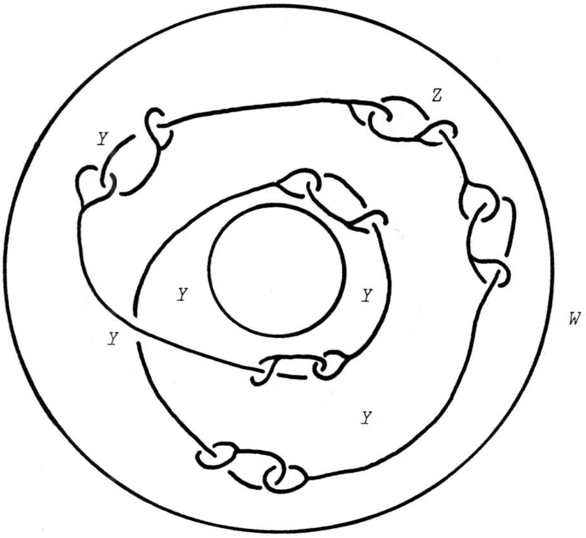

Fig. 3.2.

Proof. Suppose that M' does not have the two disk property. Then there are disjoint, PL, meridional disks D_0, D_1 of T which witness the falsity, are in general position with respect to W, and have no component of intersection with $Bd\ W$ which is trivial on $Bd\ W$. Since M has the two disk property there are subdisks D_0', D_1' of D_0, D_1, respectively, so that $(D_0' \cup D_1') \subset W$, and $(D_0' \cup D_1') \cap Bd\ W = Bd\ D_0' \cup Bd\ D_1'$ -- that is, D_0' and D_1' are meridional disks of W. Consider the double cover T of W. Then one lift $\tilde{Z}$ of Z together with a lift $\tilde{Y}$ of Y encircles T once as in Lemma 3.2. By Lemma 3.2 any meridional disk of τ must intersect $\tilde{Y} \cup \tilde{Z}$. The two lifts of D_0' separate Int τ into two 3-cells B_1, B_2. If $D_0 \cap Z = \emptyset$, then $D_0 \cap Y \neq \emptyset$ and $\tilde{Z}$ is in one of B_1 or B_2. Then the lift of D_1' which intersects the other 3-cell must intersect $\tilde{Y}$ since it must intersect $\tilde{Y} \cup \tilde{Z}$ and it does not intersect $\tilde{Z}$. Therefore D_1 intersects Y and the lemma is proved.

Next we show how to replace a cube with many handles by a torus. For this purpose we need the following lemma whose proof follows from [4, Steps 6 and 7 of proof of Lemma 7]. We include a proof for completeness.

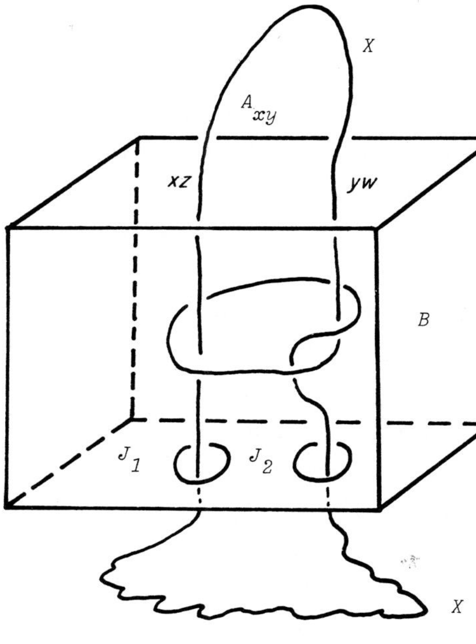

Fig. 3.3.

LEMMA 3.4. *Let X be a compact set in S^3, B be a tame 3-cell in S^3 so that $X \cap B$ is the union of two arcs embedded in B as illustrated in Figure 3.3, X contains an arc A_{xy} joining x to y in $S^3 -$ Int B, and neither simple closed curve J_1 nor J_2 is homotopic to a point in $S^3 - (X \cup$ Int $B)$. Let D be a disk in $S^3 - X$ so that Bd $D \cap B = \emptyset$, D is in general position with respect to Bd B, and each component of $D \cap$ Bd B is non-trivial in Bd $B - X$. Then there is a homeomorphism $h : S^3 \to S^3$ fixed outside a small regular neighborhood of B so that $h(D) \cap (X \cup B) = \emptyset$.*

Proof. Consider a component K of $D \cap Bd\ B$ which is innermost on D. Let D_K be the subdisk of D bounded by K. Then $D_K \cap (\text{Int } B)=\emptyset$ since each curve in $Bd\ B - X$ which is trivial in $B - X$ is also trivial on $Bd\ B - X$. This fact follows from [4, Lemma 6] in conjunction with the Loop Theorem. Therefore D_K must lie to the outside of Int B. However, K cannot separate x from y on $Bd\ B$ because then K would link a simple closed curve in $A_{xy} \cup \text{Int } B$. The hypothesis on J_1 and J_2 implies that K must separate $Bd\ B$ into two disks, one of which contains x and y and the other of which contains the other two points of $Bd\ B \cap X$.

Next we claim that every component of $Bd\ B \cap D$ is parallel to K in $Bd\ B - X$ in that it separates x and y from the other points of $Bd\ D \cap X$ and cobounds an annulus with K on $Bd\ B$ which misses $Bd\ B \cap X$. Suppose not. Then consider a curve L of $Bd\ B \cap D$ which is innermost on D among those curves not parallel to K. Then L is one boundary component of a disk with holes F in D so that each of the other boundary components of F is parallel to K. Each other boundary component of F bounds a disk in $S^3 - (X \cup \text{Int } B)$ since it cobounds an annulus on $Bd\ B - X$ with K and K bounds D_K in $S^3 - (X \cup \text{Int } B)$. Let G be a singular disk in $[S^3 - (X \cup \text{Int } B)] \cup F$ bounded by L. If F is contained in $S^3 - (X \cup \text{Int } B)$, then L bounds G in $S^3 - (X \cup \text{Int } B)$. This is contrary to hypothesis since L either separates x from y on $Bd\ B$ (and hence links the union of A_{xy} and an arc in B) or else is parallel to one of J_1, J_2. Hence, if there is an F, it is contained in $B - X$. Then L can be pushed into Int B. Use L and G to denote the pushed in L and the correspondingly adjusted G. The points x and y can be joined by an arc C_{xy} in $Bd\ B - G$ and the bottom points of $X \cap Bd\ B$ can be joined by an arc C_{zw} in $Bd\ B - G$. But L links the simple closed curve $J = (X \cap B) \cup C_{xy} \cup C_{zw}$. On the other hand L bounds G which is contained in $S^3 - J$. This contradiction allows us to conclude that all components of $Bd\ B \cap D$ are parallel to K.

There is a homeomorphism $g_1 : S^3 \to S^3$ so that g_1 restricted to the arc xz is pointwise fixed, $g_1(Bd\ B) = Bd\ B$, and $g_1(D) \cap g_1(Bd\ B)$ is contained in the lateral sides of B. There is then a further homeomorphism $g_2 : S^3 \to S^3$ fixed outside D a small neighborhood of B obtained by blowing away from the arc xz The homeomorphism $g = g_1^{-1} \circ g_2 \circ g_1$ establishes the lemma.

LEMMA 3.5. Let T be a solid torus and M be a collection of subsets of T with the two disk property one element of which is a cube with n handles H. Then the collection M', obtained from M by replacing H by a simple closed curve J which is embedded in H as illustrated in Figure 3.4, also has the two disk property.

Proof. Let D_0 and D_1 be disjoint meridional disks of T in general with respect to $\bigcup_{i=1}^{n-1} B_i$ where the B_i's are 3-cells shown in Figure 3.4. Suppose $D_0 \cap J = \emptyset$. Consider $D_0 \cap (\bigcup_{i=1}^{n-1} Bd\ B_i)$. First find new disjoint disks D_0 and D_1 so that the new D_i $(i = 0,1)$ does not intersect any element of M' which the original $D_i (i = 0,1)$ did not and so that each component of $D_0 \cap (\bigcup_{i=1}^{n-1} Bd\ B_i)$ is non-trivial in $\bigcup_{i=1}^{n-1} Bd\ B_i - X$. By Lemma 3.4, D_0 and D_1 can be moved so that the new D_0 misses $\bigcup_{i=1}^{n-1} B_i$. All of this is done without creating new intersections with the other elements of M'. Hence the new D_0 misses a core of H and can be pushed off H. Since M has the two disk property, D_0 and D_1 must intersect some common element of M other than H. Since that element belongs to M', M' has the two disk property.

The next two lemmas show a method by which a cube with many handles can be replaced by a pair of eyeglasses.

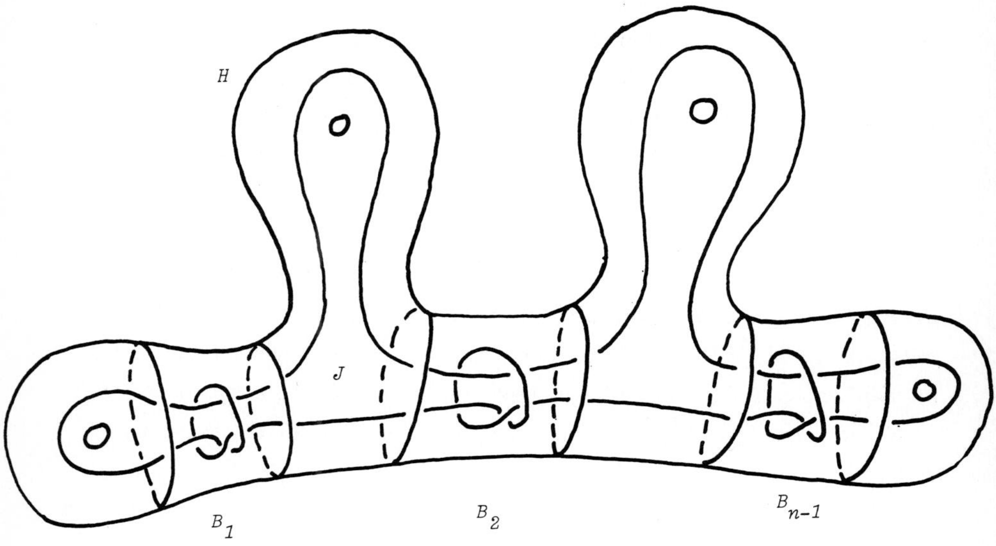

Fig. 3.4.

LEMMA 3.6. Consider the simple closed curve J and eyeglasses L illustrated in Figure 3.5. Then J is not homotopically trivial in $S^3 - L$.

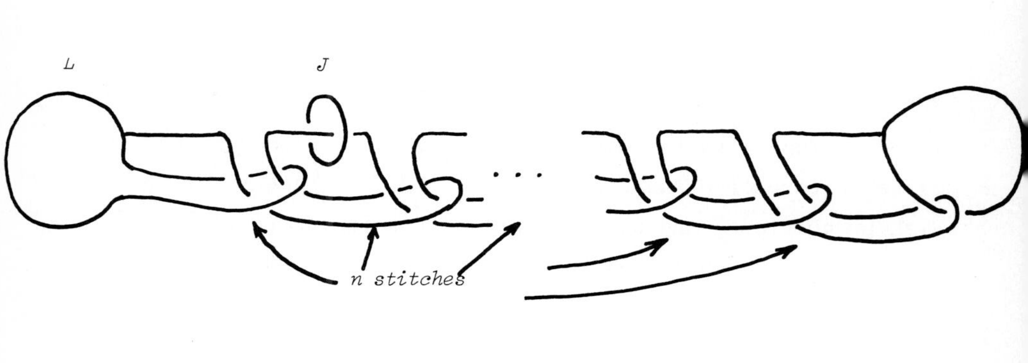

Fig. 3.5.

Proof. The proof is by induction on n, the number of stitches. Note that if J were homotopic to a point in $S^3 - L$, then J would bound a real disk D in $S^3 - L$ by Dehn's Lemma. Consider the case $n = 1$. (See Figure 3.6.)

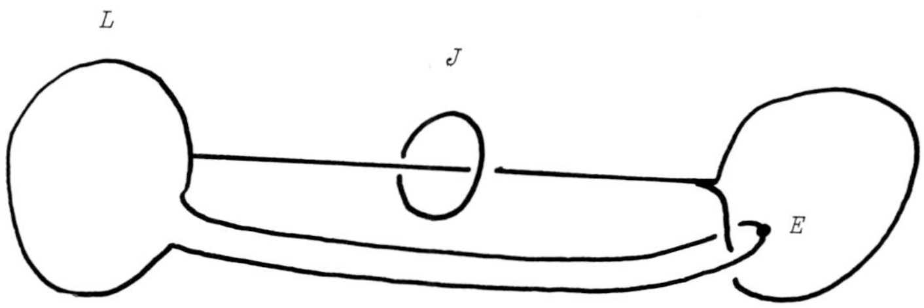

Fig. 3.6.

If J bounds a real disk in D in $S^3 - L$, consider all the general position intersections with the disk E illustrated in Figure 3.6. Trivial curves of $(E \cap D)$ in $E - L$ are removed. Any intersection innermost on D when pushed to one side of E then bounds a disk missing $L \cup E$. This is impossible since any such curve links a simple closed curve in $L \cup E$.

Suppose we have proved the lemma for k stiches and we are given the $k + 1$ case. Consider intersections with disk E as before. As before, an innermost curve of $D \cap E$ when pushed off E cannot be pushed down (see Figure 3.7) because then it would

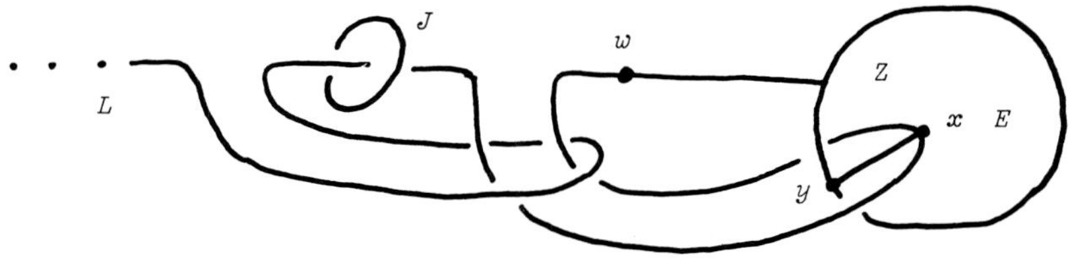

Fig. 3.7.

link $yxwzy$. On the other hand if it is pushed up, the right hand end of the eyeglass could be replaced by the new curve $yxwzy$ and induction could be applied to finish the proof.

LEMMA 3.7. Let H be a standardly embedded cube with n handles which is one element of a collection M with the two disk property. Then the collection M', obtained by replacing the element H of M by the eyeglasses G embedded in H as illustrated in Figure 3.8, still has the two disk property.

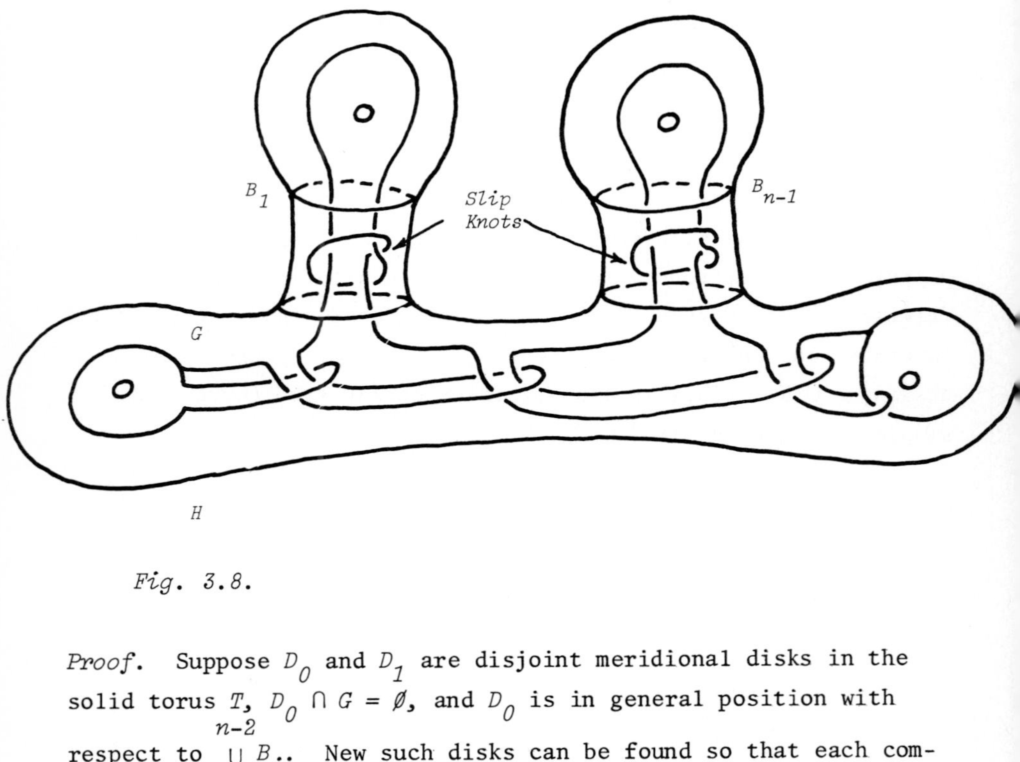

Fig. 3.8.

Proof. Suppose D_0 and D_1 are disjoint meridional disks in the solid torus T, $D_0 \cap G = \emptyset$, and D_0 is in general position with respect to $\bigcup_{i=1}^{n-2} B_i$. New such disks can be found so that each component of $D_0 \cap (\bigcup_{i=1}^{n-2} Bd\ B_i)$ is non-trivial in $\bigcup_{i=1}^{n-2} Bd\ B_i - G$ and the new D_i $(i=0,1)$ do not interest any element of M' which the old ones did not. Note that since H is standardly embedded and those slip knots can be seen to be unknotted in S^3, Lemma 3.6 applies and can be used to show that the hypotheses to Lemma 3.4 are satisfied. Use Lemma 3.4 to find new meridional disks D_0 and D_1 so that $D_0 \cap (G \cup (\bigcup_{i=1}^{n-2} B_i)) = \emptyset$ and each new D_i $(i=0,1)$ does not intersect any element of M' which the old D_i did not. Now D_0 misses a core of H. Hence, since M has the two disk property, D_0 and D_1 must intersect a common element of M other than H. That element is also in M', so M' has the two disk property.

IV. THE MAIN EXAMPLE

There is a null sequence of disjoint cellular arcs $\{A_i\}$ in S^3 such that the decomposition space $S^3/\{A_i\}$ is not homeomorphic to S^3. In fact the decomposition G is describable by a sequence $\{M_i\}_{i\in\omega}$ where each M_i $(i \geq 1)$ consists of 2^i standardly embedded solid double tori. Of course these double tori are tangled up with each other.

The defining sequence $\{M_i\}_{i\in\omega}$ for G is produced inductively. Let M_0 contain a single element, a solid torus T. Let collection M_1 consist of two solid double tori embedded in T as illustrated in Figure 2.1. For each $i \geq 2$, M_i is obtained from M_{i-1} by replacing each solid double torus in M_{i-1} by two solid double tori in such a manner that the following conditions are satisfied.

1. M_i consists of 2^i standardly embedded solid double tori.

2. Each element of M_i is contained in the interior of a 3-cell contained in an element of M_{i-1}.

3. For each element of M_{i-1}, one of the two solid double tori of M_i which it contains has diameter less than $1/i$.

4. Let C be a solid double torus in M_i. Then C is divided into a chain consisting of two solid tori at the ends and 3-cells in between. (See Figure 4.1) Call those ordered links $\{c_i\}_{i=1}^{n_c}$. Then each c_i has diameter less than $1/i$.

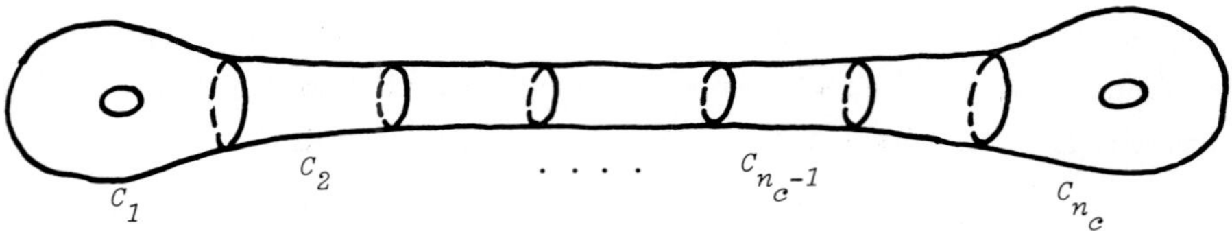

Fig. 4.1.

5. Suppose E is a double torus belonging to M_i which is contained in a double torus C belonging to M_{i-1}. Let $\{e_i\}_{i=1}^k$ and $\{c_j\}_{j=1}^n$ be the links of E and C respectively. Then E runs straight through C in the following sense. If for some i and j, $e_i \cap c_j \neq \emptyset$, then for each $r < i$, $e_r \cap c_{j+2} = \emptyset$.

6. M_i has the two disk property.

PROPOSITION 4.1. If G is a decomposition of E^3 described by a sequence of collections $\{M_i\}_{i\in\omega}$ satisfying the above properties, then G is an upper semi-continuous decomposition of S^3 into points and a null sequence of cellular arcs for which S^3/G is not homeomorphic to S^3.

Proof. Property 2 guarantees that each element of G is cellular. Property 3 implies that the non-degenerate elements of G form a null sequence. Properties 4 and 5 imply that the elements of G are points and arcs. Property 6 implies by Theorem 2.1 that S^3/G is not homeomorphic to S^3.

It remains only to construct the defining sequence $\{M_i\}_{i\in\omega}$ which satisfies Properties 1-6.

Description of the M_i's. Recall that collection M_1 lies in the solid torus T as illustrated in Figure 2.1. One can choose these double tori so that they can be divided into links satisfying Property 4.

Lemma 3.3 together with Lemma 3.1 implies that M_1 has the two disk property.

For each $i \geq 1$, M_{i+1} is obtained from M_i in three steps as follows.

Step 1. Replacing double tori by tori. For each solid double torus H in M_i find a torus W in H embedded as illustrated in Figure 4.2.

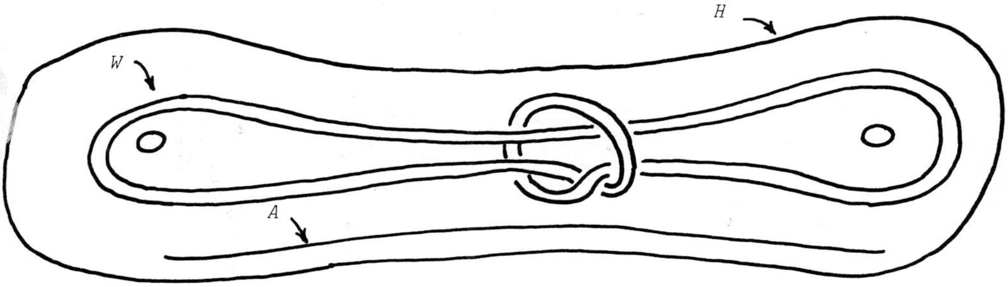

Fig. 4.2.

By Lemma 3.5, the collection M' obtained from M by making the above replacements also has the two disk property.

Step 2. Replacing each torus by a double torus and a cube with many handles. Each element of M_i' is a solid torus W contained in a solid double torus H which is an element of M_i.

Find an arc A in Int $H - W$ so that A intersects each link of H in one subarc of A. In W put a chain of small eyeglasses, one called Z, the others called $\{Y_i\}$, so that each eyeglass intersects at most two links of H and so that the eyeglasses are embedded in W as described in Lemma 3.3. For each eyeglass Y_i in W find an arc E_i joining that eyeglass to A as illustrated in Figure 4.3 so that E_i lies in one link of H, $E_i \cap A$ is one point, and so that the set Y, equal to $A \cup (\cup E_i) \cup (\cup Y_i)$ can be isotoped in H down near A. Therefore Y is contained in a cell contained in H. A regular neighborhood of Y is a standardly embedded cube with handles. By Lemmas 3.3 and 3.1, the collection M_i'', obtained from M_i' by replacing each solid torus W of M_i' by regular neighborhoods of the two objects Y and Z, still has the two disk property.

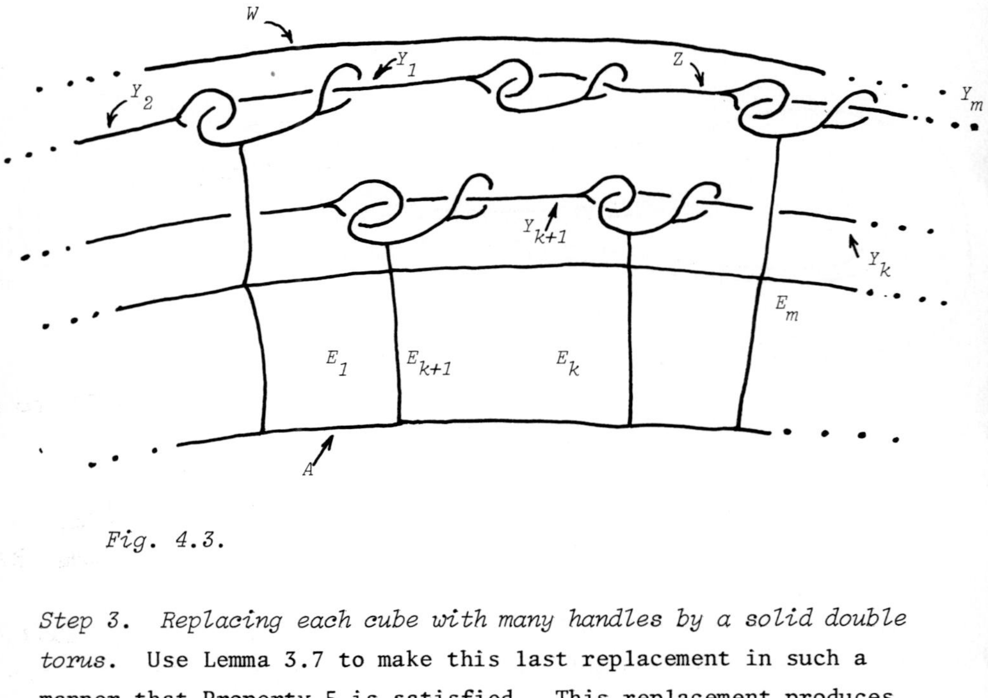

Fig. 4.3.

Step 3. Replacing each cube with many handles by a solid double torus. Use Lemma 3.7 to make this last replacement in such a manner that Property 5 is satisfied. This replacement produces a collection M_{i+1} which satisfies all six Properties so the example is complete.

REFERENCES

1. Armentrout, Steve, Decompositions of E^3 with a compact 0-dimensional set of nondegenerate elements, Trans. Amer. Math. Soc., 123 (1966), 165-177.
2. Bing, R. H., A decomposition of E^3 into points and tame arcs such that the decomposition space is topologically different from E^3, Ann. of Math., 65 (1957), 484-500.
3. __________, Upper semicontinuous decompositions of E^3, Ann. of Math., 65 (1957), 363-374.
4. __________, Necessary and sufficient conditions that a 3-manifold be S^3, Ann. of Math., 68 (1958), 17-37.
5. __________, A wild surface each of whose arcs is tame, Duke Math. J., 28 (1961), 1-16.
6. __________, Point-like decompositions of E^3, Fund. Math., 50 (1962), 431-453.
7. Everett, Dan, Shrinking countable decompositions of E^3 into points and tame cells, (these Proceedings).
8. Gillman, David S., and Martin, Joseph M., Countable decompositions of E^3 into points and point-like arcs, Notices Amer. Math. Soc., 10 (1963), 74.
9. Starbird, Michael, Cell-like 0-dimensional decompositions of E^3, Trans. Amer. Math. Soc. (to appear).
10. Starbird, Michael and Woodruff, Edythe P., Decompositions of E^3 with countably many nondegenerate elements, (these Proceedings).
11. Woodruff, Edythe P., Decomposition spaces having arbitrarily small neighborhoods with 2-sphere boundaries, Trans. Amer. Math. Soc., (to appear).
12. __________, Decomposition of E^3 into cellular sets, (these Proceedings).

SPECIAL REPRESENTATIONS FOR 3-MANIFOLDS

Joan S. Birman[1]
Jerome Powell[1]

Department of Mathematics
Columbia University
New York, N.Y.

Using techniques from the link calculus, it is shown that each closed, oriented 3-manifold admits a very special type of Heegaard decomposition which is also a very special surgery. The associated sewing map is the restriction to the Heegaard surface of a homeomorphism of a handlebody which fixes a complete system of meridian discs. The groups of isotopy classes of all such maps is closely related to the classical pure braid group. The associated Heegaard diagram consists of a pair of curves $(\vec{x},\vec{y})$ which have the following property: there is an associated diagram $(\vec{x}, \vec{z})$ for S^3 such that $(\vec{y}, \vec{z})$ also defines S^3 and the intersection matrices $\vec{x} \cap \vec{z}$ and $\vec{y} \cap \vec{z}$ are each the identity matrix. These diagrams have a great deal of interesting structure, including certain self-duality and finiteness properties.

I. INTRODUCTION

It is well-known that each closed, orientable 3-manifold M^3 may be represented by two different methods:

[1]*This work was partially supported by NSF Grant #MCS76-08230.*

ISBN 0-12-158860-2

Method 1: Choose a handlebody A which is imbedded in the 3-sphere S^3 in a standard manner, so that the closure of the complement, $S^3 - A$, is again a handlebody. Remove A from S^3 and resew to obtain $M = M^3$. The resulting decomposition of M as $A \cup A'$, where $A' = M - A$, is a Heegaard splitting of M. The splitting may be specified by a "diagram", i.e. by two families $\vec{x}$ and $\vec{x}'$ of simple closed curves on ∂A, where $\vec{x}$ (respectively $\vec{x}'$) bounds a complete system of meridinal discs in A (respectively A'). We denote this diagram by the symbol $(\partial A, \vec{x}, \vec{x}')$.

Method 2: Let L be a link, i.e. the embedded image of n pairwise-disjoint copies of S^1 in S^3. Remove pairwise disjoint tubular neighborhoods V_i $(i = 1,\ldots,n)$ of the components of L from S^3 and resew to obtain M. The resulting description of M as $(S^3 - \bigcup_{i=1}^{n} V_i) + (\bigcup_{i=1}^{n} V_i)$ is a "surgery" description.

In general, one might say that the difficulties in using either of these representations of 3-manifolds as a tool in studying the entire class is that there are too many of them. This paper continues an effort, begun in [2], to specialize diagrams and surgeries by placing additional structures on them. Our ultimate goal is to associate to each M^3 a *unique* diagram which would also (from another point of view) yield a unique surgery for each M^3. We have not achieved that goal, but we do obtain diagrams and surgeries which have some very restrictive properties.

The concepts of a "special surgery" and "special diagram" were introduced in [2]. A link $L \subset S^3$ is special if it is a subset of a graph G_L which has regular neighborhood U in S^3 such that

(1) The genus n of U is equal to the number of components of L, and

(2) The closure of $S^3 - U$ is homeomorphic to U.

A surgery is special if its link L is special. A diagram $(\partial A, \vec{x}, \vec{x}')$ is special if there is an associated diagram $(\partial A, \vec{x}, \vec{z})$ for S^3 such that $(\partial A, \vec{x}', \vec{z})$ is also a diagram for S^3. There is a correspondence between the special diagrams and special surgeries which represent a given M^3 and a natural way to pass from one to the other. It was proved in [2] that each M^3 may be represented by a special surgery or diagram.[1]

In this paper we continue the work described above and study still more special representations. Our philosophy is roughly this: Surgery representations of 3-manifolds are very flexible, because they may be altered very easily by the moves of the link calculus [3] however such alterations are in general quite unmotivated. On the other hand, Heegaard splittings are associated with a very deep structure, namely the group-theoretic structure of the set of gluing maps, under composition of mappings. That group is, however, very difficult to study directly. After working with mapping class groups of surfaces for some period of time, we were led to conjecture the result which appears as Theorem 4.6 of this paper, however we could not prove it via techniques from surface topology. The link calculus yielded what turned out to be a very easy proof, and at the same time revealed additional structure which we had not expected. Thus our general approach has been to use the mapping class group to tell us what to prove and the link calculus as a method of proof.

Here is an outline of this paper, Section 2 sets forth notation and conventions. In Section 3 we introduce framed links of class i $(i = 1,2,3,4)$ with the special links defined earlier $\supset$ class 1 and class $i-1 \supset$ class i for $i = 2,3,4$. We then prove in Theorem 3.1 that each M^3 may be represented by a framed link of

[1] *The definition in* [2] *also required that* $(\partial A . \vec{x}', \vec{z})$ *satisfy additional requirements, however the seemingly weaker requirement given above is equivalent to the stronger concept.*

class i for each $i = 1,2,3,4$. Theorem 3.4 shows that there are only finitely many framed links of class 4 with a fixed number of components. In Section 4, which contains the main results of this paper, we consider the diagrams associated to class i framed links. Some of the properties of these diagrams are delineated in Theorem 4.2, Corollary 4.4, Theorem 4.5, 4.6, 4.7, 4.9 and 4.10. Here are examples of the type of result we prove: Corollary 4.4 asserts that the Heegaard diagram associated to a class 1 link has a geometric intersection matrix which coincides with the corresponding geometric intersection matrix for the dual diagram. Theorem 4.5 shows that the associated Heegaard gluing maps are in a very special subgroup of the homeomorphism group of a handlebody. Theorem 4.9 asserts that for each genus n there are only finitely many Heegaard diagrams which are associated to class 4 framed links (Recall that each M^3 may be represented by a class 4 framed link).

Since we are constructing special surgery representation for 3-manifolds, we might expect them to yield very special 4-manifolds with given M^3 boundary. This is a topic for future research. This and other interesting open problems are discussed in Section 5.

II. DEFINITIONS, NOTATION, CONVENTIONS

We begin by introducing notation and defining terms. The symbols $A, A', B, \ldots$ denote handlebodies and $\partial A\ \partial A', \partial B, \ldots$ denote closed, orientable surfaces, which may (or may not) bound natural handlebodies $A, A', B, \ldots$. A system of properly imbedded discs in a handlebody A is a *complete system of discs* for A if A, cut open along these discs, is a 3-cell; a system of pairwise disjoint simple closed curves in ∂A is a *complete system of curves* for ∂A if ∂A cut open along these curves is a 2-sphere.

A *Heegaard splitting* of a closed, orientable, connected 3-manifold M is a decomposition of M as a union of handlebodies, $M = A \cup A'$, where $A \cap A' = \partial A = \partial A'$. The *genus* of the splitting is n = genus A = genus A' = rank $\pi_1 A$ = rank $\pi_1 A'$. The manifold M is determined up to homeomorphism by a *diagram* $D = (\partial A, \vec{x}, \vec{y})$, where $\vec{x} = \{x_1, \ldots, x_n\}$ and $\vec{y} = \{y_1, \ldots, y_n\}$ are complete systems of curves in $\partial A = \partial A'$ which bounds discs in A and A' respectively. The manifold M determined by the diagram D will be denoted $M(D)$. A diagram D is a *standard diagram for* S^3 if the curves $\vec{x}$, $\vec{y}$ satisfy the following intersection properties: $x_i \cap y_j = \emptyset$ if $i \neq j$ and $x_i \cap y_i$ = one point, $1 \leq i, j \leq n$. We will denote this by the symbol $\vec{x} \cap \vec{y} = \mathrm{Id}$.

We will sometimes add or drop subscripts which denote genus or super scripts which denote dimension. Thus A may be written A_n and M may be written M^3.

A *framed link* $L = (\vec{w}, \vec{f})$ is a system of $2n$ pairwise disjoint simple closed curves $(\vec{w}, \vec{f}) = \{w_1, \ldots, w_n, f_1, \ldots, f_n\}$ in S^3, where each f_i lies on the boundary of a tubular neighborhood V_i of w_i in S^3 and meets each meridian of ∂V_i once. Therefore each pair w_i and f_i cobound an annulus A_i in V_i. The curves $\vec{w}$ are the *components* of L and the curves $\vec{f}$ are the *framings*. The linking numbers $\vec{\ell}_i = \ell k(w_i, f_i)$ with respect to a fixed (say a right-handed) orientation of S^3 determine $\vec{f}$ uniquely up to isotopy in $S^3 - \vec{w}$, and sometimes we will specify these linking numbers $\vec{\ell} = \bigcup_{i=1}^{n} \ell_i$ instead of the curves $\vec{f}$. A framed link determines a 3-manifold $M(L)$, via surgery, in the following manner: cut out the solid tori V_i, $1 \leq i \leq n$, from S^3 and resew, identifying a meridian in V_i with the curve $f_i \subset \partial(S^3 - V_i)$. Thus each f_i bounds a disc in the surgered manifold $M(L)$. The *4-manifold* $W_4(L)$ *associated to* L is the 4-manifold obtained by attaching 2-handles to the boundary of the 4-ball D^4 along tubular neighborhoods of the curves of L, using their framings.

A link $\vec{w}$ in $S^3 = E^3 \cup \{\infty\}$ is said to be represented as a *2m-plat* (with respect to a height function on the t-axis) if it is imbedded in E^3 in such a way that it has m local maxima and m local minima with respect to the t axis, when E^3 is parametrized by rectangular coordinates (r,s,t). It is a *pure plat* if it is a $2m$-plat and if it also has m components. See Figure 2.1. It is

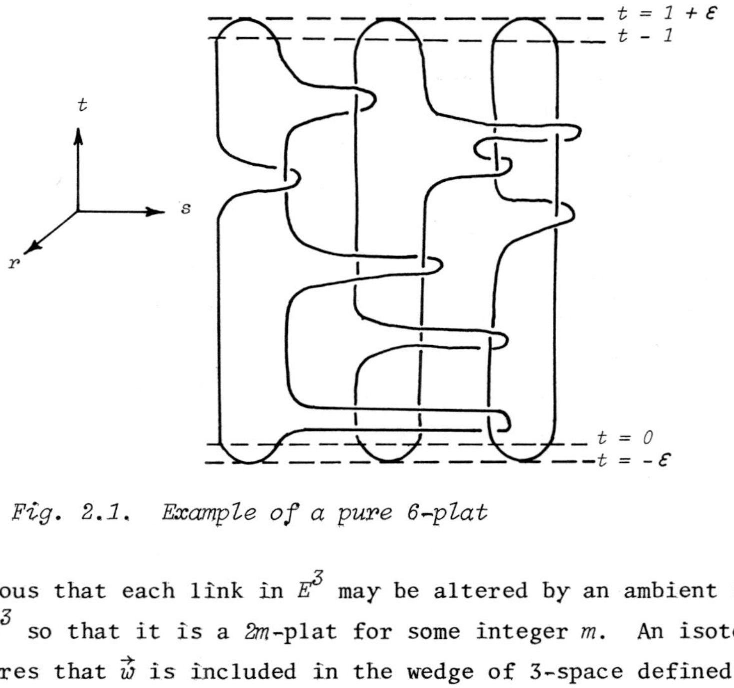

Fig. 2.1. Example of a pure 6-plat

obvious that each link in E^3 may be altered by an ambient isotopy of E^3 so that it is a $2m$-plat for some integer m. An isotopy ensures that $\vec{w}$ is included in the wedge of 3-space defined by the inequalities $1 + \varepsilon \geq t \geq -\varepsilon$, and also the local maxima (and minima) are located in the intersection of the planes $t = 1 + \varepsilon$ (and $t = -\varepsilon$) and $r = 0$. We may further assume that the arcs of $\vec{w} \cap \{(r,s,t) \mid 1 + \varepsilon \geq t \geq 1$ and $0 \geq t \geq -\varepsilon\}$ lie in the plane $r = 0$. Then the region of $\vec{w}$ which is between the planes $t = 1$ and $t = 0$ is a *geometric braid* on $2m$ strings, and the projection of $\vec{w}$ onto the $s - t$ plane determines a well-defined element in the classical braid group. We will assume that the reader is familiar with that group (see [1] for a general reference). There is a natural homomorphism from the braid group to the group of permutations on $2m$ letters, and its kernel is the *pure braid group* P_{2m}. We will refer to the elementary braids $\sigma_i (i = 1,\ldots,1m-1)$ and A_{ij} $(1 \leq i < j \leq 2m)$ pictured in Figure 2.2

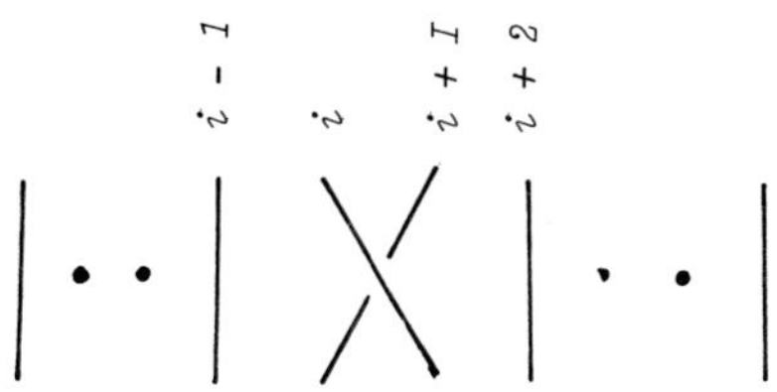

2-2(a) Elementary braid G_i

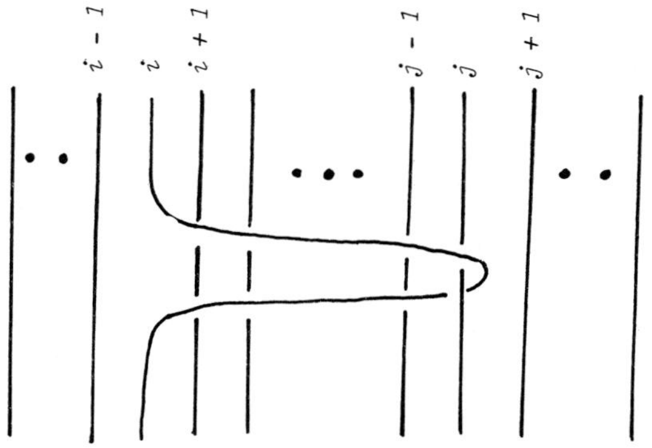

2-2(b) Elementary pure braid A_{ij}

Fig. 2.2. Generators of the braid group and pure braid group.

as *standard generators* for the braid group and for the pure braid group. Note that each element in the braid group determines a $2m$-plat and each element in the pure braid group determines a $2m$-pure plat; these are defined by identifying the strings of a geometric representative in pairs at the top and bottom of the braid, joining the $(2i-1)^{st}$ string to the $2i^{th}$ string, $i = 1,\ldots,m$.

It will be convenient to think of a 3-dimensional handlebody A_n as a 3-ball D^3 to which n 1-handles $D^1 \times D^2$ have been attached. The ith *handle* of A_n (respectively ∂A_n) is the ith copy $(D^1 \times D^2)^{(i)}$ (respectively $(D^1 \times \partial D^2)^{(i)}$) regarded as a subset of A_n (respectively ∂A_n). Its *feet* are the discs $(\partial D^1 \times D^2)^{(i)}$ (respectively circles $(\partial D^1 \times \partial D^2)^{(i)}$). The *sphere part* of ∂A_n is the subset $\partial A_n - \bigcup_{i=1}^{n} (D^1 \times \partial D^2)^{(i)}$ of ∂A_n. The ith *meridian* of ∂A_n is the curve $(\frac{1}{2} \times \partial D^2)^{(i)}$.

If $L = \bigcup_{i=1}^{n} (w_i, f_i)$ is a framed link, the *4-manifold associated to* L is the manifold $W^4(L)$ obtained by attaching n 2-handles to the boundary of the 4-ball D^4, along the circles of L. The attaching maps are defined by the framings (see [3]). Note that L is not oriented since an orientation since an orientation for $W^4(L)$ and for $M^3(L) = \partial W^4(L)$ is obtained by extending a fixed orientation on D^4 over $W^4(L)$.

A Heegaard splitting $A_n \cup B_n$ of S^3 will be said to be *standard* if A_n is a 3-ball with n unknotted, unlinked handles attached to it. Let $A_n \cup B_n$ be such a splitting. Let

$$\overline{H} = \overline{H}(n) = \{\overline{h} : \partial A_n \to \partial A_n \mid \overline{h} \text{ is an orientation-preserving homeomorphism}\}$$

$$\overline{A} = \overline{A}(n) = \{\overline{h} \in \overline{H}(n) \mid \overline{h} \text{ extends to a homeomorphism of } A_n\}$$

$$\overline{B} = \overline{B}(n) = \{\overline{h} \in \overline{H}(n) \mid \overline{h} \text{ extends to a homeomorphism of } A_n\}.$$

Let H, A, B be the reductions of these groups mod isotopy, where the natural map $\overline{H} \to H$ sends $\overline{h} \to h$.

If c is a simple closed curve on ∂A_n, let $\tau_c \in H(n)$ denote the isotopy class of a Dehn twist about c. Note that $\tau_c \in A(n)$ if c bounds a disc in A_n and $\tau_c \in B(n)$ if c bounds a disc in B_n.

III. REPRESENTATIONS OF 3-MANIFOLDS BY SPECIAL FRAMED LINKS

The concept of a special framed link was defined in Section 1. In this section we will introduce several classes of framed links (all of which will be shown to be special framed links), and we will prove that each M^3 can be represented by members of each such class. These results will be applied in Section 4, below, to Heegaard theory.

A framed link $L = (\vec{w}, \vec{f}) = (\vec{w}, \vec{\ell})$ will be said to be

Class 1 if it satisfies the property

(i) $\vec{w}$ may be represented as a pure plat;

Class 2 if it is class 1 and also

(ii) the framings ℓ_i are all even integers;

Class 3 if it is class 2 and also

(iii) $\vec{w}$ is a pure plat with respect to a height function on the t-axis (see Figure 2.1) and also with respect to a height function on the s-axis, when $\vec{w} \subset E^3 = S^3 - \{\infty\}$, and E^3 is parameterized by rectangular coordinates (r, s, t);

(iv) for any two components w_i, w_j the geometric and algebraic linking numbers coincide, and there are 0, $+1$ and -1;

(v) the 4-manifold $W^4(L)$ associated to L is parallelizable and has index 0 or 8.

Class 4 if it is class 3 and if

(vi) the framings are all 0, $+2$ or -2.

An example of a framed link of class 3 is given in Figure 3.1.

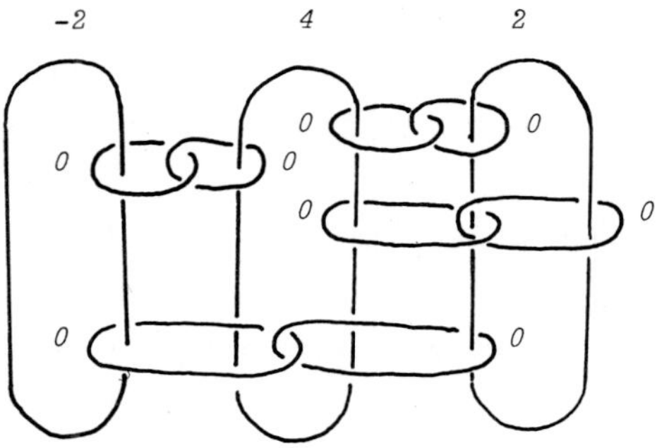

Fig. 3.1. Example of a class-3 framed link

THEOREM 3.1. Each closed, orientable 3-manifold M^3 is M(L) for some framed link L of class 1, 2, 3 or 4. Moreover, a constructive procedure exists to convert any framed link into one of class 1, 2, 3 or 4 by using the link calculus [3].

Remark. The fact that each M^3 admits a representation as $M(L)$ where L satisfies (i), and (iv) follows from the construction on page 538 of [4] together with remarks on page 279 of [5]. However, that proof does not yield (ii), (iii) (v), or (vi) and it also is non-constructive. An interesting connection between our proof and the methods in [4] will be given in Theorem 4.7 and Remarks 4.8, below.

Proof of Theorem 3.1. We will prove that $M = M(L)$ where L is class 3 or 4. Since Class 1 $\supset$ class 2 $\supset$ class 3, the theorem will follow.[1]

[1] *Since class i-1$\supset$class i the reader may wonder why we introduced classes 1, 2 and 3? Our reason is that we are not sure whether the particular properties of class 4 links are the most appropriate for an ultimate normal form, hence we wished to indicate larger classes which might be useful for future investigations.*

We begin with an arbitrary surgery representation of $M = M(L)$, where L is a framed link. Using a constructive procedure due to Kaplan [10] this framed link presentation may be altered to a new one $L^{(1)} - (\vec{w}^{(1)}, \vec{f}^{(1)})$ such that $M(L^{(1)}) = M(L)$ and also all of the framings $\ell_i^{(1)}$ are even integers. We may further assume that $\vec{w}^{(1)}$ is represented as a $2m$-plat with respect to a height function on the t-axis, and that the plat has been altered by isotopy so that it determines a well-defined element in the braid group B_{2m}, where the local maxima and minima and braid strings are numbered to correspond to increasing s-coordinate. We assert:

LEMMA 3.2. The link $\vec{w}$ has a $2m$-plat representation in which the local maxima and local minima are ordered so that those belonging to w_i have smaller s-coordinates than those belonging to w_{i+1} for each $i = 1, \ldots, n-1$.

Proof of Lemma 3.2. We may alter the plat by isotopy at the top and bottom in the manner indicated in Figure 3.2 (a) and (b) to achieve any desired permutation of the local maxima or minima.

We now consider the braid associated to $\vec{w}^{(1)}$, after the modifications of Lemma 3.2. It will be helpful to think of the different components as having different colors, e.g. red, blue, green,... . Then, after the modifications of Lemma 3.2 the strings of the braid will be arranged in monochromatic groups, e.g. $2m_1$ red strings, followed by $2m_2$ blue strings, $2m_3$ green strings ... at the top and at the bottom of the braid. The action of the permutation group Σ_{2m} restricts to an action on each monochromatic grouping, hence we may associate to our modified plat the n permutations $\pi_1, \ldots, \pi_n$ belonging to the n components of w. Each π_i is a permutation on m_i symbols, $1 \leq i \leq n$.

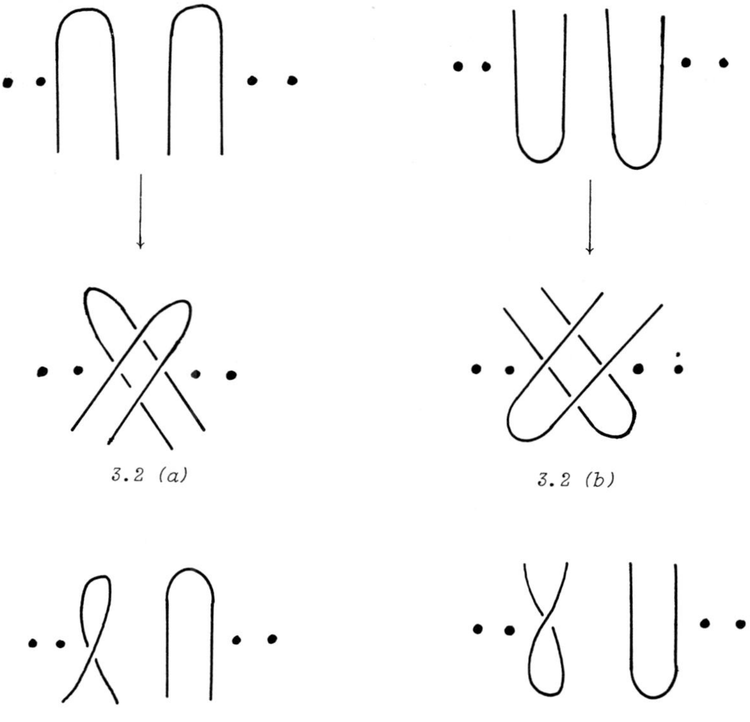

3.2 (a) 3.2 (b) 3.2 (c) 3.2 (d)

Fig. 3.2. Moves that take 2m plats → 2m plats

LEMMA 3.3. We may without loss of generality assume that $\pi_i = (23)(45) \ldots (2m_i-2, 2m_i-1)$.

Proof of Lemma 3.3. We proceed by induction on m_i. If $m_i = 1$, then π_i = identity or (12); if $\pi_i = (12)$ we may apply the modification of Figure 3.2 (d) to achieve π_i = identity. For $m_i > 1$, if $\pi_i(1) \neq 1$ or 2 we may alter the plat by a move of the type illustrated in Figure 3.2 (b) to achieve $\pi_i(1) = 1$ or 2, followed by a move of the type illustrated in Figure 3.2 (d), if necessary, to achieve $\pi_i(1) = 1$. Hence we may, without loss of generality, assume that $\pi_i(1) = 1$. Now, if $\pi_i^{-1}(2) \neq 3$ or 4, a move of the type illustrated in Figure 3.2 (a) will achieve $\pi_i^{-1}(2) = 3$ or 4, and a move of the type illustrated in 3.2 (c), if necessary, will yield $\pi_i^{-1}(2) = 3$. Induction on m_i completes the proof of the lemma.

To continue the proof of the theorem we assume that $(\vec{w}^{(1)}, \vec{f}^{(1)})$ has been altered in the manner indicated in Lemmas 3.2 and 3.3. Then the $2m$ braid associated to $\vec{w}^{(1)}$ may be taken to be a product of a sequence of canonical permutation braids $\sigma_2\sigma_3 \cdots \sigma_{2m_i-2}$, one for each component (cf. Figure 2.2a), and a pure braid on $2m$ strings. The strings of the pure braid are arranged in an ordered sequence of monochromatic groups. The braid itself may be assumed to be a product of the elementary braids A_{ij} (see Figure 2.2b), where in general i and j need not have the same color. We now apply the moves of the link calculus (see [3]) to $(\vec{w}^{(1)}, \vec{f}^{(1)})$, introducing new components in order to alter the picture in the manner illustrated in Figure 3.3 (a). After a finite number of such alternations, we will achieve a new framed link $L^{(2)} = (\vec{w}^{(2)}, \vec{f}^{(2)}$, with $M(L^{(2)}) = M(L^{(1)})$ which admits a projection of the type illustrated in Figure 3.4. Note that $\vec{w}^{(2)}$ is now a pure-plat with respect to a height function on the s-axis, i.e. $\vec{w}^{(2)}$ has property (i).

After a rotation of 90° about the r axis, we may assume that $\vec{w}^{(2)}$ is a pure-plat with respect to a height function on the t-axis, and after isotopy (as before) that $\vec{w}^{(2)}$ is in standard position, so that it defines an element in the pure braid group P_{2q}, where q is the number of components in $\vec{w}^{(2)}$. As before, that pure-braid is a product of the elementary pure braid generators A_{ij}. Hence we may apply the moves in Figure 3.3 (a), as before, to achieve a new framed link $L^{(3)} = (\vec{w}^{(3)}, \vec{f}^{(3)})$ which is a pure plat with respect to height functions on *both* the s and t axis. Thus we have achieved condition (iii), and a few moments

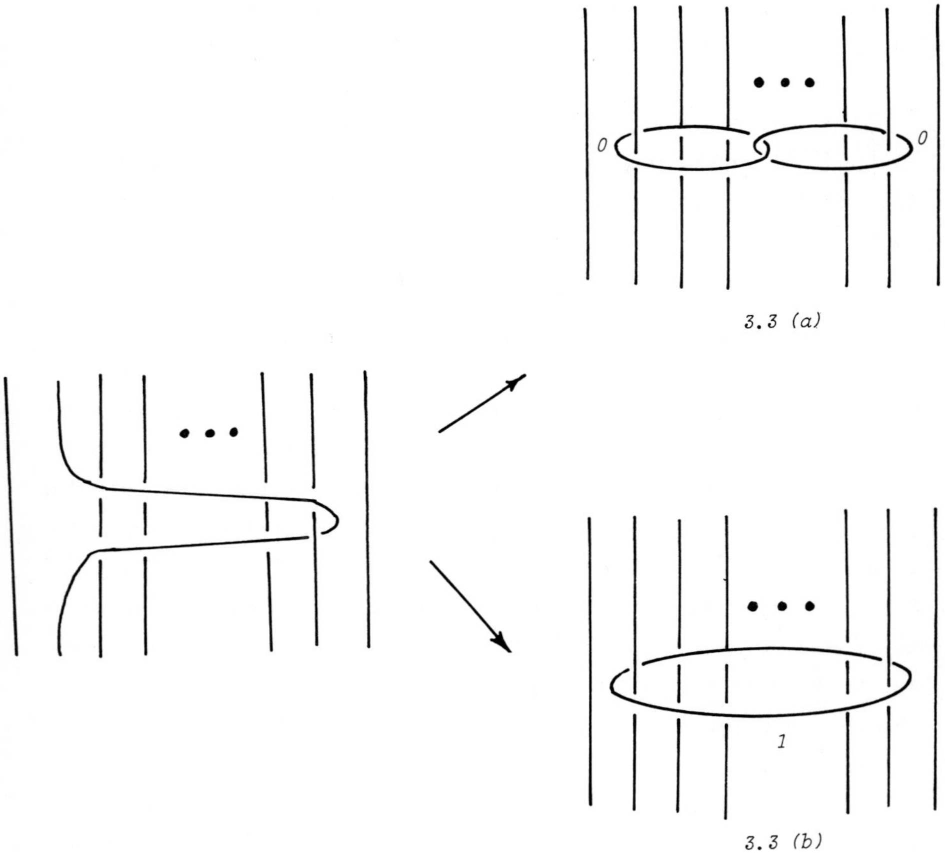

3.3 (a)

3.3 (b)

Fig. 3.3. Moves of the link calculus

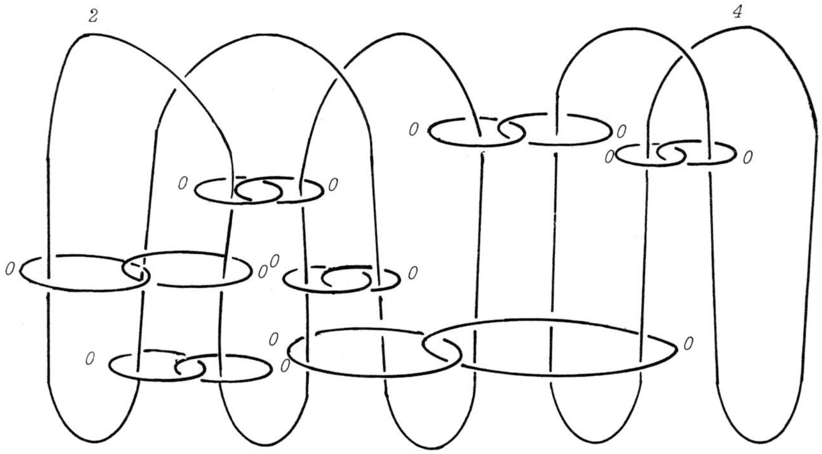

Fig. 3.4.

thought indicate that we have also achieved (iv). Finally, since the new components introduced by our moves have framing zero, and since the original framings are unaltered by our moves, we have also achieved (ii).

We will now establish (v). Let $W^4(L^{(3)})$ be the 4-manifold associated to $L^{(3)}$. Since the framings are all even the manifold W^4 is parallelizable and the bilinear form associated to $H^2(W^4)$ is type II. By theorem 5.1 on page 24 of [11] its index must then be a multiple of 8. Let $L^* = (\vec{w}^*, \vec{f}^*)$ be any framed link which defines a 2-handle presentation of the Kummer surface. The procedure which we used above to alter the framed link presentation for M^3 may also be applied to L^*, hence we may without loss of generality assume that L^* satisfies conditions (i), (ii), (iii), (iv). Now, the index of the Kummer surface is ± 16 and its boundary is homeomorphic to S^3, hence if we augment $L^{(3)}$ by adding an appropriate number of copies of $\pm L^*$ we will achieve a new framed link $L = (\vec{w}, \vec{f})$ which satisfies properties (i) - (v), i.e. L is class-3 and $M(L^{(3)}) = M(L)$.

Small modifications in the proof will now yield a proof that L may (alternatively) be chosen to be class 4. First, note that S^3 may be defined by surgery on a link (call it $\pm T$) of two unknotted, simply-linked components, one of which has framing 0 and the other ± 2 (see [10] or [2]). If a component L_i has framing $\ell_i \neq 0, +2, -2$, then $|\ell_i|$ may be reduced by introducing $\pm T$ and replacing L_i with its band sum with $\pm T$. This does not change $M(L)$, because $M(\pm T) = S^3$. Each such band sum causes a pair of components to link with linking number ± 2, however, such linking may be modified to ± 1 by the procedure used earlier (see Figure 3.3 (a).) Induction on $|\ell_i|$ and on the number of components with framing different from 0, ± 2 completes the proof.[1]

THEOREM 3.4. For each integer n there are a finite number of class-4 framed links L of n components.

Proof of Theorem 3.4. Since the components of L are a pure plat, they are represented by an element in pure braid group on 2n strings. The generators of that group are the elementary braids A_{ij}, hence, L is defined by a braid word $W = \prod_{x=1}^{r} A_{i_x j_x}^{\varepsilon_x}$. Since the geometric intersection numbers between w_i and w_j are all 0 or 1 (property (iv)) each generator $A_{ij} (1 \leq i < j \leq 2n)$ appears in W at most once, hence r is bounded. For fixed r there are only a finite number of possibilities, depending on the order of the individual letters A_{ij}^{ε} in the braid word W, hence there are only finitely many admissible W. The framings are all 0, $+2$ or -2, hence, there are a finite number of possibilities for L.

[1]*Acknowledgement: The last step in this proof was suggested to us by remarks of Professor M. A. Ŝtanko, during a private conversation in December 1977. Professor Ŝtanko also described to us, at that time results which he had obtained independently, which intersect some of those in this section. His results are announced withouts proofs, in* [15].

IV. THE HEEGAARD DIAGRAMS AND SEWINGS ASSOCIATED TO CLASS 1, 2, 3 AND 4 FRAMED LINKS.

In this section we study some of the properties of the Heegaard diagrams and sewings associated to framed links of class 1,2,3 and 4. The main results are Theorem 4.2, Corollary 4.4, Theorems 4.5, 4.6, 4.7, 4.9 and 4.10. To begin, we develop a method for associating a Heegaard diagram D to each class-1 framed link. From the construction it will be clear that each class-1 framed link is special, in the sense defined in Section 1 and in [2]. We will also develop the inverse construction, which allows us to pass from a class-1 diagram to an associated class-1 framed link.

THEOREM 4.1. Each class-1 framed link L determines a Heegaard diagram D such that $M(D) = M(L)$.

Proof of Theorem 4.1. Let $M^3 = M(L)$, where $L = (\vec{w},\vec{f})$ is class-1. The first step in our proof will be to associate to $\vec{w} \subset S^3$ a Heegaard splitting $A_n \cup B_n$ of the 3-sphere S^3. We will then use the Heegaard surface $A_n \cap B_n = \partial A_n$ to define a Heegaard diagram for $M(L)$.

If L is class 1, its components $\vec{w}$ will define a pure plat with respect to a height function on (say) the t axis. The first step will be to alter $\vec{w}$ to a new representation as an n-bridge link. To begin, we push the local maxima down a little bit and the local minima up a little bit until $\vec{w}$ is the union of

<u>n arcs $\vec{\alpha}$</u> in the intersection of the plane $t = 0$ $r = 0$.

<u>n arcs $\vec{\beta}$</u> in the intersection of the plane $t = 1$, $r = 0$ which lie directly above the n arcs of $\vec{\alpha}$.

<u>$2n$ arcs $\vec{\gamma}$</u> which join the corresponding points of $\partial\vec{\beta}$ and $\partial\vec{\alpha}$, also each arc γ_i meets each intermediate plane $t = t_0$, $0 \leq t_0 \leq 1$, in exactly one point.

We next construct a homeomorphism $h : E^3 \to E^3$ which is supported in $t \geq 1$ and $t \leq -1$, also $h(\partial\vec{\alpha}) = \partial\vec{\alpha}$, also $h(\vec{\gamma})$ consists of vertical arcs joining $\partial\vec{\beta} = \partial h(\vec{\beta})$ to $\partial\vec{\alpha} = \partial h(\vec{\alpha})$. This may be accomplished by untwisting the braid strings $\vec{\gamma}$ at the expense of twisting the arcs $\vec{\alpha}$ in the plane $t = 0$ (see Theorem 5.4 of [1]), keeping $\vec{\beta}$ fixed. Then $h(\vec{w}) = h(\vec{\alpha}) \cup h(\vec{\beta}) \cup h(\vec{\gamma})$ is a bridge representation of $\vec{w}$. We now join the n arcs $h(\vec{\alpha})$ by $n-1$ edges $\vec{e}$ in the plane $t = 0$ to form a connected tree $\vec{e} \cup h(\vec{\alpha})$. Let $G = \vec{e} \cup h(\vec{w})$. Then G has a regular neighborhood A_n which is a 3-cell (a regular neighborhood of the tree $\vec{e} \cup h(\vec{\alpha})$) with n unknotted and unlinked handles (regular neighborhoods of the bridges) attached. Clearly A_n determines a Heegaard splitting $A_n \cup B_n$ of S^3.

Let $\bigcup_{i=1}^{n} V_i$ be pairwise disjoint tubular neighborhoods of the components $\bigcup_{i=1}^{n} w_i$ of L, and choose curves $f_i \subset \partial V_i$ to define framings in the components. We may without loss of generality assume that (i) A_n is obtained from $\bigcup_{i=1}^{n} V_i$ by adding $n-1$ connecting beams, and (ii) the curves f_i are so placed on ∂V_i that they avoid the discs where these beams are attached, so that $\bigcup_{i=1}^{n} f_i \subset \partial A_n$. Since each torus ∂V_i split open along f_i is a cylinder, it follows that ∂A_n split open along $\vec{f}$ is a punctured sphere, hence $\vec{f}$ is a complete system for ∂A_n.

Next we choose a second system of curves $\vec{x} = \{x_1, \ldots, x_n\} \subset \partial A_n$ which bounds a complete system for A_n. The curves $\vec{x}$ will be select so that each x_i encircles the ith handle once in the manner indicated in Figure 4.1. Let $D = (\partial A_n, \vec{x}, \vec{f})$. Then D is a Heegaard diagram for a 3-manifold $M(D)$. It is *the diagram associated to the class-1 framed link* L (with connecting edges $\vec{e}$). Remark: a different choice of $\vec{e}$ will in general lead to a different diagram.

It remains to prove that $M(D) = M(L)$. Now, the manifold $M(D)$ (respectively $M(L)$ is obtained from S^3 by cutting out the handlebody A_n (respectively the solid tori $\bigcup_{i=1}^{n} V_i$) and resewing so that the curves f_i bound discs. Since A_n is obtained from $\bigcup_{i=1}^{n} V_i$ by adding $n-1$ connecting beams, it follows that $M(D)$ may be obtained from $M(L)$ by cutting out those beams and resewing. Each connecting beam is a 3-cell and surgery on a 3-cell does not alter the topological type of a 3-manifold, hence $M(D) \cong M(L)$. This completes the proof of Theorem 4.1.

To develop properties of the diagram D we examine the association $L \to D$ in somewhat greater detail. The reader is referred to Section 2 for the definition of a standard handlebody A_n in S^3, and for our conventions about ∂A_n, its "sphere part" and its "handles", their "feet" and their "meridians". Using that terminology, we see that each curve x_i which was selected in the proof of Theorem 4.1 decomposes into the union of two arcs: an arc $x_i^{(1)}$ on the ith handle and an arc $x_i^{(2)}$ on the sphere part (cf. Figure 4.1). Note that a curve on the sphere part of ∂A_n goes "under the ith handle" precisely when it intersects $x_i^{(2)}$ once.

THEOREM 4.2. The diagram $D = (\partial A_n, \vec{x}, \vec{f})$ associated to a class-1 framed link has the following properties:

(i) There is an associated standard diagram $D' = (\partial A_n, \vec{x}, \vec{z})$ for S^3 such that $D'' = (\partial A_n, \vec{f}, \vec{z})$ is also a standard diagram for S^3.

(ii) Each curve f_i in the diagram D goes over the ith handle once and spends the rest of its time on the sphere part of ∂A_n, i.e. it does not go over the jth handle for any $j \neq i$. Also, when f_i is on the sphere part of ∂A_n it does not pass under the ith handle.

Proof of Theorem 4.2. Let $\vec{z} = \{z_1, \ldots, z_n\}$ be the meridian of the handles of ∂A_n. Then, by construction, $D' = (\partial A_n, \vec{x}, \vec{z})$ is a

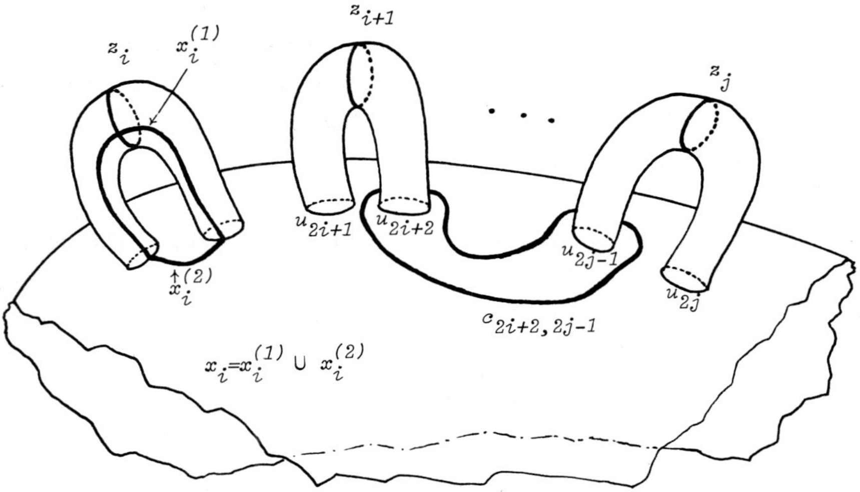

Fig. 4.1. Some curves and arcs on a standard handlebody

standard diagram for S^3. Since each curve f_i is a framing curve for a component w_i of the link, and since z_i bounds a meridian disc in a tubular neighborhood V_i of w_i, it follows that the geometric intersection matrix $\| \, |f_i \cap z_j| \, \|$ is the identity matrix, hence D'' is standard, hence $M(D'') = S^3$. This proves (i). A few moments thought should convince the reader that (ii) is just another way of saying (i).

Remark. Theorem 4.2 generalizes the main result of [2]. The diagrams D' and D'' are both standard, in the notation of that paper.

Definition. We will say that a Heegaard diagram D is class-1 if it satisfies the properties given in Theorem 4.2.

PROPOSITION 4.3. Each Class 1 Heegaard diagram determines a Class-1 framed link.

Proof of Proposition 4.3. Using the associated diagram D' for S^3, we may regard the Heegaard surface ∂A_n in D as a Heegaard surface in $S^3 = A_n \cup B_n$. The curves $\vec{f}$ may then be pushed off the surface transversally into A_n (or B_n) to new curves $\vec{w}$ which we will take as the components of our link. The Heegaard surface determines the framings on $L = (\vec{w},\vec{f})$. The curves $\vec{z}$ in D'' bound meridian discs in tubular neighborhoods of the w_i, and each f_i meets z_i once (also $f_i \cap z_j = \emptyset$ if $i \neq j$). The framed link is class 1 because it is represented as an n-bridge link, hence it also may be represented as a $2n$-plat, where n is the number of components. The manifold $M(L)$ is homeomorphic to $M(D)$, because $M(D)$ is the diagram associated to L.

COROLLARY 4.4. If D is class-1 and L satisfies property (iv), then the geometric intersection matrix $\| \, |f_i \cap x_j| \, \|$ for $D = (\partial A_n, \vec{x}, \vec{f})$ will coincide with the corresponding matrix $\| \, |x_i \cap f_j| \, \|$ for the dual diagram $D^* = (-\partial A_n, \vec{f}, \vec{x})$.

Proof of Corollary 4.4. By construction the intersection matrix $\| \, |f_i \cap x_j| \, \|$ coincides with the linking matrix $\| \ell k(w_i, w_j) \|$, where the diagonal entries in the latter are taken to be the framings ℓ_i. The fact that the matrices for D and D^* coincide then follows from the symmetry of linking numbers!

Remark. The symmetry noted in Corollary 4.4 has nothing whatsoever to do with the antisymmetry $|f_i \cap x_j|_0 = -|x_j \cap f_i|_0$ of algebraic intersection numbers on a surface.

To develop further properties of the diagram associated to class-1 and class-2 framed links, choose an orientation-preserving homeomorphism h such that $h(x_i) = f_i$, $1 \leq i \leq n$. This is possible because $\vec{x}$ and $\vec{f}$ are each complete systems for ∂A_n, and any two complete systems are equivalent under $H(n)$. The map h is a *Heegaard sewing;* a sewing is class i if its defining diagram (or framed link) is class i $(i = 1,2,3,4)$.

It was shown in [2] that a sewing is special if and only if the defining map h is in the subgroup $A(n)$ of $H(n)$. We will now show that for sewings of class 1 and 2 a much stronger group theoretic property holds.

THEOREM 4.5. A sewing $h \in A(n)$ is class 1 if and only if h is isotopic to a map $\overline{h}$ which satisfies $\overline{h}(z_i) = z_i$, $1 \leq i \leq n$.

Proof. For each $1 \leq i \leq n$ the curves f_i and x_i go over the ith handle once and do not cross over the jth handle if $i \neq j$. Hence $x_i \cap z_j = |f_i \cap z_j| = \delta_{ij}$, the Kronecker symbol, so that (up to isotopy) $\overline{h}(z_i) = z_i$, $1 \leq i \leq n$. Conversely, if $\overline{h}(z_i) = z_i$, $1 \leq i \leq n$, then h belongs to $A(n)$ [14] hence h defines a special sewing [2]. To see that the sewing is class-1, consider the associated diagram $D = (\partial A_n, \vec{x}, h(\vec{x}))$. The associated diagram $D' = (\partial A_n, \vec{x}, \vec{z})$ is a standard splitting for S^3, also $D'' = (\partial A_n, h(\vec{x}), \vec{z})$ is standard because $|f_i \cap z_j| = \delta_{ij}$, hence D is class-1, hence h is class-1.

The maps h of class-1 or 2 constitute a well-defined subgroup of $A(n)$ which we denote $R(n)$. We now describe its structure. The reader is referred to Figure 4.1. Let u_{2i-1} and u_{2i} denote the feet of the ith handle of ∂A_n. For each $1 \leq r < s \leq 2n$ choose a curve c_{rs} on the sphere part of ∂A_n which encircles u_r and u_s and separates them from the remaining curves u_t, $t \neq r,s$. Recall that τ_c denotes the isotopy class of a Dehn twist about a simple closed curve c on ∂A_n. Define maps

$$\hat{A}_{ij} = \tau_z^{-1} \tau_{z_i}^{-1} \tau_{c_{ij}}, \qquad 1 \leq i < j \leq 2n. \tag{4.1}$$

Note that $\hat{A}_{ij} \in A(n)$ because z_i, z_j, c_{ij} all bound disks in A_n. Let $\hat{P}_{2n}$ be the subgroup of $A(n)$ generated by the maps $\hat{A}_{ij}$, $1 \leq i < j \leq 2n$, and let T_k denote the infinite cyclic group generated by τ_{z_k}, $1 \leq k \leq n$, and let $2T_k$ denote its subgroup generated by τ_{z_k}. Let

$$(4.2) \qquad \tilde{R}(n) = \hat{P}_{2n} \oplus T_1 \oplus \ldots \oplus T_n$$

$$(4.3) \qquad R(n) = \hat{P}_{2n} \oplus 2T_1 \oplus \ldots \oplus 2T_n.$$

THEOREM 4.6.

1. *A sewing h is class 1 (respectively class 2) if and only if $h \in \tilde{R}(n)$ (respectively $R(n)$).*
2. *The group $\hat{P}_{2n}$ is isomorphic to the pure mapping class group of the 2n-punctured sphere. Its generators $\hat{A}_{ij}$ may be interpreted as geometric braids, and visualized as in Figure 2.2 (b).*

Proof. Suppose, first that h is class-1. By Theorem 4.6 the map $\bar{h}$ restricts to a map of ∂A_n, cut open along $\vec{z}$, hence defines an element in the mapping class group of a sphere with $2n$ discs removed. The structure of that group is investigated in Chapter 4 of [1]. It is the direct sum of the free abelian group of rank $2n$ generated by the isotopy classes of twists about the boundary curves and the pure mapping class group $\hat{P}_{2n}$ of the sphere with $2n$ points removed. Generators for $\hat{P}_{2n}$ are given in Theorem 4.5 and Lemma 1.8.2 of [1] and these are realized in our situation by the maps which we call $\hat{A}_{ij}$. The group $\hat{P}_{2n}$ is a homomorphic image of the pure braid group P_{2n} introduced in Section 1 (the braid group of the plane or disc) under the mapping $A_{ij} \to \hat{A}_{ij}$ (cf. Figure 2.2 (b)). The map $\hat{A}_{ij}$ may also be visualized geometrically as a sliding of the ith foot of a handle along a path which encircles the jth foot and separates it from all other feet of other handles (see [9]). Thus the subgroup of $A(n)$ which consists of the isotopy classes of all maps which fix the z_i pointwise is isomorphic to the direct sum of $\hat{P}_{2n}$ and the free abelian group freely generated by $\tau_{z_1},\ldots,\tau_{z_n}$. It remains to show that class 2 sewings are associated to the subgroup $R(n)$ of $\tilde{R}(n)$, and this is an immediate consequence of the fact that the framings on a class-2 framed link are even integers, hence each curve f_i intersects x_i an even number of times, hence $\bar{h}$ restricted to the ith handle can be chosen to be isotopic to some power of $\tau^2_{z_i}$.

Our next result produces an explicit sewing h satisfying the properties of Theorem 4.6 for each class-1 diagram D.

THEOREM 4.7. Let $f'_i = \tau_{z_i}(f_i)$, $1 \leq i \leq n$. *Then* $h = (\prod_{i=1}^{n} \tau_{f'_i})(\prod_{j=1}^{n} \tau_{x_j}\tau_{z_j})$ *is a Heegaard sewing associated to the class 1 diagram D.*

Proof. By Theorem 4.2 the curve systems $\vec{x}$, $\vec{f}$, $\vec{z}$ satisfy the intersection properties $|x_i \cap z_j| = |f_i \cap z_j| = \delta_{ij}$. It then follows from elementary properties of twist maps that

$$\text{(4.4)} \qquad \tau_{x_i}\tau_{z_i}(x_k) = x_k \text{ if } k \neq i$$
$$= z_k \text{ if } k = i$$

$$\text{(4.5)} \qquad \tau_{f'_i}(z_k) = z_k \text{ if } k \neq i$$
$$= f_i \text{ if } k = i$$

Hence $h(x_i) = f_i$, $1 \leq i \leq n$, as required.

Remark 4.8. In Theorem 4.7 the map h is expressed as a product of Dehn twists, hence we may apply the method of Lickorish [4] to produce a framed link which defines $M(h)$. We now show that this link is identical with the link associated to the diagram $D = (\partial A_n, \vec{x}, h(\vec{x}))$, using the method of Proposition 4.3. To see this, note that $h = h_2 h_1$ where $h_1 = \prod_{i=1}^{n} \tau_{x_i} \tau_{z_i}$ and $h_2 = \prod_{i=1}^{n} \tau_{f_i'}$. From equation (4.5) we see that $M(h_1) = S^3$. The procedure given by Lickorish for finding a framed link from a Heegaard sewing assumes, implicitely, some fixed reference sewing for S^3, e.g. h_1. Hence, in Lickorish's notation, $M(h)$ is defined by h_2, not $h_2 h_1$. His procedure then is to push the curves f_i' off the Heegaard surface to obtain a link in S^3. Since the push-off of f_i' into A_n is isotopic to the push-off of f_i into A_n, the same link is obtained as was obtained earlier, in Proposition 4.3. The framing which is obtained by our method is $\ell k(f_i, w_i)$, when f_i, w_i are regarded as curves in S^3. The framing which is obtained by the procedure given in [4] is ± 1 *under the assumption that* $\ell k(f_i', w_i) = 0$. In the more general situation considered here, the framing must be modified to $\pm 1 + \ell k(f_i', w_i) = \ell k(f_i, w_i)$, hence the two methods yield identical framed links.

The remarks above expose a subtle aspect of both constructions, namely that the framed link which is obtained from D depends not only upon the expression of the sewing h as a product of Dehn twists, but also on the choice of a reference sewing h_1 for S^3. It is clear that, using the methods of [4], different reference sewings will yield distinct framed links. In our construction, part of this ambiguity has been avoided by the requirement that h_1 determine a *standard* diagram $D' = (\partial A_n, x, h_1(x))$ and also that $D'' = (\partial A_n, h_2(x), h_1(x))$ be standard, however there are some very delicate questions involved here.

THEOREM 4.9. *For each genus n there are a finite number of diagrams associated to class-4 links.*

Proof. This is an immediate consequence of Theorem 3.4.

THEOREM 4.10. Let $D = (\partial A_n, \vec{x}, \vec{f})$ *be a class-1 diagram. Then, if any pair of curves* x_i, f_i *are removed from the diagram the remaining curves* $(x_1, \ldots, \hat{x}_i, \ldots, x_n, f_1, \ldots, \hat{f}_i, \ldots, f_n)$ *will lie on a subset of the Heegaard surface* ∂A_n *which has genus* $n-1$*. These can be used to define a new class-1 diagram* D_i *for a new 3-manifold* $M(D_i)$ *on an appropriate closed surface* ∂A_{n-1} *obtained from* ∂A_n *by "pinching off a handle".*

Proof. Examine the corresponding property for the framed link.

V. QUESTIONS AND SPECULATIONS

In our view, one of the most interesting things that we learned from the work reported here is that class-2 sewings constitute a group, which is in fact a very familiar group (see Theorem 4.6), and also a very restrictive subgroup of the mapping class group of a surface. A fascinating question is whether there might be a ultimate group or monoid of 3-manifolds, with multiplication defined as it is here by composition of sewings. To produce such a group it will be necessary to place further restrictions on sewings, because multiplication of class-2 sewings is not well defined on the equivalence class of all class-2 sewings which define the same M^3. Thus our first question concerns the equivalence relation on special sewings.

To make the question precise, recall first that if $h, h \in H(n)$, and if $M(h) = M(h')$, then the classical Reidemeister-Singer theorem, translated here into the language of sewings, tells us that there are "stable versions" h_s, h'_s of h, h', with $h_s, h_{s'} \in H(n+m)$, $m \geq 0$, such that $h'_s \in Bh_sB$, where $B = B(n+m)$. Call a sewing "special if $h \in A(n)$. (This is consistent with [2]). What is the equivalence relation on special sewings?

We conjecture:

(i) If $h, h' \in A(n)$, with $M(h) = M(h')$, then there are stable versions $h_s, h'_s \in A(n+m)$, $m \geq 0$, such that $h'_s \in (A \cap B)h_s(A \cap B)$, where $A \cap B = A(n+m) \cap B(n+m)$.

Remark. There are "special" 4-manifolds $W^4(h)$, $W^4(h')$ associated to h, h' (see [2]) and conjecture (i) implies stable equivalence between these, hence the stabilization involved in going from h, h' to h_s, h'_s necessarily involves overcoming index obstructions.

We were not able to prove (i), however, it is not necessary to establish (i) in order to study the equivalence relation on class-2 sewings. Suppose that (i) is true. First, we remark that the group $A \cap B$ was studied by the second author in [12]. Second, we note that if L is a class-2 link, and if $h \in R$ is the associated class-2 sewing, then any $h' \in (A \cap B)h$ will be special, and the link L' associated to h' will be equivalent to L. Finally, we note that if $h \in R$ is associated to L, then there exist elements[1] $f \in A \cap B$, $f \notin R$, such that $h'' = f^{-1}hf \in R$, however the link L'' associated to h'' may be inequivalent to L.

Consider the question of whether there might be a well-defined multiplication of 3-manifolds induced by composition of Heegaard sewings? We speculate on the following interesting question, which could be tackled quite independently of (i), or of the equivalence relation on class-2 sewings.

(ii) Is there a *normal* subgroup $N(n) \subset A(n)$ such that each $M^3 = M(h)$ for some n and some $h \in N(n)$?

Observe that (ii) may not be intractible, for the following reasons: we have been able to show that each M^3 is $M(h)$ for some class-3 sewing h. Now, class-3 $\subset$ class-2, however class-3

[1]*We have produced explicit examples, however, they are too complicated to describe here.*

sewings do not constitute a group. Thus, it seems possible to achieve a good deal of additional specialization beyond class-2, hence it does not seem unreasonable to ask whether (ii) is possible?

In a somewhat different direction, we speculate on the self-dual property established in Corollary 4.4. It suggests the question: can each M^3 be represented as $M(D)$ where D is self-dual? Now, we note that this is impossible, for the following reasons: If M^3 admits a self-dual diagram, then M^3 admits an orientation-reversing involution which interchanges the two Heegaard handlebodies associated to the splitting. However, there exist closed, orientable M^3 which admit no periodic homeomorhpisms [13]. Hence, we are led to ask the weaker question:

(iii) What are necessary and sufficient conditions for a manifold M^3 to admit a self-dual Heegaard diagram? What special properties do these manifolds enjoy?

Finally, we ask: in what precise ways are the 4-manifolds associated to our diagrams and links special? How are two such 4-manifolds related? Is there a special version of the link calculus which relates them?

REFERENCES

1. Birman, Joan S., Braids, links and mapping class groups, Annals of Mathematics Studies #82, Princeton University Press, 1974.
2. _______________, "Special Heegaard disgrams for 3-manifolds", Topology, to appear.
3. Kirby, R., "A calculus for framed links", Inventiones Math., to appear.
4. Lickorish, W.B.R., "A representation of orientable combinatorial 3-manifolds", Annals of Math., 76 (1962), pp. 531-538.
5. Rolfson, Dale, Knots and links, Publish or Perish, Inc., 1976.
6. Montesinos, Jose M., "Surgery on links and double branched covers of S^3" in Knots, Groups and 3-manifolds, Editor L.P. Neuwirth, Annals of Math. Studies #84, pp. 227-260, Princeton University Press 1975.

7. Hilden, Hugh, "Generators for two groups related to the braid group", Pac. J. of Math. 59, No. 2, 1975, pp. 475-486.
8. Fox, Ralph H., "A note on branched cyclic coverings of spheres", Rev. Mat. Hisp.-Amer. 32 (1972), pp. 158-166.
9. Suzuki, Shinichi, "On homeomphisms of a 3-dimensional handlebody", Can. J. Math. XXXIX, No. 1, 1977, pp. 111-124.
10. Kaplan, Steve J., "Constructing framed 4-manifolds with given almost-framed boundary", preprint.
11. Milnor, J. and Husemoller, D., Symmetric bilinear forms, Ergibnisse der mathematik, Band 73, Springer-Verlag, 1973.
12. Powell, Jerome. "Homeomorphisms of the 3-sphere that leave a Heegaard handlebody invariant", to appear.
13. Raymond, Frank and Tollefson, Jeffrey, "Closed 3-manifold which admit no periodic maps", Trans AMS 221, No. 2, 1976.
14. McMillan, D. R. Jr., "Homeomorphisms on a solid torus", Proc. AMS 14, No. 3, June 1963, pp. 386-390.
15. Ŝtanko, M. A., "Even surgeries in 3-manifolds", Proc. VI Topological Conference of Soviet Union, Tbilisi, October 2-7, 1972, p. 133 (in Russian).

SHRINKING COUNTABLE DECOMPOSITIONS OF E^3 INTO POINTS AND TAME CELLS

Dan Everett

Department of Mathematics
University of Georgia
Athens, Georgia

We show that if G is an upper semicontinuous decomposition of E^3 into points and countably many tame cells, then $E^3/G = E^3$. This generalizes a result of Bing (Annals of Math, Vol. 65 (1956), p. 363), who proved the case in which each nondegenerate element of G is a tame arc. Our result has been obtained independently by Micheal Starbird and Edythe P. Woodruff (these Proceedings).

I. INTRODUCTION

In [1] R. H. Bing proved that if G is an upper semi-continuous decomposition of E^3 whose only nondegenerate elements are countably many tame arcs, then E^3/G is homeomorphic to E^3. In this paper we extend Bing's result to the case where G has countably many nondegenerate elements, each of which is a tame arc, disk, or 3-cell in E^3. The principal tools of the proof are the following two theorems:

THEOREM. (R. L. Moore [4]*): Suppose G is an upper semicontinuous decomposition of E^2 such that no decomposition element separates E^2. Then $E^2/G = E^2$.*

ISBN 0-12-158860-2

THEOREM (Edythe P. Woodruff): Suppose G is an upper semicontinuous decomposition of E^3 *such that each point of* E^3/G *has arbitrarily small neighborhoods U such that FR U is a 2-sphere missing* $\pi(G^*)$. *Then* $E^3/G = E^3$.

Our plan of attack is as follows: first we consider any countable upper semicontinuous decomposition of E^3 all of whose nondegenerate elements are tame 3-cells. Choose any neighborhood of any element of G; within this neighborhood we can find a tame 2-sphere S^2 which contains the given element of G in its interior. In section 2 we show how to push the 1-skeleton of a triangulation of S^2 off the nondegenerate elements of G, and in section 3 we show how to push the 2-simplexes of S^2 (whose boundaries now miss the nondegenerage elements of G) off a single nondegenerate element of G. Section 4 contains the convergence arguments needed to iterate this procedure and reembed S^2 missing all the nondegenerate elements, thus completing the proof of this special case (Theorem 1). The general theorem (Theorem 2) is obtained in section 5 by "inflating" each nondegenerate element of G to a tame 3-cell and applying Theorem 1.

We use the following notation: if G is an upper semicontinuous decomposition of a manifold M, then G^* will denote the union of the nondegenerate elements of G. The decomposition space projection $M \to M/G$ will be called π. (If $X \subset M$, then $\pi^{-1}\pi(X)$ is the *saturation* of X, the union of all decomposition elements which meet X. If $\pi^{-1}\pi(X) = X$, then X is called *G-saturated*.) The decomposition space M/G is a metric space whose metric will be called d. If X is a subset of a metric space and ε is a positive number, then $N(X, \varepsilon)$ will denote the ε-neighborhood of X.

The author would like to thank Duane Loveland and David Wright for their helpful conversations. Also, an equivalent result has been obtained independently by Michael Starbird and Edythe P. Woodruff [6].

II. PUSHING A 1-COMPLEX OFF G^*

LEMMA 1. Let A be an arc in an open 2-manifold M; G a 0-dimensional upper semicontinuous decomposition of M; p a point of $A-G^*$; and V a neighborhood of p in M. Then V contains two points of $A-G^*$ which are not separated in V by any element of G.

Proof. By the upper semicontinuity of G, there is a neighborhood W of p such that $\pi^{-1}\pi(W) \subset V$. Let $q \neq p$ be another point of $A \cap W \cap (M-G^*)$. Let H be the 0-dimensional upper semicontinuous decomposition of M whose nondegenerate elements are the elements of G which separate p from q. (If no such elements exist, we are done.) To see that $H^* \cup \{p,q\}$ is closed in M, consider an element g of G which does not separate p from q, and let B be an arc from p to q in $M-g$. Then any element of G which lies sufficiently close to g misses B and so does not separate p from q. It follows easily that H is upper semicontinuous.

If g is an element of G, define $U(g)$ to be the union of the bounded components of $M-G$. If no such components exist, let $U(g) = \emptyset$. Let $a \neq p,q$ be a point of $A \cap W \cap (M-G^*)$; then a is not a limit point of H^*. Partition the nondegenerate elements of H into disjoint sets R and S, an element g of H belonging to R if and only if $a \in U(g)$. By the upper semicontinuity of H and the fact that a is not a limit point of H^*, there is an element g_0 of R such that $g_0 \subset U(g)$ for each $g \in R - \{g_0\}$. See Figure 1.

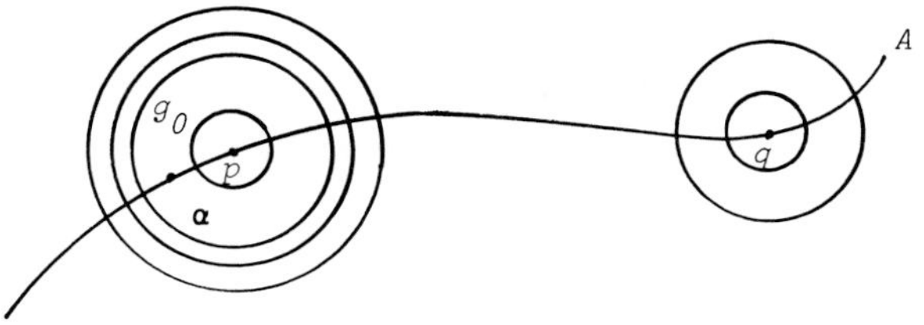

Fig. 1.

Let K' be the collection $\{g \cup U(g) \mid g \in G \text{ and } g \subset U(g_0)\}$. Let K be the set of elements of K' which are maximal with respect to inclusion. The upper semicontinuity of G and the fact that S is closed ensure that each element of K' lies in a unique element of K. In fact, K is a 0-dimensional upper semicontinuous decomposition of $U(g_0)$. Choose two points of $A \cap U(g_0)-K^*$; then these points satisfy the conclusion of the lemma, because each element of G in $U(g_0)$ lies in an element of K and no element of K separates V.

LEMMA 2. Suppose that D is a disk, H is a 0-dimensional upper semicontinuous decomposition of D such that $H^* \cap \partial D = \emptyset$, and A is an arc in D which meets ∂D exactly in its end points. Then there is a homeomorphism h of D, fixing ∂D, such that $h(A) \cap H^* = \emptyset$.

Proof. Let $A \cap \partial D = \{a,b\}$. As in the proof of lemma 1, for each $g \in H$ let $U(g)$ be the union of all the components of $D-g$ which do not contain ∂D. Let $K' = \{g \cup U(g) | g \in H\}$ and let K be the set of maximal elements of K' with respect to inclusion. Then K is an upper semicontinuous decomposition of D, none of whose elements separates D. Therefore Moore's theorem shows that $D/K = D$.

Let B be an arc in D/K from $\pi(a)$ to $\pi(b)$ which meets $\partial D/K$ exactly in $\{\pi(a),\pi(b)\}$. Write $\pi(H^*) = H_1 \cup H_2 \cup \dots$ where H_i is the image in D/K of those elements of H whose diameters are at least 2^{-i}. Then each H_i is closed in D/K. Let $f_1{:}B \to D/K-H_1$ be a reembedding, $f_2{:}B \to D/K-H_2$ be another reembedding close to f_1, etc. With care the sequence $f_1, f_2, \dots$ may be made to converge to a reembedding $f{:}B \to D/K-H^*$ which fixes $\{\pi(a),\pi(b)\}$ and takes $B-\{\pi(a),\pi(b)\}$ into $\pi(\text{Int}D)$. Then $\pi^{-1}f(B)$ is an arc in $D-H^*$ which joins a to b; and there is a homeomorphism h of D, fixed on ∂D, taking A to $\pi^{-1}f(B)$.

LEMMA 3. Suppose that S^2 *is a 2-sphere; K is a finite graph in* S^2*; L is a subcomplex of K such that K has no free vertices which do not lie in L; and G is a 0-dimensional upper semicontinuous decomposition of* S^2 *such that* $L \cap G^* = \emptyset$. *Then there is a homeomorphism h of* S^2 *which fixes L and such that* $h(K) \cap G^* = \emptyset$.

Proof. It is sufficient to consider the case where $Cl(K-L)$ is an arc with both end points in K.

Let U be the component of S^2-L which contains $K-L$. Then U is an open 2-cell, and there is a map $f : B^2 \to S^2$ such that:

f takes Int B^2 homeomorphically onto U;

there is an arc A in B^2 which meets ∂B^2 exactly in its end points, such that $f(A) = Cl\ (K-L)$. See Figure 2.

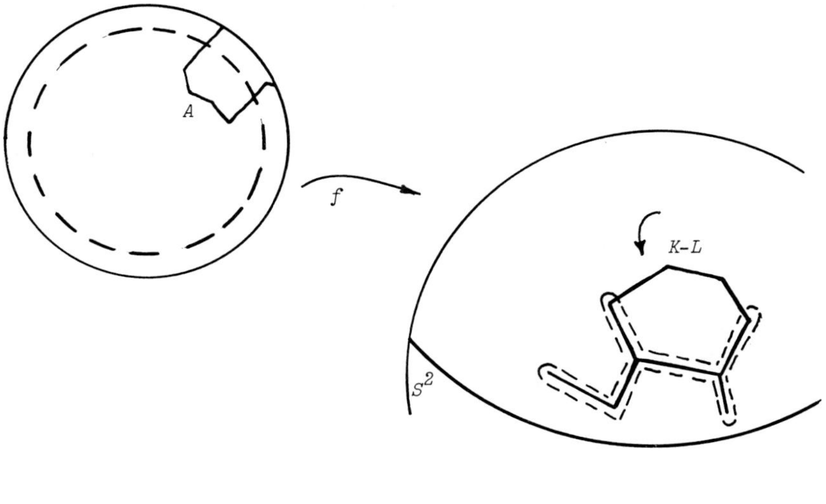

Fig. 2.

Use lemma 2 to find a homeomorphism g of B^2, fixed on B^2, such that $g(A) \cap f^{-1}(G^*) = \emptyset$. Then the required homeomorphism of S^2 is given by

$$h(X) = \begin{cases} fgf^{-1}(x) & \text{if} \quad x \in S^2 - L; \\ x & \text{if} \quad x \in L. \end{cases}$$

ADDENDUM TO LEMMA 3. If V is a G-saturated neighborhood of L, we may require that $f(K) \subset V$.

LEMMA 4. Suppose that G is an upper semicontinuous decomposition of E^3 into points and tame 3-cells; g_o is an element of G; S^2 is a tame 2-sphere in E^3 whose interior contains g_o; and T is the 1-skeleton of a triangulation of T. Then there is a homeomorphism h of S^2 such that $h(T) \cap G^* = \emptyset$, and no element of G separates $h(T)$ from g_o in $S^2 \cup \mathrm{Int}\, S^2$.

Proof. Let $B = S^2 \cup \mathrm{Int}\, S^2$. Let H be the 0-dimensional upper semicontinuous decomposition of S^2 whose nondegenerate elements are the components of $g \cap S^2$, where g is a nondegenerate element of G ([3, Thms 3-31, 3-39]). Let G_1 be the collection of elements of G which separate B, and let H_1 be the decomposition of S^2 whose nondegenerate elements are the components of $g \cap S^2$ which bound components of $g \cap B$ which separate B, for elements g of G_1. For each nondegenerate element g of H_1 let $U(g)$ be the union of all components of $S^2 - g$ which are separated from g_o in B by the element of G which contains g (Figure 3).

As in the proofs of lemmas 1 and 2, let $K' = \{g \cup U(g) \mid g \in H_1\}$ and let K be the upper semicontinuous decomposition of S^2 whose nondegenerate elements are the maximal elements of K' with respect to inclusion. Let p be the decomposition space projection map $S^2 \to S^2/K$. By Moore's theorem, S^2/K is a 2-sphere. Furthermore, there is an upper semicontinuous decomposition $p(H)$ of S^2/K

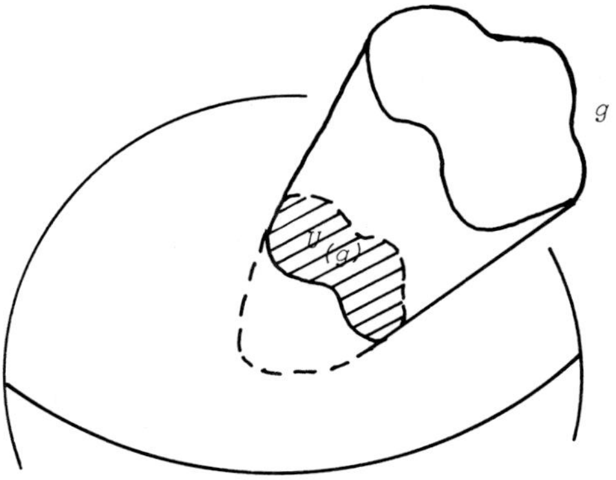

Fig. 3.

defined by $p(H) = \{p(g) | g \in H\}$. (Note that perhaps $p(g_1) = p(g_2)$ for $g_1 \neq g_2$.) By lemma 1 there are points a and b of $S^2/K-p(H)^*$ which are not separated by any element of $p(H)$. Let A be an arc in $S^2/K-p(H)^*$ joining a to b; such an arc may be constructed by the methods of lemma 2. Choose points c, d of $A-p(K^*)$. Then $p^{-1}(c)$ and $p^{-1}(d)$ are points of S^2-H^* which are not separated in S^2 by any elements of H. Therefore the techniques of Lemma 2 may be used once more to construct an arc A' in S^2-H^* joining $p^{-1}(c)$ and $p^{-1}(d)$. By construction, no element of G_1 separates A' from g_o in B; hence, no element of G separates A' from g_o in B.

We complete the proof by letting h_1 be a homeomorphism of S^3 which takes some 1-simplex of T onto A', and h_2 be a homeomorphism of S^2 which pushes T off G^* fixing A' (from lemma 3). Then $h = h_2h_1$ is the required homeomorphism.

ADDENDUM TO LEMMA 4. If T' is a subcomplex of T such that no element of G separates T' from g_o in $S^2 \cup$ Int S^2, $T' \cap G^* = \emptyset$, and V is a neighborhood of T' in S^2, then we may require that h fix T' and that $h(T) \subset V$.

III. PUSHING 2-CELLS OFF ELEMENTS OF G

Throughout this and the next section G is an upper semicontinuous decomposition of E^3 into points and tame 3-cells.

Definition. An ordered pair of embeddings (f, f') of a triangulated 2 sphere S^2 into E^3 is said to have *property A* if and only if, whenever σ and σ' are distinct 2-simplexes of S^2 and g is an element of G which meets both $f'(\sigma)$ and $f'(\sigma')$, then g meets both $f(\sigma)$ and $f(\sigma')$.

Property A seems to be the key to the proof of the main results of this paper, for the following reason. In Figure 4 below we have drawn the first two stages of the defining sequence for a "peculiar decomposition" introduced by Bing [2]. This decomposition has only countably many nondegenerate elements, each of which is cellular in E^3, but the decomposition space is not E^3. Figure 4 shows part of a triangulated 2-sphere S^2 in E^3. There

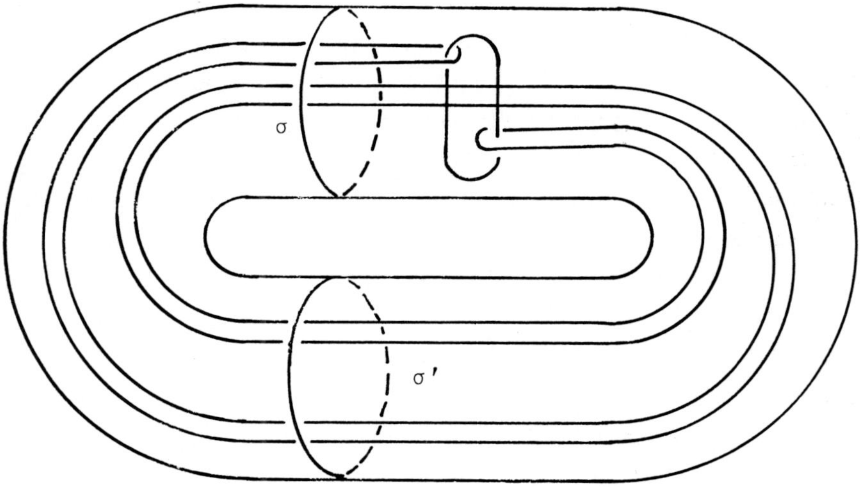

Fig. 4.

is a reembedding of S^2 in E^3, fixed on the 1-skeleton of S^2, which takes S^2 off those nondegenerate elements of the decomposition which meet S^2; but this reembedding does not satisfy property *A*.

LEMMA 5. *Suppose that D is a nondegenerate element of G, U is a neighborhood of D, S^2 is a triangulated 2-sphere, and $f: S^2 \to E^3$ is a tame embedding such that the image of the 1-skeleton of S^2 under f misses D.*

Then there is a PL embedding $f': S^2 \to E^3 - D$ such that

i) $f' = f$ on the component of $S^2 - f^{-1}(U)$ which contains the 1-skeleton;

ii) (f, f') satisfies property A.

Proof. Using a small "general position" homeomorphism of E^3, fixed off U, we may assume that the intersection $f(S^2) \cap \mathrm{Bd}\, D$ is the union of finitely many pairwise disjoint simple closed curves, each of whose inverse images under f lies in the interior of some 2-simplex of S^2. Let $f_o = f$ and suppose a PL embedding $f_i: S^2 \to E^3$ has been constructed. We will show how to construct a PL embedding $f_{i+1}: S^2 \to E^3$ so that $f_{i+1}(S^2)$ has fewer intersections with S^2 and (f_i, f_{i+1}) satisfies property *A*. Since property *A* defines a transitive relation among functions, this will prove the lemma.

Step 1. Let $S_1, \ldots, S_n$ be the components of $\mathrm{Bd}\, D \cap f_i(S^2)$ whose inverse images under f_i are inmost on some 2-simplex of S^2. Let $I(S_j)$ be the image under f_i of $f_i^{-1}(S_j)$ together with the interior of $f_i^{-1}(S_j)$ in the 2-simplex of S^2 which contains it. Each S_j bounds two disks in $\mathrm{Bd}\, D$, which we call $C_1(S_j)$ and $C_2(S_j)$. For each j we will choose one of $C_1(S_j)$ and $C_2(S_j)$ to be called $C(S_j)$, and the 3-cell in E^3 bounded by $C(S_j) \cup I(S_j)$ we will call E_j. Here is the inductive procedure for choosing $C(S_j)$:

If $I(S_j)$ is not contained in D, choose $C(S_j)$ so that E_j does not contain D (see Figure 5).

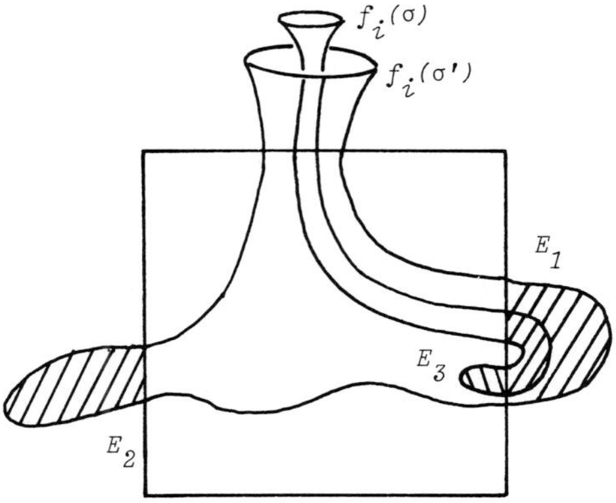

Fig. 5.

If $I(S_j) \subset D$ and $I(S_j)$ is not contained in E_k for any $k < j$, choose $C(S_j) = C_1(S_j)$.

If $I(S_j) \subset D$ and $I(S_j) \subset E_k$ for some $k < j$, choose $C(S_j)$ to be the one of $C_1(S_j)$, $C_2(S_j)$ which is contained in E_k. Then $E_j \subset E_k$. This selection process assures that if any two of the E_j's intersect, one is a subset of the other.

Step 2. Let $I = \{j \in \{1, \ldots, n\} | E_j$ is not a subset of D or of any E_k, $k \neq j\}$. For each $j \in I$ choose a neighborhood U_j of E_j so that $U_j \cap f(S^2)$ is connected, and the U_j's are pairwise disjoint. For each $j \in I$ construct a homeomorphism h_j of E^3 which is fixed off U_j and takes $U_j \cap f_i(S^2)$ into Int D. Let h be the composition of all the h_j's (if $I = \emptyset$, let $h - id$) and let $g_i = hf_i$. Then (f_i, g_i) satisfies property A, since $g_i(S^2) - f_i(S^2) \subset$ Int D. See figure 6.

Step 3. By step 2 we may assume that $E_j \subset D$ for each j. Choose m so that E_m is an innermost one of the E_j's, and let σ be the 2-simplex of S^2 which contains $f_i^{-1}(E_m)$. Using a bicollar on *Bd D* we can find a neighborhood $V \subset U$ of E_m whose intersection

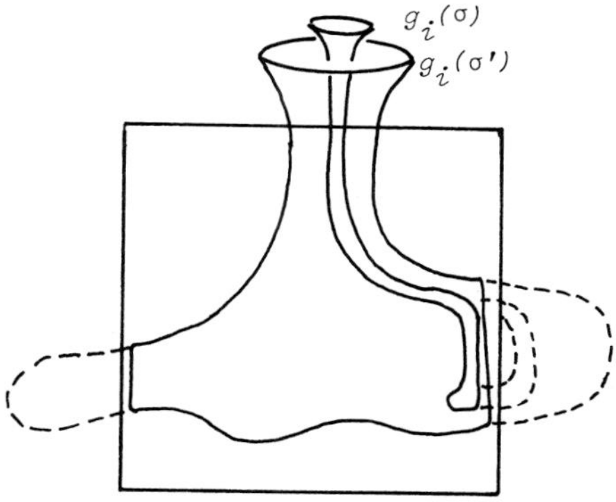

Fig. 6.

with $f_i(S^2)$ is a disk containing S_m. Choose $\eta > 0$ so that $N(C(S_m), \eta) \subset V$, and let W be a neighborhood of D so that $W \cap V \cap f_i(S^2)$ is a disk and $\pi^{-1}\pi(W) \subset N(D, \eta)$. Let $f_{i+1}: S^2 \to E^3$ be a PL embedding which is identical with f_i on $S^2 - f_i^{-1}(W \cap V)$ and sends $f_i^{-1}(W \cap V)$ to a disk K parallel to $C(S_m)$ in $W \cap V - D$.

Figure 7 shows why (f_i, f_{i+1}) (more precisely, (g_i, f_{i+1})) has property A. Suppose that g is an element of G which meets both $f_i(\sigma)$ and $f_i(\sigma')$, where σ and σ' are distinct 2-simplexes of S^2 and $f_i^{-1}(S_m) \subset \sigma$. We claim that g meets both $f_i(\sigma)$ and $f_i(\sigma')$. This could only be in doubt if $g \cap K = \emptyset$; but then since $\pi^{-1}\pi(K) \subset N(D, \eta)$, it must be that g intersects the annulus $(V \cap f_i(\sigma)) - D$, proving our claim and the lemma.

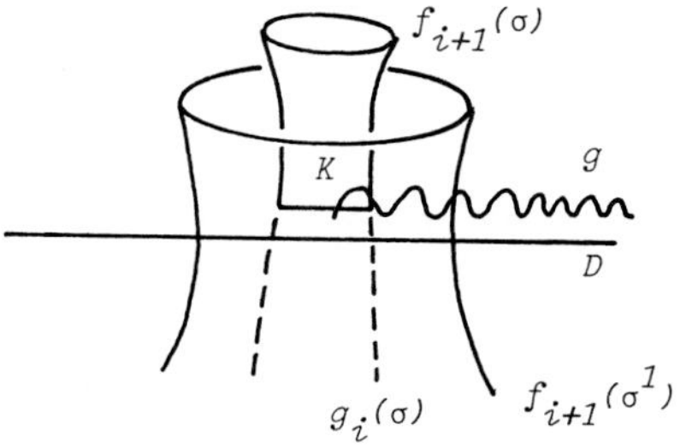

Fig. 7.

ADDENDUM TO LEMMA 5. There are pairwise disjoint disks $D_1, \ldots, D_k$ *in* $S^2 - T^1$ *such that*

$f'(\mathrm{Fr}D_i) \cap G^* = \emptyset$ *for each* i;

$f' = f$ *on* $S^2 - \cup D_i$;

$f'(D_i) \subset U$ *for each* i.

Proof of Addendum. Let $K_1, \ldots K_k$ be the disks whose interiors are the components of $f'(S^2) - f(S^2)$. By construction $\pi^{-1}\pi(K_j) = \emptyset$ if $i \neq j$, and $\pi^{-1}\pi(K_i) \subset U$ for each i. Let K be the decomposition of $f'(S^2)$ whose nondegenerate elements are $K_1, \ldots, K_n$, and let $p : f'(S^2) \to f'(S^2)/K$ be the decomposition space projection map. Using the techniques of section 2, construct pairwise disjoint simple closed curves $C_1, \ldots, C_k$ in $p(S^2 \cap U - G^*)$ such that the C_i's bound pairwise disjoint disks $\{D'_i\}$ in $p(S^2 \cap U)$ with $p(K_i) \subset D'_i$ for each i. Then the disks $D_i = f'^{-1}\, p^{-1}\, (D'_i)$ satisfy the addendum.

IV. DECOMPOSITIONS OF E^3 INTO POINTS AND TAME 3-CELLS

Recall that G is an upper semicontinuous decomposition of E^3 into points and (countably many) tame 3-cells, with decomposition space projection map $\pi : E^3 \to E^3/G$.

Definition. A decomposition H of E^3 is called a *closed subdecomposition of* G if the nondegenerate elements of H form a closed subset of the nondegenerate elements of G (in the topology of E^3/G). It follows immediately that H is upper semicontinuous.

LEMMA 6. Suppose that H *is a closed subdecomposition of* G, S^2 *is a triangulated 2-sphere,* $f : S^2 \to E^3$ *is a tame embedding which takes the 1-skeleton* T' *of* S^2 *into* $E^3 - G^*$, U *is a neighborhood of* H^*, *and* $\varepsilon > 0$. *Suppose further that* $E_1, \ldots, E_k$ *are disks in* S^2 *and* $V_1, \ldots, V_m$ *are* G*-saturated open sets in* E^3 *such that*

$Fr V_i \cap Fr V_j = \emptyset$ *whenever* $i \neq j$;

$Fr E_i \cap Fr E_j = \emptyset$ *whenever* $i \neq j$;

$f(Fr E_i) \cap G^* = \emptyset$ *for each* i; *and*

for each i, $f(E_i) \subset V_j$ *for some* j.

Then there are pairwise disjoint disks $D_1, \ldots D_n$ *in* $S^2 - T^1$, *G-saturated open sets* $U_1, \ldots U_m$ *in* E^3, *and a tame embedding* $f' : S^2 \to E^3 - H^*$ *such that:*

$f'(S^2) - f(S^2) \subset U_1 \cup \ldots \cup U_m$;

for each i, $f'(D_i) \subset U_j$ *for some* j;

$f'(Fr D_i) \cap G^* = \emptyset$ *for each* i;

$\operatorname{diam} \pi(U_i) < \varepsilon$ *for each* i;

for any i *and* j, *if* $U_i \cap V_j \neq \emptyset$ *then* $U_i \subset V_j$;

for any i *and* j, *if* $D_i \cap E_j \neq \emptyset$ *then* $D_i \subset E_j$;

$Fr U_i \cap Fr U_j = \emptyset$ *if* $i \neq j$;

(f, f') *has property* A.

Proof. Let $g_1, g_2, \ldots$ be the nondegenerate elements of H, and let U_1 be a G-saturated open neighborhood of g_1 such that $\operatorname{diam} \pi(U_1) < \varepsilon$. Also, if $g_1 \subset V_j$ for some j, choose U_1 to be a subset of V_j; otherwise, choose U_1 so that $U_1 \cap V_j = \emptyset$ for each j. By lemma 5 and its addendum there are pairwise disjoint disks $D_{11}, \ldots D_{1K(1)}$ in $S^2 - T^1$ and a tame embedding $f_1 : S^2 \to E^3 - g_1$ such that:

$f_1 = f$ on $S^2 - \cup D_i$;

$f_1(D_{1i}) \subset U_1$ for each i; and

(f, f_1) has property A.

Since $Fr\, E_j \cap G^* = \emptyset$ for each j, we may choose U_1 to be so close to g_1 that $FrE_1, \ldots, FrE_k, FrD_{11}, \ldots FrD_{1k(1)}$ are pairwise disjoint. Let $W_1 \subset U_1$ be a G-saturated open neighborhood of g_1 such that $f_1(S^2) \cap W_1 = \emptyset$.

Notice that f_1 satisfies the hypotheses of the lemma, with $\{E_1, \ldots, E_k\}$ replaced by $\{E_1, \ldots, E_k, D_{11}, \ldots D_{1k(1)}\}$ and $\{V_1, \ldots, V_k\}$ replaced by $\{V_1, \ldots, V_k, U_1\}$. Therefore we can proceed inductively to define tame embeddings $f_2, f_3, \ldots$ of S^2 into E^3; collections of pairwise disjoint disks $\{D_{2j}\}, \{D_{3j}\}, \ldots$; G-saturated open subsets $U_2, U_3, \ldots$ of E^3; and G-saturated open neighborhoods $W_2, W_3, \ldots$ of $g_2, g_3, \ldots$ such that:

$f_i(FrD_{ij}) \cap G^* = \emptyset$ for each i and j;

for $i' > i$, either $U_{i'} \subset U_i$ or $U_{i'} \cap U_i = \emptyset$;

for $i' > i$ and any j, j', either $D_{i'j'} \subset D_{ij}$ or $D_{i'j'} \cap D_{ij} = \emptyset'$;

$f_i(S^2) \subset E^3 - W_i$;

for $i' > i$, $U_{i'} \cap W_i = \emptyset$ (if $g_{i'} \subset W_i$, let $U_{i'} = \emptyset$);

diam $\pi(U_i) < \varepsilon$ for each i;

for $i' \geq i$ and each j, $f_{i'}(D_{ij}) \subset U_i$ and $f_{i'}(E_j) \subset V_j$;

$f_{i+1}(S^2) - f_i(S^2) \subset U_{i+1}$ for each i;

(f_i, f_{i+1}) has property A.

These properties assure that for $i' \geq i$, $f_{i'}(S^2) \cap W_i = \emptyset$. We now use this fact to show that for some N, $f_N(S^2) \cap H^* = \emptyset$. If this is not so, then there is an increasing sequence $i(1), i(2), \ldots$ such that $f_{i(j)}(S^2) \cap g_{i(j+1)} \neq \emptyset$ for each j. Since H is a closed subdecomposition of G and $\pi f_i(S^2) \subset N(\pi f(S^2), \varepsilon)$ for each i, there is a nondegenerate element g_m of H and a subsequence $\{k(j)\}$ of $\{i(j)\}$ such that $\{g_{k(j)}\} \to g_m$. But for large enough j, $g_{k(j)} \subset W_m$ and $k_{(j)} > m$, so $f_{k(j-1)}(S^2) \cap g_{k(j)} = \emptyset$, a contradiction. Therefore there is some N such that $f_N(S^2) \cap H^* = \emptyset$; let $f' = f_N$.

THEOREM 1. Let G be an upper semicontinuous decomposition of E^3 into points and tame 3-cells. Then $E^3/G = E^3$.

Proof. By Woodruff's theorem, quoted in the introduction, it is enough to consider an element g_o of G and a G-saturated open neighborhood U of g_o, and show that $U-G^*$ contains a 2-sphere which contains g_o in its interior.

Let $f:S^2 \to U$ be a PL embedding of a 2-sphere into U such that $g_o \subset \text{Int } f(S^2)$. Let $T_1, T_2, \ldots$ be triangulations of S^2 such that T_{i+1} is a subdivision of T_i for each i and mesh $T_i \to 0$. Let T_i^1 denote the carrier of the 1-skeleton of T_i. Choose a point $p \in f(S^2) - G^*$ which is not separated from g_o in Int $f(S^2)$ by any element of G. By lemma 4 there is a homeomorphism h of E^3 which takes $f(S^2)$ to $f(S^2)$ and such that

$$p \in hf(T_1^1) \qquad \text{and}$$

$$hf(T_1^1) \cap G^* = \emptyset$$

Let $g_1 = hf$, and let M be a G-saturated compact continuum in Int $f(S^2) \cup \{p\}$ which contains p and g_o. (To construct M, draw an open arc from p to g_o in Int $f(S^2) - G^*$). If $g \in G$ and $g \cap S^2 \neq \emptyset$, let $U_o(g)$ denote the union of all the components of $f(S^2)-g$ which do not contain p. Let G_1 be the closed subdecomposition of G whose nondegenerate elements are the elements g of G such that either

1) diam $g \geq 1/2$;

2) diam $\pi(g \cup U_o(g)) \geq 1/4$; of

3) there are disjoint 2-simplexes σ and σ' of T_1 such that $g \cap g_1(\sigma) \neq \emptyset$ and $g \cap g_1(\sigma') \neq \emptyset$. The proof that the elements of G satisfying (1) form a closed subset of E^3/G is a standard exercise, and the corresponding proofs for (2) and (3) are similar.

By lemma 6 there are pairwise disjoint disks $D_{11}, \ldots D_{1k(1)}$ in $S^2 - T_1^1$, pairwise disjoint open subsets $U_{11}, \ldots U_{1m(1)}$ of E^3, and a tame embedding $f_1:S^2 \to E^3 - G_1^*$ such that:

$f_1 = g_1$ on $S^2 - \cup_i D_{1i}$

$f_1(FrD_{1i}) \cap G^* = \emptyset$ for each i;

for each i, $f_1(D_{1i}) \subset U_{1j}$ for some j;

diam $\pi(U_{1j}) < 1/2$ for each j;

(f, f_1) has property A.

It follows that if σ σ' are disjoint 2-simplexes of T_1, $\pi f_1(\sigma) \cap f_1(\sigma') = \emptyset$.

We pause briefly for an important technical definition. Suppose that $E{:}S^2 \to E^3$ is an embedding of a triangulated 2-sphere into E^3 such that f takes the 1-skeleton T^1 of S^2 into $E^3 - G^*$ and such that, if σ_1 and σ_2 are disjoint 2-simplexes of S^2, $\pi E(\sigma_1) \cap \pi E(\sigma_2) = \emptyset$. Define $D{:}S^2 - T^1 \to (0, \infty)$ as follows. If x lies in the interior of a 2-simplex σ of S^2, let $D(x)$ be the smaller of $d(\pi E(x), \pi(T^1))$ and $\min \{d(\pi E(x), \pi(\sigma^1) \mid \sigma^1$ is a 2-simplex of S^2 disjoint from $\sigma\}$. For any 2-simplex σ of S^2, let $W(\sigma) = \pi^{-1} (\bigcup_{x \in \sigma} N\ \pi E(x), \frac{1}{3} D(x))$. Then $W(\sigma)$ is an open, G-saturated neighborhood of $E(\sigma)$, and if σ and σ' are disjoint then $Cl\ W(\sigma) \cap Cl\ W(\sigma') = \emptyset$.

Suppose inductively that for $i = 1, 2, \ldots, n-1$ we have defined collections $\{D_{ij}\}_{j=1}^{k(i)}$ of pairwise disjoint disks in $S^2 - T_i^{\,1}$ and $\{U_{ij}\}_{j=1}^{1(i)}$ of G-saturated open subsets of E^3; closed subdecompositions G_i of G; and tame embeddings $g_i, f_i{:}S^2 \to E^3 - G_i^*$, satisfying all of the following inductive hypotheses:

IH1) $g_i(T_i^{\,1}) \cap G^* = \emptyset$;

IH2) $d(\pi g_{i+1}(x), f_i(x)) < 2^{-i}$ for each $x \in S^2$;

IH3) $g_{i+1} = f_i$ on $T_i^{\,1} + \cup_j FrD_{ij}$;

IH4) G_i contains all elements of G with diameters $\geq 2^{-i}$;

IH5) for $i' \geq i$ either $U_{ij} \cap U_{i'j'} = \emptyset$ or $U_{i'j'} \subset U_{ij}$;

IH6) for $i' \geq i$ either $D_{ij} \cap D_{i'j'} = \emptyset$ or $D_{i'j'} \subset D_{ij}$;

IH7) diam $\pi(U_{ij}) < 2^{-i}$ for each i and j;

IH8) for $i' \geq i$, $f_{i'}(D_{ij}) \subset U_{ik}$ for some k;

IH9) $f_i = g_i$ on $S^2 - \cup_j D_{ij}$;

IH10) $f_i(Fr D_{ij}) \cap G^* = \emptyset$ for each j;

IH11) $f_i f^{-1}(p) = p$;

IH12) for each $g \in G$ such that $g \cap f_i(S^2) \neq \emptyset$, either $f_i^{-1}(g \cup U_i(g)) \subset D_{ij}$ for some j, or diam $\pi(g \cup U_i(g)) < 2^{-i-1}$, where $U_i(g)$ is the union of all components of $f_i(S^2) - g$ which do not contain p;

IH13) if σ is a 2-simplex of T_i and $i' \geq i$, then $f_{i'}(\text{Int } \sigma) \subset W(\sigma)$;

IH14) for each i, $M - \{p\} \subset \text{Int } f_i(S^2)$; and

IH15) for $i' > i$, $d(\pi f_{i'}(S^2), \pi H_i^*) > \frac{1}{2} d(\pi f_i(S^2), \pi H_i^*)$.

By applying lemma 4 and its addendum to each D_{ij} we can construct a homeomorphism g of $f_{n-1}(S^2)$ such that:

$$g = id \text{ on } S^2 - \cup_{i,j} D_{ij};$$

$$g f_{n-1}(T_n' \cap (\bigcup_{i,j} D_{ij})) \cap G^* = \emptyset.$$

Applying lemma 4 and its addendum to $S^2 - \cup_{i,j} D_{ij}$, we can construct a homeomorphism g' of $f_{n-1}(S^2)$ such that:

$$g' = id \text{ on } \cup_{i,j} D_{ij} \text{ and on } T_{n-1}';$$

$$g' g f_{n-1}(T_n^1) \cap G^* = \emptyset;$$

$$d(g'(x), x) < 2^{-n} \text{ for each } x \in f_{n-1}(S^2).$$

(That this last condition is possible depends on *(IH12)*.) Let $g_n - g' g f_{n-1}: S^2 \to E^3$, and note that g_n satisfies *(IH1)*, *(IH2)*, *(IH3)*.

Let G_n be the closed subdecomposition of G whose nondegenerate elements are the elements g of G satisfying at least one of the following three properties:

diam $g \geq 2^{-n}$;

diam $\pi(g \cup U_{n-1}(g)) \geq 2^{-n-1}$; or

there are disjoint 2-simplexes σ_1 and σ_2 of T_n such that both $g_n(\sigma_1)$ and $g_n(\sigma_2)$ intersect g. (Note that our choice of G_n satisfies *(IH4)*.) Using lemma 6 we can construct pairwise disjoint disks $D_{n1},\ldots,D_{nk(n)}$ in $S^2 - T_n^{\,1}$, pairwise disjoint G-saturated open sets $U_{n1},\ldots U_{nm(n)}$ in E^3, and a tame embedding $f_n: S^2 \to E^3$ such that:

for any i, j, and k, either $U_{nk} \cap U_{ij} = \emptyset$ or $U_{nk} \subset U_{ij}$;
for any i, j, and k either $D_{nk} \cap D_{ij} = \emptyset$ or $D_{nk} \subset D_{ij}$;
diam $\pi(U_{nk}) < 2^{-n}$ for each j;
$f_n(FrD_{nj}) \cap G^* = \emptyset$ for each j;
for each j, $f_n(D_{nj}) \subset U_{nm}$ for some m;
$f_n = g_n$ on $S^2 - \cup_j D_{nj}$;

If $\sigma \in T_i$ and $f(\sigma) \cap U_{nj} \neq \emptyset$, then $U_{nj} \subset W(\sigma)$. The first three of these conditions assure that *(IH5)*, *(IH6)*, and *(IH7)* are satisfied, while the next three conditions assure that *(IH8)*, *(IH9)*, *(IH10)*, *(IH11)*, and *(IH12)* are satisfied. The last condition, which is possible since $W(\sigma)$ is open and G-saturated, assures that *(IH13)* is satisfied. The remsining two inductive hypotheses, *(IH14)* and *(IH15)*, can be satisfied by judicious use of our ability (from lemma 6) to require $f_n(S^2)$ to lie in an arbitrarily close G-saturated neighborhood of $f_{n-1}(S^2)$.

We now complete the proof of Theorem 1 by showing that the sequence $\{f_i\}$ converges to an embedding $F: S^2 \to E^3 - G^*$ such that $g_o \subset \text{Int } F(S^2)$. First, we show that $\{\pi f_i\}$ converges to a function by considering the series

$$\sum_{i=1}^{\infty} d(\pi g_{i+1}(x), \pi f_i(x))$$

and

$$\sum_{i=1}^{\infty} d(\pi f_i(x), \pi g_i(x)).$$

The convergence of the first series is a consequence of *(IH2)*. To see that the second series converges, consider a point $x \in S^2$ and a number i for which $f_i(x) \neq g_i(x)$. Then $x \in D_{ij}$ for some j, and it follows from *(IH7)* and *(IH8)* that for each $i' > i$, $d(\pi f_{i'}(x), \pi f_i(x)) < 2^{-i}$. Therefore the second series converges, so $\{\pi f_i\}$ converges to a function $F' : S^2 \to E^3/G$. A similar argument shows that F' is continuous.

To see that F' is an embedding, consider distinct points x and y of S^2. For some large enough i, x and y do not lie in closed simplexes of T_i which meet. Then *(IH3)* and *(IH13)* show that $F'(x) \neq F'(y)$. By *(IH15)*, $F'(S^2) \cap \pi(G^*) = \emptyset$ so $F = \pi^{-1} F' : S^2 \to E^3 - G^*$ is an embedding. Finally, $g_o \subset \text{Int } F(S^2)$ by *(IH14)*. This completes the proof of Theorem 1.

We remark that it is possible to use the techniques presented here to prove Theorem 1 without using Woodruff's theorem. However, the details would be even worse than what we have presented here. Here is an outline of the proof. Let $T = T_1^2 \cup T_2^2 \cup \ldots$ be the union of the 2-skeleta of a sequence of finer and finer subdiisions of E^3, such that each component of $E^3 - T$ is a point. Construct a reembedding $f: T \to E^3 - G^*$ such that each component of $E^3 - f(T)$ is an element of G. Then there is a homeomorphism from E^3/G to E^3 which takes $\pi f(x)$ to x for each $x \in T$, and takes each point of $\pi(G^*)$ to a point of $E^3 - T$.

V. COUNTABLE DECOMPOSITIONS OF E^3 INTO POINTS AND TAME CELLS.

THEOREM 2. Let G be an upper semicontinuous decomposition of E^3 having only countably many nondegenerate elements, each of which is a tame arc, disk, or 3-cell. Then $E^3/G = E^3$.

Proof. By a theorem of Schurle [5], there is an upper semicontinuous decomposition H of E^3 into points and tame 3-cells such that $E^3/G = E^3/H$. Hence, $E^3/G = E^3$. (Actually Schurle's theorem yields a somewhat stronger version of Theorem 2, in which "tame arc, disk, or 3-cell" is replaced by "cellular set with mapping cylinder neighborhood".

Figure 8 depicts the first two stages of a process which produces such a decomposition H: the tame 3-cells $f_1^{-1}(g_1)$ and $f_1^{-1} f_2^{-1}(g_2)$ are elements of H. Note that the "deflating maps" $f_1, f_2, \ldots$ may be chosen to be the final stages of small pseudo-isotopies; thus the sequence $f_1, f_2f_1, \ldots$ can be made to converge to a uniformly continuous function.

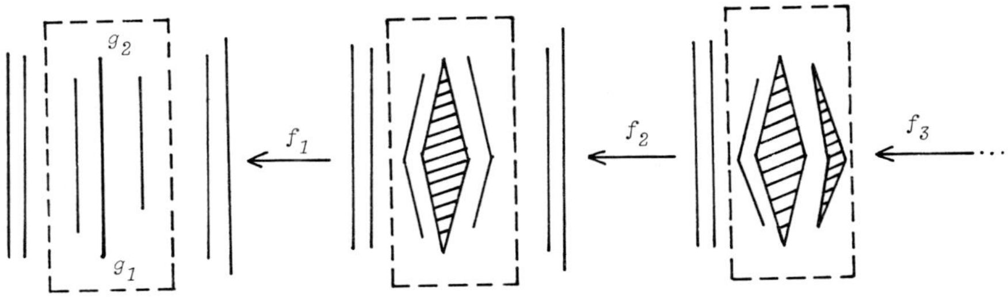

Fig. 8.

REFERENCES

1. Bing, R. H., Upper semicontinuous decompositions of E^3, Ann. of Math 65 (1957), pp. 363-374
2. __________, A peculiar decomposition of E^3, Abstract 452, Bull. Amer. Math. Soc. 62 (1956), p. 365
3. Hocking, John G., and Young, Gail S., Topology, Reading, Mass.: Addison-Wesley, 1961; p. 137.
4. Moore, R. L., Concerning upper semi-continuous collections of continua, Trans. Amer. Math. Soc. 27 (1925), pp. 416-428.
5. Schurle, Arlo W., Decompositions of E^3 into strongly cellular sets, Ill. J. of Math 18 (1974), pp. 160-164.
6. Starbird, Michael and Woodruff, Edythe P., Decompositions of E^3 with countably many nondegenerate elements, these Proceedings.
7. Woodruff, Edythe P., Decomposition spaces having arbitrarily small neighborhoods with 2-sphere boundaries, Trans. Amer. Math. Soc. (to appear).

LOCAL SPANNING MISSING LOOPS

Gene G. Garza

Department of Mathematics
University of Georgia
Athens, Georgia

In [2], *Burgess showed that a 2-sphere in* E^3 *is flat if it can be locally spanned in its complementary domains. The definition was then modified and special cases were solved by Burgess and Loveland. Here we present still more special cases - the main one being what we call "local spanning missing loops". This essentially says that if, in each neighborhood of each point on the 2-sphere, the spanning disks may be chosen to miss small loops in these neighborhoods, then the 2-sphere is flat.*

DEFINITION 1. We say that a 2-sphere Σ can be *locally spanned in the component U of* $E^3 - \Sigma$ *missing loops* if for each $p \in \Sigma$ and each $\varepsilon > 0$ there is an open ε-*nghd* $N(p)$ and an ε-disk $D \subset N(p) \cap \Sigma$ with $p \in \overset{\circ}{D}$ such that for any loop α in $U \cap N$ and any $\delta > 0$ there is an ε-disk D' in $U \cap N$ with $D' \cap \alpha = \emptyset$ and a map f of BdD onto BdD' that moves no point more than a distance δ.

All other definitions (except for def. 2) and notation are standard (see [3]).

THEOREM 1. Suppose Σ *is a 2-sphere in* E^3 *which can be locally spanned in the component U of* $E^3 - \Sigma$ *missing loops. Then* Σ *is tame from U.*

ISBN 0-12-158860-2

Proof. We shall show that U is *1-LC* with respect to polyhedral unknotted simple closed curves (see Hempel [6]) at each point of Σ. We first show that either each point of Σ is a *1-LC* point for U or that at each point p which is not *1-LC* there exist for each $\varepsilon > 0$, ε - disks $D \subset \Sigma$ with $p \in \overset{\circ}{D}$ and $\ni BdD_\varepsilon$ lies inside a locally flat annulus on Σ and BdD_ε is also the boundary for an ε-disk D' which lies inside U except for BdD'_ε. Then we show U is also *1-LC* at p.

Let $p \in W$, the set of wilds points of Σ. Suppose U is not *1-LC* at p. As in the definition we choose a small ε-*nghd* N_p and spanning disk D_p. Choose $D_{p'}$, a slightly larger disk on Σ which contains D_p and which lies in N_p. We shall show the open annulus A bounded by $Bd(D_p \cup D_{p'})$ contains no points of W, i.e., A is locally flat. A will be locally flat if U is *1-LC* at each point of A ([3]) so we pick a point $q \in A$ and show that U is *1-LC* at q. Choose a *nghd* N_q and spanning disk $D_q \subset A$ for $q \ni N_q \cap D_p = \emptyset$. Choose disjoint spherical *nghds* $B_{r_1}(p)$ and $B_{r_2}(q) \ni B_{r_2}(q) \subset N_q, B_{r_2}(q) \cap \Sigma \subset D_q, B_{r_1}(p) \subset N_p$, and $B_{r_1}(p) \cap \Sigma \subset \overset{\circ}{D}_p$. Let α_p be an arbitrary polyhedral unknotted loop in $B_{r_1}(p) \cap U$ and α_q a similar loop in $B_{r_2}(q) \cap U$. Let D_{α_p} and D_{α_q} be polyhedral disks bounded by α_p and α_q (respectively) which lie inside $B_{r_1}(p)$ and $B_{r_2}(q)$. These disks are disjoint and our task is to show that they do not hit Σ, i.e., after some small adjustment. If they hit Σ, they do so inside D_q and D_p and of course inside $D_{p'}$, if we assume D_q and N_q are chosen small enough. Thus by using the side approximation theorem ([7]) we may join D_{α_q} and D_{α_p} by an arc which is then fattened up into a third polyhegral disk $D_0 \subset U \cap N_p$ which intersects each disk in only a small subarc of its boundary. Let ℓ_p, ℓ_q, L_p and L_q denote the subarcs of α_p and α_q with $\ell_p \cup L_p = \alpha_p$ and $\ell_q \cup L_q = \alpha_q$ where ℓ_p and ℓ_q denote the subarcs which form part of the boundary of D_0. We note that the loop $P = BdD \cup L_p \cup L_q - \ell_p - \ell_q$ lies inside N_p and D_p can therefore be spanned by a disk D'_p which misses P. We must show that D'_p can also be assumed to miss $\ell_p \cup \ell_q$. To do this we choose D'_p to miss P and the associated map $f_{D_p}: BdD_p \to BdD'_p$ to move points much less than $d(D_{\alpha_q} \cup D_{\alpha_p} \cup D_Q, BdD_p)$, $d(D^c_p, B_{r_1}(p) \cap D_p)$ and $d(D_p, D_q)$. We may then assume general position between D'_p and $D_{\alpha_p} \cup D_{\alpha_q} \cup D_Q$ with the intersection being a finite disjoint union circles. Looking at an intersection circle G which is intermost on D'_p we eliminate G by cutting and pasting in the usual manner with the only problem occurring when G hits $\ell_p \cup \ell_q$. If it hits only one way use a 'slice' or subdisk of D_Q to eliminage G. So suppose it hits both. Let G' denote the disk on $D_{\alpha_p} \cup D_{\alpha_q} \cup D_Q$ bounded by G. Then $G' \cap D_{\alpha_q}$ and $G' \cap D_{\alpha_p}$ is a finite collection C of disk bounded by subarcs of G and $\ell_p \cup \ell_q$. Using this and a spanning disk for D_q we may adjust G until it no longer hits $\ell_p \cup \ell_q$. Thus we may span missing $\alpha_q \cup BdD_Q \cup \alpha_p$.

Now that D'_p misses both ℓ_p and ℓ_q it must separate either α_p or α_q from Σ in D_{α_p} or D_{α_q} (this follows by a simple linking argument if f_{D_p} moves points a sufficiently small distance). If it separates α_p, it is clear that α_p bounds inside N_p which would be a contradiction since α_p was arbitrary and we are assuming U is not *1-LC* at p. So D'_p must separate α_q from Σ in D_{α_q}. Choose a polyhedral spanning disk D'_q ([1]) for D_q which misses α_q and $\ni$ the associated map f_{D_q} moves points less than $d(\Sigma, D'_p)$. This will insure that D'_q and D'_p do not intersect along their boundaries. Note that if D'_q separates α_q from Σ in D_{α_q}, U is *1-LC* at q. So suppose not. In this case however, D'_p and D'_q must intersect and in fact α_q is contained within an open subset of $N_q \cap U$ which is bounded by a subset of $D'_p \cup D'_q$. Cutting and pasting on D'_p shows that this subset may be assumed to be a polyhedral sphere inside N_q so that α_q is still null homotopic in N_q.

Thus we have shown that if p is not *1-LC*, then there exist for each $\varepsilon > 0$, ε-disks $D_\varepsilon \subset \Sigma$ with $p \in \overset{\circ}{D}$ and $\ni$ BdD_ε lies inside a locally flat annulus A_ε on Σ. To see that BdD_ε is also the boundary for a polyhedral ε-disk D' which lies in $U \cap N_p$ except for its boundary, we repeat the previous argument except that α_q will now be the 1-level of a small (polyhedral) collar for BdD_ε which lies inside $U \cap N_p$ except for the *0*-level which is of course just BdD_ε. We connect α_q and α_p again and choose a spanning disk D'_p which misses $\alpha_q \cup BdD_\varepsilon \cup \alpha_p$ just as before. Then, as before, D'_p will separate α_q from Σ (but this time inside the collar $BdD_\varepsilon \times [0,1]$). Using this and by cutting and pasting we form D'_ε whose boundary is BdD_ε. To complete the proof we need we need only the following:

LEMMA. If U is either 1-LC at each point or "capped" as above, then U is 1-LC everywhere and hence $\bar{U}$ is a 3-cell, i.e., Σ is tame from U.

Proof. Let α be an unknotted loop in a small *nghd* of p. Then a bounds a singular disk D_α in only a slightly larger *nghd* of p. Furthermore D_α is polyhedral mod a *0*-dimensional set C with C containing all the singularities of D_α and $D_\alpha \cap \Sigma = C$ ([4]). We may assume that C misses the boundaries of all of the caps for each "capped" point since these boundaries each lie inside a locally flat annulus. Thus we may adjust D_α so that it only hits Σ at *1-LC* points. But then these intersections are easily removed by the definition of *1-LC* since C is *0*-dimensional.

This completes the proof of the lemma and the proof of the theorem.

DEFINITION. Σ is said to be *sequentially 0-ULC locally spanned at the point* $p (p \in \overset{\circ}{D})$, if (in the sequence of spanning disks $\{D_i\}_{i=1}^{\infty}$ and maps $f_i : BdD \to BdD_i$ moving points no more than $1/i$), for $\varepsilon > 0$ $\exists$ a $\delta > 0$ and an $N \ni$ two points on D_n, $n \geq N$, which are less than δ distance apart can be joined by an ε-arc on D_n.

We can now prove three theorems concerning sequential spanning.

THEOREM 2. Sequentially 0-ULC locally spanned and weakly flat imply flatness.

Proof. The object is of course to show U is 1-LC with respect to unknotted loops at each point $p \in \Sigma$ where U is a complimentary domain of Σ. Let $\varepsilon' > 0$ given. Let D be a spanning disk containing p in its interior and $\ni D \subset N_{\varepsilon'/3}(p)$. Choose a spherical *nghd* $B_r(p) \ni B_r(p) \cap \Sigma \subset \mathring{D}$ and $B_r(p) \subset N_{\varepsilon'/3}(p)$. Let $\varepsilon = r/8$. Then $\exists$ a $0 < \delta < r/8$ and an $N \ni$ two δ-points on D_n, $n \geq N$, can be joined by an ε-arc on D_n. Now let α be an unknotted loop in $B_\delta(p) \cap U$. α bounds D_α in U and we must show how to adjust D_α so that it is a small, possibly singular, disk. Choose D_n, $n \geq N$, $\ni d(f_n, id)$ is much smaller than $d(\Sigma, D\ )$. If $D_n \cap \alpha = \emptyset$, we may easily complete the argument. So suppose $D_n \cap \alpha \neq \emptyset$. We assume general position so that the intersection $D_n \cap D_\alpha$ is a finite collection of disjoint circles and arcs. We eliminate circles by replacing subdisks of D_α with subdisks of D_n and pushing off in the usual manner. We observe that the part of D_α below D_n is already small (it is clear what is intended by the word below since D_n is two sided ([9])). So consider the innermost situation which is that of a subdisk K of D_α bounded by a subarc α' of α and an arc β from the intersection, lying above D_n. This disk may however be large. So we look at a preimage of D_α, i.e., D'_α where $h : D'_\alpha \to D_\alpha$ is a homeomorphism. In $h^{-1}(K)$ we draw an arc β' whose endpoints are the endpoints of $h^{-1}(\beta)$ and lies otherwise inside $h^{-1}(\mathring{K})$. Let β'' be an ε-arc on D_n joining the endpoints of β. We redefine h on $h^{-1}(K)$ to be the same on the boundary, to map β' to β'', to map the region between β' and $h^{-1}(\beta)$

into D_n (by the Tietze extension theorem). We redefine h on the remaining part by observing that $\alpha' \cup h(\beta')$ is a loop inside $B_{r/2}(p)$ and bounds a singular disk H there. But since $\alpha' \cup h(\beta')$ lies above (general position slightly) D_n we may cut off H on D_n by the extension theorem again. The remainder of $h^{-1}(K)$ is then mapped to this singular disk. Note that all of this is accomplished inside $N_{\varepsilon'}(p)$. A slight variation of this takes care of the remaining cases so that we may conclude that U is *1-LC* at p and therefore at each point. Thus Σ is flat.

Only minor changes prove the following two theorems.

THEOREM 3. Sequentially 0-ULC locally spanned and simply connected complimentary domains imply flatness.

THEOREM 4. Sequentially 0-ULC locally spanned and locally free imply flatness.

Proof. Use the disk in Lemma 12 of [5].

REFERENCES

1. R. H. Bing, Approximating surfaces with polyhedral ones, Ann. of Math. (2) 65 (1957), 456-483
2. C. E. Burgess, Characterizations of tame surfaces in E^3, Trans. Amer. Math. Soc. 114 (1965), 80-97.
3. ______________, and J. W. Cannon, Embeddings of surfaces in E^3, Rocky Mt. J. Math. 1 (1971), #2, 259-344.
4. J. W. Cannon, Singular side approximations for 2-spheres in E^3, Illinois J. Math. 18 (1974) 27-36.
5. W. T. Eaton, Side approximations in crumpled cubes, Duke Math. J. 35 (1968), 707-719.
6. J. P. Hempel, Free surfaces in E^3, Trans. Amer. Math. Soc. 141 (1969), 263-270.
7. F. M. Lister, Simplifying intersections of disks in Bing's side approximation theorem, Pacific J. Math. 22 (1967) 281-295.
8. L. D. Loveland, Tame surfaces and tame subsets of spheres in E^3, Trans. Amer. Math. Soc. 123 (1966), 355-368.
9. T. B. Rushing, Two-sided submanifolds, flat submanifolds and pinched bicollars, Fund. Math., 74 (1972) 73-84.

A PROBLEM OF RUSHING

O. G. Harrold

Department of Mathematics
Florida State University
Tallahassee, Florida

Let α and β be tame arcs in three-space, $\mathbb{R}^3$. Suppose $\alpha \cap \beta$ is a singleton and is an end-point of each of them. If $\alpha \cup \beta$ is wild, it fails to be locally tame only at the point $\{P\} = \alpha \cap \beta$. Such arcs have been called *mildly* wild.

A problem of T. B. Rushing (Georgia Topology Conference of 1973) [5] asks; does there exist a tame arc γ meeting $\alpha \cup \beta$ only at $\{P\}$, which is an end-point of γ such that both $\alpha \cup \gamma$ and $\beta \cup \gamma$ are tame?

1. The following example, due to DeWitt Sumners, shows that the answer is sometimes yes. The example is essentially Example 1.4 of Fox-Artin [2] with a segment γ attached as indicated in Figure 1.

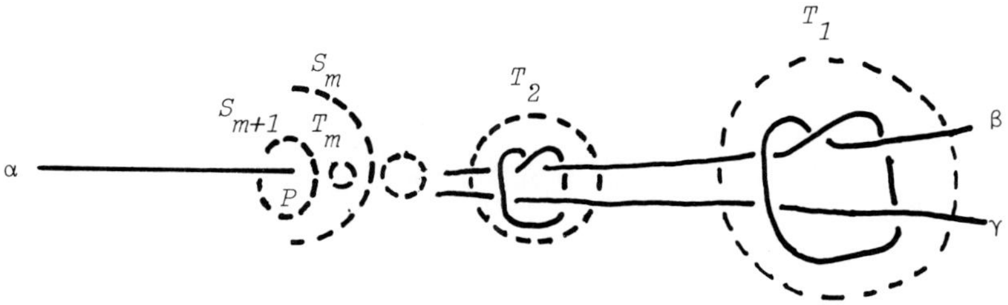

Fig. 1.

ISBN 0-12-158860-2

The segment α is tame. β is tame (folk-lore), and $\alpha \cup \beta$ is wild by Fox-Artin.

The arc γ is chosen to "go through the center" of each successive knot diagram formed by β. Clearly γ is tame since it has 1 - 1 projection into a plane.

Now, to show γ has the desired properties we note first $\alpha \cup \beta \cap \gamma$ is an end-point p of γ. Second, $\alpha \cup \gamma$ is tame since it has a 1 - 1 projection into a plane. It remains to show, third, that $\beta \cup \gamma$ is tame.

The following three diagrams will show how to perform a *canonical* operation. Then, by a sequence of such operations we will obtain a map that is a homeomorphism such that the image of $\beta \cup \gamma$ will clearly be tame, and hence $\beta \cup \gamma$ is tame.

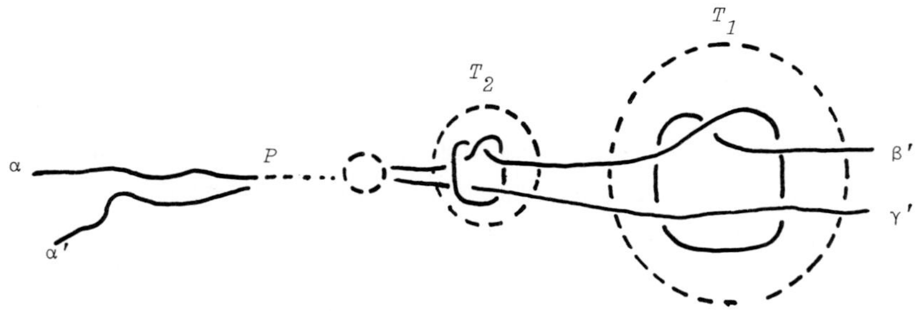

Fig. 2.

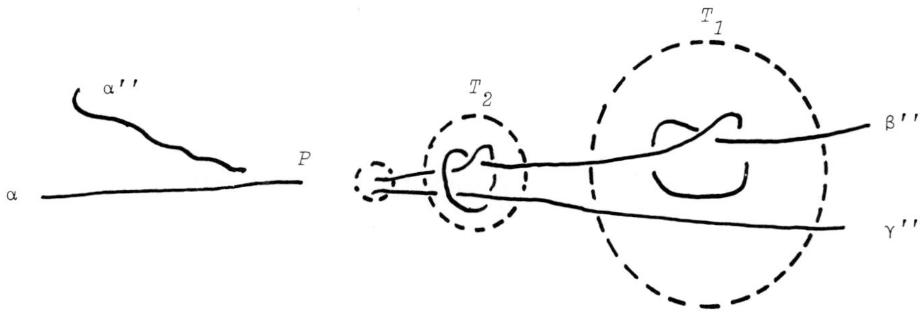

Fig. 3.

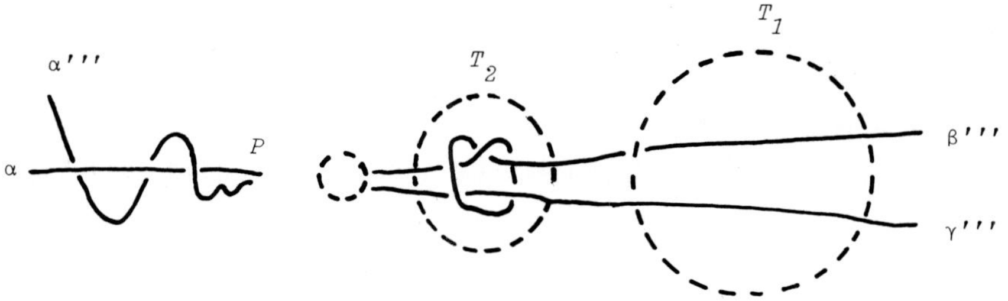

Fig. 4.

The canonical operation has two parts. First we need some notations. Let T_n be the nth indicated sphere counting from the right of the diagram. Let S_n be a sphere centered at p that meets $\alpha \cup \beta$ in a pair of points and $\alpha \cup \gamma$ in a pair of points. By the construction of $\alpha \cup \beta$ this is possible. By construction of γ, this is also true for $\alpha \cup \gamma$.

Let h_n be a homeomorphism on three-space that squeezes S_{n+1} almost to a point, then this shrunken sphere is brought "through the hole" in T_n so that a consecutive pair of over-crossings in the diagram of $\beta \cup \gamma$ become under-crossings, and, then the shrunken sphere is returned to its original position (See Figures 1 & 2, $n = 1$).

The curves marked α', β' and γ' are the images of α, β and γ, respectively, under h_n. Then, by isotopies on three-space, the identity outside a neighborhood of T_n we pass successively to Figures 3 and 4. If the net result of these three motions is denoted by $g_n(x)$, for any $x \neq P$, there is an integer N such that for $n \geq N$, $g_n(x) = g_N(x)$. Set

$$g = \lim_{n\to\infty} g_n g_{n-1} \cdots g_1(x).$$

Then g is a well-defined continuous, 1 - 1 map on $\mathbb{R}^3$ and hence is a homeomorphism. If α^*, β^* and γ^* denote the images of α, β, and γ under g, then α^*, β^* and γ^* (or α, β and γ) satisfy the conditions of Rushing's problem.

2. Let α and β be tame arcs in R^3 meeting only at P, an end-point of each of them. Suppose the arc $\alpha \cup \beta$ is wild (mildly wild in some terminology). We restrict $\alpha \cup \beta$ further to be LPU (locally peripherally unknotted) at P, i.e., there are arbitrarily small topological 2-spheres enclosing P which meet $\alpha \cup \beta$ in a pair of points. (such an arc has been called a Wilder arc [3]).

THEOREM. If α and β are tame arcs in R^3 and $\alpha \cap \beta = \{P\}$ is an end-point of each of α and β and if $\alpha \cup \beta$ is LPU *at P, then there is an arc γ that is tame and such that $\alpha \cup \gamma$ and $\beta \cup \gamma$ meet only at P, an end-point of γ, and $\alpha \cup \gamma$ and $\beta \cup \gamma$ are tame.*

The notation for S_n and T_n is as above. The segment of β joining T_n and T_{n+1} is denoted by β^*_n and the part of β in T_n is denoted by β_n.

The set β_n is a finite polygonal arc and modulo its boundary $(= x_n \cup x'_n)$ is a simple closed curve β'_n. (Not to be confused with the β' of Figure 2.) To each tame simple curve there is associated a certain Gordian number $\mu(\beta'_n) = \mu_n$, the minimal number of interchanges of over-crossings to under-crossings in a knot diagram in order to obtain the unknot [6].

The task before us is now to define γ and this is done by first defining the part of γ in T_n, denoted by γ_n.

The arc γ_n is to be a tame, unknotted arc in T_n, disjoint to β_n plus other requirements which we now explain. Let δ_n be a tame unknotted arc in T_n from a point on BdT_n near x_n to another point on BdT_n near x'_n that is splittable with respect to β_n. (Two sets H and K in a cube C are called splittable if there is a disk D such that H and K lie in different components of $C \setminus D$.) The arc γ_n is obtained from δ_n by making μ_n simple twists about a segment of $x_n'x_n$. An interval γ_n^* is adjoined connecting BdT_n and BdT_{n+1}. It may be arranged so that

$$Cl\{\bigcup_1^\infty \gamma_n \cup \bigcup_1^\infty \gamma_n^*\} = \gamma$$

is an arc (unknotted) joining x_1' and p.

For each n let k_n be a homeomorphism on $\mathbb{R}^3$ (active only on a neighborhood of T_n) such that $k_n(\gamma_n)$ is a straight line segment joining a pair of points of BdT_n and assume diam $T_n < \frac{1}{2^n}$.

Then $k = \lim_{n\to\infty} k_n k_{n-1} \ldots (x)$ exists, is 1 - 1 on $\mathbb{R}^3$ and leaves α fixed.

Then $\alpha = k(\alpha)$ and $k(\alpha) \cup k(\gamma) = k(\alpha \cup \gamma)$ has a 1 - 1 projection on the horizontal plane, hence

$$\alpha \cup \gamma$$

is tame.

Next, we define a homeomorphism on $\mathbb{R}^3$ designed to show that $\beta \cup \gamma$ is tame.

For each $k = 1, \ldots, \mu_n$ a finite set of three motions (isotopies) is carried out as on page 79, one is a "through the hole" motion as described on page 79. After such a set of three we arrive at a curve β_n' one step nearer the unknot and to an arc γ_n' whose linking number with β_n' is $\mu_n - 1$. After μ_n such sets of three homeomorphisms we arrive at the unknot $\gamma_n^{\mu_n - 1}$ from one point BdT_n to another point of BdT_n whose projection on the horizontal plane has at most a finite number of points in common with the projection of $\beta_n^{\mu_n - 1}$.

Assuming the "through the hole" isotopy is clear, we explain the second and third motions which will free the projections of the images of β_n and γ_n of common points.

LEMMA. Given two polyhedral splittable arcs H and K in a closed, polyhedral 3-cell B, there is a polyhedral 2-cell E in B such that $\partial E \subset \partial B$, int $E \cap \partial B = \square$ *with H and K in different components of B/E. If, further, H and K are unknotted, it may be arranged that their projections on the horizontal plane are each simple, i.e., 1-1.*

Thinking of B as being symmetric with respect to the origin and as a cube 2 units in a side with its faces in the co-ordinate planes $x = \pm 1$, $y = \pm 1$, $z = \pm 1$, an application of the Schoenflies theorem permits us to strengthen the conclusion of the above Lemma so that if $z = 0$ is the horizontal plane, then E lies in the plane $x = 0$.

The application of the strengthened Lemma, for $n = 1, 2, \ldots$ corresponds to application of the second and third homeomorphisms on page 3 which reduces the number of crossing points in the diagram of β_n and the elimination of the finite set of points common to the projection of β_n and γ_n on the horizontal plane (See Figures 3 and 4.)

THE INDUCTIVE HYPOTHESIS. For each $\mu_n = 1, 2, \ldots$ *there is a homeomorphism* g_{μ_n} *on* $\mathbb{R}^3$, *moving no point more than* ε_n, *leaving the point p fixed such that*

$$(*) \quad \bigcup_{k=1}^{n} \underline{\text{proj}}[g_{\mu_n}(\gamma_1 \cup \cdots \cup \gamma_{\mu_k} \cup \gamma_1^* \cup \cdots \cup \gamma_{\mu_k}^*)] \cap$$

$$\bigcup_{k=1}^{n} [\underline{\text{proj}}\ g_{\mu_n}(\beta_1 \cup \cdots \cup \beta_{\mu_k} \cup \beta_1^* \cup \cdots \cup \beta_{\mu_k}^*)] = \square, \quad \textit{where}$$

$$\sum_1^\infty \varepsilon_k < \infty.$$

For $\mu_n = 1$, the proof follows the procedure of the above example. Let y be the point on the knot diagram such that if over and under-crossings are reversed, one obtains the unknot β_n' instead of β_n. The point that traverses γ_n (in reverse order) from x_n towards x_n' is very close to β_n until in a neighborhood of a point that projects onto y. Then there is a "through the hole" motion g_n which reverses the character of the crossings at y, then the shrunken sphere S_{n+1} follows γ_n in (positive) order back to x_n. Finally the image of S_{n+1} is carried back to its original position. Taking $H = \beta_n'$ and $K = \gamma_n'(=\delta_n)$ we note these are unknotted, splittable arcs in T_n and the Lemma may be applied. Hence there is a homeomorphism g_1, on $\mathbb{R}^3$ such that

$$\text{proj } g_1(\gamma_n' \cup \gamma_n^*) \cap \text{proj } g_1(\beta_n' \cup \beta_n^*) = \square .$$

Assuming $(*)$ holds for $j = 1,\ldots,m-1$, let $y_1,\ldots,y_m$ be the points on the knot diagram which if the crossings are reversed, replaces β_n by the unknot. Recall that the arc γ_n traversed (in reverse order) from x_n towards x_n' is very close to β_n until in a neighborhood of a point that projects into y_m. There is a "through the hole motion" g_n^m which reverses the character of the crossings near y_m then follows γ_n in (positive) order back to x_n. Finally, the shrunken inage of S_{n+1} is carried back to its original position. We now look at $g_n^m(x_n x_n')$ and $g_n^m(\gamma_n^{m-1})$ and apply the inductive hypothesis to obtain $g_{\mu_{m-1}}$ and the relation $(*)$ for $j = m - 1$. By composition of this homeomorphism and g_n^m we find g_{μ_m} and $(*)$ for $j = m$.

If the composition of all homeomorphisms at stage n is denoted by $h_n(x)$, then for any positive number ε there is an integer N such that for any $n \geq N$, $h_n(x) = x$. Then

$$h(x) = \lim_{n\to\infty} h_n h_{n-1} \cdots h_1(x)$$

is continuous, 1-1 on $\mathbb{R}^3$ and is therefore a homeomorphism. On account of *(*)*, it follows that

$h(\beta)$ and $h(\gamma)$

have no common points in their projection onto the horizontal plane other than p. Further, their projections are 1-1 and hence $h(\beta) \cup h(\gamma) = h(\beta \cup \gamma)$ is tame. The arcs α, β and γ satisfy the conditions of Rushing's problem.

REFERENCES

1. J. W. Alexander, Topological Invariants of Knots and Links, Trans. Amer. Math. Soc., Vol. 30 (1928), pp. 275-306.
2. Ralph H. Fox and Emil Artin, Some wild cells and spheres in three-dimensional spaces, Ann. of Math. Vol. 49 (1948), pp. 979-990.
3. R. H. Fox and O. G. Harrold, The Wilder arcs, Topology of Manifolds, Univ. of Georgia, 1962, pp. 184-187.
4. O. G. Harrold, Locally tame curves and surfaces in 3-dimensional manifolds, B.A.M.S., Vol. 63 (1957), p. 298.
5. T. B. Rushing, Proc. Georgia Topology Conference, 1973, p. 54.
6. H. Wendt, Die gordische Auflosungen von Knoten, Math. Zeit, Vol. 42 (1937), pp. 680-696.

AN ADDITIVE INDEX THEOREM FOR CERTAIN WILDLY EMBEDDED ARCS AND CURVES

O. G. Harrold

Department of Mathematics
Florida State University
Tallahassee, Florida

The notion of penetration index of an embedded arc or curve was introduced several years ago by Alford and Ball, Harrold and others [1], [4]. If p is an interior point of an arc J, the penetration index $P(p,J)$ is defined to be the inf cardinal $S(p,\varepsilon) \cap J$ over all topological spheres of diameter less than ε containing p in its interior. Clearly such a number may be defined for any embedded graph or even any embedded regular curve in the sense of Urysohn-Menger. Also, if p is an endpoint of an arc, such a number may be defined (and even $=\infty$). Suppose p is an interior point of an arc J and $J\backslash p = J_a \cup J_b$, where J_a and J_b are separated. Then $\overline{J}_a$ and $\overline{J}_b$ are arcs and there is a penetration index of $\overline{J}_a$ at p defined. If J is tame, pen $\overline{J}_a$ at p + pen $\overline{J}_b$ at p = pen index of J at p $(=2)$. If $\overline{J}_a$, $\overline{J}_b$ are tame, but J is wild at p, it may happen that pen $J_a(p)$ + pen $J_b(p)$ < pen $J(p)$. Lomonaco gave the first example of this phenomena [7]. In case the inequality does *not* hold we say J has the additive index property at p (denote by AIP). A simple closed curve is called locally periphally unknotted (LPU) if it has penetration index 2 at each point. A simple closed curve j is called locally unknotted at x (LU) if there is a disk whose boundary contains a

ISBN 0-12-158860-2

neighborhood of x relative to J. These concepts are independent. Examples exist of curves being LPU at all points but LU nowhere and vice-versa [5]. Although both properties are local, one is strictly local (LPU) and the other is medial. The theorem below gives a partial relation.

THEOREM 1. If J is a simple closed curve in three space, that is a union of two tame arcs, then J has the additive index property at each point where J is LU.

Remark. The proof of the above theorem, applied to the case of two tame arcs having a single endpoint in common says that if p is the common endpoint of a standard interval J_a and if J_b is a tame arc chosen so that the AIP does not hold at p for $J = J_a \cup J_b$, then J is locally knotted at p (as well as being locally peripherally knotted).

We begin by observing that the AIP for $J = \overline{J}_a \cup \overline{J}_b$ at p together with the facts that $\overline{J}_a$ and $\overline{J}_b$ are each tame permits us to conclude that $\overline{J}_a \cup \overline{J}_b$ is LPU at p. Now, precisely as in [4], given $\varepsilon > 0$ one constructs a polyhedral 2-sphere enclosing J and lying in the ε-neighborhood of J. Or, if J is regarded as a subset of a single closed curve C that is locally tame modulo p, there is a polyhedral torus T enclosing C and lying in the ε-neighborhood of C. The following theorem will imply Theorem 1.

THEOREM 2. Let C be a simple closed curve, $C = C_a \cup C_b$ where C_a, C_b are tame arcs having only endpoints a and b in common. If C is locally unknotted at $a(b)$ and has the AIP at $a(b)$, then C is locally tame at $a(b)$.

Proof. Consider the point a. Since C is LU at a, there is a disk whose boundary contains a neighborhood N of a in C excluding the point b. Since C_a and C_b are tame and C is AIP at a we may diminish N if necessary to obtain a neighborhood of a in C that is LU and LPU at all point. Hence by [6], C is locally tame at a. If C is LU and AIP at each of a and b, by Bing's result [2], C is tame.

COROLLARY. Let p be an isolated wild point of a simple closed curve J. If J fails to have the AIP at p, then J is both locally knotted and locally peripherally knotted at p.

APPLICATION. Let A, B be tame arcs, $A \cap B = \{p\}$, an endpoint of each. Suppose there is a tame arc C meeting $A \cup B$ only at p. If $A \cup C$ and $B \cup C$ are locally unknotted at p, then $A \cup C$ and $B \cup C$ are tame.

Proof. Recall $A \cup C$ is locally tame mod p. The arc $A \cup C$ is locally unknotted at each point and must have the AIP at p. That is, $A \cup C$ is also locally peripherally unknotted at each point. Hence $A \cup C$ is tame, similarly for $B \cup C$.

REFERENCE

1. W. R. Alford and B. J. Ball, Some almost polyhedral wild arcs, Duke Math. Journal, Vol. 30 (1963), pp. 33-38.
2. R. H. Bing, Locally tame sets are tame, Annals of Math., Vol. 59 (1954), pp. 145-158.
3. R. H. Fox and E. Artin, Some wild cells and spheres in 3-space, Annals of Math., Vol. 49 (1948), pp. 979-990.
4. O. G. Harrold, The enclosing of simple arcs and curves by polyhedra, Duke Math. Journal, Vol. 21, No. 4, pp. 615-622, December, 1954.
5. O. G. Harrold, Locally unknotted sets in 3-space, Yokohama Math. Journal, Vol. 21, No. 1, 1973, pp. 47-60.
6. O. G. Harrold, H. C. Griffith, and E. E. Posey, A characterization of tame curves in 3-space, Trans. Amer. Math. Soc., Vol. 79 (1955), No. 1, pages 12-34.
7. S. J. Lomonaco, Uncountably many mildly wild non-Wilder arcs, Proc. Amer. Math. Soc., Vol. 19 (1968), pp. 859-898.
8. D. G. Stewart, Cellular subsets of the 3-sphere, Trans. Amer. Math. Soc., Vol. 114, No. 1, January 1965, pp. 10-22.

SEIFERT FIBERED SPACES IN 3-MANIFOLDS

William Jaco
Peter B. Shalen

Department of Mathematics
Rice University
Houston, Texas

This is an exposition of results by the authors which describe, up to homotopy, certain maps of Seifert fibered spaces into a rather general 3-manifold M, in terms of a canonical system of embedded Seifert fibered spaces in M. Applications are given to the study of 3-manifold groups and to the classification problems for 3-manifolds.

Let M be a closed, connected, orientable 3-manifold. The following strong version of the Sphere Theorem combines results of Kneser, Papakyriakopoulos, Whitehead and Milnor. *There exist finitely many disjoint, non-contractible, pairwise non-parallel, embedded 2-spheres in* M*, whose homotopy classes generate* $\pi_2(M)$ *as a* $\pi_2(M)$*-module; and modulo the Poincaré conjecture, these 2-spheres are unique up to ambient homeomorphism.* Thus all "singular 2-spheres" in M, i.e. maps of S^2 into M, may be described, up to homotopy, in terms of a geometric picture in M.

A special case of the main theorem we shall discuss here gives an analogous description (up to homotopy) of "singular tori" in M, i.e. maps of $S^1 \times S^1$ into M, in the case that M is a "Haken manifold"--that is, a compact, irreducible, orientable,

ISBN 0-12-158860-2

connected 3-manifold containing some two-sided incompressible surface. In this case it is easy to see that the only interesting maps of $S^1 \times S^1$ into M are those that induce monomorphisms of fundamental groups; we shall call these *non-degenerate singular tori*. The description we refer to has also been obtained by K. Johannson, and there is a good deal of overlap between his work and the results we shall discuss.

The strong version of the sphere theorem referred to above gives a great deal of information about fundamental groups of compact 3-manifolds, for example that they are finite free products of torsion-free groups and finite groups. It also provides (in a slightly refined version) a reduction of the classification problem for compact, oriented 3-manifolds to the classification problem for compact, *irreducible*, 3-manifolds. Similarly, it will be seen that our main theorem gives information about the fundamental groups of Haken manifolds; and that it provides a reduction of the classification problem for Haken manifolds to a more special classification problem, and one that now seems quite tractable.

We shall work in the piecewise-linear category.

One cannot hope to describe the non-degenerate singular tori in the Haken manifold M in terms of disjoint embedded tori in M. For example, if $M = S^1 \times T$, where T is a compact, orientable surface, one can in general obtain quite complicated maps of $S^1 \times S^1$ into M as products of the identity with maps of S^1 into T. A similar construction is possible when M is a *fibered manifold* in the classical sense of Seifert. (In scissors-and-paste terms, a Seifert fibered manifold M is obtained from an S^1-bundle over a surface by attaching solid tori to boundary components, in such a way that the fibers in the boundary of the S^1-bundle do not bound discs in the solid tori. In the case where the bundle is a product $S^1 \times T$, for example, the complicated singular tori in $S^1 \times S^1$ may be regarded as singular tori in M, and many of these will be non-degenerate in M.)

The fibered manifolds were classified by Seifert himself, and it is not hard to describe, up to homotopy, all the maps of $S^1 \times S^1$ into an arbitrary Siefert fibered manifold. The following theorem includes a description of all non-degenerate singular tori in an arbitrary Haken manifold in terms of non-degenerate singular tori in Seifert fibered manifolds. However, the theorem is a good deal more general than this; it describes maps whose domains are themselves Seifert fibered manifolds rather than $S^1 \times S^1$. Note that by specializing it to the particular Seifert fibered manifold $S^1 \times S^1 \times I$, which is homotopy-equivalent to $S^1 \times S^1$, one gets the description of non-degenerate maps promised above.

THEOREM. Let M be a Haken manifold. Then there exist disjoint Seifert fibered manifolds $\Sigma_1, \ldots, \Sigma_k \subset \mathring{M}$, *whose boundary components are incompressible in M, such that for any Seifert fibered manifold S with non-cyclic fundamental group, and any map* $f : S \to M$ *such that* $f_\# : \pi_1(S) \to \pi_1(M)$ *is a monomorphism, f is homotopic in M to a map whose image is contained in some* Σ_i. *We may (clearly) choose the* Σ_i *so that for* $i = j$ *the inclusion map* $\Sigma_i \to M$ *is never homotopic in M to a map of* Σ_i *into* Σ_j. *Given this additional property, the set* $\{\Sigma_1, \ldots, \Sigma_k\}$ *is unique up to ambient isotopy.*

(Since a Haken manifold is irreducible, we do not need the Poincaré conjecture to guarantee uniqueness.)

A more general, "relative" version of this theorem will be stated below. First we shall discuss some applications of the "absolute" theorem.

L. P. Neuwirth has asked for a description of all the *roots* of an element $g \neq 1$ of a knot group, i.e. solutions of $x^n = g$, $n \in \mathbb{Z}$. For instance, in the torus knot group $G = \langle a,b : a^p = p^q \rangle$, where p and q are relative prime integers >1, the element $g = a^p$ has as roots a and b -- and also all conjugates of a and b, since

g lies in the center of G. In particular the roots of g do not all lie in a single cyclic subgroup of G; we shall express this by saying that g has *non-trivial root structure in* G. As a second example, each cabled knot space contains a submanifold with incompressible boundary (a "cable space") whose fundamental group has the presentation $\langle a,b:ab^na^{-1} = b^n\rangle$ for some $n > 1$. Clearly $g = b^n$, which may be regarded as an element of the cabled knot group, has non-trivial root structure.

A knot space is a Haken manifold, and a root of an element of a group is clearly in the centralizer of the element; thus one can generalize Neuwirth's question by asking for a description of the centralizers of elements in the fundamental group of a Haken manifold M. The above theorem, combined with results of Waldhausen and Scott-Shalen, provides such a description. *Suppose, for convenience, that* M *contains no embedded Klein bottles.* Let $\Sigma_1, \ldots, \Sigma_k$ be given by the above theorem. Then *every non-abelian subgroup of* $\pi_1(M)$, *which is the centralizer of an element of* $\pi_1(M) - \{1\}$, *is conjugate to one of the subgroups* $\zeta_i = Im(\pi_1(\Sigma_i) \to \pi_1(M))$, $1 \leq i \leq k$. (These subgroups are well-defined up to conjugacy.) It also turns out that every *non-cyclic* subgroup of $\pi_1(M)$, which is the centralizer of an element of $\pi_1(M) - \{1\}$, is conjugate to a *subgroup* of one of the ζ_i; one can determine all the abelian, non-cyclic centralizers, and they are all free abelian of rank 2 or 3.

In particular, the centralizer of any element of $\pi_1(M)$ is finitely generated; and up to conjugacy, only finitely many non-abelian subgroups of $\pi_1(M)$ can arise as centralizers of elements. (These corollaries are true even if M contains Klein bottles, although the analogues of the previous statements are more technical.)

Returning to Neuwirth's problem, we find (with a little effort) that the only Seifert fibered manifolds that can be embedded in a knot space M so that their boundary components are incompressible are; the torus knot spaces, the cable spaces, and products of compact, planar surfaces with S^1. (These are all indeed Seifert fibered.) A knot space can, in fact, contain a large number of disjoint submanifolds of these types; for example, the given knot could be a sum of several torus knots and several cabled knots. In any case, it is not hard to conclude that, up to conjugacy, the only elements of the given knot group that have non-trivial root structure are the images under inclusion of the powers of the elements a^p and b^n in the respective fundamental groups of the torus knot spaces and cable spaces Σ_i. (Here we have denoted the generators of these groups in terms of the above presentations.) Furthermore, the only roots of these elements are the ones that lie in the subgroups $\mathrm{Im}(\pi_1(\Sigma_i) \to \pi_1(M))$, and it is easy to write these all down. As a corollary to this solution of Neuwirth's problem, one may state that any element g of a knot group has two roots x and y, not necessarily distinct, such that every root of g is a power of x or of y.

Another consequence of the general result on centralizers is that a relation of the form $ab^pa^{-1} = b^q$, $|p| \neq |q|$, cannot hold between elements a and b of the fundamental groups of a Haken manifold unless $b = 1$. This generalizes the earlier result that the Baumslag-Solitar groups $\langle a,b : ab^pa^{-1} = b^q \rangle$, $|p| \neq |q|$, are not 3-manifold groups. (Distinct proofs of this latter fact have been given by Jaco, Heil, Kawauchi, Hempel and Thurston; each proof illustrates an important property of 3-manifold groups.)

We now turn to the application of our main theorem to the classification problem for Haken manifolds. We define a 3-manifold M to be *simple* if every incompressible torus in M is parallel to a component of ∂M. The basic result here is the following "splitting Theorem." *Let M be a Haken manifold. Then there exists a disjoint family $\{T_1, \ldots, T_m\}$ of incompressible tori in $\mathring{M}$, unique up to ambient isotopy, such that (i) each component of the manifold obtained by splitting M along $T_1 \cup \ldots \cup T_m$ is either Seifert-fibered or simple, and (ii) no proper subfamily of $\{T_1, \ldots, T_m\}$ satisfies condition (i).* (Note that

$T_1 \cup \ldots \cup T_m$ does not always separate M; but the T_i are two-sided since M is orientable, and so the operation of splitting is well defined.)

The uniqueness of $\{T_1, \ldots, T_m\}$ up to ambient isotopy is the most significant part of the splitting theorem.

The connection between the T_j and the Σ_i of the main theorem is easily described. The components of the $\partial\Sigma_i$ are incompressible tori in ∂M. It may happen that certain distinct components of $\partial\Sigma_1 \cup \ldots \cup \partial\Sigma_k$ are parallel in M; for example, some Σ_i may be homeomorphic to $S^1 \times S^1 \times I$. However, it follows easily from the properties of the Σ_i stated in the main theorem that no three components of $\partial\Sigma_i$ can be parallel. We may obtain the family $\{T_1, \ldots, T_m\}$ from the set of components of $\partial\Sigma_1 \cup \ldots \cup \partial\Sigma_k$ simply by discarding one component from each pair of parallel components.

To apply the splitting theorem to the classification problem let us consider -- for convenience -- a closed manifold M whose components are Haken manifolds. Let M' denote the manifold obtained by splitting M along $T_1 \cup \ldots \cup T_m$. Then M is a quotient space of M', and $\partial M'$ has a natural involution τ : for $p \in \partial M$, $\tau(p)$ is the unique point of ∂M which is distinct from p and is identified with p in the quotient space M. Clearly M is determined by M' and τ.

Given a pair (M', τ), where M' is a compact 3-manifold and τ is an involution of $\partial M'$, it is not hard to write down necessary and sufficient conditions for (M', τ) to arise from a Haken manifold M in the above way. For example, the components of M' are Haken manifolds, whose boundary components are tori; each of them is either Seifert-fibered or simple (or both); and τ leaves no component of $\partial M'$ invariant. (A few more weak, technical conditions are also needed because of the presence of condition (ii) in the statement of the Splitting Theorem.) It is also easy to see that two pairs of the form (M, T') come from the same M if and only if they are equivalent in a simple, natural sense. The upshot of all this is that the classification of closed Haken manifolds is reduced to the following two problems: (a) the classification of the simple Haken manifolds, and (b) the determination of the group of self-homeomorphisms of the boundary of a given simple Haken manifold which extend to self-homeomorphism of the manifold.

Thurston has shown that all simple Haken manifolds -- with the possible exception of certain closed surface bundles over S^1 -- admit hyperbolic structures. According to a theorem of Mostow's, these hyperbolic structures are unique, and the self-homeomorphisms of the manifolds are isotopic to unique hyperbolic isometries. (This implies a theorem, first proved topologically by Johannson, that the homeomorphsm groups of these manifolds are finite.) Thus the above problems (a) and (b) are equivalent to problems in hyperbolic geometry. This connection between 3-dimensional topology and hyperbolic geometry has already profoundly affected both fields; in particular, problems (a) and (b) now seem tractable.

The following result of ours, whose proof does not depend on hyperbolic geometry, gives information about the fundamental group G of any simple irreducible 3-manifold. *Any two-generator subgroup of such a group G is either (i) free (of rank ≤ 2), (ii) free abelian of rank 2 and peripheral, or (iii) of finite index in G.* This is similar to a theorem proved independently by T. Tucker.

We shall conclude by stating the general, "relative" version of our main theorem. We *define a 3-manifold pair* to be a pair (M,T), where M is a 3-manifold and $T \subset \partial M$ is a 2-manifold. We call the 3-manifold pair (M,T) a *Haken pair* if M is a Haken manifold and T is compact and incompressible. An S^1-pair is a 3-manifold pair (S,F) where S is a Seifert fibered manifold and F is saturated in some Seifert fibration of S. (In terms of the above scissors-and-paste description of Seifert fibered manifolds, this means that F is a union of fibers in the fiber bundle from which S was constructed.) An *I-pair* is a 3-manifold pair (S,F), where S is an I-bundle over a compact surface and F is the associated ∂I-bundle. A *Seifert pair* is a finite disjoint union of S^1-pairs and I-pairs.

Now let (S,F) be a connected Seifert pair -- that is, an S^1-pair or an I-pair; let (M,T) be a Haken pair, and let $f : (S,F) \to (M,T)$ be a map of pairs. We shall say that f is *degenerate* if either (i) $\mathrm{Im}(\pi_1(S) \to \pi_1(M)) = \{1\}$, or (ii) $\mathrm{Im}(\pi_1(S) \to \pi_1(M))$ is cyclic and $F = \phi$, or (iii) S has a Seifert fibration in which F is saturated and such that the restriction of f to a fiber is homotopic in M to a constant map. Then our general theorem may be stated as follows.

Let (M,T) be a Haken pair. Then there is a Seifert pair (Σ,Φ), with $\Sigma \subset M$, $\Phi \subset T$, such that for each connected Seifert pair (S,F) and each non-degenerate map $f : (S,F) \to (M,T)$, f is homotopic as a map of pairs to a map f' such that $f'(S) \subset \Sigma$ and $f'(F) \subset \Phi$. Furthermore, we may choose (Σ,Φ) so that

(i) $\Sigma \cap \partial M = \Phi$,

(ii) the frontier of Σ in M is incompressible,

(iii) no component of this frontier is "parallel rel boundary" to a surface in T,

and (iv) there is no component σ of Σ such that the inclusion map $(\sigma,\varphi) \to (M,T)$ (where $\varphi = \sigma \cap \Phi$) is homotopic as a map of pairs to a map j such that $j(\sigma) \subset \Sigma-\sigma$, $j(\varphi) \subset \Phi-\varphi$.

Finally, given these additional properties, (Σ,Φ) is unique up to ambient isotopy.

In the case $\Phi = \phi$, this is a slightly strenthened version of the above "absolute" theorem. One consequence of the properties of (Σ,Φ) stated in the theorem is a description up to homotopy, of "singular annuli," i.e. maps from the pair $(S^1 \times I, S^1 \times \partial I)$ to (M,T), analogous to the above description of singular tori. Let us define a singular annulus to be *non-degenerate* if it induces a monomorphism from $\pi_1(S^1 \times I)$ to $\pi_1(M)$, and is not homotopic rel $S^1 \times \partial I$ to a map sending $S^1 \times I$ into T. Then every non-degenerate singular annulus is homotopic (as a map of pairs) to a map sending $S^1 \times I$ into Φ.

This relative theorem must actually be stated in order to *prove* our absolute theorem: the proof involves induction on the length of a "hierarchy" for the given Haken manifold. The relative theorem can also be used to refine many of the above consequences of the absolute theorem. The manifold Σ of our relative theorem has been studied independently by Johannson; it is his "characteristic manifold" in the case where M is boundary-irreducible and $T = \partial M$. Johannson has discovered some remarkable properties of the characteristic manifold, which are described in his paper in this volume.

ON EXOTIC HOMOTOPY EQUIVALENCES OF 3-MANIFOLDS

Klaus Johannson

Fakultät für Mathematik
Universität Bielefeld
West Germany

A homotopy equivalence between 3-manifolds (compact, orientable, irreducible) will be called exotic if it cannot be deformed into a homeomoprhism. It is well-known that such homotopy equivalences exist between 3-manifolds with non-empty boundaries. A theorem will be stated which can be considered as a classification of exotic homotopy equivalences between 3-manifolds, and some aspects of its proof will be discussed.

I. INTRODUCTION

We are concerned with the study of exotic homotopy equivalences between 3-manifolds, that is with those homotopy equivalences which cannot be deformed into a homeomorphism. Here a 3-manifold is always compact, orientable and irreducible. Since Waldhausen [11] has shown that there are no exotic homotopy equivalences between closed 3-manifolds which are sufficiently large, we shall restrict ourselves to 3-manifolds with non-empty boundary. To exclude uninteresting phenomena we suppose that, in addition, the inclusion of every boundary component induces a monomorphism of the fundamental groups (such a 3-manifolds is called "boundary-incompressible").

ISBN 0-12-158860-2

In contrast to the closed, sufficiently large 3-manifolds one finds exotic homotopy equivalences in 3-manifolds with boundaries. One type of example can be found in knot spaces (e.g. granny and square knot [1]). Another type of example is provided by the Seifert fibre spaces. To see the latter, observe that the homeomorphy-type of a Seifert fibre space (different from solid torus and S^1-bundle over the Möbius band) is given by the homeomorphy-type of the orbit surface together with a certain number of pairs (α_i, β_i) of coprime integers $0 < \beta_i < \alpha_i$ (which themselves correspond to the exceptional fibres) [7,9], while the homotopy-type is given by the homotopy- and orientability-type of the orbit surface together with the integers α_i only [10].

To describe our main result on exotic homotopy equivalences we first have to define the "characteristic submanifold". There are several ways of doing this, the most convenient one is the following:

Definition 1.1. Let M be a 3-manifold. Then a submanifold V of M is called a *characteristic submanifold* if the following holds:

1. The inclusion of every component of $(\partial V - \partial M)^-$ induces a monomorphism of the fundamental groups.
2. Each component X of V admits either a fibration as I-bundle, with projection $p : X \to B$, such that

$$\partial X \cap \partial M = (\partial X - p^{-1}\, \partial B)^- \quad ,$$

or as Seifert fibre space, with fibre projection $p : X \to B$, such that

$$\partial X \cap \partial M = p^{-1} p(\partial X \cap \partial M).$$

3. If W is any union of components of $(M - V)^-$, then $V \cup W$ is *not* a submanifold with 2.

4. If W' is any submanifold in M with 1. and 2. above, then W' can be properly isotoped into V.

THEOREM 1.2. Let M be a 3-manifold with non-empty boundary (as given above). Then the characteristic submanifold of M exists and is unique, up to ambient isotopy.

By the way, we note that this also holds for closed, sufficiently large 3-manifolds (in this case the characteristic submanifold consists of Seifert fibre spaces only). The characteristic submanifold has certain nice properties which are useful in the study of mappings of annuli, tori, I-bundles and Seifert fibre spaces into 3-manifolds, and which can be applied to the study of fundamental groups of 3-manifolds (for more details we refer to [4]). Furthermore, our main result on homotopy equivalences may now be formulated as follows:

THEOREM 1.3. Let M_1 and M_2 be two 3-manifolds with non-empty boundaries. Let V_1 and V_2 be the characteristic submanifolds in M_1 and M_2, respectively. Then every homotopy equivalence $f : M_1 \to M_2$ can be deformed so that afterwards

1. $f(V_1) \subset V_2$ *and* $f(M_1 - V_1)^- \subset (M_2 - V_2)^-$,
2. $f|V_1 : V_1 \to V_2$ *is a homotopy equivalence, and*
3. $f|(M_1 - V_1)^- : (M_1 - V_1)^- \to (M_2 - V_2)^-$ *is a homeomorphism.*

COROLLARY 1.4. If every incompressible annulus in M_1 is boundary-parallel, then there are no exotic homotopy equivalences $f : M_1 \to M_2$.

Complete proofs of these results will appear in [4]. For more background material the reader should consult Waldhausen's summary [12]. Here we are mainly interested in giving some indications why theorem 1.3. has a chance of being true.

II. EXOTIC HOMOTOPY EQUIVALENCES OF BALLS

To get an understanding of the phenomenon of the existence or non-existence of exotic homotopy equivalences, we now generalize the concept of a "manifold with boundary" to that of "manifold with boundary-pattern". This is given as follows:

Let M be a compact n-manifold, $n \leq 3$. A *boundary-pattern* for M consists of a set $\underline{\underline{m}}$ of compact connected $(n-1)$-manifolds in ∂M, such that the intersection of any two of them consists of $(n-2)$-manifolds, the intersection of any three consists of $(n-3)$-manifolds, and so on. The elements of $\underline{\underline{m}}$ are called *bound sides* of $(M,\underline{\underline{m}})$, while the components of $(M - \cup_{G \in \underline{\underline{m}}} G)^-$ are called the *free sides* of $(M,\underline{\underline{m}})$. By $\overline{\underline{\underline{m}}}$ we denote the set of all the bound and free sides of $(M,\underline{\underline{m}})$.

An *admissible map* $f : (N,\underline{\underline{n}}) \to (M,\underline{\underline{m}})$ is a map $f : N \to M$ satisfying

$$\underline{\underline{n}} = \cup_{G \in \underline{\underline{m}}} \{\text{components of } f^{-1}G\} .$$

An *admissible homotopy* is a continuous family of admissible maps. Having defined "admissible homotopy" one also has defined "admissible isotopy" and "admissible homotopy equivalence".

Finally, let N be a codimension zero submanifold of $(M,\underline{\underline{m}})$ such that $N \cap \partial M$ is a codimension zero submanifold of ∂M, in general position with respect to $\underline{\underline{m}}$. Then the boundary-pattern of N, given by

$$\underline{\underline{n}} = \{G | G \text{ is component either of } (\partial N - \partial M)^- \text{or of } N \cap H, \text{ where } H \in \underline{\underline{m}}\}$$

will be refered as the *proper boundary pattern of N*.

The concept of manifolds with boundary-patterns and admissible maps takes into account the fact that the possibility of an exotic homotopy equivalence depends really on the existence of a boundary. This will be illustrated with the following examples. Here the homeomorphy-type (not the admissible homeomorphy-type) of the manifolds involved is trivial, namely a 2-disc or a 3-ball.

If $(D,\underline{\underline{d}})$ is a 2-disc (with boundary-pattern) with at least four free sides, then it is easy to construct admissible homotopy equivalences $(D,\underline{\underline{d}}) \to (D,\underline{\underline{d}})$ which are not admissibly homotopic to a homeomorphism. Since every surface (not necessarily orientable) with non-empty boundary can be obtained from some $(D,\underline{\underline{d}})$ by attaching bound sides pairwise, we find exotic homotopy equivalences of arbitrary surfaces with boundary.

If M is a ball, we have a much greater variety of possible boundary-patterns for M. Hence one should expect that there can be found a lot of exotic admissible homotopy equivalences between 3-balls with boundary-patterns. However, in fact, the contrary is true. Indeed, there are 3-balls with arbitrarily many free sides, but which do not admit any exotic admissible homotopy equivalence at all (this gives already a first hint to the fact that there are no exotic homotopy equivalences in the complement of the characteristic submanifold).

In order to explain this phenomenon more closely, we first have to put some restrictions on our choice of boundary-patterns. For this define an *i-faced disc*, $i \geq 1$, to be a 2-disc with i bound sides, $1 \leq i \leq 3$, and no free sides. Then the boundary-pattern of a 3-manifold $(N,\underline{\underline{n}})$ is called *useful* if every admissibly embedded i-faced disc, $1 \leq i \leq 3$, is parallel to a disc D from ∂N such that $D \cap \bigcup_{G \in \underline{\underline{n}}} \partial G$ is the cone on $\partial D \cap \bigcup_{G \in \underline{\underline{n}}} \partial G$. Observe that, by the loop-theorem [8], a 3-manifold N is boundary-incompressible if and only if $\{\partial N\}$ is a useful boundary-pattern for N. As we shall see later the notion of 3-manifolds with boundary-patterns is the appropriate concept, for our purpose, to work with. A 3-manifold $(N,\underline{\underline{n}})$ with useful boundary-pattern is called *simple* if its admissible homeomorphy-type cannot be simplified by splitting $(N,\underline{\underline{n}})$ along an admissibly embedded 4-faced disc (square), annulus (whose boundary-pattern consists of all the boundary components), or torus. To be more precise, we say a manifold $(N',\underline{\underline{n}}')$ is obtained from $(N,\underline{\underline{n}})$ by splitting along an admissibly embedded surface $(F,\underline{\underline{f}})$, $F \cap \partial N = \partial F$, if $N' = (N - U(F))^{-}$, for some regular neighborhood $U(F)$ of F, and if $\underline{\underline{n}}'$ is the proper boundary-pattern of $N' \subset N$. Then a 3-manifold $(N,\underline{\underline{n}})$ with useful boundary-pattern is a simple 3-manifold if at least one component of $(N',\underline{\underline{n}}')$ is an I- or S^1-bundle over an i-faced disc, $1 \leq i \leq 4$, or annulus, provided $(N',\underline{\underline{n}}')$ is obtained from $(N,\underline{\underline{n}})$ by splitting at a square, annulus, or torus.

The following proposition tells us that the presence of non-trivial squares is the only obstruction for deforming an admissible homotopy equivalence between balls into a homeomorphism. On the other hand, all such homotopy equivalences can be obtained by a finite number of flips along non-trivial squares.

PROPOSITION 2.1. Let $(M_1,\underline{\underline{m}}_1)$ and $(M_2,\underline{\underline{m}}_2)$ be two balls. Suppose that $\overline{\underline{\underline{m}}}_j$, $j = 1,2$, is a useful boundary-pattern, and that $(M_1,\overline{\underline{\underline{m}}}_1)$ is a simple ball. Then there exists no exotic admissible homotopy equivalence $f : (M_1,\overline{\underline{\underline{m}}}_1) \to (M_2,\overline{\underline{\underline{m}}}_2)$.

Proof. The boundary-patterns $\underline{\underline{m}}_j$ and $\overline{\underline{\underline{m}}}_j$, $j = 1,2$, induce canonical cell complexes C_j and $\overline{C}_j$: the 2-cells of these are the bound sides of $(M_j,\underline{\underline{m}}_j)$ resp. $(M_j,\overline{\underline{\underline{m}}}_j)$, the 1-cells are the bound sides of the 2-cells resp., and the 0-cells are the bound sides of the 1-cells resp.

Define K_j and $\overline{K}_j$ as the dual complexes of C_j and $\overline{C}_j$ resp. Since $\overline{\underline{\underline{m}}}_j$ is a useful boundary-pattern and since $(M_1,\overline{\underline{\underline{m}}}_1)$ is supposed to be a simple ball, the following facts are easily verified:

1. $\overline{K}_j$ is a triangulation of the 2-sphere ∂M_1.
2. Every simple closed and simplicial curve in $\overline{K}_1$ which consists of three (resp. four) 1-simplices is the boundary of a 2-simplex (resp. the link in $\overline{K}_1$ of some 0-simplex in $\overline{K}_1$).

3. K_1 is obtained from $\overline{K}_1$ by removing a finite number of disjoint open stars of 0-simplices in $\overline{K}_1$.

We shall call the simplices of K_j *"bound"* and those of $\overline{K}_j - K_j$ *"free"*.

Observe that the above three facts characterize the complexes $K_j \subset \overline{K}_j$ which are induced by simple balls $(M_j, \underline{\underline{m}}_j)$. Hence, conversely, there is a great variety of simple balls. However, we shall show that every simplicial isomorphism $\varphi : K_1 \to K_2$ extends to a simplicial isomorphism $\varphi : \overline{K}_1 \to \overline{K}_2$. This would prove our proposition, for every admissible homotopy equivalence $f : (M_1, \underline{\underline{m}}_1) \to (M_2, \underline{\underline{m}}_2)$ defines a simplicial isomorphism $\varphi : K_1 \to K_2$ in the obvious way.

The proof is by contradiction. Therefore assume that there is at least one free 0-simplex x_1 such that

$$\varphi(\text{link}(x_1, \overline{K}_1)) \neq \text{link}(y, \overline{K}_2), \text{ for all } y \in \overline{K}_2^{(0)} \quad .$$

Since φ is a simplicial isomorphism, $\varphi(\text{link}(x_1, \overline{K}_1))$ is a simple closed curve which is simplicial in K_2. By the Jordan curve theorem, this curve splits the 2-sphere ∂M_2 into two discs D_1' and D_2'.

By our assumption, it follows from 3. that each disc D_j', $j = 1, 2$, contains at least one bound 1-simplex t_j' with $t_j' \cap \partial D_j' \neq \emptyset$ and which meets $\partial D_j'$ in points. We may suppose that t_1' and t_2' have a common point z_1', for the proof in the other case is similar.

Since φ is a simplicial isomorphism, there must be precisely one 0-simplex z_1 in $\text{link}(x_1, \overline{K}_1)$ which is mapped under φ into z_1'. Moreover, by our supposition, there are two bound 1-simplices t_1, t_2 in star $(z_1, \overline{K}_1)$ such that $\varphi(t_j)$, $j = 1, 2$, is a 1-simplex in D_j' which meets $\partial D_j'$ in points. By 3., we may suppose without loss of generality, that both t_1 and t_2 lie in the link of some free 0-simplex x_2.

Considering the curve $\varphi(\text{link}(x_2,\overline{K}_1))$ we find that $\text{link}(x_1,\overline{K}_1)$ and $\text{link}(x_2,\overline{K}_1)$ must meet themselves in at least two different points z_1 and z_2. But this leads to a contradiction, either to 1. or to 2.

Hence we have a contradiction to our assumption that φ does not extend to $\overline{\varphi}$ and this completes the proof of 2.1.

III. INDICATION OF THE PROOF OF THE MAIN THEOREM

In this paragraph we would like to outline how the result on balls can be linked with the study of homotopy equivalences in general. The program to be formulated here is the burden of the proof of theorem 1.3., and will be carried out in detail in [4] (see also [12]).

First of all one defines a characteristic submanifold for irreducible 3-manifolds with useful boundary-patterns (modelled after that in the absolute case). Again the characteristic submanifold exists and is unique, now up to admissible ambient isotopy. Moreover, one has to prove that every admissible homotopy equivalence $f : (M_1,\underline{\underline{m}}_1) \to (M_2,\underline{\underline{m}}_2)$ can be split at the characteristic submanifolds, that is that f can be admissibly deformed, so that afterwards

$$f|V_1 : V_1 \to V_2 \text{ and } f|(M_1 - V_1)^- : (M_1 - V_1)^- \to (M_2 - V_2)^-$$

are admissible homotopy equivalences, where V_i, $i = 1,2$, is the characteristic submanifold of $(M_i,\overline{\underline{\underline{m}}}_i)$ and where V_i and $(M_i - V_i)^-$ are endowed with the proper boundary-patterns.

Now, given any surface F in $(M_2,\underline{\underline{m}}_2)$, one cannot expect in general that an admissible homotopy equivalence $f : (M_1,\underline{\underline{m}}_1) \to (M_2,\underline{\underline{m}}_2)$ can be admissibly deformed, so that $f^{-1}F$ is connected (counterexamples are easily constructed in dimension 2 already). However, it is a matter of fact that for *simple* 3-manifolds $(M_2,\underline{\underline{m}}_2)$ with $\overline{\underline{\underline{m}}}_2 \neq \emptyset$ it can be proved that there is at least one (non-separating) surface F_2 such that $f^{-1}F_2$ is connected, up to admissible homotopy, for every admissible homotopy equivalence $f : (M_1,\underline{\underline{m}}_1) \to (M_2,\underline{\underline{m}}_2)$. Moreover, one can show that f can be admissibly deformed, so that afterwards

$$f|f^{-1}U(F_2) : f^{-1}U(F_2) \to U(F_2) \text{ and}$$

$$f|(M_1 - f^{-1}U(F_2))^- : (M_1 - f^{-1}U(F_2))^- \to (M_2 - U(F_2))^-$$

are admissible homotopy equivalences, where $U(F_2)$ is a regular neighborhood of F_2 and where the manifolds involved are endowed with their proper boundary-patterns.

These two splitting theorems for admissible homotopy equivalences, together with the observation that the complement of the characteristic submanifold is a simple 3-manifold, lead us to the definition of a *great hierarchy*. A great hierarchy for a 3-manifold $(M,\emptyset) = (M_1,\underline{\underline{m}}_1)$ is a sequence $(M_1,\underline{\underline{m}}_1)$, $(M_2,\underline{\underline{m}}_2),\ldots$ of 3-manifolds such that

$M_{2i} = (M_{2i-1} - V_{2i-1})^-$, for all $i \geq 1$, where V_{2i-1} is the characteristic submanifold in $(M_{2i-1},\underline{\underline{m}}_{2i-1})$,

$M_{2i+1} = (M_{2i} - U(F_{2i}))^-$, for all $i \geq 1$, where F_{2i} is a surface chosen such that it has the above splitting property, and $U(F_{2i})$ is a regular neighborhood of F_{2i} .

The boundary-patterns in question are the proper ones. By a result of Haken [2], such a sequence always has an end, i.e. more precisely, there is an integer n such that $(M_n,\underline{\underline{m}}_n)$ consists of simple balls.

Now, let us be given any homotopy equivalence $f : M \to M'$ between 3-manifolds with boundaries. Then we define a useful boundary-pattern of M and M' by setting $\underline{\underline{m}} = \emptyset$ and $\underline{\underline{m}}' = \emptyset$. Now, we fix a great hierachy $M' = M_1'$, M_2' , ..., M_n' for M'. Then, with the help of the above splitting theorems, we obtain a sequence of admissible homotopy equivalences $f_i : (M_i, \underline{\underline{m}}_i) \to (M_i', \underline{\underline{m}}_i')$, and at last we end up with admissible homotopy equivalences between simple 3-balls. Next, we may apply the result of Chapt. 2, and, to complete the proof of 1.3., it then finally remains to work up the hierarchy again. For this purpose we have to prove:

> If f_{2i+2} can be admissibly deformed into a homeomorphism, then also f_{2i}, for all $i \geq 1$.

This finishes the indication of the proof.

IV. APPENDIX

Having established theorem 1.3., one can push the study of homotopy equivalences of 3-manifolds a bit further still. As an immediate consequence of the theorem we obtain that the homotopy type of any sufficiently large 3-manifold contains at most finitely many sufficiently large 3-manifolds, up to homeomorphy (see [6, problem 3.3.]). Furthermore, if M is a 3-manifold with boundary, then let $\underline{\underline{H}}_M$ denote the gruppoid defined by

obj $\underline{\underline{H}}_M$ = 3-manifolds homotopy equivalent to M
mor $\underline{\underline{H}}_M$ = homotopy classes of homotopy equivalences.

Now, suppose that the components of the characteristic submanifold of M admit fibrations as required in 2. of 1.1. but without any exceptional fibres. Then, with the help of theorem 1.3., one can prove that $\underline{\underline{H}}_M$ is finitely generated. This makes use also of the theorem that the homeotopy group of any simple 3-manifold is finite [5]. To show the latter one uses the theory of characteristic submanifolds to reduce the problem to a surface problem, namely the conjugacy problem of the homeotopy groups of surfaces with boundary, which was recently solved by Hemion [3].

REFERENCES

1. Fox, R. H., On the complementary domains of a certain pair of inequivalent knots, Ned. Akad. Wetensch., Indag. Math. 14, (1952) 37-40.
2. Haken, W., Über das Homöomorphieproblem der 3-Mannigfaltigkeiten I, Math. Z. 80 (1962) 89-120.
3. Hemion, G., On the classification of homeomorphisms of 2-manifolds and the classification of 3-manifolds, preprint (1976).
4. Johannson, K., Homotopy equivalences of 3-manifolds with boundary, to appear as Springer Lecture Note.
5. Johannson, K., On the mapping class group of sufficiently large 3-manifolds, preprint (1978).
6. Kirby, R. (ed.), Problems in low dimensional manifold theory, to appear.
7. Seifert, H., Topologie dreidimensionaler gefaserter Räume. Acta Math. 60 (1933) 147-238.
8. Stallings, J., On the loop theorem, Ann. of Math. 72 (1960) 12-19.
9. Waldhausen, F., Eine Klasse von 3-dimensionalen Mannigfaltigkeiten I, II, Invent. Math. 3 (1967) 308-333, 4 (1967) 87-117.
10. Waldhausen, F., Gruppen mit Zentrum und 3-dimensionale Mannigfaltigkeiten, Topologie 6 (1967) 505-517.
11. Waldhausen, F., On irreducible 3-manifolds which are sufficiently large, Ann. of Math. 87 (1968) 56-88.
12. Waldhausen, F., Recent results on sufficiently large 3-manifolds, Proc. Symp. Pure. Math., vol. 32, Amer. Math. Soc., Providence, R. I., to appear.

EIGHT FACES OF THE POINCARÉ HOMOLOGY 3-SPHERE

R. C. Kirby

Department of Mathematics
University of California
Berkeley, California

M. G. Scharlemann[1]

Department of Mathematics
University of California
Santa Barbara, California

We give eight different descriptions of the Poincaré homology sphere, and show that they do define the same 3-manifold. The definitions are: (1) plumbing on the E_8 graph, (2) surgery on the E_8 link, (3) the link of the singularity $z_1^2 + z_2^3 + z_3^5 = 0$, (4) S^3/I^* where I^* is the binary icosahedral group, (5) the dodecahedral space, (6) the Seifert bundle, (7) surgery on the trefoil knot, (8) the p-fold cover of the (q,r)-torus knot, for $\{p,q,r\} = \{2,3,5\}$.

The dodecahedral space of Poincaré was established long ago as a manifold of unusual interest, both because it was the first example of a homology sphere which is not a sphere and also because it lies in a class of three manifolds closely related to the Platonic solids. Interest in the manifold has increased in recent

[1] *Supported in part by NSF Grant MCS 7607181.*

ISBN 0-12-158860-2

years because of its surprisingly diverse applications to problems in topology (see e.g. [19, §2], [20], [16], [17], [21]). Part of the explanation for its usefulness is the large number of ways, discovered over the years, to describe the dodecahedral space. It is our aim in this paper to collect the most useful of these and verify at an elementary level that all do define the same 3-manifold. This paper arose from seminar notes in 1973 (and we thank L. Siebenmann for a substantial contribution to that seminar). We apologize for the untimely delay in appearance of this exposition, and remind the reader that since 1973, two excellent works, [12] and [14], have appeared which include parts of this paper.

I. EIGHT DESCRIPTIONS

Description 1 (Plumbing). Let $p : T^4 \to S^2$ be the contangent disk bundle over $S^2 = CP^1$ (this is just the tangent disk bundle with the opposite orientation so that the Euler characteristic is -2). Over any cell B^2 in S^2 the bundle is trivial so there is a commutative diagram

$$\begin{array}{ccc} p^{-1}(B^2) & \xrightarrow{\varphi} & B^2 \times B^2 \\ & {\scriptstyle p}\searrow \quad \swarrow{\scriptstyle proj_1} & \\ & B^2 & \end{array}$$

where φ is a diffeomorphism.

Two copies T_1 and T_2 of T can be "plumbed" together by identifying, for any $(x,y) \in B^2 \times B^2$, the points $\varphi_1^{-1}(x,y)$ and $\varphi_2^{-1}(y,x)$. The fibers of the first bundle over B^2 correspond to trivial sections of the second bundle over B^2.

Let P^4 be the result of plumbing together 8 copies of T as follows:

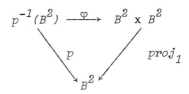

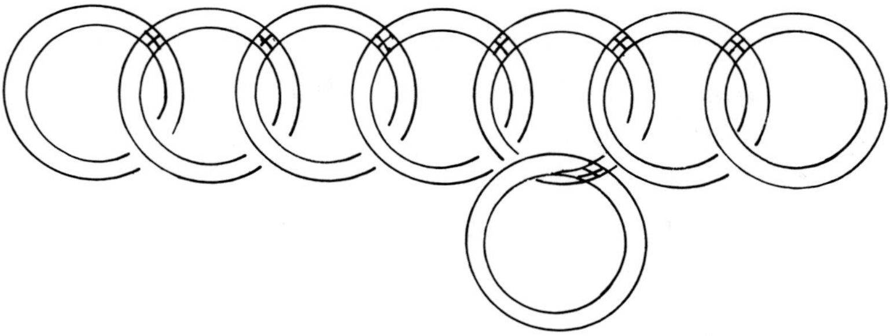

Fig. 1.

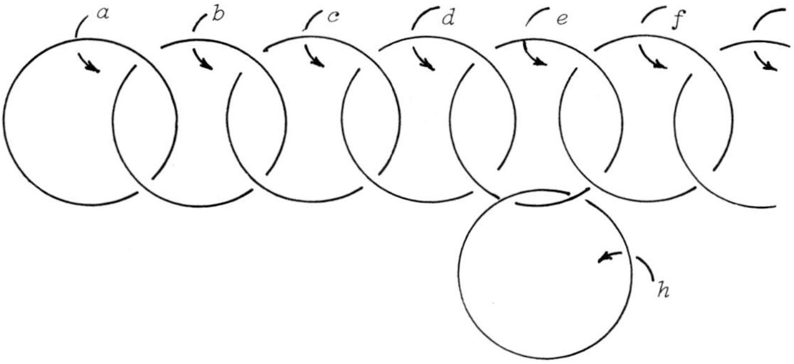

Fig. 2.

After rounding corners, P becomes a smooth 4-manifold. The first description of the dodecahedral manifold will be ∂P^4. In some of the future descriptions, we will recover not only ∂P^4, but P^4 as well.

Description 2 (Surgery on a link). Consider the link Λ of 8 circles in S^3, drawn in Figure 2. Each circle can be assumed planar in $R^3 = S^3 - \infty$, so each has an obvious trivialization of its normal disk bundle (choose one normal vector field in the plane, one orthogonal to the plane). The trivialization τ which we choose, however, is one obtained from the first by rotating the normal disk at each $\theta \in S^1$ by an angle 2θ. Attach $B^2 \times B^2$ to B^4 by $\tau : \partial B^2 \times B^2 \to S^3 = \partial B^4$. Since we may regard B^4 as the trivial 2-disk bundle over a 2-disk whose boundary is the attaching circle, the result is a 2-disk bundle over S^2. Since the framing chosen differs from the standard framing by 2 full left handed twists, the bundle has Euler characteristic -2, and so is the cotangent bundle of $S^2 = CP^1$.

Adjoin 8 copies of $B^2 \times B^2$ to B^4, one to each circle in Λ using the trivialization τ. The boundary of the resulting manifold is the second description of ∂P^4. In fact, since a pair of linking circles in Λ bound 2-disks in B^4 which intersect at just one point, it may be seen by inspection that descriptions 1 and 2 are equivalent, indeed that the 4-manifold just described is P^4.

With this description it is easy to compute $\pi_1(\partial P^4)$ using the calculus of Crowell and Fox [2]. The group $\pi_1(S^3 - \Lambda)$ is generated by loops around each circle a, b, c, d, e, f, g, h with relations for each crossing, $ab = ba$, $bc = cb$, $cd = dc$, $de = ed$, $ef = fe$, $fg = gf$, $eh = he$. The 8 copies of $B^2 \times \partial B^2$ attached to $S^3 - \Lambda$ provide 8 more relations $1 = a^2b = ab^2c = bc^2d = cd^2e = de^2fh = ef^2g = fg^2 = eh^2$. By substitution $e = a^5 = g^3 = h^{-2}$ and $h^{-1} = ag$, so we have generators a and g with $a^5 = g^3 = (ag)^2$.

This group $\{x,y;\ x^3 = y^5 = (yx)^2\}$, has an independent history, and is known in the literature (for reasons which will become clear) as the binary icosahedral group I^*. I^* is the only finite group which can occur as the fundamental group of a homology 3-sphere [8].

Description 3 (Link of a singularity). Let $f : C^3 \to C$ be the complex polynomial $f(z_1, z_2, z_3) = z_1^2 + z_2^3 + z_3^5$. $f^{-1}(0)$ is a complex variety which is non-singular except where $\partial f/\partial z_j = 0$ for all $j = 1,2,3$. Evidently the only singular point is the origin $z_1 = z_2 = z_3 = 0$. The intersection of the unit 5-sphere about the origin with this variety will also be shown to be ∂P^4.

Description 4 (The quotient S^3/I^*). The icosahedron is a regular solid with twenty faces, thirty edges and twelve vertices. It is the dual complex to the dodecahedron. The group I of isometries of the icosahedron (or dodecahedron) centered at the origin is naturally a subgroup of $SO(3)$, the group of orthogonal rotations of R^3.

Let $SU(2)$, the unitary transformations of C^2, act on C^2 on the right, that is, if $u = \begin{pmatrix} a & b \\ -\bar{b} & \bar{a} \end{pmatrix}$, $a\bar{a} + b\bar{b} = 1$, then $u(z,w) = (z,w)\begin{pmatrix} a & b \\ -\bar{b} & \bar{a} \end{pmatrix}$. This action of $SU(2)$ on C^2 commutes with complex multiplication, taking lines to lines, so it defines an action on $CP^1 = S^2 = C^1 \cup \infty$. If $u = \begin{pmatrix} a & b \\ -\bar{b} & \bar{a} \end{pmatrix} \in SU(2)$, then for $z \in C^1 \cup \infty$, $u(z) = (z,1)\begin{pmatrix} a & b \\ -\bar{b} & \bar{a} \end{pmatrix} = (az - \bar{b},\ bz + \bar{a}) = \dfrac{az - \bar{b}}{bz + \bar{a}}$. Hence u gives a linear fractional transformation of $C^1 \cup \infty$. If we identify $C^1 \cup \infty$ with S^2 by stereographic projection, these transformations map onto $SO(3)$. This map $q : SU(2) \to SO(3)$ defines a covering projection which is 2-fold since $q^{-1}(\text{identity}) = \begin{pmatrix} 1 & 0 \\ 0 & 1 \end{pmatrix} \cup \begin{pmatrix} -1 & 0 \\ 0 & -1 \end{pmatrix}$. (topologically this is the map $S^3 \to RP^3$.)

The lift of I to $SU(2)$ is denoted I^*; we will later show that I^* is the group $\pi_1(\partial P^4)$ calculated above. Since $SU(2)$ considered as a map $R^4 \to R^4$ preserves distance from $\{0\}$, $SU(2)$ acts on S^3. (Later we will show that in fact $SU(2)$ is S^3.) The quotient of S^3 by I^* is the fourth description of ∂P^4. Indeed, we will show there is a homeomorphism $C^2/I^* \to f^{-1}(0)$ above which is biholomorphic off of zero.

Description 5 (Poincare's). The dual of the icosahedron, the dodecahedron, is a regular solid with twelve faces, thirty edges and twenty vertices (see for example, [3, p. 11]). Identify opposite faces of the dodecahedron by the map which pushes each face through the dodecahedron and twists it $2\pi/10 = 36^\circ$ about the axis of the push in the direction of a right-hand screw. This identification is consistent along the edges (see [18]) and the quotient space is a 3-manifold (this requires some checking along the edges). This 3-manifold is ∂P^4.

Description 6 (Seifert bundle). ∂P is a Seifert bundle over S^2 with three exceptional fibers of Seifert invariant $(2,1)$, $(3,1)$ and $(5,1)$ and cross-section obstruction -1. Equivalently ∂P may be obtained by surgery with appropriate framings on any three "anti-Hopf" circles in S^3.

Here we outline what this means (see [13], Chapt. 1). Let M be an oriented 3-manifold with a smooth circle action. Each orbit α has a neighborhood diffeomorphic to $S^1 \times B^2$ (α corresponds to $S^1 \times 0$), with slices $s \times B^2$, $s \in S^1$, being taken to slices. The orbit is principal if $S^1 \times b$ is also an orbit for all $b \in B^2$. The orbit is exceptional with Seifert invariant $(n,1)$ if its neighborhood could be obtained from the principal orbit case by cutting $S^1 \times B^2$ at some slice $s \times B^2$, rotating the slice $2\pi/n$, and then gluing back together (assume the orbit followed by the slice gives the orientation of M). Thus an orbit near an exceptional orbit goes n times parallel to the exceptional orbit, and once around it; S^1 pushes a point *on* the exceptional orbit n times around the orbit. If $d_n : S^1 \times \partial B^2 \to S^1 \times \partial B^2$ is a diffeomorphism represented by the matrix $\begin{pmatrix} n & -1 \\ 1 & 0 \end{pmatrix}$, then $d_n(S^1 \times b)$ is the typical orbit near an exceptional orbit. The action of $\theta \in S^1$ near the orbit is given by $\theta(d_n(s,b)) = d_n(\theta \cdot s, b)$.

Denote the quotient space M/S^1 by $\overline{M}$. If there are only principal and exceptional orbits, then $\overline{M}$ is an oriented 2-manifold and away from the exceptional orbits M is an S^1 fiber bundle over $\overline{M}$. A cross-section to the action on the boundary of a tubular neighborhood of an $(n,1)$ orbit is given by $d_n(s \times \partial B^2)$. There is an obstruction in $H^2(\overline{M}, \pi_1(S^1)) = Z$ to extending these cross-sections to a cross-section to the circle action over all of M-(exceptional orbits). Choose a sigh for this obstruction as follows:

Let (j,k) denote a path going j times around S^1 and k times around ∂B^2. If, for some exceptional orbit, we choose as a cross-section not $d_n(0,1)$ but $d_n(c,1)$, we say the obstruction changes by c. In particular, for some c, the obstruction to extending $d_n(-c,1)$ vanishes; we call c the cross-section obstruction.

The cross-section obstruction may also be defined as follows. Let $\alpha : S^1 \times B^2 \hookrightarrow M$ be a tubular neighborhood of a *principal* orbit $\alpha(S^1 \times 0)$ so that the action at $\theta \in S^1$ is given by $\theta(s,b) = (\theta \cdot s, b)$. A cross-section for the action is $\alpha(0,1)$ then c is the integer such that $\alpha(-c,1)$ extends to a cross-section of all other principal orbits, a cross-section which coincides with $d_n(0,1)$ near the exceptional orbits.

As an example, let $M = S^3$ be the unit sphere around the origin in C^2. One action of the circle on S^3 is given by $\lambda(z,w) = (\lambda z, \bar{\lambda} w)$, $\lambda \in S^1 \subset C$, $(z,w) \in S^3 \subset C^2$. All the orbits are principal; indeed the quotient map is the "anti-Hopf" fibration $\bar{H} : S^3 \to S^2$, which is conventionally oriented so that the Euler class is $+1$. (This convention is motivated by the theory of complex manifolds, in which the natural action of S^1 on D^4, given by $\lambda(z,w) = (\lambda z, \lambda w)$, $(z,w) \in D^4 \subset C^2$, may be lifted to an action on the Hopf bundle by "blowing up" the origin in D^4 replacing the origin by a 2-sphere whose normal bundle has Euler class -1).

In general for M a bundle over $\bar{M}$ with no exceptional orbits and cross-section obstruction c, the Euler class is $-c$. Here, in particular, is how to see that the cross-section obstruction for the anti-Hopf circle action on S^3 is -1. Regard S^3 as the union $(S^1 \times B^2)_1 \cup_f (S^1 \times B^2)_2$ of two solid tori by a homeomorphism $f : (S^1 \times \partial B^2)_1 \to (S^1 \times \partial B^2)_2$ whose matrix is $\begin{pmatrix} 0 & 1 \\ 1 & 0 \end{pmatrix}$. Let $(S^1 \times B^2)_i$ have zero-section $S^1 \times \{0\}$ corresponding to the axis $w = 0$ for $i = 1$ and $z = 0$ for $i = 2$. Let $\theta \in S^1$ act on $S^1 \times B^2$ by $\theta(s,b) = (\theta s, b)$. The anti-Hopf action of the circle on S^3, restricted to $(S^1 \times B^2)_1$, is then $\alpha\,\theta\,\alpha^{-1}$, where $\alpha : S^1 \times B^2 \to (S^1 \times B^2)_1$ is the homeomorphism whose matrix is $\begin{pmatrix} 1 & 0 \\ -1 & 1 \end{pmatrix}$. Then $f\alpha(1,1) = (0,1)$ which extends to the cross-section $(\times B^2)_2$ over $(S^1 \times B^2)_2$. Hence $c = -1$.

To construct ∂P^4, remove three orbits from S^3 (labeled $\alpha_2, \alpha_3, \alpha_5$) and sew them back in using d_2, d_3 and d_5. More precisely, construct

$$(S^3 - (\alpha_2 \cup \alpha_3 \cup \alpha_5)) \cup_{g_2} S^1 \times B^2 \cup_{g_3} S^1 \times B^2 \cup_{g_5} S^1 \times B^2$$

where

$$g_i : S^1 \times (B^2 - 0) \to S^1 \times (B^2 - 0)$$

is $d_i \times id_{(0,1]}$ (we consider $B^2 - 0$ to be $\partial B^2 \times (0,1]$) and the domain of g_i is identified with a neighborhood of α_i, $i = 2,3,5$ by taking $S^1 \times b$ to an anti-Hopf circle for each $b \in B^2 - 0$. See Figure 3.

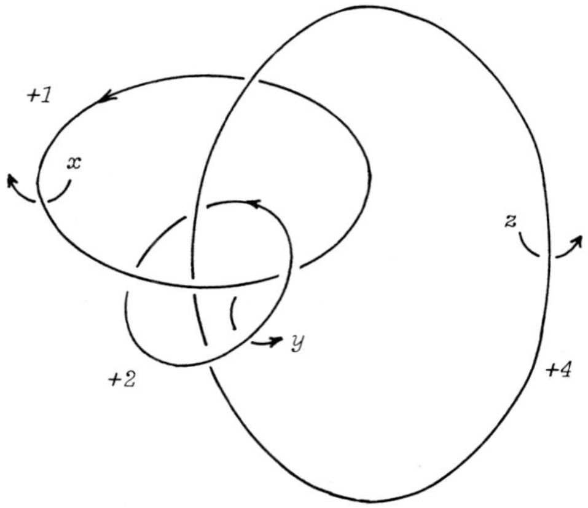

Fig. 3.

Clearly this describes the above Seifert bundle, and, simultaneously, a surgery on a link in S^3. Once again it is possible to calculate the fundamental group knot-theoretically. Let L be the link of 3 Hopf circles (Figure 3); then
$\pi_1(S^2 - L) = \{x,y,z | x = y^{-1}z^{-1} x zy,\ y = z^{-1}x^{-1} yxz,\ z = x^{-1}y^{-1} zyx\}$.
Each surgery kills the element corresponding to $d_n^{-1}(s, \partial B^2)$, which is a curve going once along α_n and winding around α_n n times compared to an anti-Hopf circle. Thus in Figure 2, the curves winding around n-1 times are killed and we add the relations

$1 = x^{-1}zy = y^{-2}xz = z^{-4}yz$. Then $x = zy$ and $x^2 = y^3 = z^5$, so the fundamental group is again I^*.

Description 7 (surgery on the trefoil knot). Surgery on the left handed trefoil knot L (Figure 4) with framing -1 gives ∂P^4 (this trefoil knot is called left handed because the crossings correspond to a left handed screw). A knot bounds a smooth orientable surface in S^3, which determines a normal vector field to the knot (tangent to the surface) and hence a framing (or trivialization) for the normal bundle. This is the zero framing, and framing n comes from twisting the 0-framing n times in a right handed direction. If we push the trefoil knot off itself using the framing, we get a curve homotopic to the dotted curve c in Figure 4. A presentation for $\pi_1(S^3 - L)$ is $\{a, b, c \mid ab = bc = ca\}$. Surgery kills the class represented by c, so we add the relation $baca^{-2} = 1$. Since $c = b^{-1}ab = aba^{-1}$, we have $(ab)^3 = (ab)(bc)(ca) = (abc)^2 = (a^2b)^2 = a^2(ba^2ba^{-3})a^3 = a^2(baca^{-2})a^3 = a^5$ so the group is $I^* = \{a, b \mid a^5 = (ab)^3 = (a^2b)^2\}$.

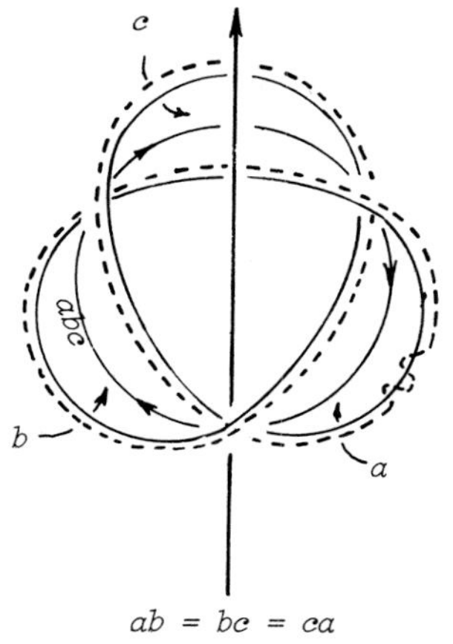

Fig. 4.

Description 8 (A branched cover). ∂P^4 is the 5-fold branched covering over the right handed trefoil knot $(= (2,3)$ torus knot). Similarly it is the 2-fold branched covering of the $(3,5)$ torus knot and the 3-fold branched covering of the $(2,5)$ torus knot.

Here is a brief description of the n-fold branched cover of a knot K in S^3. K has a trivial normal bundle. We will show that for some trivialization of the normal bundle, $T : S^1 \times B^2 \to S^3$, $T(S^1 \times 0) = K$, there exists a map $f : S^3 - K \to S^1$, unique up to homotopy, such that $fT|S^1 \times \partial B^2 = p_2 : S^1 \times \partial B^2 \to \partial B^2$. Let E denote the total space of the normal circle bundle of K.

Recall that for any space X, there is a natural isomorphism $[X,S^1] = [X,K(Z,1)] \cong H^1(X,Z)$. Since, for F a fiber of the normal circle bundle to K, inclusion induces an isomorphism $H^1(S^3 - K;Z) \cong H^1(F;Z)$, it also induces an isomorphism $[S^3 - K, S^1] \to [F, S^1]$. Thus a generator f of $[S^3 - K, S^1]$ carries F to S^1 by a degree one map. But varying the framing $T : S^1 \times B^2 \to S^3$ changes the degree of the composite $(S^1 \times b) \xrightarrow{T} E \xrightarrow{f} S^1$, $b \in \partial B^2$, by multiples of (degree $f|F) = 1$; hence we may choose T so that $(S^1 \times b) \xrightarrow{T} E \xrightarrow{f} S^1$ is zero. But the map $p_2 : S^1 \times \partial B^2 \to \partial B^2 \cong S^1$ is also of degree one on $F = (s \times \partial B^2)$ and degree zero on $(S^1 \times b)$. Since the two maps are homotopic on the 1-skeleton of E, they are homotopic on E. Thus we may take T so that $fT|S^1 \times \partial B^2 = p_2$.

To define the n-fold branched covering space Σ_n of K, let V be the bundle over $S^3 - K$ induced by f and the n-fold covering of S^1;

$$\begin{array}{ccc} V & \longrightarrow & S^1 \\ \downarrow{\scriptstyle \nu} & & \downarrow{\scriptstyle n} \\ S^3 - K & \xrightarrow{f} & S^1 \end{array}$$

The end of V is homeomorphic to $S^1 \times S^1 \times R$, so we can sew K back in to obtain the manifold Σ_n. The projection ν defines a map $\Sigma_n \xrightarrow{\nu} S^3$ which is a homeomorphism on $\nu^{-1}(K)$ and an n-fold covering elsewhere.

II. EQUIVALENCE OF THE DESCRIPTIONS

We will prove the following equivalences:

$$\begin{array}{ccccccccc} (5) & \Longleftrightarrow & (4) & \Longleftrightarrow & (3) & \Longleftrightarrow & (6) & \Longleftrightarrow & (1) \\ & & & & \Updownarrow & & \Updownarrow & & \Updownarrow \\ & & & & (8) & \Longleftrightarrow & (7) & \Longleftrightarrow & (2) \end{array}$$

While several arguments are long, they are designed to be self-contained. No deep theorems are required.

A few words are necessary about orientations. A complex manifold has a unique orientation ($SU(n)$ is connected) and we use this fact to determine a preferred orientation for ∂P^4. The variety $z_1^2 + z_2^3 + z_3^5 = 0$ is the cone on ∂P^4 and if we require (traditionally) that the first vector of its unique orientation be an outward pointing normal to ∂P^4, then we have oriented ∂P^4. If the singularity is resolved, we get the complex manifold int P^4 which corresponds to plumbing disk bundles with Euler characteristic -2.

If two complex linear subspaces in C^n intersect at a point, then together they must give the unique orientation of C^n, so algebraically their intersection must be $+1$. Thus the Hopf circles in S^3 (which are the intersections of complex lines in C^2 with S^3), must have linking number $+1$  ; also R^3 has the usual right handed orientation. The variety $z_1^2 + z_2^3 = 0$ (or $z_1^2 = z_2^3$) in C^2 meets S^3 in the right handed trefoil knot ($2,3$ torus knot).

On the other hand, $-\partial P^4$ bounds a complex manifold, the handle body obtained by attaching a 2-handle to B^4 along the right handed trefoil knot with framing $+1$. This complex manifold union P^4 is $CP^2 \# \#^8(-CP^2)$.

The most useful reference is [13] which contains proofs (buried in more general theorems) of the equivalences *(1)<=>(3)<=>(4)<=>(6)*; our proof of *(3)<=>(8)* is taken from [12], which also contains a somewhat different proof of *(3)<=>(4)*. A proof of *(5)<=>(4)<=>(3)<=>(1)* for a more general class of manifolds can be found in [4]. [14] contains equivalences *(2)<=>(7)<=>(8)*.

Equivalence of descriptions (4) and (5): First we describe how to picture S^3/I^*, then provide a proof that indeed S^3/I^* is the dodecahedron with opposite sides identified. Imagine the following circular chain of dodecahedra: place one dodecahedron on one face on the table, and then place nine more on it to form a tower with each dodecahedron rotated $\pi/5$ around the vertical axis compared to the one just below it; then identify top and bottom. Take another copy of this circular chain and place it adjacent to the first, at a slant of 36°, and winding once around the first, like a pair of Hopf circles. In this way wind five circular chains about the first one. In R^3 these do not fit perfectly together, but in S^3 they do. Note that it makes no difference which way they wind. Take two copies of this and sew them together the way one sews together solid tori to get S^3.

Thus S^3 is decomposed into 120 dodecahedra whose centers can be taken to be the elements of I^*. These elements permute the dodecahedra; in particular there is an element of I^* which pushes our original dodecahedron up one in the tower, identifying bottom and top of the dodecahedron. Similar "towers" through the other ten faces lead us to identify all opposite pairs of faces.

In order to prove that the fundamental domain of I^* is indeed the dodecahedron requires an analysis of how $SU(2)$ acts on S^3. We assume that $SU(2)$ acts on C^2 on the right, that is, if $u = \begin{pmatrix} a & b \\ -\bar{b} & \bar{a} \end{pmatrix}$, $a\bar{a} + b\bar{b} = 1$, then $u(z,w) = (z,w)\begin{pmatrix} a & b \\ -\bar{b} & \bar{a} \end{pmatrix}$. Identify

$SU(2)$ with the unit 3-sphere in C^2 by taking $\begin{pmatrix} a & b \\ -\bar{b} & \bar{a} \end{pmatrix}$ to (a,b). Thus S^3 acts on itself; we want a simple geometric picture of this action.

The complex lines in C^2 intersect only at the origin, so the complex lines intersected with S^3 give a decomposition (foliation) of S^3 into circles. (These circles are also the orbits under the circle action $\lambda(z,w) = (\lambda z, \lambda w)$, $\lambda \in S^1 \subset C$.) By sterographic projection identify S^3 with $R^3 \cup \infty$, with coordinates (r,s,t) on R^3. Assume the complex line $w = 0$ (the z-axis) intersects S^3 in $(r\text{-axis} \cup \infty) = S_1$ and the line $z = 0$ intersects S^3 in the unit circle S^3 in the (s,t) plane; in particular $(1,0,0,0)$ in S^3 goes to $(0,0,0)$ in R^3. The other complex lines intersect S^3 in the following kinds of circles; the complement of $S_1 \cup S_2$ in S^3 is a union of "concentric" tori (since S^3 is the join of S_1 and S_2). Each torus is the union of disjoint circles obtained from the -45° lines in the square by identifying opposite sides of the square in the orientation preserving way. In particular, we orient these circles continuously so that S_2 is oriented consistently with the usual orientation of the (st)-plane and S_1 has the same orientation as the z-axis. (If we think of S^2 as the unit disk in the (st)-plane with S_2 collapsed to a point, then each circle intersects S^2 exactly once; this defines the Hopf map $H : S^3 \to S^2$.)

Armed with this picture, the action of $\begin{pmatrix} \lambda & 0 \\ 0 & \bar{\lambda} \end{pmatrix} \in SU(2)$, $\lambda \in S^1 \subset C$, is easy to see. It twists the circle S_1 by an angle λ, and the circle S_2 by the angle $\bar{\lambda}$; the orbits of the induced action on the tori are perpendicular to those of λ.

There is, for each circle S in S^3 and point p in S, a copy of R^3 in R^4 perpendicular to S at p. The intersection with S^3 of this perpendicular R^3 will be called the perpendicular sphere at p in S. Clearly $\begin{pmatrix} \lambda & 0 \\ 0 & \bar{\lambda} \end{pmatrix}$ carries the perpendicular sphere at p in S_1 to the perpendicular sphere at λp.

In general, $n = \begin{pmatrix} a & b \\ -\bar{b} & \bar{a} \end{pmatrix}$ can be described similarly. It is the product of rotations through θ in the *real* plane L_u spanned by

$(1,0)$ and (a,b) and its orthogonal complement L'_u. This can be seen as follows. If $a = \alpha + i\beta$ and $b = \gamma + i\delta$ then the embedding $i : SU(2) \to SO(4)$ gives

$$i(u) = \begin{pmatrix} \alpha & \beta & \gamma & \delta \\ -\beta & \alpha & -\delta & \gamma \\ -\gamma & \delta & \alpha & -\beta \\ -\delta & -\gamma & \beta & \alpha \end{pmatrix}$$

with $\alpha^2 + \beta^2 + \gamma^2 + \delta^2 = 1$.

With respect to the basis

$$(1,\ 0,\ 0,\ 0),\quad \left(0,\ \frac{\beta}{\mu},\ 0,\ \frac{-\nu}{\mu}\right),$$

$$\left(0,\ \frac{\gamma}{\mu},\ \frac{\delta}{\omega},\ \frac{\beta\gamma}{\mu\nu}\right),\quad \left(0,\ \frac{\delta}{\mu},\ \frac{-\gamma}{\nu},\ \frac{\beta\delta}{\mu\nu}\right),$$

where $\mu = \sqrt{1-\alpha^2}$ and $\nu = \sqrt{\gamma^2+\delta^2}$,

$$i(u) \cdot \begin{pmatrix} \alpha & \sqrt{1-{}^2} & 0 & 0 \\ -\sqrt{1-\alpha^2} & \alpha & 0 & 0 \\ 0 & 0 & \alpha & -\sqrt{1-\alpha^2} \\ 0 & 0 & \sqrt{1-\alpha^2} & \alpha \end{pmatrix}$$

and $\cos\theta = \alpha = \mathrm{Re}(a)$. Thus u (or any other element of $L_u \cap S^3$) defines a Hopf-like decomposition of S^3 into circles.

Let S_u be that circle in S^3 which in R^3 is the line $t \cdot \left(\frac{\beta}{\mu}, \frac{\gamma}{\mu}, \frac{\delta}{\mu}\right)$, $t \in R$. Here $\left(\frac{\beta}{\mu}, \frac{\gamma}{\mu}, \frac{\delta}{\mu}\right)$ is the image of $(1,0,0)$ under the change of basis, so S_u is the image of S_1. Accordingly, u carries spheres perpendicular to S_u to other such spheres.

We now embed I^* in $SU(2)$ as described above. Place the dodecahedron with center at $0 \in R^3$ so that the barycenter of a face is tangent to $S^2 \cong C \cup \infty$ at the point $(1,0,0)$ in R^3, which corresponds to the point 0 in C under stereographic projection. Twist the dodecahedron by $2\pi/5$; this element g_o corresponds in C to multiplication by λ^2, where $\lambda = e^{\pi i/5}$, or, equivalently, to the linear fractional transformation whose matrix is $\begin{pmatrix} \lambda & 0 \\ 0 & \bar{\lambda} \end{pmatrix}$. Thus g_o is covered by $\begin{pmatrix} \lambda & 0 \\ 0 & \bar{\lambda} \end{pmatrix}$ in $SU(2)$. We have examined the action of such an element on S^3; it maps the sphere perpendicular to $e^{-\pi i/10} \in S_1$ to that perpendicular to $e^{\pi i/10}$. Thus the Z_{10} subgroup of $SU(2)$ covering the rotations of the dodecahedron about $(1,0,0)$ has a fundamental domain the region lying between these two perpendicular spheres--a lens shaped region (which gives its name to the Lens space S^3/Z_{10}).

Now let $z = u + iv$ be the barycenter of another face of the dodecahedron tangent to $S^2 = C \cup \infty$. Let $\rho = \sqrt{1 + z\bar{z}} = \sqrt{1 + (u^2 + v^2)}$ and note that the linear fractional transformation carrying 0 to z is given by the matrix

$$A = \begin{pmatrix} \frac{1}{\rho} & \frac{-\bar{z}}{\rho} \\ \frac{z}{\rho} & \frac{1}{\rho} \end{pmatrix} .$$

The $2\pi/5$ twist about this barycenter, denoted g_z, then corresponds to a matrix $A\begin{pmatrix} \lambda & 0 \\ 0 & \bar{\lambda} \end{pmatrix}A^{-1}$ in $SU(2)$. Let $\lambda = e + if$, and denote, as above, the entries in the matrix image of $A\begin{pmatrix} \lambda & 0 \\ 0 & \bar{\lambda} \end{pmatrix}A^{-1}$ in $SO(4)$ by $\alpha, \beta, \gamma, \delta$. An easy calculation shows that $\alpha = e$, $\beta = f/\rho^2(2 - \rho^2)$, $\gamma = \frac{2fv}{\rho^2}$, $\delta = \frac{2fu}{\rho^2}$.

As before, let S_z be a circle in S^3 such that $S_z \cap R^3 = t(\beta/\mu, \gamma/\mu, \delta/\mu)$. Then g_z carries spheres perpendicular to S_z to other such spheres. In this case $\mu = \sqrt{1 - e^2} = f$

$$\left(\frac{\beta}{\mu}, \frac{\gamma}{\mu}, \frac{\delta}{\mu}\right) = \left(\frac{2-\rho^2}{\rho^2}, \frac{2v}{\rho^2}, \frac{2\mu}{\rho^2}\right) .$$

But the coordinates of z in R^3 under stereographic projection are

$$\left(\frac{2-\rho^2}{\rho^2}, \frac{2u}{\rho^2}, \frac{2v}{\rho^2}\right) .$$

Comparing the two vectors in R^3 we deduce (after the orthogonal rotation which switches the last two coordinates) that rotation of the dodecahedron about the barycenter at z lifts in $SU(2)$ to the same action as rotation about the barycenter at 0, except the axis in R^3 of the translation of S^3 *now points through* z *instead of through* $(1,0,0)$. Thus the axis of the (lens-shaped) fundamental domain of g_z passes through z. The intersection of all the fundamental domains of all rotations about barycenters of faces is then the intersection of those lenses whose axes point in the direction of the barycenters. But this intersection is precisely the fundamental domain of I^*, since it is easy to see that all elements of I are compositions of rotations about barycenters of faces. But the intersection of these lenses is clearly the dodecahedron. Furthermore, we have seen that the action of I^* on S^3 identifies opposite sides of the lenses (hence of the dodecahedron) with a $\pi/5$ twist. This completes the proof.

Incidentally, it is possible to explicitly calculate generators and relations for I^* by using description 5) for ∂P, as in [18]. This is then a roundabout proof that $I^* = \{(x,y) \mid x^3 = y^5 = (xy)^2\}$.

Equivalence of descriptions (3) and (4). The proof is a medley of [12] and [10]. Our aim is to find a homeomorphism $P : C^2/I^* \to f^{-1}(0)$ where $f : C^3 \to C$ is $f(z_1,z_2,z_3) = z_1^2 + z_2^3 + z_3^5$. We will find three homogeneous polynomials $p_1, p_2, p_3 : C^2 \to C^1$, and define $\overline{P} = (p_1,p_2,p_3) : C^2 \to C^3$. We must show that

(i) $\overline{P}$ is invariant under action by I^* so that $\overline{P}$ defines $P : C^2/I^* \to C^3$.

(ii) $p_1^2 + p_2^3 + p_3^5 = 0$ so that image $(P) \subset f^{-1}(0)$.

(iii) $d\overline{P}$ has rank 2 on $C^2 - 0$ and thus $P\ (C^2/I^*)-0$ is a covering map.

(iv) $\overline{P}^{-1}(\text{point}) = 120$ points, so P is one-to-one and a homeomorphism.

$$10 \text{ elements} \quad \pm \begin{pmatrix} \varepsilon^{3\mu} & 0 \\ 0 & \varepsilon^{2\mu} \end{pmatrix}$$

$$10 \text{ elements} \quad \pm \begin{pmatrix} 0 & -\varepsilon^{2\mu} \\ \varepsilon^{3\mu} & 0 \end{pmatrix}$$

$$50 \text{ elements: } \pm \frac{1}{\sqrt{5}} \begin{pmatrix} -\varepsilon^{3(\mu+\omega)}(\varepsilon-\varepsilon^{4}) & \varepsilon^{3(\omega-\mu)}(\varepsilon^{2}-\varepsilon^{3}) \\ \varepsilon^{3(\mu-\omega)}(\varepsilon^{2}-\varepsilon^{3}) & \varepsilon^{-3(\mu+\omega)}(\varepsilon-\varepsilon^{4}) \end{pmatrix}$$

$$50 \text{ elements: } \pm \frac{1}{\sqrt{5}} \begin{pmatrix} \varepsilon^{3(\mu-\omega)}(\varepsilon^{2}-\varepsilon^{3}) & \varepsilon^{-3(\mu+\omega)}(\varepsilon-\varepsilon^{4}) \\ -\varepsilon^{3(\mu+\omega)}(\varepsilon-\varepsilon^{4}) & -\varepsilon^{3(\omega-\mu)}(\varepsilon^{2}-\varepsilon^{3}) \end{pmatrix}$$

This yields the polynomials

$$p_3 = -(1728)^{1/5}\, z_1 z_2 (z_1^{10} + 11 z_1^5 z_2^5 - z_2^{10})$$

$$p_1 = (z_1^{30} + z_2^{30}) + 522(z_1^{25} z_2^5 - z_1^5 z_2^{25}) - 10005(z_1^{30} + z_1^{10} z_2^{20})$$

$$p_2 = -(z_1^{20} + z_2^{20}) + 228(z_1^{15} z_2^5 - z_1^5 z_2^{15}) - 494 z_1^{10} z_2^{10} .$$

The reader may verify directly that $p_1^2 + p_2^3 + p_3^5 = 0$, but the following argument is both more elegant and requires no calculation.

Consider the complex vector space V of homogeneous polynomials of degree 60; V has dimension 61, with basis $z_1^i z_2^{60-i}$, $i = 0,\dots,60$. There is a 2-dimensional subspace $W = \{\lambda p_3^5 + \mu p_1^2\}$, for $\lambda,\mu \in C$. Given a barycenter (a,b) of a face of the icosahedron, the annihilator A of the 1-dimensional subspace $az_2 - bz_1 = 0$ has dimension 60. Thus $\dim(W \cap A) \geq 1$, so for some fixed λ and μ, not both zero, $(\lambda p_3^5 + \mu p_1^2)(z_1, z_2) = 0$ if $az_2 - bz_1 = 0$.

The orbits of points in CP^1 under the action of I are of four types:

(1) the 12 vertices of the icosahedron

(2) the 30 barycenters of edges

(3) the 20 barycenters of faces

(4) orbits containing 60 points.

Since p_1, p_2, and p_3 are invariant under I^*, it follows that the zeroes of $\lambda p_3^5 + \mu p_1^2$ must consist of complex lines through entire orbits. Suppose there are ω_i orbits of type i, $i = 1,2,3,4$, in the zeroes of $\lambda p_3^5 + \mu p_1^2$, multiplicities included. Then degree $(\lambda p_3^5 + \mu p_1^2) = 60 = 12\omega_1 + 30\omega_2 + 20\omega_3 + 60\omega_4$. Since $\lambda p_3^5 + \mu p_1^2$ is zero on the complex line through a barycenter of a face, it follows that $\omega_3 \neq 0$, but then $\omega_3 = 3$ and $\omega_1 = \omega_2 = \omega_4 = 0$. Thus $\lambda p_3^5 + \mu p_1^2$ has the same zeroes as p_2^3, so $\lambda p_3^5 + \mu p_1^2 = \nu p_2^3$. We redefine p_3 to be the old p_3 divided by $\lambda^{1/5}$ and so on, so that $p_1^2 + p_2^3 + p_3^5 = 0$. We have now satisfied properties (i) and (ii).

To show that $\overline{P} : C^2 \to f^{-1}(0)$ or $P : C^2/I^* \to f^{-1}(0)$ is locally bihilomorphic off zero, it suffices to prove that the matrix

$$d\overline{P} = \begin{pmatrix} \dfrac{\partial p_1}{\partial z_1} & \dfrac{\partial p_2}{\partial z_1} & \dfrac{\partial p_3}{\partial z_1} \\[2em] \dfrac{\partial p_1}{\partial z_2} & \dfrac{\partial p_2}{\partial z_2} & \dfrac{\partial p_3}{\partial z_2} \end{pmatrix}$$

has rank 2 everywhere. Note that

$$\Delta_2 = \text{determinant} \begin{pmatrix} \dfrac{\partial p_1}{\partial z_1} & \dfrac{\partial p_3}{\partial z_1} \\ \\ \dfrac{\partial p_1}{\partial z_2} & \dfrac{\partial p_3}{\partial z_2} \end{pmatrix}$$

is homogeneous of degree 40 and is invariant under I^*; therefore it can be zero only on the complex lines through the orbit of points determined by the twenty faces of the icosahedron. The other determinants are zero only on the lines through the vertices or barycenters of edges, so $d\overline{P}$ has rank 2 except on zero.

The map $\overline{P}$ is proper since p_1, p_2 and p_3 are polynomials. Therefore $\overline{P}(C^2 - 0)$ is closed in $f^{-1}(0) - 0$. It is also open because $\overline{P}$ is locally a homeomorphism. Since $f^{-1}(0) - 0$ is connected, $\overline{P}$ and P are onto. Thus $P : C^2/I^* - 0 \to f^{-1}(0) - 0$ is a covering space.

Finally we must show that $\overline{P}^{-1}$(point) has 120 points; then P is one-to-one and so a homeomorphism. Since order $(I^*) = 120$, $\overline{P}^{-1}$(point) > 120 points. Consider $(a,b,0) \in f^{-1}(0)$, (a,b) a vertex of the icosahedron. $p_3(z_1,z_2) = 0$ has the usual solution, 12 lines. On each line, p_1 and p_2 restrict to polynomials in one variable of degree 30 and 20 respectively. The solutions of $z^{30} = a$ and $z^{20} = b$ are the vertices of a regular 30-gon and 20-gon respectively. So there are at most 10 common solutions on each of the 12 lines, so $\overline{P}^{-1}(a,b,0) \leq 120$ points. This finishes the construction of the homeomorphism $P : C^2/I^* \to f^{-1}(0)$.

The required homeomorphism $Q : S^3/I^* \to f^{-1}(0) \cap S^5$ is defined as follows. For each $x \in S^3$, let $R_x \subset C^2 - 0$ be the ray through x. If $y \in I_*(x)$ then $R_y \in I_*(R_x)$; thus $P(R_x)$ is well defined. Furthermore, since p_1, p_2 and p_3 are homogeneous polynomials, as the distance from the origin to a point t on R_x increases, so does the distance from the origin to $P(t)$. Thus $P(R_x) \cap S^5$ is one point, which we call $Q(x)$. Q is clearly smooth and one-to-one; it is onto since P is.

Equivalence of descriptions (3) and (6). We will define a circle action on the link L of the singularity $z_1^2 + z_2^3 + z_3^5 = 0$, an action which gives L the required Seifert manifold structure. For $\gamma \in S^1 \subset C$, let $\gamma(z_1,z_2,z_3) = (\gamma^{15} z_1, \gamma^{10} z_2, \gamma^6 z_3)$. Clearly this circle action on C^3 leaves L invariant.

The orbits of S^1 are principal if all $z_i \neq 0$, for if $\gamma(z_1,z_2,z_3) = (z_1,z_2,z_3)$ then $\gamma^6 = \gamma^{10} = \gamma^{15} = 1$, so $\gamma = 1$. The exceptional orbits are the three orbits $z_1 = 0$, $z_2 = 0$, $z_3 = 0$.

S^1 acts on the orbit $z_3 = 0$ by $\gamma(z_1,z_2,0) = (\gamma^{15},\gamma^{10} z_2, 0)$ so $Z_5 \subset S^1$ acts trivially on the orbit. Furthermore if $\omega \in S^1$ satisfies $\omega^5 = 1$, then ω acts on a disk perpendicular to the orbit via complex multiplication by $(\omega)^6 = \omega$. Thus the orbit is exceptional of type $(5,1)$. Similarly $z_2 = 0$, $z_1 = 0$ are exceptional orbits of type $(3,1)$ and $(2,1)$.

Next we show that any Seifert manifold M whose only exceptional orbits are of type $(2,1)$, $(3,1)$ and $(5,1)$ and which is an integral homology sphere is a Seifert manifold with quotient space S^2 and cross-section obstruction -1. This will complete the proof, for we know from the equivalence of 3 and 5 that $\pi_1(L) = \{x,y\colon x^3 = y^5 = (yx)^2\}$ so $H_1(L) = 0$.

First note that if the quotient space $\overline{M}$ of M by the circle action is of genus g, then the first homology of the (trivial) circle bundle obtained by deleting the exceptional orbits has rank $2g + 3$. Sewing back the 3 exceptional orbits can at most decrease the rank to $2g$, because sewing in a copy of $S^1 \times B^2$ adds only the relation corresponding to $s \times \partial B^2$. Thus $g = 0$ and $\overline{M} = S^2$.

To calculate the cross-section obstruction b, construct a presentation for $\pi_1(M)$ as follows. A cross-section for the circle bundle away from the exceptional orbits is a 3-punctured sphere with fundamental group $\{q_1,q_2,q_3;\ q_1q_2q_3 = 1\}$, where each q_i is a path around an exceptional orbit. Let $L \in \pi_1(M)$ be represented by a principal orbit.

Let (j,k) denote the path in $S^1 \times \partial B^2$ which goes j times around S^1 and k times around ∂B^2. Attach the exceptional orbits by d_i, $i = 2,3,5$. We may assume that $d_2(-c,1) = q_1$, $d_3(0,1) = q_2$, $d_5(0,1) = q_3$. But $d_i(1,i) = (0,1)$ which is null-homotopic in $S^1 \times B^2$. Thus adding the exceptional orbits introduces the relations $d_i(1,i) = 0$. Thus $q_1^2 = d_2(-2c,2) = d_2(-2c-1,0) = h^{-2c-1}$, $q_2^3 = d_3(0,3) = d_3(-1,0) = h^{-1}$, $q_3^5 = d_5(0,5) = d_5(-1,0) = h^{-1}$. Thus $\pi_1(M) = \{q_1,q_2,q_3,h; q_1^2\, h^{2c+1} = q_2^3\, h = q_3^5\, h = [q_i,h] = q_1q_2q_3 = 1\}$. Eliminate q_1 by $q_1 = (q_2q_3)^{-1}$ and h by $h = q_2^{-3}$ and abelianize to obtain

$$H_1(M) = \{q_2,q_3; 2q_3 + (6c+5)q_2 = 5q_3 - 3q_2 = 0\}.$$

Thus $H_1(M)$ is of order

$$\det \begin{vmatrix} 6c+5 & 2 \\ -3 & 5 \end{vmatrix} = 30c + 25 + 6.$$

Since $H_1(M) = 0$, $c = -1$.

Equivalence of descriptions (6) and (1). Examine the structure of the plumbing construction P^4, 8 copies of the cotangent disk bundle T plumbed together as shown:

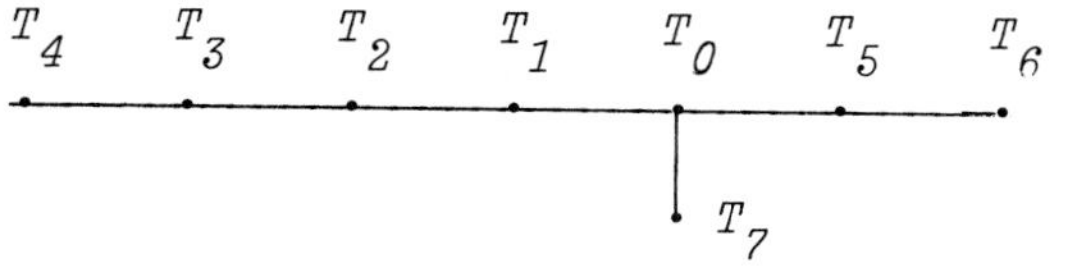

This may be viewed as plumbing 3 "arms" of length 4, 2, and 1, to the central T_0. That part of ∂P^4 lying in each arm has a very simple description. In particular, $T_1 \cap \partial P^4$ is an S^1 fiber bundle over S^2 with neighborhoods of two fibers removed. But removing two fibers leaves a trivial bundle, so $T_1 \cap \partial P^4 \cong S^1 \times S^1 \times I$ for $i = 1,2,3,5$, and $T_j \cap \partial P^4 = S^1 \times B^2$ for $j = 4,6,7$. Thus each arm, consisting, for example of $T_1 \cup T_2 \cup T_3 \cup T_4$, intersects in a copy of $S^1 \times B^2$. Hence ∂P^4 is obtained from ∂T_0 by removing three tubular neighborhoods $(S^1 \times B^2)_i$ of fibers in ∂T_0 and, to each $(S^1 \times \partial B^2)$ boundary component, attaching a copy of $S^1 \times B^2$ by some (linear) attaching map $g_i : (S^1 \times \partial B^2)_i \to S^1 \times \partial B^2$.

There is a natural circle action on ∂T_0, in which the circle acts on each fiber by rotation. Remove $(S^1 \times B^2)_i$, $i = 1,2,3$ and attach three copies of $S^1 \times \partial B^2$ by $\{g_i\}$. The action on $(S^1 \times B^2)_i$ extends linearly over each attached $S^1 \times B^2$, so the circle action extends over ∂P^4. We will verify that this circle action gives the required Seifert manifold structure by calculating g_i.

LEMMA 1. The attaching map $g_i : (S^1 \times \partial B^2)_i \to S^1 \times \partial B^2$ for an arm of length $m \geq 0$ is given by the matrix

$$\begin{pmatrix} 2 & 1 \\ -1 & 0 \end{pmatrix}^m = \begin{pmatrix} m+1 & m \\ -m & 1-m \end{pmatrix} .$$

Proof. This is certainly the case for $m = 0$, that is, for ∂T_0 itself. The proof is by induction. Suppose it is true for an arm of length $m \geq 0$.

Let T_m denote the copy of T at the end of the arm to which we plumb T_{m+1}. Since T_{m+1} has Euler class -2, it is made from two charts, $(B^2 \times B^2)^1$ and $(B^2 \times B^2)^2$, and we identify $(B^2 \times \partial B^2)^1$ with $(B^2 \times \partial B^2)^2$ by extending the map $(\partial B^2 \times \partial B^2)^1 \to (\partial B^2 \times \partial B^2)^2$ given by the matrix $\begin{pmatrix} 1 & 2 \\ 0 & -1 \end{pmatrix}$ linearly across $(B^2 \times \partial B^2)^1$. Plumb T_{m+1} in along $(B^2 \times B^2)^2$. Then a copy $(S^1 \times B^2)^3$ in ∂T_m is identified with $(B^2 \times \partial B^2)^1$ by switching factors, so the attaching map $(S^1 \times \partial B^2)^3 \to (\partial B^2 \times \partial B^2)^1$ is given by the matrix $\begin{pmatrix} 0 & 1 \\ 1 & 0 \end{pmatrix}$. But by induction hypothesis, the map $(S^1 \times \partial B^2)_i \to (S^1 \times \partial B^2)^3$ is given by $\begin{pmatrix} 2 & 1 \\ -1 & 0 \end{pmatrix}^m$. Thus the attaching map $(S^1 \times \partial B^2)_i \to (S^1 \times \partial B^2)^2$ is given by

$$\begin{pmatrix} 1 & 2 \\ 0 & -1 \end{pmatrix}\begin{pmatrix} 0 & 1 \\ 1 & 0 \end{pmatrix}\begin{pmatrix} 2 & 1 \\ -1 & 0 \end{pmatrix}^m = \begin{pmatrix} 2 & 1 \\ -1 & 0 \end{pmatrix}^{m+1},$$

proving the lemma.

LEMMA 2. Each arm of length m attached to T_0 adds an exceptional orbit of type $(m+1, 1)$ and decreases the cross-section obstruction by one.

Proof. The proof is based on the matrix identity

$$\begin{pmatrix} m+1 & m \\ -m & 1-m \end{pmatrix} = \begin{pmatrix} 1 & 0 \\ -1 & 1 \end{pmatrix}\begin{pmatrix} m+1 & -1 \\ 1 & 0 \end{pmatrix}\begin{pmatrix} 1 & 1 \\ 0 & 1 \end{pmatrix}$$

Thus from Lemma 1 g_i is the composition of three automorphisms $(S^1 \times \partial B^2)\hookleftarrow$. The first, represented by $\begin{pmatrix} 1 & 1 \\ 0 & 1 \end{pmatrix}$ has no effect on the circle action, but changes the cross-section $(0,1)$ to $(1,1)$. The second automorphism is just d_{m+1}. The original cross-section obstruction c is the obstruction to extending the cross-section $d_{m+1}(1,1)$. Then the obstruction of extending $d_{m+1}(0,1)$ is $c - 1$. Thus the composition

$$\begin{pmatrix} m+1 & -1 \\ 1 & 0 \end{pmatrix}\begin{pmatrix} 1 & 1 \\ 0 & 1 \end{pmatrix}$$

represents the addition of an exceptional orbit of type $(m+1, 1)$, and a decrease by 1 in the cross-section obstruction. The third automorphism, represented by $\begin{pmatrix} 1 & 0 \\ -1 & 1 \end{pmatrix}$ extends to an automorphism $S^1 \times B^2 \hookleftarrow$ and thus has no effect on the homeomorphism type. This proves the lemma.

Since ∂P^4 is obtained from T_Q, which has cross-section obstruction 2 (Euler class -2), by attaching arms of length 1, 2, and 4, it follows that the cross-section obstruction is -1, and ∂P^4 has exceptional orbits of type $(2,1)$, $(3,1)$ and $(5,1)$.

Equivalence of descriptions (1) and (2). The equivalence of descriptions (1) and (2) follows from the definitions; see definition of (2) above.

Equivalence of descriptions (3) and (8). We sketch the proof of Milnor [12]. The torus knot of type $(2,3)$ is the knot which wraps around the standard torus in R^3, twice in one direction and three times in the other. In other words it is the image in $S^3 \subset C^2$ of the circle $S^1 \subset C$ under the map $t \to \frac{1}{\sqrt{2}}(t^3, t^2)$. This is the intersection of S^3 and the variety $\{(z_1, z_2) \in C^2 \mid z_1^2 = z_2^3\}$.

Let L be the link of the singularity $z_1^2 + z_2^3 + z_3^5 = 0$, and M be the 5-fold branched cover of the trefoil knot.

Evidently $V = \{(z_1, z_2, z_3) \in C^3 - 0 \mid z_1^2 + z_2^3 + z_3^5 = 0\}$ is the 5-fold branched cover of $C^2 - 0$ along $B = \{(z_1, z_2) \in C^2 - 0 \mid z_1^2 + z_2^3 = 0\}$. Indeed the projection $(z_1, z_2, z_3) \to (z_1, z_2)$ is a 5-fold cover away from $z_1^2 + z_2^3 = 0$ corresponding to the 5 roots of $z^5 \neq 0$, but is a homeomorphism when $z_1^2 + z_2^3 = 0 = z_3^5$. R^+ acts on V and $C^2 - 0$ by $t(z_1, z_2, z_3) = (t^{1/2} z_1,\ t^{1/3} z_2,\ t^{1/5} z_3)$ and $t(z_1, z_2) = (t^{1/2} z_1,\ t^{1/3} z_2)$. The action commutes with projection. Since each orbit of R_+ intersect L precisely once, $V/R_+ \cong L$. Similarly $C^2 - 0/R_+ \cong S^3$ and the induced map $L \to S^3$ is a branched 5-fold cover over $B/R_+ \cong$ trefoil knot.

Equivalence of descriptions (2) and (7). We need to show that doing surgery on the framed link Λ of (2) gives the same 3-manifold as doing surgery on the left handed trefoil knot using the -1 framing.

The first author shows that two framed links Λ and Λ' yield the same 3-manifold if and only if they are related by a series of link operations of two kinds [9]:

$\mathcal{O}_1$: Add to or subtract from a link an unknotted circle with framing ± 1, which is separated from the other circles by an embedding S^2 in S^3.

$\mathcal{O}_2$: Given two components γ_0 and γ_1 of an oriented, framed link, push γ_1 off itself, using its given framing, to obtain γ_1'. Join γ_0 and γ_1 by a strip $b : I \times I \hookrightarrow S^3$ such that $b(I \times I) \cap \gamma_i = b(i \times I)$, $i = 0,1$. Then substitute for γ_0 and γ_1 the circles γ_1' and $\gamma_0 \#_b \gamma_1 = \gamma_0 \cup \gamma_1 \cup b(I \times \partial I) - b(\partial I \times I)$.

The framing for γ_1' is the same as that for γ_1; that for $\gamma_0 \#_b \gamma_1$ is the sum of the framings of γ_0 and γ_1 plus or minus twice the linking number of γ_1 with γ_0. The sign is plus if and only if $b(I \times I)$ can be oriented consistently with γ_0 and γ_1.

The full strength of this theorem is unnecessary. Here we use only the "easy" part, that if two links are related by $\mathcal{O}_1$ and $\mathcal{O}_2$, then the corresponding 3-manifold are homeomorphic. Indeed, $\mathcal{O}_1$ corresponds to taking connected sum with or splitting off a copy of the complex projective plane $\pm CP^2$, with one of its orientations, from the trace of the surgery. This follows immediately from the fact that $\pm CP^2$ - *(4-disk)* is the Hopf disk bundle over S^2 with Euler class ± 1.

$\mathcal{O}_2$ corresponds to sliding the 2-handle attached along γ_0 (in the trace of the surgery) across the 2-handle along γ_1.

LEMMA 3. *If we change a portion of a framed link as in Figure 5 below, then the 3-manifold resulting from surgery is not changed.*

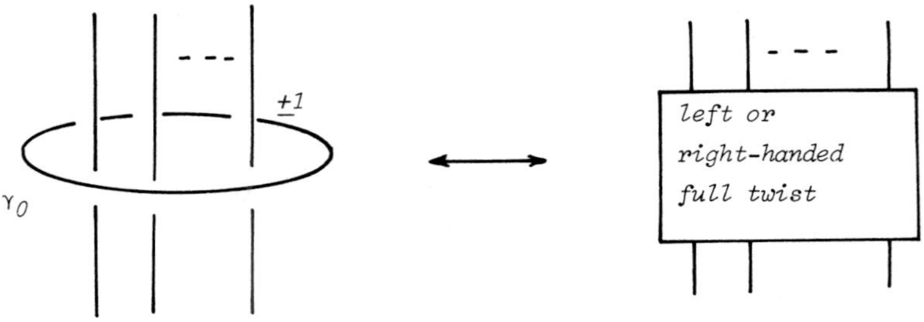

Fig. 5.

If γ has framing n in the left case, then it will have framing $n \pm (\ell(\gamma_0, \gamma))^2$ in the right case.

The proof, given in [9], is a straightforward application of $\mathcal{O}_1$ and $\mathcal{O}_2$, and can be worked out easily by the reader for one or two strands through γ_0.

By a series of applications of Lemma 3 we change the framed link Λ to the *-1* trefoil knot. First we introduce three unknots with *+1* framing ($\mathcal{O}_1$) and then slide the end circles of Λ over them ($\mathcal{O}_2$) to get

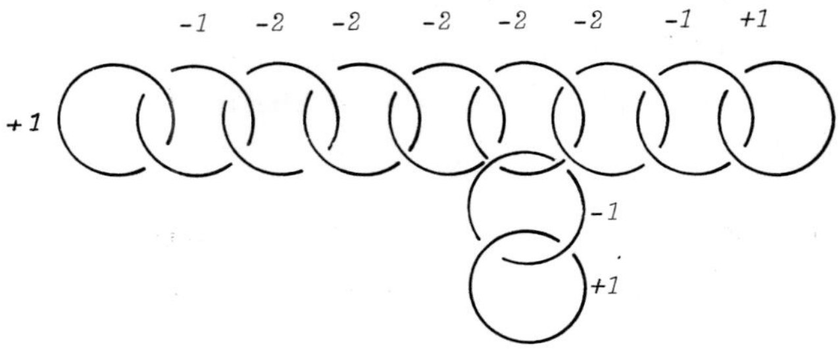

Fig. 6.

Then we remove the *-1* circles, using the lemma until we get

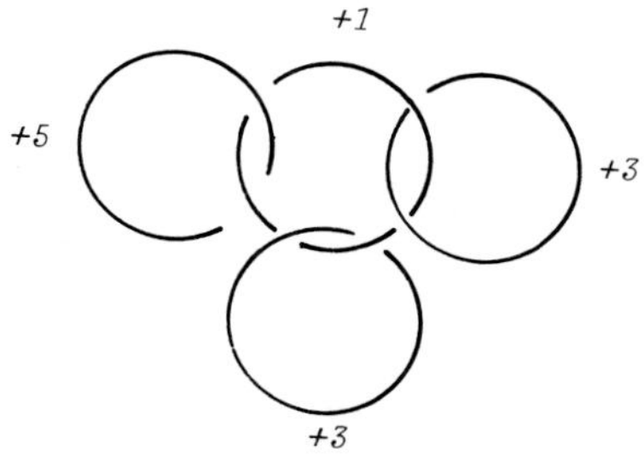

Fig. 7.

Removing the *+1* circle, we get

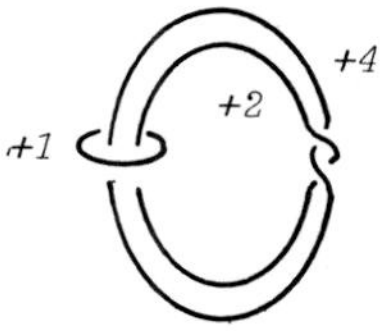

Fig. 8.

Blowing down the *+1* circle gives

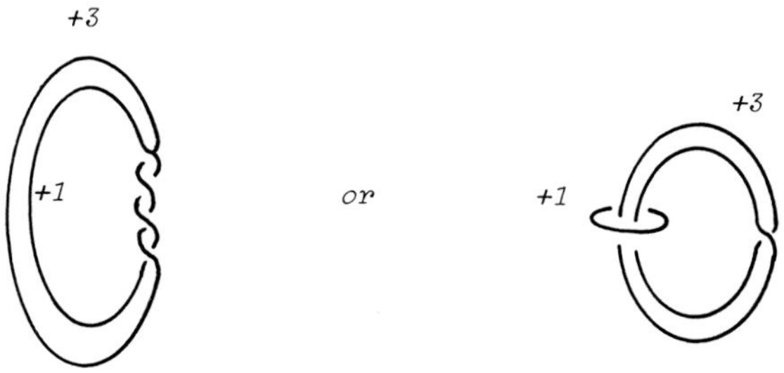

Fig. 9.

And one last application of the lemma finishes the proof.

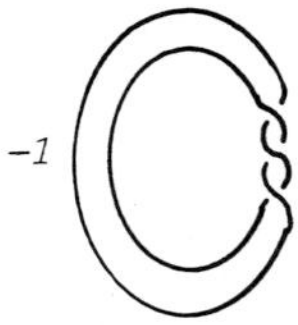

Fig. 10.

Equivalence of descriptions (7) and (8): We will find a handle body or surgery description of the 5-fold branched cover of the right handed trefoil knot K. First we perform a surgery on an unknot J with framing 1, so that S^3 is the result. But if the unknot is chosen appropriately, K is unknotted in the new S^3 (see Figure 11). It is easy to see that the 5-fold branched cover is still S^3, but the curve J lifts to 5 copies, $J_1,\ldots,J_5$, whose framings can be calculated from the formula

$$\ell\left(\sum_{i=1}^{5} J_i, \sum_{i=1}^{5} J_i\right) = \sum_{i,j=1}^{5} \ell(J_i,J_j) = 5\,(J,J) = 5$$

which implies by symmetry that

$$\sum_{i=1}^{5} \ell(J_1,J_i) = 1.$$

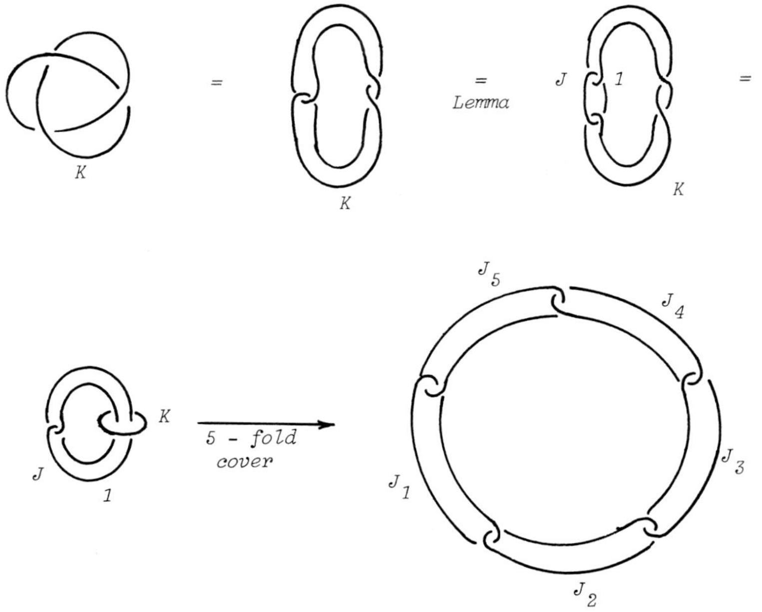

Fig. 11.

We see by inspection that $\ell(J_1, J_2) = \ell(J_1, J_5) = 1$ and $\ell(J_1, J_3) = \ell(J_1, J_4) = 0$ so that $\ell(J_1, J_1) = -1$ and hence $\ell(J_i, J_i) = -1$. The point now is that if surgery on J in S^3 gives K in S^3, then surgery on $J_1, \ldots, J_5$ gives the 5-fold branched cover of K. To see this surgery, we apply Lemma 3 several times, first to say, J_1 and J_4, obtaining

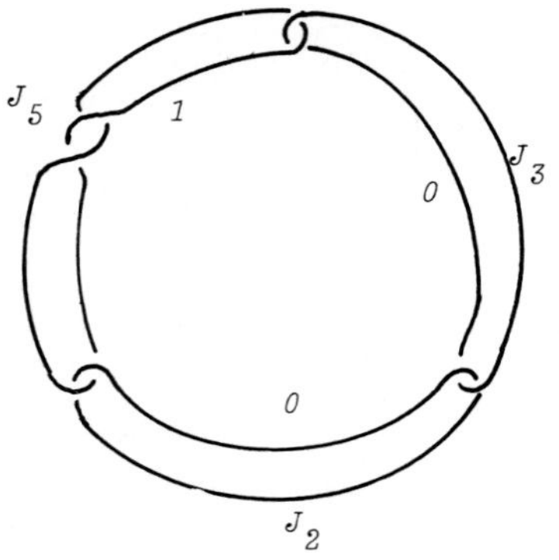

Fig. 12.

then to J_5,

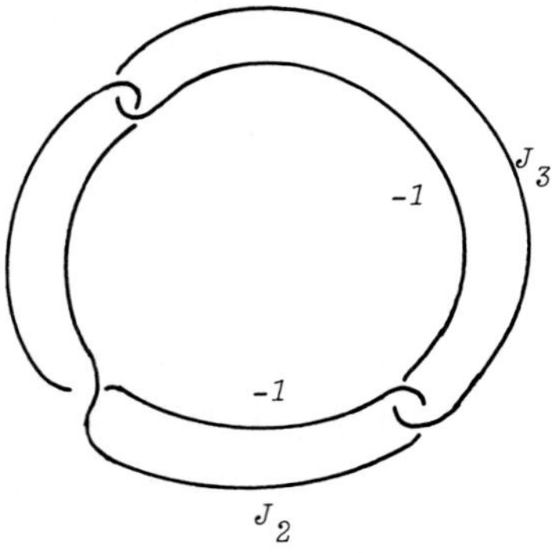

Fig. 13

and finally to, say, J_3.

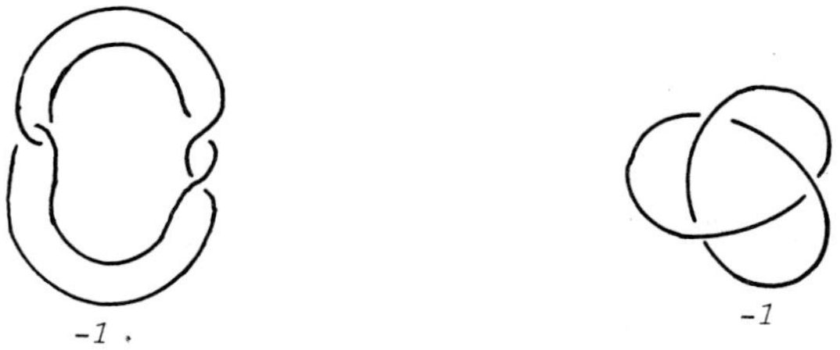

Fig. 14.

Equivalence of descriptions (6) and (2) or (7). We have seen in description (6) that the Seifert surface is equivalent to surgery on 3 anti-Hopf circles (Figure 3) with framings 1, 2, and 4. But exactly this framed link turns up while showing that descriptions (2) and (7) are equivalent.

III. OTHER POSSIBILITIES

The equivalences proven here are not necessarily the most direct paths between two points. The interested reader would find it quite rewarding to construct shortcuts. Here are some which exist in the literature.

The equivalence of (1) and (3) can be seen by resolving the singularity $z_1^2 + z_2^3 + z_3^5 = 0$. In fact, the minimal resolution of $z_1^2 + z_2^3 + z_3^5 = 0$ is a complex manifold homeomorphic to P^4. The resolution of the singularity is explicitly done in [11, p. 23-27]. Perhaps a more piquant approach, through, is to use the more general theorems on resolution of singularities found, for example in [6], [7], [1], [15], to discover the connection between the resolution of those singularities corresponding to the platonic solids (e.g., the dodecahedron) and the Dynkin diagrams used to classify semi-simple Lie groups.

In [16] the second author sketches a proof of $(6) \Longleftrightarrow (7)$ by studying the circle action on S^3 given by $\gamma(z_1, z_2) = (\gamma^3 z_1, \gamma^2 z_2)$. The trefoil knot $z_1^2 + z_2^3 = 0$ is an orbit of this action.

A Heegard splitting for ∂P^4 is drawn on page 19 of [5] and on page 245 of [14]. The latter shows that his Heegard splitting coincides with our description (7). The reader can construct his own Heegard splitting via, say, the 2-fold branched cover of the $(3,5)$-torus knot K. Note that we can decompose S^3 into two $B^2 \times I$'s such that for each one, we have $(B^2 \times I) \cap (S^3, K) \cong (B^2 \times I, ((-\frac{1}{2}, 0) \times I) \cup ((0,0) \times I) \cup ((\frac{1}{2}, 0) \times I))$; the 2-fold branched cover of $B^2 \times I$ over 3 unknotted strands is the solid 2-holed torus, i.e. $B^3 \cup$ (two 1-handles). What remains is to "see" the homeomorphism by which the two are glued together.

Description (8) for ∂P^4 can be extended to give a definition of P^4 as a p-fold cover of B^4 branched over a certain Seifert surface of the (q,r)-torus knot, $\{p,q,r\} = \{2,3,5\}$. The surface is obtained by pushing into B^4 the fiber of the map $S^3 - K \to S^1$ given by $(z,w) \to (z^q+w^r)/|z^1+w^r|$, $(z,w) \in C^2$. A Seifert surface for the $(3,5)$-torus knot is drawn in Figure 15. Its double branched cover is exactly Figure 2.

Fig. 15.

If we take the usual Seifert surface for the right trefoil knot, Figure 16, push it into B^4 and take the 5-fold branched cover, we get Figure 17. S. Akbulut and J. Harer pointed out this description; it occurs naturally as a complex submanifold of the Kummer surface. It is not hard to slide 2-handles over 2-handles to get from Figure 17 to Figure 2. We leave a description of the 3-fold cover to the reader.

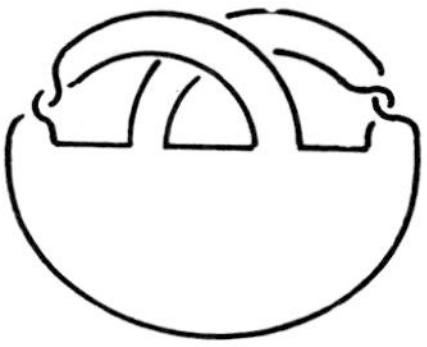

Fig. 16.

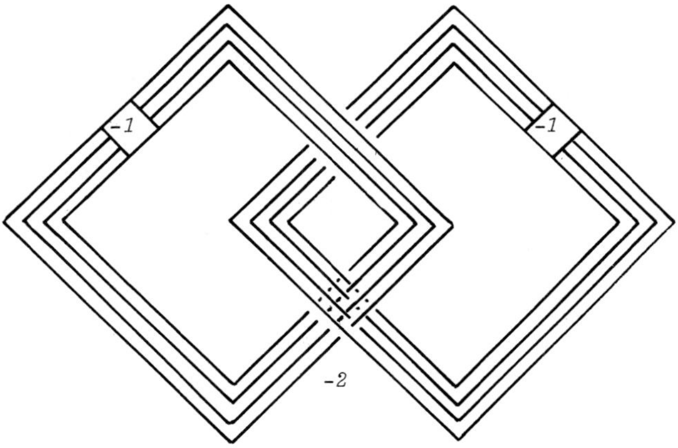

Fig. 17. The -1 means one full left-handed twist. Each circle has framing -2.

REFERENCES

1. Brieskorn, E., Die Auflösung der rationalen Singularitäten holomorpher Abbildungen, Math. Annalen, 178 (1968), 225-270.
2. Crowell, R., and Fox, R., Introduction to knot theory, Boston, Ginn, 1963.
3. Coxeter, H., Regular complex polytopes, Cambridge Univ. Press, 1974.
4. Duval, P., Homographies, Quaternions and Rotations, Oxford Univ. Press, 1964.

5. Hempel, J., 3-Manifolds, Princeton U. Press, 1976.
6. Hirzebruch, F., The topology of normal singularities of an algebraic surface (d'apres Mumford), Séminaire Bourbaki, 15e annés, 1962/63, no. 250.
7. ______, Topological methods in algebraic geometry, Springer-Verlag, 1966.
8. Kervaire, M., Smooth homology spheres and their fundamental groups, Trans. Amer. Math. Soc., 144 (1969), 67-72.
9. Kirby, R., A calculus for framed links in S^3, Inv. Math. 1978.
10. Klein, F., Lectures on the icosahedron and the solution of equations of the fifth degree, New York, Dover, 1956.
11. Laufer, H., Normal two-dimensional singularities, Ann. of Math. Study no. 71, Princeton University Press, 1971.
12. Milnor, J., On the 3-dimensional Brieskorn manifold, $M(p,q,r)$, in Knots, Groups and 3-manifolds, edited by L. P. Neuwirth, Princeton University Press, 1975, pp. 175-226.
13. Orlik, P., Seifert manifolds, Lecture notes in mathematics, Springer-Verlag, New York, 1972.
14. Rolfsen, D., Knots and Links, Publish or Perish, Boston, 1976.
15. Samelson, H., Notes on Lie algebras, Van Nostrand, 1969.
16. Scharlemann, M., Constructing strange manifolds with the dodecahedral space, Duke Math. Journal, 43 (1976), 33-40.
17. ______, Thoroughly knotted homology spheres, Houston Journal of Math., 3(1977), 271-283.
18. Seifert, H., and Threlfall, W., Lehrbuch der Topologie, Chelsea, 1947.
19. Seibenmann, L., Topological manifolds, Actes, Congres Intern. Math. 1970, v.2, 133-163.
20. Stein, E., A smooth action on a sphere with one fixed point, thesis, Princeton University, 1975.
21. Zeeman, E. C., Twisting spun knots, Trans. Amer. Math. Soc., 115 (1965), 471-495.

DECOMPOSING ANALYTIC SURFACES

Richard Mandelbaum

Department of Mathematics
University of Rochester
Rochester, N. Y.

The article which follows arose out of a series of lectures on the Topology of Analytic Surfaces given at the University of Georgia and the University of California, Berkely in the summer of 1977. In those lectures we discussed some of the algebraic and analytical tools used by geometers in classifying algebraic and analytic surfaces and then described the work of B. Moishezon and myself in studying the topology of such surfaces. The lectures were delivered to an audience of topologists and any tools needed from algebraic geometry were explained (a bit sloppily) and introduced when needed. This article maintains the spirit of the lectures and we therefore apologize in advance to any algebraic geometers who read it. We have expanded the original lectures somewhat and have included some new unpublished material in the later sections. We have also included a brief introduction to some new work of Moishezon on Surfaces of General Type (See [27]*).*

I. DECOMPOSING 4-MANIFOLDS

One of the basic problems of topology is the characterization of manifolds by means of algebraic invariants. In the case of

ISBN 0-12-158860-2

compact manifolds of dimension less than three there exists a complete classification. In fact such manifolds are completely determined by their fundamental groups. In high dimensions the undecidability of the word problem for finitely-presented groups precludes any complete algebraic classification. We thus are led to simplify our problem by restricting our attention to compact simply-connected manifolds. In dimension three our characterization question then becomes the Poincare Conjecture. In high dimensions (≥ 5) the work of Browder-Novikov-Wall, [3], [32], [38], on surgery theory shows that the homotopy type and Pontrayagin classes determine a simply connected manifold up to a finite number of possibilities, and Sullivan [37] has shown that simply-connected Kaeller manifolds are so determined by their *real* cohomology ring and Pontrayagin classes! This leaves the case of $n = 4$. We then have the following result of Whitehead [42] as sharpened by Milnor [23].

THEOREM 1.1. (Whitehead [42]) Suppose M_1, M_2 are compact simply connected smooth 4-manifolds. Let L_{M_1}, L_{M_2} be the bilinear forms on $H^*(M_1)$, $H^*(M_2)$ induced by the cup-product pairing. Then M_1 is homotopy-equivalent to M_2 if and only if L_{M_1} is isomorphic to L_{M_2} (in the sense of [35, 39]).

Thus in the case of such 4-manifolds their real cohomology completely determines their homotopy type. Actually as the following theorem of Wall shows it determines much more than that.

THEOREM 1.2. (Wall) (1) Suppose M_1, M_2 are compact simply-connected smooth 4-manifolds with associated cup-product forms L_{M_1}, L_{M_2}. Then M_1 is h-cobordant to M_2 if and only if L_{M_1} is isomorphic to L_{M_2} ; (2) Furthermore if M_1 is h-cobordant to M_2 then there exists an integer k such that

$$(*) \quad M_1 \# k(S^2 \times S^2) \text{ if diffeomorphic to } M_2 \# k(S^2 \times S^2).$$

The existence of k above is established in a highly non-constructive fashion and, even in principle, no bound whatever is obtained on the integer k. The h-cobordism conjecture in dimension 4 (i.e. for 5-dimensional h-cobordisms between 4-dimensional manifolds) would of course then be that $k = 0$ in (*) above. Not knowing this result to be true we can initially attempt to at least find upper bounds on k. For this purpose it is more convenient to use the following alternate formulation of (*).

Let P denote CP^2 and let Q denote CP^2 with orientation opposite to the usual. Then it can be observed that

(**) $(S^2 \times S^2) \# P$ is diffeomorphic to $(P \# Q) \# P$

(We recall [39] that $P \# Q$ is diffeomorphic the unique non-trivial S^2-bundle over S^2. The proof of (**) is based on the same geometric reasoning which shows that $T^2 \# RP^2 = K^2 \# RP^2$, where T^2 is the 2-torus and K^2 the 2-dimensional Klein bottle).

Thus using (**) we can reformulate (*) to give us (read '=' as 'is diffeomorphic to').

ALTERNATE THEOREM 1.3. *Let M_1, M_2 be simply-connected compact smooth 4-manifolds. Suppose M_1 is h-cobordant to M_2. Then there exist integers $k_1 \geq 0$, $k_2 \geq 0$ such that*

$$(***) \quad M_1 \# k_1P \# k_2Q = M_2 \# k_1P \# k_2 Q.$$

In fact, for any simply connected compact smooth 4-manifold there exist integers $\ell_1 \geq 0$, $\ell_2 \geq 0$ such that

$$(\ddagger) \quad M \# \ell_1P \# \ell_2Q = aP \# bQ$$

where $a = \ell_1 + \frac{1}{2}(b_2(M) + \sigma(M))$, $b = \ell_2 + \frac{1}{2}(b_2(M) - \sigma(M))$. ($b_2(M)$, $\sigma(M)$, the 2nd Betti-number and signature of M respectively).

If $\ell_1 = \ell_2 = 0$ in (‡) above then we shall call M completely decomposable. Although it is not unreasonable to conjecture that one can always take $k_1 = k_2 = 0$ in (***), a similar conjecture for (‡) (i.e. all such 4-manifold are completely decomposable) is clearly false. In fact any spin-manifold (such as $S^2 \times S^2$) can never be completely decomposable. To see this clearly we digress momentarily to discuss the classification of bilinear forms.

As bilinear forms arising from cup-product pairings on manifolds are symmetric and unimodular we restrict our attention to such forms. Choosing a basis for $H_2(M,\mathbb{Z})$ we can represent L_M by a symmetric unimodular matrix A_M and we note that isomorphism classes of L_M's are in 1-1 correspondance to congruence classes of matrices A_M.

We can immediately divide symmetric bilinear forms into two types. (See [35])

A form L is of type II if its associated quadratic form Q_L takes on only even values (which is equivalent to some associated matrix A_L having only even entries on its diagonal). Otherwise we say L is of type I.

As L is unimodular it is of course non-singular and we can also distinguish those L which are definite ($rk\ L = |\text{signature } L|$) from those which are indefinite ($rk\ L > |\text{signature } L|$).

The indefinite forms are completely classified. In fact let $\pm I$ be the form with matrix representative (± 1), U the form with matrix representative $\begin{vmatrix} 0 & 1 \\ 1 & 0 \end{vmatrix}$, and $\pm E_8$ the form with matrix representative

$$\pm \begin{pmatrix} 2 & 1 & 0 & 0 & 0 & 0 & 0 & 0 \\ 1 & 2 & 1 & 0 & 0 & 0 & 0 & 0 \\ 0 & 1 & 2 & 1 & 0 & 0 & 0 & 0 \\ 0 & 0 & 1 & 2 & 1 & 0 & 0 & 0 \\ 0 & 0 & 0 & 1 & 2 & 1 & 0 & 1 \\ 0 & 0 & 0 & 0 & 1 & 2 & 1 & 0 \\ 0 & 0 & 0 & 0 & 0 & 1 & 2 & 0 \\ 0 & 0 & 0 & 0 & 1 & 0 & 0 & 2 \end{pmatrix}$$

Then we have

(1) If L is of type I and indefinite then

$$L \simeq a\, I \oplus b(-I)$$

where
$$a = [rk(L) + \sigma(L)]/2$$
$$b = [rk(L) - \sigma(L)]/2$$

(2) If L is of type II then

$$\sigma(L) \equiv 0 \pmod{8}$$

and if L is indefinite

$$L \sim a\, E_8 \oplus b\, U$$

where
$$a = \sigma(L)/8$$
$$b = (rk(L) - |\sigma(L)|)/2$$

and $rk(L)$ is the rank of L and $\sigma(L)$ the signature of L. (A complete classification of definite forms is not known. See [25, 35] for further details). Now suppose M is a smooth simply-connected compact 4-manifold with cup-product form L_M. Then it is easily seen that L_M is of type II if and only if M is a spin manifold (i.e., its second Steifel-Whitley class $W_2(M)$ vanishes). For such manifolds a further condition on the class of L_M is given by the following remarkable theorem of Rohlin [34].

THEOREM 1.4. (Rohlin) Let M be a <u>smooth</u> 4-manifold with $W_2(M) = 0$. *Let* L_M *be the cup-product form of M on* $H^*(M,Q)$. *Then* $\sigma(L_M) \equiv 0 \pmod{16}$.

Thus a further necessary condition that a form L of type II be the cup-product form of a smooth 4-manifold is that $\sigma(L) \equiv 0 \pmod{16}$ and we in fact have that if $L = L_M$ for an appropriate spin 4-manifold M. Then

$$(2') \quad L \sim a'(E_8 \oplus E_8) \oplus b\, U$$

where $a' = a/2$ and a, b are as in (2) above.

We thus have necessary conditions for a symmetric unimodular form L to be the form of a 4-manifold M. Are these conditions sufficient? Here the situation is very murky indeed!

Using our previous notation for CP^2 and $\overline{CP}^2$ we have:

$$L_P \cong \langle 1\rangle, \ L_Q \cong \langle -1\rangle, \ L_{S^2 \times S^2} \cong \begin{pmatrix} 0 & 1 \\ 1 & 0 \end{pmatrix} .$$

Then if L_M is indefinite and of type I we have that

$$M \sim aP \# bQ; \ a,b \text{ as in (2)}$$

and $\sim$ is to be read as 'h-cobordant to' .

If L_M is of type II and indefininte then were there to exist a simply-connected smooth manifold V with $W_2(V) = 0$ and $(V) = b_2(V) - 16$ we could assert that

$$m \sim a' V \# b(s^2 \times S^2); \ a',b \text{ as in (2')}$$

However no such manifold is known to exist!! In fact no 4-manifolds with definite intersection form other than $\#_{i=1}^{n} P$ or $\#_{i=1}^{n} Q$ are known to exist!!

The simplest 4-manifold with $|\sigma(M)| = 16$ known has $\sigma(M) = -16$, $b_2(M) = 22$. The Kummer surface is such a manifold and by a theorem of Kodaira is diffeomorphic to the projective algebraic variety $V = V_4$; the non-singular quartic hypersurface in CP^3 with equations in homogeneous coordinates z_i $i = 0, \ldots 3$ in CP^3

$$V_4 : z_9^4 + z_1^4 + z_2^4 + z_3^4 = 0$$

Thus the most that can be said about 4-manifolds whose intersection forms are indefinite and of type II. If a', b are as in 2' and $b \geq 3|a'|$ then

(3) $M \sim_k a' V_4 \# b'(S^2 \times S^2)$ where $b' = b - 3|a'|$

In any event it is clear that a spin manifold can never be completely decomposable. However, if M is an arbitrary simply-connected compact 4-manifold then either $M \# P$ or $\overline{M} \# P$ has an indefinite intersection form of type I. (where $\overline{M}$ denotes the manifold obtained by reversing the orientation of M).

If $M \# P$ is completely decomposable we shall say that M is almost completely decomposable. (ACD)

We can then conjecture that:

CONJECTURE I. *Let M be a smooth simply-connected compact 4-manifold. Then either M or $\overline{M}$ is almost completely decomposable.*

Before tackling this conjecture we recall a few lessons from the classification theory of compact orientable 2-manifolds.

We recall that every such 2-manifold F_g is topologically just a connected sum $\overset{n}{\underset{i=1}{\#}} (T^2)$; of 2-Torii (where the empty sum F_0 is defined to be S^2). Now although this classification can be done in a purely topological fashion, one of the beautiful things about orientable 2-manifold is that every such manifold is diffeomorphic to a non-singular algebraic curve in CP^3. (In fact every such manifold is diffeomorphic to a hypersurface in either $S^2 \times S^2$ or $P \# Q$.)

Thus the topology of orientable 2-manifolds can be studied by algebraic or analytic methods and it can in fact be shown that if M is a compact complex 1-dimensional manifold, then $M \approx F_h$

where h is the complex dimension of the space of holomorphic 1-forms on M.

Now if $n = 4$ it is of course *not* the case that all 4-manifolds, even all simply connected 4-manifolds, admit complex structures. For example neither S^4, nor Q, nor $S^2 \times S^2 \# S^2 \times S^2$ admit such structures. However the compact algebraic surfaces do provide a wide variety of examples of simply-connected 4-manifolds.

Furthermore, as a consequence of work of Bogomulov, Miyaoka and Yau [1, 26, 44] all simply connected analytic surfaces other than CP^2 have indefinite intersection forms. It is therefore reasonable to conjecture that

CONJECTURE I'. Let V be a simply-connected compact analytic surface.

Then V is almost completely decomposable!

The bulk of the succeeding article will discuss what progress has been made on Conjecture I' and related questions. The first step in our attach on this conjecture will be a discussion of the diffeomorphic classification of simply-connected analytic surfaces and some related analytic invariants.

II. ANALYTIC SURFACES

To begin our study of analytic surfaces we recall the fundamental classification theorem of Enriques-Kodaira for the simply-connected case.

THEOREM 2.1. Suppose M is a simply connected compact complex surface. Then there exists a non-singular projective algebraic surface V such that V is diffeomorphic to M and one of the following three possibilities hold.

a) V is rational

b) V is elliptic

c) V is of general type

REMARK 1. If we omit the assumption that M is simply connected then there exist two additional categories:

d) Ruled Surfaces

e) Abelian Varieties.

Topologically elements of class d) are of the form $X \# nQ$ where X is an S^2 bundle over some orientable surface F_g, while elements of class e) are all topologically of the form $T \# nQ$ where T is a 4-Torus.

Without discussing the whole theory of classification of surfaces we shall mention some topological invariants which help characterize the various classes above. In order to do this we first introduce the operations of 'blowing up' and 'blowing down' and minimal model. We initially discuss the operation of 'blowing up' a point in an analytic surface. The more general operation of 'blowing up' a subvariety of codimension ≥ 2 we defer until somewhat later.

A. 'Blowing Up and Down'

Let V be an analytic surface and let $p \in V$. Let U be some coordinate neighborhood of p in V and let (x,y) be local coordinates on V with $x(p) = y(p) = 0$. We may suppose without loss of generality that $D_2 = \{q \,|\, |x(q)|^2 + |y(q)|^2 < 1\}$ is contained in U.

Let $\tilde{D} \subset D_2 \times CP^1$ be the submanifold defined by pairs $\{((x,y),\ [t_0,t_1]) \mid xt_1 = yt_0\}$. Let π be the restrictions to $\tilde{D}$ of the canonical projection of $D_2 \times CP^1$ onto its first factor.

Then $\pi : \tilde{D} \to D_2$ is a holomorphic map onto D_2 such that if $(x,y) \neq (0,0)$ then $\pi^{-1}(x,y)$ is the unique point $((x,y), [x:y])$. Thus $\pi | D - \pi^{-1}(0,0) \to D_2 - (0,0)$ is biholomorphic. On the other hand

$$\pi_1^{-1}(0,0) = ((0,0), [t_0,t_1]) \approx CP^1.$$

We can now define

$$\tilde{V} = V - \overline{D}_1 \cup_\pi \tilde{D},$$

where D_1 is the obvious subdisc of D_2 and $(x,y) \in D_2 - \overline{D}_1$ is identified with $\pi^{-1}(x,y) \in \tilde{D}$. Clearly $\tilde{V}$ is a complex manifold since π is biholomorphic over $D_2 - \overline{D}_1$.

Let $L = \pi^{-1}(0,0) \subset \tilde{V}$. Then it is readily verifiable that the self intersection of L as a 2-cycle in $\tilde{V}$ is -1. Then $\tilde{D}$ is diffeomorphic to a Hopf-disc bundle over S^2 and we can think of $\tilde{D}$ as $Q - B^4$ (i.e. $\overline{CP}^2$ - a closed 4-ball). Thus topologically $\tilde{V} = V - B^4 \cup Q - B^4 = V \# Q$. We say $\tilde{V}$ is V blown up (holomorphically) at the point p and we have seen topologically that this is the same as taking a connected sum with Q at a ball centered at $p \in V$.

We note that the operation $V \# P$ can similarly be interpreted as an anti-holomorphic 'blowing up' by simply using local coordinates (x^1, y^1) in the above construction with $x^1 = x;\ y^1 = \overline{y}$. This has the effect of producing an analagously defined L^1 with $L^1 \cdot L^1 = +1$. If V is compact then we see immediately that the effect of a holomorphic (anti-holomorphic) blowing up of a point $p \in V$ produces a manifold $\tilde{V}$ (resp. $\tilde{V}^1$) with

1) $b_2(\tilde{V}) = b_2(V) + 1 \qquad (b_2(\tilde{V}^1) = b_2(V) + 1)$
2) $\sigma(\tilde{V}) = \sigma(V) - 1 \qquad (\sigma(\tilde{V}^1) = \sigma(V) + 1)$

Now $\tilde{V}$ is a complex manifold and an easy calculation shows that its first Chern class satisfies $c_1^2(\tilde{V}) = c_1^2(V) - 1$. $\tilde{V}^1$ however does not admit even an almost-complex structure. This can be seen as follows: We recall that if M is a compact smooth orientable 4-manifold then M admits an almost complex structure if there exists a cohomology class $C \in H^2(M,\mathbb{Z})$ whose reduction mod 2 is $W_2(M)$ and such that $C^2(M) = 2\chi(M) + 3\sigma(M)$, where χ,σ are the Euler number and signature of M. [33] Now if V is a complex manifold then clearly $2\chi(\tilde{V}^1) + 3\sigma(\tilde{V}^1) = 2\chi(V) + 3\sigma(V) + 5$. Furthermore if $K \in H_2(V,\mathbb{Z})$ is dual to $C_1(V)$ then any candidate for $C \in H^2(\tilde{V}^1,\mathbb{Z})$ would of necessity have a dual of the form $K^* + sL^1$ where $K^* \in H_2(\tilde{V}^1,\mathbb{Z})$ is such that $\pi_*(K^*) = K$ and $K^{*2} = K^2 = c_1^2$. Furthermore, since $K^1 + sL^1$ mod 2 is dual $W_2(\tilde{V}^1)$ we must have that s is odd. But $(K^* + sL^1)^2 = K^{*2} + s^2 = C_1^2(V) + s^2 = 2\chi(V) + 3\sigma(V) + s^2$ and since S is an odd integer we can't have $S^2 = 5$. Thus $\tilde{V}^1$ does not admit an almost complex structure.

(Note that $V^{\#} = V \# 2P$ always *does* admit an almost complex structure if V does. This is because the analogues in condition on $V^{\#}$ would be of the form $(K^* + S_1L^1 + S_2L^{11})^2 = C_1^2 + 10$ with S_1, S_2 odd and $K^{\#2} = C^2$. But taking $S_1 = 3$ $S_2 = 1$ gives us the requisite element of $H_2(V^{\#},\mathbb{Z})$ and thus by duality the requisite cohomology class C. Similarly if V has an almost complex structure then $V \# n(S^2 \times S^2)$ has an almost complex structure if and only if n is even.)

In certain cases, we can reverse the process of 'blowing up'. We have the following criteria due to Castelnuevo-Kodaira.

CASTELNUEVO CRITERION. Let V be a complex surface and suppose S is a complex one dimensional submanifold of V with self-intersection (as a homology class) equal to -1.

Then if S is homeomorphic to a sphere, there exists a complex surface W and a point $p \in W$ such that V is biholomorphic to W blown up at p. (Topologically then V is of course just $W \# Q$).

The Castelnuevo criterion is topologically just a condition allowing us to decompose a given 4-manifold as a connected sum.

Interpreted topologically we can in fact generalize it as follows;

Let V be a 4-manifold and S an embedded sphere with $S \cdot S^1 = \pm 1$. Then there exists a 4-manifold W with $V = W \# \begin{Bmatrix} P \\ Q \end{Bmatrix}$ respectively.

We say W is obtained by blowing V down along S.

In the analytic case we call S an exceptional curve (of the first kind).

It is then easily seen that given any complex surface V^1 there exists a surface V without any exceptional curves such that $V^1 = V \# nQ$. (In fact V^1 is biholomorphic to some $V \# nQ$). Such a V is called a (relatively) minimal surface. (Note V is not unique. For example $P \# 2Q = S^2 \times S^2 \# Q$ and $S^2 \times S^2$ is minimal).

For topological purposes it suffices to consider only minimal surfaces and our criteria for distinguishing between types of complex surfaces will be designed to distinguish between 'minimal models' of such surfaces.

B. Analytic Invariants

Suppose V is a complex manifold. Then by $\mathcal{O}_V$ we will mean the sheaf of holomorphic functions on V. (This is also called the structure sheaf of V. It is defined by means of the presheaf U open in $V \mapsto \mathcal{O}_U = \{$ring of holomorphic functions on the open set $U\}$. We need to also consider the sheaf of nowhere vanishing holomorphic functions and we denote this by $\mathcal{O}_V^*$. Closely related to the holomorphic functions are the meromorphic functions. These are not defined on all of V as holomorphic functions are and properly speaking are not functions on V at all. Nevertheless we define: (following Whitney [43]).

Definition. A meromorphic function on V is a pair (H,f), where H is a dense open subset of V and f is a holomorphic function on H such that for every $p \in V$ there is a neighborhood U of p and holomorphic functions φ, ψ in U such that $Z_\psi = \{X \in U | \psi(X) = 0\}$ is nowhere dense in U and $f = \varphi/\psi$ in $(U \cap H) - Z_\psi$.

Note: In fact we can always extend f to be holomorphic in some maximal open set $H_f \subset V$ with complement P_f a nowhere dense subvariety of V. We will always identify f with its extensions. We call P_f the set of poles of f or polar set of f. If f is meromorphic on V and $z \in H_f$, then since f is holomorphic on H_f we can easily define $f(z) \in \mathbb{C}$ and speak of the value of f at z.

If f doesn't vanish identically on H_f we define $Z_f = P_{f^{-1}}$ where f^{-1} is clearly meromorphic on V if f is. Similarly if f is not identically constant on an open subset of V we can define $Z_f(a) = P_{(\frac{1}{f-a})}$. (Alternatively $Z_f(a) = \{p \in V | \exists$ a sequence $p_n \in H_f \to p$ with $f_{(p)} \to a\}$. Here $Z_f(\infty) = P_f$ in the obvious fashion).

We call $Z_f \cap P_f = \cap_{a \in \bar{C}} Z_f(a)$, the indeterminacy set of f and denote it by I_f. (Note that in contrast to the case of one-dimenisonal complex manifolds, where a meromorphic function is always assigned the value ∞, at its pole set, in the higher dimensional case we can assign f the value ∞ only on $P_f - I_f$. No values can be assigned on I_f and thus again in contra-distinction to the one-variable case a meromorphic function isn't even well-defined on all of V as a mapping to CP^1).

We can also define meromorphic functions on V more directly in a sheaf-theoretic way to obtain the sheaf M_V of meromorphic functions on V (We can also get the corresponding sheaf M_V^* of not-identically zero meromorphic functions.)

M_V is constructed as follows: For $x \in V$, let M_x be the ring of formal quotients of the integral domain $\mathcal{O}_x$ = {ring of germs of holomorphic functions at x} = stalk of $\mathcal{O}_V$ at x. Let $M_V = \bigcup_{x \in V} M_x$. M_V carries the natural structure of the sheaf associated to the presheaf $U \mapsto M_U$ = quotient ring of $\mathcal{O}_V$, for U open in V).

In this construction $\mathcal{O}_V$ (resp $\mathcal{O}_V^*$) is a natural subsheaf of M_V (resp M_V^*) and O_a a subring of M_a. (The sections of M_V over U (Which contains, but is, in general, strictly larger than M_V)) are the meromorphic functions).

If f is meromorphic on V then it can be shown that Z_f, $Z_f(a)$, P_f, I_f are analytic subsets of V with $\dim Z_f = \dim Z_f(a) = \dim P_f = \dim V-1$ and $\dim I_f \leq \dim V-2$. We shall be particularly interested in the codimension 1 subvarieties of V and to that end define

$$\mathrm{Div}(V) = \text{free abelian group on subvarieties } Z \subset V \text{ of codimension } 1.$$

We call elements of Div (V), divisors on V (more properly, Weyl divisors on V). There is a partial order $\geq$ on Div (V) given by $D \geq 0, D = \Sigma n_i Z_i$ if $n_i \geq 0$ all i. Then $D \geq E$ if $D - E \geq 0$. Now if f is a meromorphic function then we can associate to f a divisor $(f) \in$ Div (V) as follows: Let $Z_1, \ldots Z_t$; $P_1, \ldots, P_r$ be the irreducible components of Z_f, resp. P_f. To each Z_i, resp. P_j we can associate the order of f at Z, $\mathrm{ord}_Z f \in \mathbb{Z}$ in the obvious fashion. (For details see [30])

$$\text{Then } (f) = \Sigma\, \mathrm{ord}_{Z_i}(f) \cdot Z_i + \Sigma\, \mathrm{ord}_{P_j}(f) \cdot P_j$$
$$= \Sigma\, \mathrm{ord}_{Z_i}(f) \cdot Z_i - \Sigma |(\mathrm{ord}_{P_j}(f))| \cdot P_j$$

Letting $M_V = \Gamma(V, \mathcal{M}_V)$=the field Of all meromorphic functions on V(with $M_V^* = M_V - (0)$) we get a map $M_V^* \xrightarrow{\mathcal{D}}$ Div (V) by $f \to (f)$. We call the image of $\mathcal{D}$ the subgroup of principal divisors of V. The quotient group Div $(V)/\mathcal{D}M_V^*$ is one of the key groups encountered in studying algebraic and analytic geometry. It is called the Picard group of V.

If $D_1, D_2 \in$ Div (V) define the same element in Pic (V) (i.e. $D_1 - D_2 = (f)$, for some $f \in M_V^*$) we say D_1 is linearly equivalent to D_2.

We are particularly interested in the relation between divisors on V and holomorphic line bundles (i.e. one-dimensional complex vector bundles) on V. Let $G = C^\infty$ or holomorphic. We recall that if $\xi \overset{\varphi}{\to} V$ is a G-line bundle over V then there exists a covering $\{V_i\}$ of V and isomorphisms $\varphi^{-1}(U_i) \xrightarrow[\eta_1]{\sim} U_i \times \mathbb{C}$ with φ restricting to the natural projection over U_i. Furthermore it is required that $\varphi_i\varphi_j^{-1} : U_i \cap U_j \times \mathbb{C} \to U_i \cap U_j \times \mathbb{C}$ be linear on fibers, so there are G-maps $V_{ij} : U_i \cap U_j \to \mathbb{C}^*$ such that $\eta_i\eta_j^{-1}$: $(X,V) \to (X \cdot V_{ij}(X) \cdot V)$. The correspondence $(\xi \overset{\varphi}{\to} V) \to (V_{ij})$ defines a map 'line bundles on V^1' $\to$ cocycles in $Z^1((U_i),G^*)$ (where G^* is the sheaf $C_V^{\infty *}$ or $\mathcal{O}_V^*$). The correspondence so defined maps isomorphic bundles to equivalent cycles and thus $H^1(V,\mathcal{O}^*)$ (resp. $H^1(V,C^{\infty *})$) can be identified with the group of equivalence classes of holomorphic (resp. smooth) line bundles on V. (The group operation $(\alpha_{ij}) \cdot (\beta_{ij}) = (\alpha_{ij}\beta_{ij})$ in $H^1(V,\mathcal{O}^*)$ corresponds to tensor multiplication of line bundles). (Note that a similar correspondence can be constructed for arbitrary rank-n vector bundles. Then $H^1(V,G^*)$ is the *set* of cohomology classes with coefficients in the sheaf $G = GL(n,\mathcal{O}^*)$ or $GL(n,C^{\infty *})$ (See [6] or [15]).

In particular we shall henceforth think of line bundles as represented by cocycles $(V_{ij}) \in Z^1(U,\mathcal{O}^*)$ for some covering $\mathcal{U} = (U_i)$ of V.

Now to relate line bundles to divisors we note that if V is a manifold and Z is a positive (Weyl) divisor on V then for any point $p \in V$ there exists an open set $U \ni p$ and a holomorphic function $f \in O_U$ such that $\dot{Z} \cap U = Z_f = \{x \in U | f(x) = 0\}$. (Note that this is not necessarily true if V is singular. For example if $V = \{(x,y) \in \mathbb{C}^2 | y^2 = X^3 + X^2\}$ and $Z = (0,0) \in V$. Then Z can't be locally defined by the vanishing of *one* function f). Now any divisor D can be written as the difference of positive divisors, $Z_1 - Z_2$ and thus we can associate a meromorphic function $f \in M_U^*$ to $D \cap U$ such that $D \cap U = (f)$.

Furthermore if we multiply f by $u \in O_U^*$ we see that $(f_u) = (f) = D \cap U$. Thus we can define a (Cartier) divisor D on V by means of an open covering $\{U_i\}$ of V and an assignment of an equivalence class $(f_i) \in M_v^*/\mathcal{O}_v^*$ to $\{U_i\}$. We thus can obtain a sheaf $\mathcal{D}_V = M_V^*/\mathcal{O}_V^*$ of divisors on V. If V is nonsingular then $\Gamma(V, \mathcal{D}_V) \simeq$ Div (V) and we can identify Cartier Divisors with Weyl Divisors in a natural fashion. (Again we point out that this is not generally true for V singular).

Now the exact sequence $0 \to \mathcal{O}_V^* \to M_V^* \xrightarrow{P} \mathcal{D}_V \to 0$ defining the divisor sheaf gives rise to the exact cohomology sequence

$$0 \to H^0(V, \mathcal{O}_V^*) \to H^0(V, M_V^*) \xrightarrow{P} H^0(V, \mathcal{D}_V) \xrightarrow{\delta^*} H^1(V, \mathcal{O}_V^*) \to H^1(V, M_V^*)$$

But a fundamental result of Kodaira [15] says that for V compact and non-singular, $H^1(V, M_V^*) = 0$. Thus $H^1(V, \mathcal{O}_V^*) \simeq H^0(V_1 \mathcal{D}_V)/p^*H^0(V, M_V^*) =$ Div $V/\mathcal{D}M_V^* =$ Pic (V). In particular then, isomorphism classes of line bundles on V are in 1-1 correspondence with linear equivalence classes of divisors.

(Explicitly given a divisor class $[D] \in$ Pic (V). We can represent D by $f_i \in H^0(V, M_V^*/\mathcal{O}_V^*)$. Thus since on $U_i \cap U_j$ we have $(f_i) = (f_j)$, the quotient f_j/f_i defines an element of $\mathcal{O}_{U_i \cap U_j}^*$. The correspondence $\{f_i\} \to \{\sigma_{ij} = f_j f_i^{-1}\}$ defines an element of $H^1(V, \mathcal{O}_V^*)$ and thus a line bundle on V).

We note that if ξ is any holomorphic vector bundle on V then we can form in the obvious way the sheaf of holomorphic cross sections of ξ, $\mathcal{O}_V(\xi)$. (This is defined locally by $f_i \in O_{U_i}$ such that $f_i = \xi_{ij} f_j$ in $U_i \cap U_j$ where (ξ_{ij}) represents ξ in $H^1(V, \mathcal{O}_V^*)$) Now since $\xi_{ij} \in \mathcal{O}_{U_i \cap U_J}^*$ we see that on $U_i \cap U_j$, $(f_i) = (f_j)$. Thus if $f \in \Gamma(V, \mathcal{O}_V^*(\xi))$ is a global holomorphic cross-section of ξ it makes sense to speak of the divisor (f) of f.

In particular suppose T_V^* is the holomorphic cotangent bundle of V. We can pick a basis $dz_1, \ldots dz_n$ for $\Gamma(V, T_V^*)$ $(n = \dim_{\mathbb{C}} V)$. This basis generates the holomorphic 1-forms on V. Forming the alternating product bundles $\Lambda^k T_V^*$ we obtain the vector spaces $\Gamma(V, \Lambda^k T_V^*)$ of holomorphic k-forms on V. Taking $k = n$ we form $\Lambda^n T_V^*$. As is well known for any $x \in V$, $\dim_{\mathbb{C}}(\Lambda^n T_V^*)_x = 1$ and $\Lambda^n T_V^*$ is thus a line-bundle on V. It is called the canonical line-bundle on V and denoted by $K_V = \Lambda^n T_V^*$.

Its cross-sections are the holomorphic n-forms which can be locally represented as $w_U = f \cdot dz_1 \wedge \ldots \wedge dz_n$; $f \in \mathcal{O}_U$, U open in V. Then we can associate a divisor (w) to any $w \in \Gamma$ as described previously. Since it is immediate that any two n-forms define linearly equivalent divisors we can speak of the divisor class associated to the line bundle K_V. This divisor class is called the canonical divisor class on V and any of its members is called a canonical divisor. By 'abuse of notation' any canonical divisor is denoted by K_V. We note that

$$C_1(V) = C_1(T_V) = -C_1(T_V^*) = C_1(K)$$

where C_i is the ith Chern class.

We now let $q = \dim_{\mathbb{C}} H^1(V, \mathcal{O}_V)$, $p_g = \dim_{\mathbb{C}} H^2(V, \mathcal{O}_V) = \dim_C H^0(V, \mathcal{O}_V(K))$.

We call q the 'irregularity' of V and p_g its geometric genus. Note that $H^0(V, \mathcal{O}_V(K))$ is the space of holomorphic cross-sections of K = 'holomorphic 2-forms' on V. Similarly $H^1(V, \mathcal{O}_V)$ can be identified with the space of holomorphic 1-forms on V. Surfaces with $p_g = 0$ are called regular surfaces.

The fundamental theorem relating the analytic invariants to topological ones is the Riemann-Roch theorem and its corollary the Noether formula.

The Noether formula says that

$$C_1^2[V] + C_2[V] = 12(1 - q(V) + p_g(V))$$

Now $C_2[V]$ in this case is just the Euler number $\chi = C_2[X] = \sum_{i=0}^{4} (-1)^i b_i(V)$ and $C_1^2[V]$ is equal to $K \cdot K$, the self intersection of any canonical divisor on K.

If V is Kaehler we then have $b_1 = 2_q$ and we also have available the Hodge-Index theorem a consequence of which is that if b_2^+, b_2^- are the number of positive and negative entries in a diagonalization of the bilinear form L_V on $H_2(V,Q)$ associated to V then $b_2^+ = 2p_g + 1$.

We thus obtain the following formulas relating the topological and analytical invariants of V. $b_2 = 12\,P_g - 8q + 10 - C_1^2$
$b_2^+ = 2P_g + 1$, $b_2^- = 10P_g - 8q + 9 - C_1^2$, $\sigma = C_1^2 + 8q - 8(P_g + 1)$
$C_1^2 = 2X + 3\sigma$.

If V is simply connected then its homotopy type is completely determined by p_g, K^2 and $C(K)$ mod 2. We have $b_2 = 12\,P_g + 10 - C_1^2$, $\sigma = C_1^2 - 8(P_g + 1)$ and V is type II if $C(K)$ mod $2 = 0$.

A recent result of S. Yau [44] is that $C_1^2 < 3C_2$ for all complex surfaces other than CP^2 (with its usual complex structure) where $C_1^2 = 3C_2$.

It is then easily verified that

1) All complex manifolds not homotopy equivalent to CP^2 have indefinite intersection forms.

2) All complex spin manifolds are homotopy equivalent to connected sums of the form $aV_4 \# b(S^2 \times S^2)$.

(It is an appealing conjecture that all simply-connected complex manifolds not homotopy equivalent to CP^2 in fact satisfy $C_1^2 \leq 2C_2$. This is *not* true for non-simply connected manifolds as can be seen from examples of [8].)

C. Fundamental Definitions

In the beginning of this chapter we quoted a theorem which divided all simply connected compact complex surfaces into 3 categories. We have as yet not defined these categories or characterized them in a topological fashion. We will proceed to correct this lacunae.

We need one additional definition.

Definition. Let V be a complex surface and K its canonical line bundle. Let nK denote the n-fold tensor product of K with itself. $(nK = \underbrace{K \otimes K \ldots \otimes K}_{n\text{-times}})$ and let $\mathcal{O}_V(nK)$ be the sheaf of holomorphic cross-sections of the bundke nK. Then $P_n(V) = P_n = \dim_{\mathbb{C}} H^1(V_1 \mathcal{O}_V(nK))$ and we call P_n the nth pluri-canonical genus of V.

We note that $p_g(V) = P_1(V)$

We recall one additional theorem

THEOREM 2.2 (Siegel) Let V be a compact complex n-manifold. Then M_V = {meromorphic functions on V} is an algebraic function field over $\mathbb{C}$ with tr. deg $M_V \leq n$.

If V is projective algebraic then tr. deg $M_V = n$

If $M_V \simeq \mathbb{C}(z_1, \ldots z_n)$ then we say V is rational.

Thus:

An algebraic surface V is rational if $M_V \simeq \mathbb{C}(z_1, z_2)$.

We have the following classification theorem for rational surfaces due to Castelnuevo.

THEOREM 2.3. (Castelnuevo-Andreotti) Let V be an algebraic surface. Then V is rational if $q = P_2 = 0$. In this case V is topologically either S^2 x *S^2 or $P \# nQ$.*

Note that all rational surfaces are certainly almost completely decomposable and if of type I are in fact completely decomposable.

(It is not difficult to show that if $V = S^2 \times S^2$ or $V = P \# mQ$ then $q(V) = P_n(V) = 0$ for all $n \geq 1$. It is tempting to conjecture in fact that $q(V) = P_g(V) = 0$ implies that V is rational, but this conjecture is clearly false as there exist many non-rational surfaces with $q = P_g = 0$ but with $H_1(V, \mathbb{Z})$ a torsion group. [For example the Enriques Surface with $H_1(X) = \mathbb{Z}_2$ which is elliptic or the Godeaux Surface with $H_1(V) = \mathbb{Z}_5$, which is of general type]. However until recently no example was known of a complex surface with $P_g = H_1(X, \mathbb{Z}) = 0$ which wasn't rational. In fact Severi conjectured that all such surfaces were rational. In [4] Dolgachev in fact constructs simply connected elliptic surfaces with $P_g = 0$, which are not rational. It is still conjectured that if V is a simply-connected surface of general type then $P_g(V) \neq 0$.)

Definition. Let V be an analytic surface. Then V has a pencil of elliptic curves if there exists a compact Riemann Surface V and a holomorphic map $\Phi : V \to R$ of V onto R such that there exist a finite number of points $X_1, \ldots X_n$, so that if $X \in R - \{X_1, \ldots X_1\}$ then $f^{-1}(X)$ is an elliptic curve (i.e. is topologically a torus). We shall say V is an elliptic surface if it admits a pencil of elliptic curves. If V is rational and also admits such a pencil we shall call V a rational elliptic surface, but not call it an elliptic surface.

If V is a non-rational elliptic surface then its minimal model is also elliptic. However if V is a rational elliptic surface its minimal model is either topologically $S^2 \times S^2$ or CP^2, which do not admit holomorphic maps as above.

If V is an elliptic surface and none of its fibers in an elliptic pencil $\Phi : V \to R$ contain exceptional curves of the first kind then V is said to be minimally elliptic. Note that every elliptic surface V admits a minimally elliptic model V^* such that $V = V^*$ blown up at a finite number of points.

(For non-rational elliptic surfaces being minimally elliptic coincides with being minimal in the usual sense, however there exists a rational elliptic surface which is minimally elliptic but not minimal.)

We then have the following criterion for elliptic surfaces.

THEOREM 2.4. Let V be a analytic surface with (relatively) minimal model V^* then V is an elliptic surface if

$$C_1^2(V^*) = K_{V^*}^2 = 0$$

What are the elliptic surfaces like topologically. We point out that as can be seen from [19, 14] there exist infinite families of elliptic surfaces U_P, such that any two surfaces X_1, X_2 in U_P are homotopy equivalent to one another but are not known to be diffeomorphic. Thus the class of elliptic surfaces could conceivably provide us with an infinite number of simply-connected smooth 4-manifolds in the same h-cobordism class, but not diffeomorphic to each other! (In fact the class of elliptic surfaces is the only class in which such a phenomena could occur!). It is thus clear that the situation is somewhat complicated. We will discuss elliptic surfaces in greater depth later when we shall show that all simply connected elliptic surfaces are almost completely decomposable.

Lastly we turn to surfaces of general type. Essentially, in the simply-connected case at least, these surfaces consist of everything which remains after removing the rational and elliptic surfaces.

More constructively we have the following

CRITERION 2.5. (Moishezon, Enriques) Let V^* be an analytic surface with (relatively) minimal model V.

Then V is of general type if,

1) *for some $n > 0$, $P_n(V) > 0$ and $C_1^2(V) = K_V^2 > 0$ or equivalently*
2) *for some $n^1 > 0$, $P_{n^1}(V^*) > 0$ and V^* does not admit a pencil of elliptic curves.*

Note that the condition 'for some n, $P_n(V) > 0$' is equivalent to demanding that V^* not be rational.

Very, very little is known about surfaces of general type and we shall also discuss what is known in greater depth later.

D. Decomposability of Surfaces

Let us now return to our basic conjecture that all simply connected compact analytic surfaces are almost completely decomposable. We will attack this conjecture by looking at certain classes of analytic surfaces and prove that these classes are almost completely decomposable.

The classes we will attack are:

1) Non-singular hypersurfaces of CP^3
2) Complete Intersection Surfaces
3) Hypersurface sections of relatively simple simply-connected 3-folds and branched cyclic covers.
4) Elliptic Surfaces
5) Surfaces with pencils of hyperelliptic curves
6) General simply connected surfaces of general type.

The first 3 classes contain all 3 types of simply connected algebraic spaces and we will indicate what is known about each class as we come to it. We begin with the hypersurfaces of CP^3

III. THE HYPERSURFACES OF CP^3

We recall that the hypersurfaces of CP^3 are precisely those subsets defined by the vanishing of a homogeneous polynomial on $\mathbb{C}^4$. That is if $f : \mathbb{C}^4 \to \mathbb{C}$ is a polynomial function. Then

$$V(f) = \{(X_0 : X_1 : X_2 : X_3) \in \mathbb{C}P^3 | f(X_0, X_1, X_2, X_3) = 0\}$$

is called a hypersurface of CP^3. $V(f)$ is non-singular if the matrix $df : \mathbb{C}^4 \to \mathbb{C}^4$, given by $\left(\frac{\partial f}{\partial X_0}, \frac{\partial f}{\partial X_1}, \frac{\partial f}{\partial X_2}, \frac{\partial f}{\partial X_3} \right)$ is non-singular.

If deg $f = n$ we shall say $V(f)$ is of degree n.

We note the following

ASSERTION 1. Let V be a non-singular hypersurface of CP^3. *Then V is simply connected.*

Proof. The sharp version of Lefschetz's first theorem as proven in [24] for example asserts that:

THEOREM. Let W be a algebraic variety of complex dimension k which lies in the complex projective space $\mathbb{C}P^n$. *Let H be a hyperplane in* CP^n *which contains the singular points (if any) of W. Let* $W_H = W \cap H$. *Then the inclusion map* $W_H \to W$ *induces isomorphisms* $\pi_r(W_H) \simeq \pi_r(W)$ *for* $r < k-1$. *(Note that H need not be non-singular.)*

Now in our case, since $\mathbb{C}P^3$ is simply connected, it suffices to produce an embedding of $\mathbb{C}P^3$ into some $\mathbb{C}P^N$ in which $V = CP^3 \cap H$, for some hyperplane H of $\mathbb{C}P^N$. In fact we can do more than this. For any pair of positive integers (n,k) there exists an embedding $\mathbb{C}P^k \hookrightarrow CP^{N(k)}$, $N_n(k) = \binom{n+k}{k} - 1$, unique up to projective linear transformations of $CP^{N(k)}$. This is the Veronese embedding of degree n which we can describe as follows:

Let (n,k) be a pair of positive integers. Let $\mathrm{Par}_k(n) = \{(i_0, i_1, \ldots i_k) \mid i_j$ is a non-negative integer and $\sum_0^k i_j = n\}$ The $\mathrm{Par}_k(n)$ is a finite set of cardinality $\binom{n+k}{k} = N_n(k) + 1$. Let $\alpha = (\alpha_0, \ldots \alpha_{N_n(k)})$ be any ordering of the elements of $\mathrm{Par}_k(n)$. Then corresponding to α there exists an embedding $f_\alpha : \mathbb{C}P^k \hookrightarrow \mathbb{C}P^{N(k)}$ defined by

$$X = (X_0 : X_1 : \ldots : X_k) \to (X^{\alpha_0} : X^{\alpha_1} : \ldots X^{\alpha_{N(k)}})$$

(where $X^{\alpha_j} = X_0^{i_0} X_1^{i_1} \ldots X_k^{i_k}$ where $\alpha_j = (i_0, \ldots i_k) \in \mathrm{Par}_k(n)$.)

Note that different orderings α, β give rise to embedding f_α, f_β which clearly differ by elements in $PL(N(k), \mathbb{Z})$. Then if V is a hypersurface of degree n in $\mathbb{C}P^k$ it is defined by the zero-locus of some homogeneous form F^n of degree n in variables $X_0, \ldots X_k$ say. But the above embedding transforms any n-form F^n in $\mathbb{C}P^k$ into a linear form H_F on $\mathbb{C}P^{N(k)}$ and thus $V = f_\alpha(\mathbb{C}P^k) \cap \{H_F = 0\}$ in $\mathbb{C}P^{N(k)}$.

(Note that we have in fact shown that hypersurfaces of any $\mathbb{C}P^k$, $k > 3$ are simply connected).

Using the Veronese embedding above we can in fact easily show the following.

ASSERTION 2. Let (n,k) be a pair of positive integers. Let V, W be non-singular hypersurfaces of CP^k of degree n. Then V is diffeomorphic to W.

Proof. Let $f : \mathbb{C}P^k \to \mathbb{C}P^N$ be a Veronese embedding of degree n. Let

$$V = \{(X, a) \in \mathbb{C}P^k \times \mathbb{C}P^N | a \cdot f(X) = 0\}$$

(where we take $a \cdot b = \Sigma a_i b_i$ as scalar product in $\mathbb{C}^{N+1}$ so the zero locus $a \cdot b = 0$ is well defined on $\mathbb{C}P^N \times \mathbb{C}P^N$).

Let $\Phi = p_2|V$ and $\pi = p_1|V$ where the p_i are the obvious projection maps.

Now by our previous discussion on the relation between n-forms on $\mathbb{C}P^k$ and linear forms in $\mathbb{C}P^N$ it is clear that if $V = V(F)$ for some n-form F then $V \simeq \Phi^{-1}(a)$, for some $a \in \mathbb{C}P^N$ defined by the linear form H_F corresponding to F. In fact $V = \pi(\Phi^{-1}(a)$ which we denote by $V(a)$.

In particular then the family $(V, \Phi, \mathbb{C}P^N)$ contains as fibers all degree n hypersurfaces of $\mathbb{C}P^k$.

Where are the critical points of Φ?

It is a straightforward verification that $(z,a) \in V$ is a critical point of Φ iff $V(a)$ has a singularity at z.

Then letting A be the set of critical values of Φ we obtain by Bertini's Theorem that A is a proper analytic subvariety of CP^N. In particular then $B = CP^N - A$ is connected. Let $V^1 = \Phi^{-1}(B)$ and $\varphi = \Phi|V^1$. Then (V^1, φ, B) is in fact a differentiable fiber bundle whose fibers run through all the non-singular hypersurfaces of degree n in $\mathbb{C}P^k$. In particular any two such are diffeomorphic.

The beauty of Assertion 2 for our purposes is that it lets us pick any equation of degree n, provided it is non-singular, to generate the topological model of the hypersurface of degree n in $\mathbb{C}P^3$. We denote such a hypersurface by V_n

Either using specific equations or by more general techniques of algebraic geometry we can produce the following table of information concerning the hypersurfaces in $\mathbb{C}P^3$ (See Table I).

TABLE I

Degree	P_g	b_2	σ	K	*Homotopy Type*
1	0	1	1	$-3E_1$	P
2	0	2	0	$-2E_2$	$S^2 \times S^2$
3	0	7	-5	$-E_3$	$P \# 6Q$
4	1	22	-16	0	V_4
5	4	53	-35	E_5	$9P \# 44Q$
6	10	106	-64	$2E_6$	$4V_4 \# 9(S^2 \times S^2)$
7	20	187	-105	$3E_7$	$41P \# 146Q$
8	35	302	-160	$4E_8$	$10V_4 \# 41(S^2 \times S^2)$
9	56	457	-231	$5E_9$	$113P \# 344Q$
10	84	658	-320	$6E_{10}$	$20V_4 \# 109(S^2 \times S^2)$
m	$\frac{1}{6}(m-1)(m-2)(m-3)$	$m(m^2-4m+6)-2$	$\frac{1}{3}m(4-m^2)$	$(m-4)E_m$	$k_m^1 P \# \ell_m Q$ if m odd; $a_m^1 V_4 \# b_m^1 (S^2 \times S^2)$ if m even

where $k_m^1 = (\frac{m}{3})(m^2-6m+11)-1 \qquad \ell_m = \frac{m-1}{3}(2m^2-4m+3)$

$$a_m^1 = \frac{m(m^2-4)}{48} \qquad b_m^1 = \frac{13}{48}m^3 - 2m^2 + \frac{47}{12}m - 1$$

and E_m *is a hyperplane section of* V_m *(and thus a surface of genus* $\frac{1}{2}(m-1)(m-2)$ *embedded in* V_m*).*

In fact the following additional pieces of topological data are known.

1) $S^2 \times S^2$ is $A\ C\ D$. In fact as we mentioned previously $S^2 \times S^2$ is rational and $S^2 \times S^2 \# Q = P \# 2Q$; $S^2 \times S^2 \# P = 2P \# Q$.

2) V_3 is completely decomposable. In fact V_3 is also rational and is *algebraically* just P blown up at 6 points.

3) All of the other hypersurfaces are in fact minimal and these can't be analytically decomposable. The ones of odd degree may however be topologically decomposable.

We now sketch a proof of

THEOREM 3.1. ([16]) *All non-singular hypersurfaces of* CP^3 *are almost completely decomposable.*

Procedure.

1) We first construct a 'family' of manifolds V_t, $t \in D^2$ such that V_t is non-singular for $t \neq 0$ and show that our given hypersurface V is diffeomorphic to a non-singular V_t, $t \neq 0$.

2) If we can find a family such that V_0 is the union of two manifolds X_1, X_2 intersecting transversely in the space $\cup V_t$ then we can show using the fibration theorem which will be discussed shortly that

$$(*) \quad V_t = X_1 - N(X_1 \cap X_2) \cup_{\partial N(X_1 \cap X_2)} X_2 - N(X_1 \cap X_2)$$

where $N(X_1 \cap X_2)$ is a tubular neighborhood of the intersection of X_1 and X_2 and the two pieces in the right hand side above are identified by an appropriate diffeomorphism of

$$\partial N(X_1 \cap X_2) \cap X_1 \to \partial N(X_1 \cap X_2) \cap X_2.$$

3) We analyze the decomposition (*) further. Invoking our theorem on irrational connected sums [18] we show how the union in (*) above can be modified to produce a connected sum decomposition.

4) Having picked our family in 1) and 2) so that X_1, X_2 have a simpler structure than V we use an inductive procedure to reduce problems about V_* to problems about V_1.

The Procedure outlined above is in fact the general procedure we use to establish all our results about classes 1-5 (Class 6 will be analyzed by a slight modification of the above techniques). Before seeing explicitly what happens in the case of hypersurfaces of CP^3, let us look at hypersurfaces of CP^2. These are plane algebraic curves of degree n (and genus $\frac{1}{2}(n-1)(n-2)$) and let us look at their degenerations into two components:

$$C_n = \{(X_0 : X_1 : X_2) \in \mathbb{C}P^2 | X_0^n + X_1^n + X_2^n = 0\}$$

$$C_{n+1} = \{((X_0:X_1:X_2),(\lambda:\mu)) \in CP^2 \times CP^1 | \mu(X_0^{n+1} + X_1^{n+1} + X_2^{n+1}) + \lambda(X_0 + X_1 + X_2)(X_0^n + X_1^n + X_2^n) = 0\}$$

Let $\pi : C_{n+1} \to \mathbb{C}P^1$ be the obvious map and let D_r be a disc of radius r about $(\mu=0\ \lambda=1)$ in $\mathbb{C}P^1$.

Then an elementary calculation shows that if r is sufficiently small, the only critical value of π in D_r is at the origin. Furthermore $\pi^{-1}(0)$ is essentially $C_n \cup C_1$ with transversal intersection $C_n \cap C_1 = \{P_1, \dots P_n\} = P$. Then by the fibration theorem quoted above

$$C_{n+1} \simeq C_n - N(P) \cup C_1 - N(P)$$

As can thus be easily seen from Figure 1 C_{n+1} is thus topologically C_n with $(n-1)$ handles attached. That is $C_{n+1} = C_n \# (n-1)T^2$. Continuing inductively we find $C_{n+1} = \frac{1}{2} n(n-1)T^2$ which of course is the standard formula for genus (C_{n+1}).

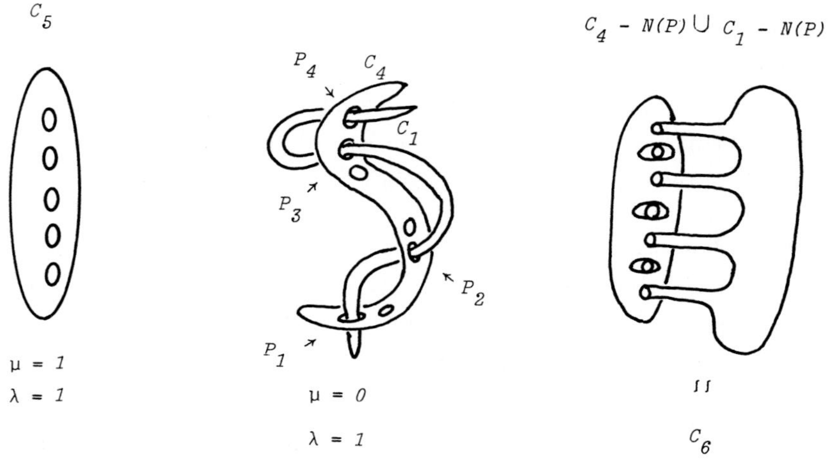

Fig. 1.

Let us now return to CP^3. Here the situation is more complicated. If we set $V_n = \{(X_i) \in CP^3 \mid \Sigma X_i^n = 0\}$ and $V_{n+1} = \{((X), (\lambda{:}\mu)) \in CP^3 \times CP^1 \mid \mu \Sigma X_i^{n+i} + \lambda(\Sigma X_i)(\Sigma X_i^n) = 0\}$ we find that V_{n+1} fails to be a manifold! In fact it has $n(n+1)$ singular points precisely at the points in $V_1 \cap V_n \cap V_{n+1}$; $\mu = 0$. Thus we can't exactly parallel the construction undertaken for algebraic curves. However, utilizing the technique of 'blowing up subvarieties' (which we shall explain shortly) we can in fact produce a manifold W_{n+1} and a holomorphic map $\pi : W_{n+1} \to CP^1$ with isolated critical value at the 'origin $(\mu'=0\ \lambda'=1)$' in CP^1 such that $\pi^{-1}(\mu=\varepsilon, |\varepsilon| \text{ small}) \approx V_{n+1}$ and $\pi^{-1}(\mu=0) = V_n \cup \sigma V_1$ with $V_n \cap \sigma V_1 \approx C_n$ and $\sigma V_1 = V_1$ blown up at $n(n+1)$ points $= P\#(n)(n+1)Q$. Then our fibration theorem gives us that

$$(*) \quad V_{n+1} \approx \overline{V_n - T(C_n)} \cup \overline{\sigma V_1 - T(C_n)}$$

(where $\partial T(C_n) \cap V_n$ is just a Circle Bundle over C_n of Euler Class $+n$ while $\partial T(C_n) \cap \sigma V_1$ is a Circle Bundle over C_n of Euler Class $-n$, and the union above is via the obvious orientation-reversing, fiber preserving, diffeomorphism between two such bundles).

We have thus reduced our algebraic problem to the following topological question.

Question? Suppose M_1, M_2 are smooth, compact 4-manifolds and $j_i : T \to M_i$; $(=1,2)$ are embeddings of an orientable surface into M_i with $T_i = j_i(T)$. Let $h : \partial T_1 \to \partial T_2$ be a diffeomorphism and suppose

$$(**) \quad V = M_1 - T_1 \cup_h M_2 - T_2$$

Describe the topology of V in the simplest possible terms!

We call a decomposition such as $(**)$ above an 'irrational connected sum' and would like to describe V in terms of standard connected sums. We then have the following theorem

THEOREM 3.2 (ICS). *Suppose M_1, M_2' are compact connected oriented 4-manifold with compact 2-submanifolds S_1, S_2' respectively. Let T_1, T_2' be tubular neighborhoods of S_1, S_2' with projection maps π, π' and set $H_1 = \partial T_1$, $H_2' = \partial T_2'$.*

Suppose $\eta : H_1 \to H_2'$ is an orientation reversing bundle morphism and set

$$\overline{V = M_1 - T_1} \cup_\eta \overline{M_2' - T_2'}$$

Suppose also that some fiber C of H_1 is <u>*inessential*</u> *as a loop in V. Then*

(1) There exists 2-discs $d_1 \subset S_1$, $d_2' \subset S_2'$ with $S_1^* = \overline{S_1 - d_1}$, $S_2'^* = \overline{S_2' - d_2'}$, $\tilde{H}_1 = H_1|S_1^*$, $\tilde{H}_2' = H_2'|S_2'^*$, $T_1^* = T_1|S_1^*$, $H_1^* = \partial T_1^*$,

$H_2'^* = \partial T_2'^*$ and η extends to a diffeomorphism $\eta^* : H_1^* \to H_2'^*$ such that if

$$V' = \overline{M_1 - T_1^*} \cup_{\eta^*} \overline{M_2' - T_2'^*}$$

then V' is diffeomorphic to either

$$V \# P \# Q \quad \text{or} \quad V \# S^2 \# S^2$$

(2) Furthermore if $M_2' = M_2 \# Q$ is obtained by blowing up some manifold M_2 at a point $P \in S_2$ with S_2' the 'strict image' of S_2 in M_2, Then

$$V \# P = \overline{M_1 - T_1^*} \cup \overline{M_2 - T_2^*} = V^*$$

where T_2^* is defined in the obvious fashion so that $T_2^* \approx T_2'^*$.

(3) If K, the 1-skeleton if S_1 in M_1, is null-homotopic in M_1. Then $V' \approx M_1 \# \chi_K(M_2')$; $V^* = M_1 \# \chi_K(M_2)$ (where $\chi_K(M_2)$ is M_2 surgered along a collection of disjoint 1-spheres obtained by pulling back K into a disjoint collection of 1-spheres $\tilde{K}$ in H_1 and transferring to $\eta(\tilde{K}) \subset M_2$ via η) Furthermore if $\eta(\tilde{K})$ is null-homotopic in M_2. Then

$$V^1 \approx M_1 \# M_2' \# \chi(S^4) \quad \text{and} \quad V^* = M_1 \# M_2 \# 2g(P \# Q)$$

where g = genus (S_1) and $\chi(S^4)$ is either $2g(P \# Q)$ or $2g(S^2 \times S^2)$.

(4) If V, M_1, M_2' are simply connected and $M_2' = M_2 \# Q$, as in (1) above: Then

$$V \# P = M_1 \# M_2 \# 2g(P \# Q)$$

Before sketching the proof of this theorem in Chapt. 5 we apply it. In the case $(*)$ above we in fact obtain

$$V_{n+1} \# P = V_n \# (V_1 \# (\eta(n+1)-1)Q) \# 2g(C_n)(P \# Q)$$

(The complete theorem is more complicated and will be described fully later)

Using $2g(C_n) = (n-1)(n-2)$ we thus obtain

$$V_{n+1} \# P = V_n \# V_1 \# (n-1)(n-2)P \# (2n^2-2n+1)\,Q.$$

Thus continuing inductively we get

$$V_{n+1} \# P = (n+1)\ V_1 \# \tfrac{1}{3}(n+1)(n^2 - 4n + 3)P \# \tfrac{n}{3}(2n^2 + 1)\ Q.$$

But $V_1 = P$ and so we get

$$V_{n+1} \# P = k_{n+1}\ P \# \ell_{n+1}\ Q$$

$$k_{n+1} = \tfrac{1}{3}(n+1)(n^2-4n+6),\ \ell_{n+1} = \tfrac{n}{3}(2n^2+1)$$

In particular then all the V_n are thus almost completely decomposable as claimed.

Let us now see how to use the above procedure to obtain similar results for Classes 2-5 above. We begin with a more detailed discussion of Step 1 and 2 above.

IV. FAMILIES OF ANALYTIC MANIFOLDS AND THEIR DEGENERATIONS

Let us begin with the fibration theorem alluded to previously. We content ourselves with stating the holomorphic version of that theorem. For the full C^∞ version see [17].

FIBRATION THEOREM [17]. *Let W be a Complex Manifold. Suppose* $f : W \to \Delta$ *is a non-constant proper holomorphic map onto a disc* $\Delta \subset \mathbb{C}$ *about the origin having a critical value only at* $0 \in \Delta$. *Suppose the zero-divisor* Z_f *of f consists of two non-singular irreducible components* A_1, A_2 *of multiplicity 1, intersecting transversely in a connected submanifold C in W. Then if* $T \to C$ *is a tubular neighborhood of C in W such that* $T_1 = T \cap A_1$ *and* $T_2 = T \cap A_2$ *are tubular neighborhoods of C in* A_1 *(resp.* A_2*)* [*Note: there will always exist such a T*] *then setting* $H_i = \partial T_i \to C$, *there exists a bundle isomorphism* $\eta : H_1 \to H_2$

of H_1 onto H_2 such that for any $\lambda \in \Delta - \{0\}$

$$\overline{A_1 - T_1} \cup_\eta \overline{A_2 - T_2} \text{ is diffeomorphic to } V_\lambda = f^{-1}(\lambda) .$$

Idea of Proof. 1) We first note that having picked T as above there will be some $\varepsilon > 0$, such that for $|\lambda| < \varepsilon$, V_λ intersects T and ∂T transversely

2) We note that $f|f^{-1}\{|\lambda| < \varepsilon\} \cap \overline{W - T}$ has no critical points and therefore conclude that

$$\overline{V_t - V_t \cap T} \approx \overline{V_0 - V_0 \cap T} = \overline{A_1 - A_1 \cap T} \amalg \overline{A_2 - A_2 \cap T}$$

3) We check that

a) $\partial(V_t \cap T) \approx \partial(A_1 - A_1 \cap T) - \partial(A_2 - A_2 \cap T)$

b) $\overline{V_t \cap T} \approx [S^1 \text{ bundle over } C] \times I$

4) Combining 2) and 3) concludes the proof.

In order to apply the fibration theorem we must be in a situation where we have a manifold W and a map $f : W \to \Delta$ of the appropriate type. However under normal circumstances the existence of such a $W = \cup\{V_2 | X \in \Delta\}$ is not immediately apparent. What is more natural is the existence of a 'linear system' of divisors of a given algebraic manifold V. [If one looks at higher codimension submanifolds of V one gets an analogous notion of 'algebraic system' of algebraic cycles on V]. We thus detour back into the realm of algebraic geometry

Linear Systems. Let H_0, H_1 be independent hypersurfaces of the same degree in CP^n. Then we can generate a family of hypersurfaces $H(\lambda:\mu) \subset CP^n$ simply by forming the linear combinations $H(\lambda:\mu) = \lambda H_1 + \mu H_0$, where if by abuse of notation we assume that the defining equation of H_i is $H_i(X_0: \ldots X_n) = 0$ then $H(\lambda:\mu)$ is given by $\lambda H_1(X) + \mu H_0(X) = 0$.

We call such a family $H(\lambda:\mu)$, a linear pencil of hypersurfaces.

Let F be the meromorphic function on CP^n defined by $F(X) = H_1(X)/H_0(X) = -\mu/\lambda$.

We then note the following

1) $Z_F = H_1$, $P_F = H_0$ and so H_1, H_2 are linearly equivalent divisors in the sense of Chapt. 3.

2) F is well defined only on $\mathbb{C}P^n - \{X|H_1(X) = H_2(X) = 0\} = \mathbb{C}P^n - H_1 \cap H_0$.

Thus $F : \mathbb{C}P^n - H_1 \cap H_0 \to CP^1$ does in fact give a holomorphic mapping with fibers of the form $H(\lambda:\mu) - H_1 \cap H_0$.

Generalizing the construction above slightly we may suppose $H_0, \ldots H_m$ are independent hypersurfaces of CP^n of the same degree and form $H(\lambda) = \Sigma\, \lambda_i H_i$. Then we call $H(\lambda)$ a linear system of hypersurfaces of CP^n. We again note that any two $H(X_\alpha)$, $H(X_\beta)$ in $H(\lambda)$ will be linearly equivalent divisors.

Now if $X \subset CP^N$ is any non-singular subvariety of CP^N we can think of the $H(\lambda)$ above as denoting hypersurface sections obtained by intersecting X with the appropriate hypersurfaces of X. We thus have linear systems of hypersurfaces sections, also denoted by $H(\lambda)$, cutting out linearly equivalent divisors of X. Also linear pencils on X will then define holomorphic maps $F' : X - X \cap H_0 \to CP^1$. (In fact in terms of our definitions of meromorphic functions we see that a linear pencil of hypersurface sections $\lambda(H_1 \cap X) + \mu(H_0 \cap X)$ defines a meromorphic function f on X with $Z_f = H_1 \cap X$, $P_f = H_0 \cap X$ and $I_f = H_1 \cap H_0 \cap X$).

To get at the broadest definition of linear system and to see how linear systems give rise to the type of maps $f : W \to \Delta$ we are interested in we must continue the discussion of divisors and meromorphic functions begun in Chapt. 4.

We thus define:

Definition. If $D = \sum_{i=1}^{t} n_i Z_i \in \mathrm{Div}(V)$, then $\mathcal{L}(D) = \mathcal{L}_V(D) = \{f \mid f$ is holomorphic on V and either $f \equiv 0$ or $(f) + D \geq 0\} =$
$\{f \mid \mathrm{ord}_{z_i}(f) \geq -n_i,\ 1 \leq i \leq t$, and $\mathrm{ord}_z(f) \geq 0$ all other $Z\}$.
Then $\mathcal{L}(D)$ is the vector space of all meromorphic functions on V with no poles other than at the z_i and at each z_i either with poles of at most a certain bounded order, or with zeroes of at least a certain order.

One of the fundamental theorems of compact analytic manifold then says that dim $\mathcal{L}(D)$ is always finite.

Closely related to $\mathcal{L}(D)$ is the projective space $|D|$ defined if $\mathcal{L}(D) \neq (0)$ as

$$|D| = \{(f) + D \mid f \in M_V^* \text{ and } (f) + D \geq 0\}$$

$$= \text{Projectivization of } L(D)$$

Then it is clear that $\dim |D| = \dim L(D) - 1$. Then

Definition. A linear system is a subset L of some $|D|$ such that $V = \{f \in M_v \mid f \equiv 0$ or $(f) + D \in L\}$ is a vector space over $\mathbb{C}$. Equivalently, L is a projective subspace of the projective space $|D|$.

The linear systems $|D|$ themselves are called complete linear systems.

If a linear system is of dimension 1 we call it a linear pencil.

We note that if $D = n_i Z_i$ is a divisor on V then $\mathrm{Supp}(D) = \cup\{Z_i \mid n_i$ in $\Sigma n_i Z_i$ is $\neq 0\}$.

Then the base points of a linear system L are defined by

$$(\text{Base points of } L) = \bigcap_{D \in L} (\text{Support of } D)$$

If $D_1, \ldots, D_k$ are linearly equivalent effective divisors so that

$|D_1| = |D_2| = \dots = |D_k|$, the linear system L spanned by them is the smallest linear subspace $L \subset |D_1|$ containing all the D_i.

Now how do we go from linear systems to mappings. We must generalize our definition of meromorphic function in order to define meromorphic mappings. So let f be a meromorphic function on V. Then define the closed graph of f

$$G_f = \text{Closure of } \{(z,f(z)) \mid f \text{ is holomorphic at } z\} \text{ in } V \times \bar{\mathbb{C}} \quad (\text{i.e. } z \in H_f)$$

(where $\bar{\mathbb{C}} = \mathbb{C} \cup (\infty) = CP^1$).

Then letting $\mathcal{J}$ be the projection of G_f to V we find

THEOREM 4.1. Under the above conditions we have

a) G_f *is a subvariety of* $V \times CP^1$

b) $\mathcal{J}$ *is holomorphic, proper and onto*

c) $\mathcal{J}$ *is biholomorphic over* H_f.

Using this Theorem as motivation we now define:

Definition. Let V,W be complex manifolds. Then f is a meromorphic mapping of V to W if and only if, $f \subset V \times W$ and there exists an open, dense subset $H \subset V$ such that

1) $f|H = \{(x,y) \in f | x \in H\}$ is holomoprhic
2) $G_f = \text{Closure}(f|H)$ in $V \times W = f$ is analytic in $V \times W$
3) If $\mathcal{J} = \pi_1 \mid V \times W \to V$ restricted to G_f then $\mathcal{J}$ is a proper map onto V.

(We note that as in the case of meromorphic functions we can always extend f to a maximal set H_f and we always identify f with its extentions. We can also show that $V-H_f$ is a proper subvariety of V and that in fact $\mathcal{J}^{-1}(V-H_f)$ is a proper subvariety of G_f. Furthermore we can prove that $\mathcal{J}$ is biholomorphic over H_f.)

We note that if $f_1 \ldots f_n$ are meromorphic functions on V then $f = (f_1 \ldots f_n)$ will define a meromorphic mapping. We say that f is a bimeromorphic mapping if f and f^{-1} are meromorphic.

(Aside: The preceding theory can be built in the algebraic category as well as in the analytic. In that case the words analytic, meromorphic are replaced by regular, rational respectively. The regular maps are those which are locally given by polynomials and the rational ones are locally quotients of polynomials.)

Before continuing let us emphasize that a meromorphic function or mapping of V to W need not be a set-theoretic single-valued function on all of V. It is single-valued on an open dense subset of V but may be "multi-valued" on proper subvarieties of V. Nevertheless f is still a relation on $V \times W$ and so $\mathrm{dmn} g$, $\mathrm{rng} f$ and $f[A] = \{(x,y) \in G_f | x \in A\}$ are well defined.

Example. Let $f = \{((x_0:x_1:x_2:),(y_0:y_1:y_2)) \in CP^2 \times CP^2 | x_0y_0 = x_1y_1 = x_2y_2\}$ Then f defines a meromorphic mapping of $\mathbb{C}P^2$ to $\mathbb{C}P^2$. In fact

1) on $\mathbb{C}P^2 - \{x_0x_1x_2 = 0\} \to \mathbb{C}P^1 - \{y_0y_1y_2 = 0\}$ f is the holomorphic map

$$(x_0:x_1:x_2) \to (\frac{1}{x_0}:\frac{1}{x_1}:\frac{1}{x_2})$$

2) On the twice punctured lines
$L'_i = \{X_i = 0\} - \{X_i \cap X_{i+1} \cup X_i \cap X_{i+2}\}$
(where we consider $i \in \mathbb{Z}_3$) $C = 0,1,2$ we have

$$f[L'_i] = Q_i = (y_0,y_1,y_2) \text{ where } y_i = 1, y_{i+1} = y_{i+2} = 0$$

3) At the points $(1:0:0), (0:1:0), (0:0:1)$, f 'blows up' each point to a line $y_i = 0$.

We have in fact that f is bimeromorphic between $\mathbb{C}P^2$ and itself. It is holomorphic on $CP^2 - \{(0:0:1),(0:1:0),(1:0:0)\}$ and biholomorphic on $\mathbb{C}P^2 - \{X_0 = 0 \cup X_1 = 0 \cup X_2 = 0\}$ to itself. (f is called the Cremona transformation)

Now suppose Z is a meromorphic mapping on V to CP^m with $Z[V] \not\subset$ any hyperplane of CP^m. Then for any hyperplane H of CP^m we can functorally obtain a divisor $Z^*[H]$ of V by assigning multiplicities correctly [to the distinct codimension 2 subvarieties of $Z^{-1}[H]$. ($Z^*[H]$ is called the total transform of H under Z). In particular then as H varies in $|H|$ we obtain a linear system $L_2 = \{Z^*[H] \mid H \in [H]\}$. This is called (naturally) the linear system determined by Z.

Now conversely if L is any linear system on V with base point set of codimension ≥ 2 and φ is a fixed isomorphism of $L \simeq CP^m$ (some m) (Recall that by definition such a φ must exist). Then L determines a meromorphic map Z_L of V to CP^m as follows:

By the definition of linear system there exists a divisor D_0 on V such that $V_L = \{f \in M_V | f \equiv 0$ or $(f) + D_0 \in L\}$ is a vector space over $\mathbb{C}$. In fact L is then Projectivization(V_i) and so φ determines an isomorphism $\Phi : \mathbb{C}^{m+1} \approx V_L$. Let f_i $i=0,\ldots m$ be the basis of V_L given by $f_i = \Phi(e_i)$, $e_i, i=0,\ldots m$ the natural basis of $\mathbb{C}^{m+1}$. $e_0 = (1,0,0\ldots)$ $e_1 = (0,1,0\ldots)$ etc.

Then $Z_L = (f_0,\ldots f_m)|$ gives us the meromorphic map defined by the linear system L.

(There is an alternate, though equivalent, way of looking at the meromorphic map associated to a linear system. We first note that if D is a divisor and $[D]$ is its associated line bundle then (D) is isomorphic to $\Gamma(V, \mathcal{O}(D) = \{\mathbb{C}$-vector space of holomorphic sections of $[D]\}$ and thus $|D| \approx P\Gamma(V,\mathcal{O}(D) =$ Projectivization of the vector space. $\Gamma(D) \approx \Gamma(V,\mathcal{O}(D))$, Thus linear systems correspond to holomorphic subspace of $\Gamma(D)$. Now if E is any line bundle and F a vector subspace of $\Gamma(V,E)$, then letting $f_0 \ldots f_m$ be a basis for F we have a meromorphic map $V \to CP^m$ given by $x \to (f_0(x),\ldots f_m(x))$ which exactly corresponds to the meromorphic map determined by the linear system corresponding to F).

What we have shown is

THEOREM 4.2. (See [30]) There is a 1-1 correspondence between

a) linear systems L of dimension m whose base points have codimension ≥ 2 together with a projective isomorphism $L \approx \mathbb{C}P^m$

b) meromorphic maps $Z : X \to CP^m$ such that $Z[X] \not\subset$ any hyperplane.

(We will use $Z \to L_z$, $L \to Z_L$ to denote the inverse operations defined above).

In particular if L is a complete linear system, so $L = |D|$ for some divisor D, we sometimes refer to Z_L as the meromorphic map determined by the divisor D. If $D = nK_V$, then the map determined by D is called the nth pluricanonical map of V. (If $n=1$ we call it the canonical map.) Using n^{th} pluricanonical maps we can give an alternate characterization of surfaces of general type. (Note that in fact as described above the pluricanonical map is determined by $|nK|$ only up to a projective transformation on its range space.)

THEOREM 4.3. Let V^* be an analytic surface with (relatively) minimal model V. Then V is of general type if and only if for some $n > 0$, the pluricanonical map f_n determined by n satisfies dim $f_n[V] \geq 2$. [*For proof; see* [28]]

Thus given a linear system L on V we can in general (i.e. if codim Base points of $L \geq 2$) get a meromorphic map $Z_L : V \to CP^m$ for some m. This is not good enough however. What we really want is some way of 'extending' or 'modifying' Z_L to get a holomorphic map $\tilde{Z}_L$ on some suitably defined 'modification' of L, whose fibers are 'in general' isomorphic to those of Z_L. In order to see how close we can come to doing this we need one more concept; that of a modification (or general blowing up) of a complex manifold. Formally we have,

Definition. Let V and X be complex manifolds and suppose $g : X \to V$ is given. Then $(X,V,\mathcal{J})$ is a modification of V (or, more informally, X is a blowing up of V (via $\mathcal{J}$) iff

a) $\mathcal{J}$ is a proper holomorphic map of X onto V

b) There are nowhere dense analytic subsets $M \subset X$, $N \subset V$ such that $f|X - M : X - M \to V - N$ is biholomorphic.

It is clear that our definition of 'blowing up' a surface at a point is simply a special case of this more general notion of modification.

Now suppose W^{n-m} is a complex submanifold of V^{n+1}. Then for any $p \in W$ there exists a coordinate neighborhood $U(z_0, \ldots z_n)$ of p in W such that $W \cap U = \{z \in U \mid z_i = 0,\ i = 0, \ldots m\}$

Let $\tilde{U}_p = \{(z,w) \in U \times \mathbb{C}P^{m-1} \mid z_i w_j - z_j w_i = 0 \quad 0 \leq i, j \leq m\}$. Patching together all the $\tilde{U}_p$ we obtain a new manifold $\tilde{V}$ and a holomorphic map $\mathcal{J} : \tilde{V} \to V$ which is a modification of V. In fact $\mathcal{J}^{-1}(W)$ is a CP^m bundle over W isomorphic to Projectivization ($N_V(W)$ = Normal bundle of W in V) and $\mathcal{J} : \tilde{V} - \mathcal{J}^{-1}(W) \to V - W$ is biholomorphic (If $m = n-1$ then $\tilde{V} = V$ and $\mathcal{J}^{-1}(W) \approx W$). We call $\tilde{V}$ the blowing up of V along W.

(Blowing up can also be defined in a purely coordinate free fashion as follows: Suppose $E \overset{\pi}{\to} W$ is a complex vector bundle and let $P(E) \overset{p}{\to} W$ be the associated complex projective bundle. Let $L \overset{\rho}{\to} P(E)$ be the tautological complex line bundle $P(E)$ (where we recall that if $(x,[\ell])$ is a point of $P(E)$, where ℓ is the complex line through the origin in the complex vector space $\pi^{-1}(x)$ and $[\ell]$ is the equivalence class of ℓ in $p^{-1}(x) = \pi^{-1}(x) - \{0\}/\mathbb{C}^*$, then ℓ is the fiber of L over $(x,[\ell])$) Then there exists a canonical map $\varphi : L \to E$ (i.e. $((x, \bar{\ell}), t) \in \rho^{-1}(x,[\ell]) \to (x,t) \in \pi^{-1}(x)$) which is biholomorphic on $L - \varphi^{-1}(W) \to E - W$ and such that $\varphi^{-1}(W)$ is a $\mathbb{C}P^r$-bundle over W (r = fiber $\dim_{\mathbb{C}} E-1$). Then L or $\varphi : L \to E$ is E blown up along W. If W is a submanifold of a complex manifold V it admits a normal bundle $N_V(W) \approx$ Tubular neighd W in V. We can the blow up W in $N_V(W)$ to get $\varphi : N \to N_V(W)$ which is biholomorphic on $\tilde{N} - \varphi^{-1}(w) \to N(w) - W$ and set $\tilde{V} = V - T(W) \cup_{\varphi} \tilde{N}$ to define the blow-up of V along W. It is readily seen that upon going to local coordinates we obtain the same definition we had previously).

We can now relate meromorphic mappings and modifications as follows:

Suppose f is a meromorphic mapping of X to Y. Let G_f be the closed graph of f and $\mathcal{T} : G_f \to X$ the obvious projection. Then clearly $(G_f, X, \mathcal{T})$ is a modification of X. In fact letting $I_f = X - H_f$ (we call I_f the indeterminancy set of f) we recall that $\mathcal{T} : G_f - \mathcal{T}^{-1}(I_f) \to X - I_f$ is biholomorphic and if I_f is a submanifold of X (i.e. a non-singular subvariety) then G_f is simply X blown up along I_f. For example if f is the Cremona transformation described previously then G_f is precisely CP^2 blown up at the three points $\{(0;0:1), (0:1:0), (1:0:0)\}$. In general though I_f has singularities. As we shall see shortly if I_f is simply the union of two submanifolds A_1, A_2 of X with $A_1 \pitchfork A_2 = Y$ (A_1 intersecting A_2 transversely in Y) then $G_f \to X$ can be factored as $G_f \xrightarrow{\mathcal{T}_1} X_{A_1} \xrightarrow{\mathcal{T}_2} X$, where $X_{A_1} \xrightarrow{\mathcal{T}_2} X$ is X blown up along A_1 and $G_f \xrightarrow{\mathcal{T}_1} X_{A_1}$ is X_{A_1} blown up along $\tilde{A}_2 = \text{Closure } \mathcal{T}_2^{-1}(A_2 - Y)$. This is a special case of a remarkable theorem of Hironaka which says that 'essentially' any meromorphic mapping can be factored through a succession of 'blowing ups' of submanifolds (as in our example above). The precise formulation involves a slight generalization of blowing up and can be found in [45]

We note that if f is meromorphic from X to Y and $\sigma_f : G_f \to Y$ is the obvious projection then σ_f is in fact a holomorphic map on the complex manifold G_f which is the 'unique extension' of the holomorphic map $f : X - I_f \to Y$ to G_f. Thus in our situation to apply our fibration theorem we will start with a linear system L on a manifold V; then go to the meromorphic map $Z_L = f$ determined by L and from there go to the holomorphic map $\sigma_f : G_f \to CP^m$ on the modification of V. Before detailing this process let us first observe that the process of going from meromorphic mapping to modification can be reversed. In fact, we note that if $\mathcal{J}: W \to V$ is a modification then $\sigma = \mathcal{J}^{-1}$ in fact defines a meromorphic mapping from V to W such that G_σ is biholomorphic to W. We also note that if f is an arbitrary meromorphic map of V to W and G_f has canonical projections $\mathcal{J}: G_f \to V$; $\sigma : G_f \to W$. Then f is bimeromorphic iff $\sigma : G_f \to W$ is a modification.

We now examine in detail the meromorphic map and consequent modification determined by a linear pencil.

Let us initially assume $X = X^n \subset P^N = CP^N$ is an n-dimensional non-singular algebraic subvariety of CP^N and suppose P_0, P_∞ are hyperplanes of CP^N, with P_0, P_∞, X intersecting pairwise transversely and $P_0 \cap P_\infty$ intersecting X transversely. (Henceforth we shall say that in such a situation P_0, P_∞, X have normal intersection). Let $X_0 = X \cap P_0$, $X_\infty = X \cap P_\infty$ and $Y = X_0 \cap X_\infty$. Then X_0, X_∞ determine a linear pencil with 'axis' Y on X. Now let φ be the meromorphic map from X to P^1 determined by the above linear system. Note that if $\tilde{X} = G_\varphi$ is the modification of X determined by φ, then $\tilde{X}$ is just X blown up along the indeterminacy set Y of φ and the projection map $f : \tilde{X} \to P^1$ is the obvious extension of the holomorphic map $X \to (P_0(X), P_\infty(X))$ of X-Y to CP^1 determined by the defining equations $p_0 = 0$, $p_\infty = 0$ of P_0, P_∞.

(In local coordinates we can produce our map as follows:

Suppose CP^N has coordinates $[Z_0:\ldots:Z_N]$. We can assume without loss of generality that $p_0 = Z_0$ and $p_0 = Z_1$ and thus the meromorphic function φ is $\varphi[Z_0,\ldots,Z_N] = [Z_0;Z_1] \in CP^1$ for $Z \in X$.

Then if

$$\tilde{P}^N = \{([Z_0,\ldots,Z_N],[t_0,t_1]) \in CP^N \times CP^1 \mid Z_0 t_1 = Z_1 t_0\}$$

with induced projection maps $\tilde{\pi} : \tilde{P}^N \to P^N$ and $\pi_1 : \tilde{P}^N \to P^1$ then $\tilde{X}$ is just $\tilde{\pi}^{-1}(X)$, and $f : \tilde{X} \to P^1$ is just $\pi_1|\tilde{X}$. We note that if $[t_0,t_1] \in P^1$, then $\tilde{\pi}$ projects $\pi^{-1}[t_0,t_1]$ biholomorphically onto the hyperplane to $Z_1 = t_1 Z_0$ in P^N and thus the points of P^1 are in 1-1 correspondence with the hyperplanes (and thus hyperplane sections) of our pencil. It is clear that $\pi = \tilde{\pi}|X$ is a holomorphic map $\tilde{X} \to X$ that maps $\pi^{-1}(X-Y)$ biholomorphically onto $X-Y$ and $\tilde{Y} = \pi^{-1}(Y)$ is biholomorphically $Y \times P^1$. Thus what we have done is blown up CP^N along $P_0 \cap P_\infty \approx CP^{N-2}$ to get $\tilde{P}^N$ and in the process blown up X^n along Y^{n-2} to get the manifold $\tilde{X}^n$. In terms of our coordinates f is just given by $f([Z_0,\ldots,Z_N),[t_0,t_1]) = [t_0,t_1])$

Now clearly if P_0, P_∞ were replaced by non-singular hypersurfaces V_0^m, V_∞^m with V_0^m, V_∞^m, X intersecting normally then the above analysis would go through unchanged and the linear pencil of hypersurfaces $\lambda V_0 + \mu V_\infty$ would induce a blowing up of X along $X \cap V_0 \cap V_\infty$ and a holomorphic map $\tilde{X} \to X$ extending the map 'V_0/V_∞' defined on $X - X \cap V_0 \cap V_\infty$. Suppose however that $X \cap V_0$, was singular. In fact suppose that V_0, V_∞ are hypersurface of CP^N (of degree m) such that $X_0 = X \cap V_0$ consists of two non-singular components A_1, A_2 such that X_∞, A_1, A_2 intersect normally. Let $S = A_1 \cap A_2$ and $C = A_1 \cap A_2 \cap X_\infty$. (Note that S, C are manifolds by our normality assumption). Then what is the modification of X determined by the pencil generated by X_0, X_∞. We note that trying to blow X up along $X_0 \cap X_\infty = A_1 \cap X_\infty \cup A_2 \cap X_\infty$ will give us a variety with singularities along C. Thus if $C \neq \emptyset$, it will not give us the nonsingular modification we are looking for. Instead we proceed as follows: (See Figure 2)

Let $B_1 = A_1 \cap X_\infty$ and $B_2 = A_2 \cap X_\infty$. Let $X' =$ X blown up along B_1 with projection $\pi_1 : X' \to X$, $A_i' =$ Closure $\pi_1^{-1}(A_i - A_i \cap B_1)$ $i = 1,2$; $X_\infty' =$ Closure $\pi_1^{-1}(X_\infty - X_\infty \cap B_1)$, $B_1' = \pi_1^{-1}(B_1)$, $B_2' =$ Closure $\pi_1^{-1}(B_2 - B_1 \cap B_2)$. Then we can verify that since B_1 intersects A_2 transversely that in blowing up X along B_1 we have simultaneously blown up A_2 along $A_2 \cap B_1 = A_2 \cap A_1 \cap X_\infty = C$. Thus $A_2' = A_2$ blown up along C. However B_1 lies in A_1 and X_∞ and thus π_1 maps A_1' and X_∞' biholomorphically onto A_1 resp. X_∞. Similarly, $B_2' \approx B_2$ while $B_1' \approx B_1 \times CP'$. We note that $A_1' \cap X_\infty' = \emptyset$ but $A_2' \cap X_\infty' = B_2'$.

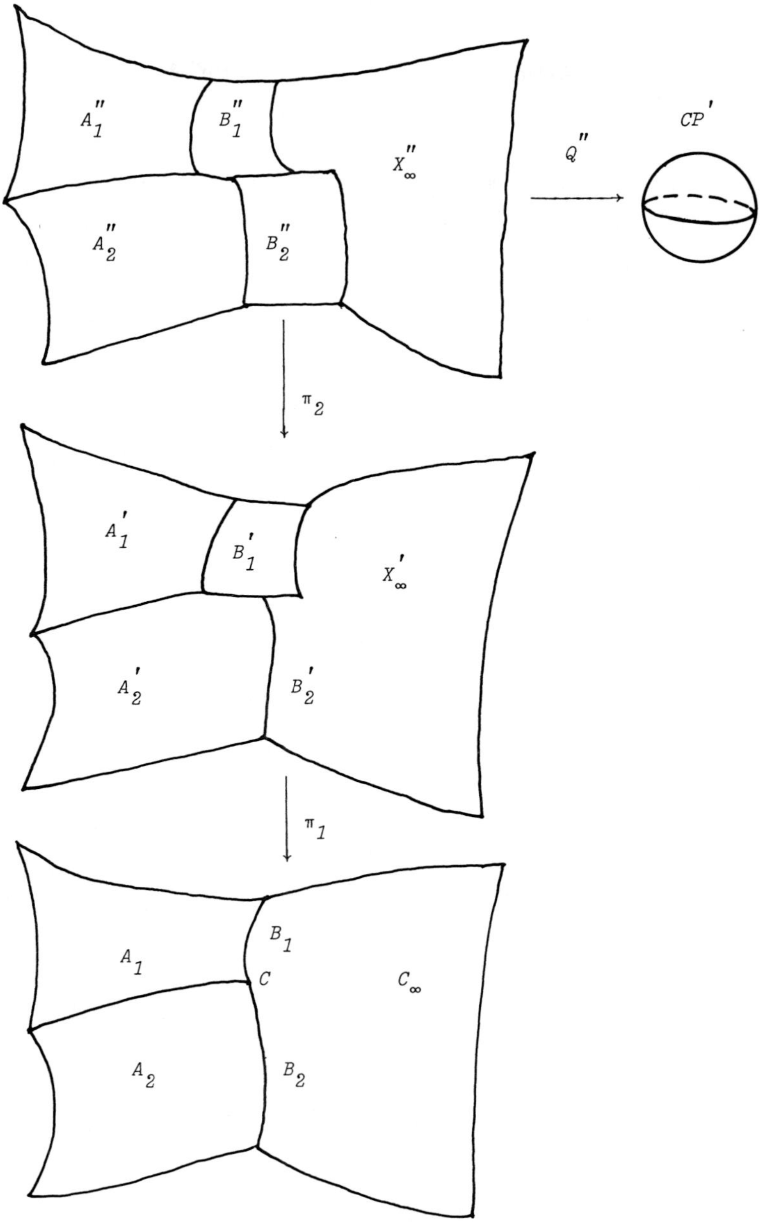

Fig. 2. Note: $A_1' = A_1$; $A_2' = A_2$ blown up at C

Thus although we can extend the meromorphic function φ defined by our linear system to a new function φ' on X', we have that in general φ' is still only meromorphic with indeterminancy set $I_{\varphi'} = B_2$. Thus we must blow up yet another time. Fortunately $I_{\varphi'} = B_2'$ is a submanifold of X' and so by blowing X' along B_2' we will get a new manifold X'' and a *holomorphic* extension $\varphi'' : X'' \to CP^1$ of φ'. More specifically letting $\pi_2 : X'' \to X'$ be the projection map defined by our blowing up and A_i'', X_∞'', B_1'', B_2'' the obvious submanifolds of X'' we can verify that $\varphi''^{-1}(0) = A_1'' \cup A_2''$; $\varphi''^{-1}(\infty) = X_\infty''$ with $A_1'' \pitchfork A_2'' \cong A_1 \pitchfork A_2 = S$ and $A_1'' \cup A_2'' \cong A_1 \cup A_2'$; $X_\infty'' \simeq X_\infty$. Lastly let us recall the analytic version of Sard's Theorem.

SARD'S THEOREM FOR COMPLEX MANIFOLDS. Let $f : M \to N$ be a surjective proper holomorphic map of complex manifolds. Then the critical values of f form a proper analytic subvariety A of N. In particular $f|M-f^{-1}(A) : M - f^{-1}(A) \to N - A$ is a holomorphic fiber bundle.

(Recall that a proper analytic subvariety A of a manifold has real codimension ≥ 2 so $N - A$ will always be connected if A is).

Applied to our pencils it says that φ will only have a finite number of isoltated critical values and that all its non-singular fibers will be diffeomorphic. Thus noting that 0 is a critical value of φ we see that there exists a disc Δ about the origin in $\mathbb{C} \subset CP^1$ such that if $f = \varphi|\varphi^{-1}(\Delta)$ and $W = \varphi^{-1}(\Delta)$ then $f : W \to \Delta$ satisfy the hypothesis of our Fibration Theorem with $f^{-1}(p) \simeq X_\infty$ for $p \neq 0$.

Summarizing our discussion above and extending it to pencils of divisors on an arbitrary complex manifold we obtain the following fibration theorem for divisors.

THEOREM 4.4. [17]. *Suppose W is a compact complex manifold and V, X_1, X_2 are closed submanifolds crossing normally in W with V linearly equivalent to $X_1 + X_2$ as divisors on W. Then there exists a complex manifold Y and a proper holomorphic map $f: Y \twoheadrightarrow \Delta$ onto a disc in $\mathbb{C}$ containing the origin with $f^{-1}(p) \approx V$ for $p \neq 0$ and $f^{-1}(0) = X_1 \cup X_2'$ with X_1, X_2' of multiplicity one and intersecting transversely in some submanifold $S' = X_1 \cap X_2'$ of Y, where X_2' is equal to X_2 blown up along $X_1 \cap X_2 \cap V$ $S \simeq X_1 \cap X_2$. In particular then $V \approx X_1 - T_1 \cup_\eta X_2' - T_2$ where T_i is a tubular neighborhood of S' in X_i and η is a bundle isomorphism of ∂T_1 onto ∂T_2.*

Let us now apply our theorem. If we first look at hypersurfaces in CP^2 it is clear that we can construct V, X_1, X_2 crossing normally and satisfying the hyperthesis of our theorem by taking $V = \{(X_0:X_1:X_2) \mid X_0^{n+1} + X_1^{n+1} + X_2^{n+1} = 0\}$
$X_1 = \{(X_0:X_1:X_2) \mid X_0^n + X_1^n + X_2^n = 0\}$ and
$X_2 = \{(X_0:X_1:X_2) \mid X_0 + X_1 + X_2 = 0\}$.
(In this case only one blowing up is in fact necessary and the manifold Y we get is essentially (C_{n+1} of the previous section). The result our fibration theorem then gives us is just the genus formula for non-singular algebraic plane curves. If we look at hypersurface in CP^3 and take V, X_1, X_2 to be defined by $\Sigma X_i^{n+1} = 0$, $\Sigma X_i^n = 0$, $\Sigma X_i = 0$, respectively our theorem produces the 'fibration' $\pi : W_{n+1} \to CP^1$ we claimed existed earlier and we obtain the decomposition $(*)$

$$V_{n+1} \simeq \overline{V_n - T(C_n)} \cup \overline{\sigma V_1 - T(C_n')}$$

where σV_1 is V_1 blown up at the $n(n+1)$ points in $V_{n+1} \cap V_n \cap V_1$.

Now suppose that V is a 2-dimensional non-singular (Scheme-theoretic) complete intersection in CP^{k+2}. That is suppose there exists k hypersurfaces, $H_i(n_i)$ of degree n_i; $(=1\ldots k$ in CP^{k+2} intersecting 'normally' in $V = V(n_1,\ldots n_k)$. [We shall assume without loss of generality that $i \leq j \Rightarrow n_i \leq n_j$].

[*Note:* We have defined a normal intersection of 3 manifolds X_1, X_2, X_3 in a complex manifold W by demanding that X_i intersects X_j transversely $i \neq j \in \{1,2,3\}$ and $X_1 \cap X_2$ intersects X_3 transversely. More generally if $X_1,\ldots X_n$ are codimension one subvarieties of a complex manifold W^{n+k} we say that $X_1,\ldots X_n$ intersect normally in W^{n+k} if $V = X_1 \cap \ldots \cap X_n$ is a non-singular subvariety of W^{n+k} such that for each point $p \in V$ there exists a coordinate neighborhood U of p in W and local coordinates $(v_1,\ldots,v_k,x_1,\ldots x_k)$ on U such that $X_i \cap U = \{z \in U | x_i(z) = 0\}$ and $V \cap U = \{z \in U | x_1(z) = \ldots = x_n(z) = 0\}$].

We claim that V can also be made the singular fiber of a 'nice' degenerating family as follows: [See 17]

1) By using some elementary commutative algebra we can find k hypersurfaces $H_i'(n_i)$ of degree n_i, $i=1\ldots k$ intersecting normally in $V = \bigcap_{i=1}^{k} H_i'$ such that

a) the intersection of the first k-1 hypersurfaces $\bigcap_{i=1}^{k=1} H_i' = W$ is non-singular and

b) there exist a hypersurface $K = K(n_k-1)$ and a hyperplane L in CP^{k+2} such that the triple $(W \cap H_k', W \cap K, W \cap L)$ satisfy the hypothesis of our Theorem 4.4.

Then symbolically we have

$$W \cap L = V(n_1,\ldots n_k);\; W \cap K = V(n_1,\ldots n_{k-1}, n_k-1)$$

$$W \cap L = V(n_1,\ldots,n_{k-1}, 1) \text{ and therefore}$$

$$V(n_1,\ldots n_k) = V(n_1,\ldots,n_k-1) - T(C(n_1,\ldots,n_k-1)$$

$$\cup\ \sigma V(n_1,\ldots,n_{k-1}, 1) - T(C'(n_1,\ldots,n_k-1))$$

where $\sigma V(n_1,\ldots n_{k-1}, 1)$ is $V(n_1,\ldots n_{k-1}, 1)$ blown up at the

$x = (\prod_{i=1}^{k} n_i)(n_k-1)$ points in $W \cap H'_k \cap K \cap L$ and $C = C(n_1, \ldots n_k-1) = W \cap K \cap L \approx C'(n_1, \ldots n_k-1)$. In particular if we now apply our irrational connected sum theorem [18] we obtain

$$V(n_1, \ldots n_k) \# P = V(n_1, \ldots, n_k-1) \# V(n_1, \ldots n_{k-1}, 1) \# x_1 P \# x_2 Q$$

where

$$x_2 = 2g(C(n_1, \ldots n_k-1)$$

and

$$x_1 = x_2 + x - 1$$

(Note that C is a non-singular algebraic curve of genus $g(C)$). Thus we can continue inductively to eventially obtain

$$V(n_1, \ldots n_k) \# P = \underbrace{V(1,1,\ldots 1) \# \ldots \# V(1,1,\ldots 1)}_{\Pi n_i \text{ times}} \# kP \# \ell P$$

for appropriate k, ℓ and since $V(1,1,\ldots 1) = P$ we have shown that every complete intersection surface is indeed almost completely decomposable *(ACD)*.

Generalizing the process whereby we showed that all complete intersections were *ACD* we can begin with an arbitrary non-singular 3-fold $W \hookrightarrow CP^N$ (some N). Then again it is always possible to find some hypersurfaces H_1, H_n, H_{n+1} of degree 1, resp. n, resp. $n+1$ in CP^N such that the hypersurface sections $X_1 = W \cap H_1$, $X_n = W \cap H_n$ and $X_{n+1} = W \cap H_{n+1}$ intersects normally in W. Then applying Theorem 4.4 we obtain

$$X_{n+1} \approx X_n - T(C) \cup \sigma X_1 - T(C')$$

where $\sigma X_1 = X_1$ blown up at the $x = n(n+1) \cdot \deg W$ points in $X_1 \cap X_n \cap X_{n+1}$ and $C = X_1 \cap X_n \simeq C'$.

If W is itself simply connected (as it was in the complete-intersection case) then by Lefschetz's Theorem so are the X_i and thus we can use our irrational connected sum theorem to obtain

$$X_{n+1} \# P \simeq X_n \# X_1 \# a_1 P \# a_2 Q$$

where as in the complete intersection case $a_2 = 2$ genus$(X_1 \cap X_n)$ and $a_1 = a_2 + X - 1$. In general $a_1 > 0$ and so we can again con- tinuing to obtain

$$X_{n+1} \# P = (n+1)\ X_1 \# k_n P \# \ell_n Q \quad \text{with} \quad k_n > 0 \text{ and } \ell_n > 0.$$

Now in the cases considered so far we were able to continue inductively until we obtained $X_1 = P$ thus showing X_{n+1} was *ACD*. If W is arbitrary then of course X_1 will in general not be CP^2. However if for whatever reason, we know that X_1 itself is *ACD* then when $k_n > 0$, X_{n+1} will also be *ACD*. Using this we obtain

THEOREM 4.5. [17] *Let W be a simply connected projective algebraic threefold. Suppose X is a non-singular hypersurface section of W of degree n which is ACD. Then if Y is any other non-singular hypersurface section of W with* deg $Y \geq n$, *then Y is ACD.*

In order to get concrete applications out of this we must first discuss how one identifies divisors on W arising as hypersurface sections of some embedding $W \hookrightarrow CP^N$. As we noted earlier every hypersurface section of W can be realized as a *hyperplane* section of the induced Verorese embedding of W. Thus it suffices to identify those divisors which are hyperplanes.

Recall that any divisor D on W determines a meromorphic map φ_D of W to some CP^N (where $N = \dim |D|$). Then D is called a *very ample* divisor if φ_D is a holomorphic embedding. D is called an ample divisor if for some $n > 0$, nD is very ample. (Note that it is easy to show that if nD is very ample so is $(n+m)\ D$ for all $m > 0$).

Now if nD is very ample with associated embedding $\varphi_n : W \to CP^N$ then in fact it is easy to check that nD is realized as a hyperplane section of $\varphi_n(W) \subset CP^N$. Conversely if H is a hyperplane section of $W \overset{i}{\hookrightarrow} CP^N$ then H defines a very ample divisor with $\varphi_N = i$.

Thus given a complex 3-fold W it is of some interest to determine if W admits any ample divisors and which ones they are. There are a few basic criteria by which one can distinguish the ample divisors. [These are due primarily to Moishezon [29] Nakai [31] See Hartshorne [7] or Kleiman [11] as a general reference.]

In particular let us consider the following case:

Suppose X is a non singular algebraic surface and C any irreducible hypersurface section of X. Let $k \geq 1$ and let E_k be the line bundle associated to the divisor kC on X. Let W_k be the completion of E_k to a projective bundle on X (i.e. $W_k = \mathrm{Proj}(E_k \oplus I)$). and let $\pi_k : W_k \to X$ be the projection map. Let X denote the zero section of W_k and set $S = \pi^{-1}(C)$.

Then using one of the criteria for very-ampleness we can check that for any $m > 0$ and $\ell \geq 0, mX + \ell S$ is a very ample divisor on X. In particular we can then always find hypersurfaces $V_{m,\ell} \in |mX + \ell S|$, $V_{m-1,\ell} \in |(m-1)X + \ell S|$ such that X, $V_{m,\ell}$, $V_{m-1,\ell}$ intersect normally in W. Then since it is not difficult to see that $V_{0,\ell}$ is just holomorphic to X blown up at a finite number of points we eventually arrive at

$$V_{m,\ell} \# P \approx mX \# k_{m,\ell} P \# k'_{m,\ell} Q$$

Hence if X is *ACD* so is $V_{m,\ell}$.

Thus we can obtain theorems of the following type:

THEOREM 4.6. *Suppose X is an almost completely decomposable non-singular algebraic surface and C is a very ample divisor on X. Let M be a k-fold cyclic covering manifold of X with branch locus $B \in |kC|$ some $k > 0$.*

Then M is almost completely decomposable.

COROLLARY 4.7. Let M be a cyclic covering manifold of CP^2. Then M is almost completely decomposable.

Proofs. The Theorem follows from a result of Wavrik [41] which shows that every such M arises as a $V_{m,\ell}$ of an appropriate projective bundle $W_k \to X$ obtained by completing the line bundle associated to B. In the corollary it suffices to observe that the branch locus of any cyclic covering of CP^2 is linearly equivalent to kL (some $k > 0$) where L is the divisor class of the canonical embedding of $CP^1 \hookrightarrow CP^2$

Now returning to (***) suppose we don't know that X_1 is almost completely decomposable. If we keep track of genus $(X_1 \cap X_m)$ in our induction process we see that by the adjunction formula of algebraic geometry we get that letting $S_m = X_1 \cap X_m$ (thought of as a submanifold of X_1) then $2g(S_m) - 2 = (K_{X_1} + S_m) \cdot S_m$ $K_{X_1} = (K_W + X_1) \cdot X_1$ is the canonical divisor on X_1 and K_W is the canonical divisor on W. Thus

$$2g(S_m) - 2 = (K_W \cdot X_1 + X_1 \cdot X_1 + X_1 \cdot X_m) \cdot X_m$$

$$= ma + m(m+1)b \text{ where } a = K_W \cdot X_1^2$$

and $$= b = X_1^3 = \deg W$$

so $$2g(S_m) = (m-1)(mb-2) + 2mg_1$$

In particular we can derive that $\ell_n; k_n > (1/3)(n-1)^3 b$.

Now even though X_1 is not known to be almost completely decomposable we do know by Wall's results that there must exist an integer $k = k(X_1)$ such that $X_1 \# k(P \# Q)$ is almost completely decomposable. Thus if $\ell_n, k_n > k$ we obtain that X_{n+1} is itself *ACD!* Calling the minimum such k above the resolving number of X_1 we obtain the following theorem and corollaries.

THEOREM 4.8 [17]. *Let W be a simply connected compact complex submanifold of* CP^N *with* $\dim_{\mathbb{C}} W = 3$. *Let* X_m *be a non-singular hypersurface section of W of degree* $m \cdot \deg W$. *Set* $b = \deg W$.

Then there exists a positive integer m_0 *such that* $m \geq m_0$ *implies* X_m *is almost completely decomposable.*

More particularly if k is the resolving number of X_1 *then* $m > \sqrt[3]{3k/b}$ *implies* X_m *is ACD. Furthermore* X_1 *is ACD implies X is ACD for all m.*

COROLLARY 4.9. Suppose X is a simply-connected non-singular algebraic surface and C is a very ample divisor on X. Let M be an n-fold cyclic covering manifold of X with branch locus $B \in |nC|$, *for some* $n > 0$.

Let $k = k(X)$ *be the resolving member of X. Then if* $b = C^2$ *and* $g = \text{genus}(C)$

1) $n > \sqrt[3]{3k/b}$ *implies M is almost completely decomposable*
2) $n > \sqrt[2]{k/g}$ *implies M is almost completely decomposable.*

COROLLARY 4.10. Suppose X is a simply connected non-singular algebraic surface and D is an ample divisor on V. Then there exists an integer $m_0 \geq 0$ *such that for any* $m > m_0$ *there exists a non-singular curve* $E_{2m} \in |2mD|$ *and for any non-singular curve* $C \in |2mD|$ *there exists a unique algebraic 2-fold covering* X_{2m} *of V with ramification locus C such that* X_{2m} *is ACD.*

This method gives us other results for which we establish some additional terminology. A field F is called an algebraic function field of two variables over C if F is a finitely-generated extension of C of transcendence degree two. Let F denote the collection of all such fields. Then for $F \in F$ there exists a non-singular algebraic surface whose field of meromorphic functions is F. We shall call any such nonsingular surface a model for F. It is then easy to see that given any two such models V_1, V_2 for F their fundamental groups are isomorphic. Thus we define the fundamental group $\pi_1(F)$ for any $F \in \mathcal{F}$ as the fundamental group of any model V for F. We then let $\mathcal{F}_0$ be the subcollection of simply-connected F in $\mathcal{F}$. For $F \in \mathcal{F}_0$ we let $\mu(F) = \inf\{k|$ a model V for F such that $V \# kP$ is completely decomposable$\}$. Using Wall's result previously mentioned, it can be seen that $\mu(F)$ is finite for any $F \in \mathcal{F}_0$. If F is a pure transcendental extension of $C, \mu(F) = 0$. If $\mu(F) \leq 1$ we shall call F a topologically normal field. We now need

Definition. Let $L, K \in \mathcal{F}$. Then L is a satisfiactory cyclic extension of K if there exist models V_L, V_K for L, resp. K and a morphism $\Phi : V_L \to V_K$ with discrete fibers whose ramification locus R_Φ is a nonsingular hypersurface section of V_K whose degree is a multiple of $\deg(\Phi)$.

We can then restate Corollary 4.10 as:

THEOREM 4.11. *Let $K \in \mathcal{F}_0$. Then there exists a satisfactory cyclic extension $L \in \mathcal{F}_0$ of K which is of degree 2 over K and topologically normal.*

In [18] it is further shown that if K itself is topologically normal then so is any satisfactory cyclic extension. These two results motivate a partial order in $\mathcal{F}_0$ defined as follows:

For $L, K \in \mathcal{F}_0$ we shall say that L is a satisfactorily resolvable extension of K if there exists a finite sequence of fields $L_0, \ldots, L_n$ in $\mathcal{F}_0$ with $L_0 = K$. L_{i+1} a satisfactory cyclic extension of L_i and $L_n = L$. We write $K < L$ if L is a satisfactorily resolvable extension of K. Then $<$ induces a partial ordering on $\mathcal{F}_0$. Our above results then say that in terms of this partial ordering every sufficiently "large" field L is topologically normal.

Lastly we mention a purely topological counterpart of Theorem 4.11.

THEOREM 4.12. Suppose X is a smooth simply-connected 4-manifold. Let $F \in H_2(X,Z)$ with $F^2 \neq 0$ and F divisible by some integer $m \geq 2$. Then there exists a smooth compact simply connected 4-manifold $\tilde{X}$ and a map $\Phi : \tilde{X} \to X$ exhibiting $\tilde{X}$ as an m-fold branched cover over X whose branch locus R is a nonsingular representative of F such that

1) *If $F^2 > 0$ then $\tilde{X} \# P$ is completely decomposable*
2) *If $F^2 > 0$ then $\tilde{X} \# Q$ is completely decomposable.*

V. IRRATIONAL CONNECTED SUMS

The lynchpin of our results on almost complete decomposability has been the irrational connected sum theorem of Sect. 3 which allowed us to change $V = M_1 - T_1 \cup M_2' - T_2'$ into $V \# P = M_1 \# M_2 \# 2g(p \# Q)$ in the notation of Theorem 3.2. In this section we will sketch a proof of that theorem.

To fix notation again let M_1, M_2' be compact connected oriented 4-manifolds. Let S be a compact oriented 2-manifold embedded in M_1, resp. M_2' with image S_1, resp. S_2' and let T_1, resp. T_2' be the obvious tubular neighborhoods. Set $H_1 = \partial T_1$ and $H_2 = \partial T_2$ and suppose $\eta : H_1 \to H_2$ is an orientation reversing bundle morphism and

$$V = M_1 - T_1 \cup_\eta M_2' - T_2'$$

We then have:

LEMMA 5.1. Suppose that some fiber C of H_1 is inessential as a loop in V. Then there exist 2-discs $d_1 \subset S_1$, $d_2' \subset S_2'$ with $S_1 = S_1 - d_1$, $S_2'^* = S_2' - d_2'$, $\tilde{H}_1 = H_1|S_1^*$, $\tilde{H}_2' = H_2'|S_2'^*$ $T_1^* = T_1|S_1^*$, $T_2'^* = T_2'|S_2'^*$, $H_1^* = \partial T_1^*$, $H_2'^* = \partial T_2'^*$ such that η extends to a *diffeomorphism*

$$\eta^* : H_1^* \to H_2'^*$$

such that if

$$V' = \overline{M_1 - T_1^*} \cup_* \overline{M_2' - T_2'^*}$$

then

V' is diffeomorphic to either
$V \# P \# Q$ or $V \# S^2 \times S^2$.

Proof. Consider Figure 3 below.

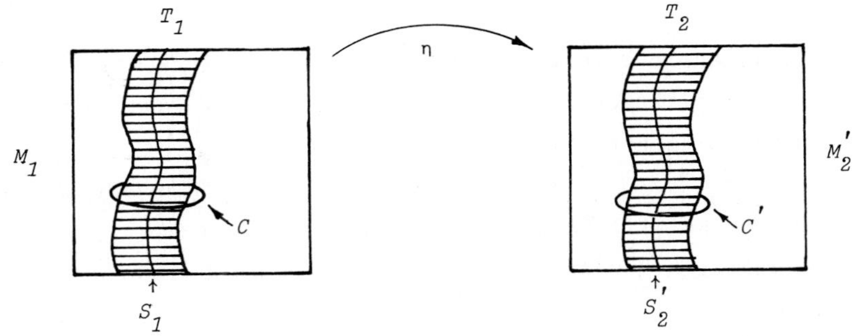

Fig. 3

Let $C' = \eta(C)$. We construct a neighborhood $N(C)$ of C in V as follows. Let $\pi: H_1 \to S_1$, $\pi' : H_2' \to S_2'$ be the obvious projection maps and set $p = \pi(C)$, $p' = \pi(C')$. Let d_1, d_2' be small discs about p, p' in S_1, S_2'. Let $N_1 = \pi^{-1}(d_1) \approx C \times D^2$, $N_2 = \pi^{-1}(d_2) \approx C \times D^2$. Using a collaring theorem we can push N_1, N_2 into M_1, M_2' to get neighborhoods $\tilde{N}_1 \approx N_1 \times I$, $\tilde{N}_2 \approx N_2 \times I$ and glue $\tilde{N}_1$ to N_2 via $\eta|N_1$ to obtain $N(C) = \tilde{N}_1 \cup_\eta \tilde{N}_2 \approx C \times D^3$. Now surger V along $N(C)$. Thus $N(C)$ is replaced by a $D^2 \times S^2$ where the framing we used in constructing $N(C)$ is precisely that which identifies $D^2 \times *$ in $D^2 \times S^2$ with a fiber $T_1 \to S_1$. That is if

$\chi(V) = V - N \cup D^2 \times S^2$ then we can decompose $\chi(V)$ as follows:

$$\chi(V) = V - N \cup D^2 \times S^2$$
$$= (M_1 - T_1 - \tilde{N}_1) \cup (M_2' - T_2' - \tilde{N}_2) \cup D^2 \times S^2$$
$$= (M_1 - T_1 - \tilde{N}_1) \cup D_1 \cup_\eta (M_2' - T_2' - \tilde{N}_2) \cup D_2$$

where we have decomposed $D^2 \times S^2$ into two pieces $D^2 \times S^2_+ \cup D^2 \times S^2_- = D_1 \cup D_2$ in the obvious fashion.

But it is easily seen that

$$(M_1 - T_1 - \tilde{N}_1) \cup D_1 \approx \overline{M_1 - T_1^*} \quad \text{while}$$
$$(M_2 - T_2' - \tilde{N}_2) \cup D_2 \approx \overline{M_2' - T_2'^*}$$

and clearly η extends to η^* as desired. Thus

$$\chi(V) = \overline{M_1 = T_1^*} \cup \overline{M_2' - T_2'^*} = V'$$

However C was null-homotopic in V so $V' = \chi(V) = V \# S^2 \times S^2$ or $V \# P \# Q$ as claimed.

We note further that if $M_2' = M_2 \# Q$ as in (2) of Theorem 3.2 then one must have that $X(V) = V \# P \# Q$ and one can check that the Q summand of M_2' can be identified with the Q summand in $X(V)$. Thus one can 'blow down' the Q on both sides of

$$X \# P \# Q = \overline{M_1 - T_1^*} \cup \overline{M_2' - T_2'^*}$$

to obtain

$$V \# P = M_1 - T_1^* \cup M_2 - T_2^* = V^*$$

using the notation of Theorem 3.2 Part (2).

In order to complete the proof of Theorem 3.2 it suffices then to prove the following:

PROPOSITION 5.2. Let M_1, M_2 be compact connected oriented 4-manifolds. Suppose K is a bouquet of 1-spheres with rk $H_1(K:\mathbb{Z}) = t$ and let K_i be the image of K in M_i under an embedding. Suppose that R_i is a regular neighborhood of K_i in M_i and $\Phi : R_1 \to R_2$ a diffeomorphism inducing the identity on $H_1(K,\mathbb{Z})$.

Let $S_i = \partial R_i$ and $\varphi = \Phi|S_i$. Suppose

$$V = \overline{M_1 - R_1} \cup_\varphi \overline{M_2 - R_2}$$

Then if K_1 is null homotopic in M_1 we have

$$V = M_1 \# \chi_K(M_2)$$

(where $\chi_K(M_2)$ is M_2 surgered along a collection of disjoint 2-spheres obtained by pulling back K_1 into a disjoint collection $\tilde{K}_1$ of 2-spheres in S_1 and transferring to $\tilde{K}_i = \varphi(\tilde{K}_i)$ in M_2)

Furthermore if $\varphi(\tilde{K}_1)$ is null-homotopic in M_2 then

$$V = M_1 \# M_2 \# tL \text{ where } L \text{ is either } P \# Q \text{ or } S^2 \times S^2.$$

Proof. 1) Since K_1 is null-homotopic in M_1 and in this codimension homotopy=isotopy we can find 2-discs D_α $\alpha = 1 \dots t$ capping off the 1-spheres in the collection K_1, such that if $B_\alpha = N(D_\alpha) - R_1$ and $B = \cup B_\alpha$ then

$$M_1 - R_1 - B = \overline{M_1 - N(K_1 \cup \cup D_\alpha)}$$
$$= \overline{M_1 - D^4}$$

(where $N(X)$ denotes a regular neighborhood of X)

2) Using the properties of Φ we have

$$R_2 \cup_\varphi B = \tilde{\Phi}(R_1 \cup B),$$

where $\tilde{\Phi}$ is Φ extended by the identity on B. Thus $R_2 \cup_\varphi B \approx D^4$.

3) We now write B_α as $D^2_\alpha \times S^2 - D^2_\alpha \times S^2_-$ (S^2_- = lower hemisphere of S^2) and obtain the following decomposition:

$$\overline{M_2 - R_2 \cup_\varphi B} = \overline{M_2 - \tilde{K}_2 \times D^3 \cup \cup_\alpha D^2_\alpha \times S^2} - \overline{R_2 - \tilde{K}_2 \times D^3 \cup \cup D^2 \times S^2}$$

But

$$\overline{M_2 - \tilde{K}_2 \times D^3 \cup \cup_\alpha D^2_\alpha \times S^2} = \chi_K(M_2^{\cdot})$$

while

$$R_2 - \tilde{K}_2 \times D^3 \cup \cup_\alpha D^2_\alpha \simeq S^2_- \times R_2 \cup \cup_\alpha B_\alpha \approx D^4 \qquad \text{by (2)}$$

4) Combining 1), 2), 3) we find

$$V = \overline{M_1 - R_1} \cup \overline{M_2 - R_2} = \overline{M_1 - R_1 - B} \cup \overline{M_2 - R_2 \cup_\varphi B}$$

$$= \overline{M_1 - D^4} \cup \overline{\chi_K(M_2) - D^4}$$

$$= M_1 \# \chi_K(M_i) \text{ as desired.}$$

5) The last part of the theorem follows from standard properties of surgery on null homotopic loops in 4-manifolds

VI. ELLIPTIC SURFACES

We now apply the methods of the last 3 sections to demonstrate:

THEOREM 6.1 [19, 27]. *Let V be a simply-connected elliptic surface. Then V is almost completely decomposable.*

To prove Theorem 6.1 we will again have to find appropriate nicely degenerating systems and try to carry through an inductive argument. To do this we first discuss Kodaira's classification of elliptic surfaces.

Thus suppose M is an elliptic surface and suppose w.l.o.g that it is minimal. We then have a holomorphic map $f : M \to R$ onto some compact Riemann Surface R such that 1) f has only a finite number of critical values $a_1, \ldots a_n$ and 2) $X \in R - \{a_1, \ldots a_n\} \Rightarrow f^{-1}(X) \approx S^1 \times S^1$ and $f^{-1}(X)$ has multiplicity one as a divisor on M.

What are the possible topological configurations for the singular fibers $f^{-1}(a_i)$?

There are two types of possibilities

I. $f^{-1}(a_i)$ is topologically a 2-torus T^2 but has multiplicity $m \geq 2$ in M. That is; for any $x \in f^{-1}(a_i)$ if z is a local parameter at $a_i \in R$ with $z(a_i) = 0$ then x has a coordinate neighborhood $U(s,t)$ such $f : M \to R$ takes the form $(s,t) \to z = s^m$ on U.

II. $f^{-1}(a_i)$ has singular points.

Fibers of Type I are called non-degenerate multiple fibers while fibers of type II are called degenerate fibers (they may in addition be multiple). A complete description of all possible degenerate fibers can be found in [12]. For simplicity we shall use the following technical trick [27].

LEMMA 6.2. *Suppose V is an elliptic surface. Then there exists an elliptic surface V', diffeomorphic to V, such that for some holomorphic map $f : V' \to R$ of V onto a compact Riemann surface R, any degenerate fiber $f^{-1}(a)$ of $f : V' \to R$ is of multiplicity one and has only one singular point P, which is an ordinary double point.*

(Recall that a singular point p is an ordinary double point if we can pick a local coordinate neighborhood $U(s,t)$ about p such that $f|U$ has the form $f(s,t) = s^2 + t^2$ (where $s(p) = t(p) = 0$).)

We call a fibering of the above type a Kodaira fibering. We now anlayze the multiple fibers in detail. To do this we must introduce the concept of a logarithmic transformation on a elliptic fibration.

Thus suppose $f : M \to R$ is an elliptic fibration and $a \in R$ is a regular value of f. Then topologically we have for some small disc D about a that $f^{-1}(D) = T^2 \times D^2$. Let C denote the curve ∂D^2 on $f^{-1}(D)$ with induced orientation. On $\partial(f^{-1}(D)) = \partial(M - f^{-1}(D)=T^3$ let Q, H_1, H_2 be simply closed curves generating $H_1(T^3)$ such that Q is homotopic to zero in $f^{-1}(D)$. Let L_m be a diffeomorphism of $\partial(f^{-1}(D) \to \partial(M - f^{-1}(D)$ which takes $C \to mQ + k_1H_1 + k_2H_2$.

We call $M - f^{-1}(a) \cup_{L_m} f^{-1}(a)$ the logarithmic transformation of M at a and write it as $M^* = L_a(m)M$. (This is of course an abreviation for $L_a(m,k_1,k_2)M$)

A fundamental result of Kodaira then is

LEMMA 6.3 [13]. *Let $M^* = L_a(m)M$ be the logarithmic transformation of M at a. Then M^* can be given the structure of an elliptic surface with fibration $f^* : M^* \to R$ such that $(f^*|M^* - f^{*-1}(a)$ 'coincides' with $f|M - f^{-1}(a)$ and on $f^{*-1}(a)$, f^* locally looks like the map $(\sigma,y) \to \sigma^m$. Thus $f^{*-1}(a)$ is a multiple fiber of multiplicity m in M^*.*

Furthermore if V^* is an arbitrary elliptic surface then $V^* = L_{a_1}(m_1)\ldots L_{a_k}(m_k)\ V$ where V is an elliptic surface with <u>no</u> multiple fibers.

By way of example let us look at the elliptic surface $E = T^2 \times S^2$ with projection $T^2 \times S^2 \to S^2$. Then by a direct computation we find that (1) if $E^* = L_a(m)E$ then E^* is diffeomorphic to $S^1 \times L$ where L is a Lens space of type (m,k) for some k relatively prime to m. (2) If $E^* = L_{a_1}(m_1)\ L_{a_2}(m_2)E$ then E^* is still diffeomorphic to $S^1 \times$ lens space $L(m,k)$ for appropriate m,k with $g.c.d(m,k) = 1$.

We now return to the simply connected case. Thus suppose V^* is a simply connected minimal elliptic surface. It is clear that if $\pi : V^* \to \Delta$ is the associated projection map then Δ must be a 2-sphere.

LEMMA 6.4. $V^* = L_{a_1}(m_1) L_{a_2}(m_2) V$ *where V is a simply connected minimal elliptic surface with no multiple fibers and g.c.d* $(m_1, m_2) = 1$. *We allow the possibility that either one or both of* m_1, m_2 *are equal to 1.*

Proof. If V^* has more than two multiple fibers we can construct a holomorphic map $R \to S^2$, R some Riemann Surface, such that if V^* has multiple fibers of multiplicity m_i at $a_i \in S^2$, $i \geq 3$, $m_i \geq 2$ then $R \to S^2$ has ramification order $m_i - 1$ at a_i for all the a_i. Pulling back V^* over R then gives a simply connected unbranched k-sheeted covering manifold W of V^* with $k > 1$ which is a contradiction. Similary if V^* has only two multiple fibers one can deduce that they can't have a common factor. Conversely if V^* is as above then a direct calculation of $\pi_1(X^*)$ shows it to be trivial provided $\pi_1(V)$ is

LEMMA 6.5. *Suppose V is a simply connected minimal elliptic surface and* $V^* = L_{a_1}(m_1) L_{a_2}(m_2) V$ *with g.c.d* $(m_1, m_2) = 1$. *Then*

1) $b_2(V) = b_2(V^*)$

2) $\sigma(V) = \sigma(V^*)$

3) $W_2(V^*) = 0 \Longleftrightarrow W_2(V) = 0$ *and* $m_1 + m_2$ *is even.*

Proof. If $f : W \to S$ is a surjective holomorphic mapping of a complex surface W onto a non-singular compact Riemann surface S with general fiber C then $\chi(W) = \chi(C)\chi(S - \cup a_i) + \Sigma\chi(f^{-1}(a_i))$ where a_i are the critical value of f and χ is the Euler number. For elliptic surfaces we have $\chi(C) = 0$ so that $\chi(W) = \Sigma\chi(f^{-1}(a_i))$. But logarithmic transformation preserve $\chi(f^{-1}(a_i))$ for all a_i so we obtain $\chi(V^+) = \chi(V)$. Since $\pi_1(V^*) = \pi_1(X) = 0$ and if V is elliptically minimal so is V^* we get using Noetlor's formula that $p_g(V) = p_g(V^*)$ and $b_2(V) = b_2(V^*)$. This implies 1) and 2). 3) is proven as in [14] or [19] using a direct calculation of the behavior of the canonial bundle K_V under logarithmic transformations. We note that using the fact that $W_2(V)$ is the mod 2 reduction of $C_1(K_V)$ we obtain $W_2(V) = 0$ if and only if $p_g(V)$ is odd.

Now if we restrict attention to elliptic surfaces without multiple fibers we have the following theorem of Kas.

THEOREM 6.6 [9]. *Let V_1, V_2 be simply connected minimal elliptic surfaces without any multiple fibers. Then V_1 is diffeomorphic to V_2 if and only if $p_g(V_1) = p_g(V_2)$.*

Now suppose W is a simply-connected minimally elliptic surface, with $pg(W) = p$. Denote any topological model for W by Wp.

We then note:

LEMMA 6.7. *1) W_{2k} is homotopy equivalent to $(4k+1) P \# (20k+9) Q$*

2) W_{2k+1} is homotopy equivalent to $(k+1)V_4 \# k(S^2 \times S^2)$

Furthermore if m,n are positive integers and $W_p(m,n)$ denotes $L_{a_1}(m)\ L_{a_2}(n)\ W_p$ then g.c.d $(m,n) = 1$ implies

3) $W_{2k}(m,n)$ is homotopy equivalent to W_{2k}

4) i) If $m+n$ is even then $W_{2k+1}(m,n)$ is homotopy equivalent to W_{2k+1}

ii) If $m+n$ is odd then

$W_{2k+1}(m,n)$ is homotopy equivalent to $(4k+3)P \# (20k+19) Q$

Remark. 1. If $p = 0$ then using the Castelnuevo criterion one can show that $L(m)\ W_0 = W_0 = P \# 9Q$ for any positive integer m. (In fact $L(m)W_0$ is rational). However if $m > 1$, $n > 1$ and *g.c.d.* $(m,n) = 1$ then $W_0(m,n)$ is *not* rational and though homotopy equivalent to $P \# 9Q$ is not known to be topologically equivalent to it. The $W_0(m,n)$ are called Dolgacev surfaces and provide conter-examples to Severe's Conjecture mentioned in Sect. 2. It is not known if any two Dolgacev surfaces with $\{m,n\} \neq \{m',n'\}$ are diffeomorphic! If $p > 0$ then $W_p(m,n)$ is known to be equal to

$W_p(m', n')$ if $\{m,n\} = \{m', n'\}$ as unordered pairs. Again if $\{m,n\} \neq \{m', n'\}$ then it is also not known whether $W_p(m,n)$ is even homeomorphic to $W_p(m', n')$. Thus, as we mentioned in Sect. 2, there could conceivably be infinitely many non-homeomorphic simply connected elliptic surfaces of the same homotopy type.

2. The existence of a W_k for each $k \geq 0$ can be demonstrated in a purely topological fashion as follows:

Let F_g be topological surface of genus g possessing an involution i_g with exactly $2g+2$ fixed points. (F_g can be constructed as a 2-sheeted branched cover of S^2 with precisely $2g+2$ distinct branch points.)

Let $V_g = F_1 \times F_{2g}/i_1 \times i_g$ with canonical projection $\pi_g : V_g \to F_g/i_g = S^2$. Then V_g is a compact complex variety with $n = 8g + 8$ singular points $p_1, \ldots, p_n$. It is easily verified that each point p_i has a neighborhood U_i which is a cone cN_i over the circle bundle $N_i \to S^2$ of Euler characteristic-2. Replacing U_i by the corresponding disc bundle T_i with boundary N_i we obtain the 4-manifold W_g. That is $W_g = V_g - \cup U_i \cup \cup T_i$.

To prove Theorem 6.1 it now clearly suffice to show that $W_k(m,n)$ is *ACD* whenever g.c.d $(m,n) = 1$. This is done as follows:

Step 1: Construct a nicely degenerating family with nonsingular fiber W_k and zero-fiber $X_0 \cup X_1$ with $X_0 \approx W_0$; $X_1 \approx W_{k-1}$ and $X_0 \cap X_1 = T^2$ embedded as a non-singular fiber of some elliptic fibrations $f_0 : X_0 \to S^2$ resp $f_1 : X_1 \to S^2$.

However $W_0 = P \# 9Q$ so using our previously developed machinery we obtain:

$$W_k \# P = W_{k-1} \# P \# 8Q \# 2(P \# Q)$$

$$= W_{k-1} \# 3P \# 10\,Q.$$

Step 2: Let $Z \to \Delta$ be the nicely degenerating family of Step 1. We construct a 3-dimensional 'logarithmic transform'.

$Z^* = L(m)\, L(n)\, Z$ of Z such that $Z^* \to \Delta$ is a nicely degenerating family with nonsingular fiber $W_k(m,n)$ and zero-fiber $X_0 \cup X_1^*$ with $X_0 \approx W_0$; $X_1 \approx W_{k-1}(m,n)$ and $X_0 \cap X_1^*$ a common non-singular fiber of some fibration of X_0, resp. X_1^*. We then get

$$W_k(m,n) \# P = W_{k-1}(m,n) \# 3P \# 10\, Q.$$

for all k,m,n.

Step 3. To complete the proof we must only show that $W_0(m,n) \# P = 2P \# 9Q$. This is done by noting that

$$W_0(m,n) = [W_0 - T^2 \times D^2] \cup [L_a(m)\, L_b(n)(T^2 \times S^2) - T^2 \times D^2]$$

$$= W_0 - T(C) \cup S^1 \times L(p,q) - T(C)$$

where C is a non-singular T^2 in W_0, resp. $S^1 \times L(p,q)$ embedded as a fiber of the corresponding elliptic fibration. Then using the Irrational Corrected Sum Theorem we obtain

$$W_0(m,n) \# P = X(S^1 \times L(p,q)) \# P \# 8Q$$

where $X(S^1 \times L(p,q)$ is $S^1 \times L(p,q)$ surgered along 2-disjoint 1-circles homotopy equivalent to generators of $\pi_1(C)$.

But $X(S^1 \times L(p,q))$ is either $P \# Q$ or $S^2 \times S^2$. (See [46] or calculate directly), and thus

$W_0(m,n) \# P = 2P \# 9Q$ as desired.

The most interesting part of the above proof is the construction of the family $Z \to \Delta$. Noting our work in Sect. 4 it suffices to construct a linear system L with generic element W_k and having appropriate degeneration properties. L is constructed as follows:

Let Σ_k be the complex analytic surface obtained by projectivization of the $\mathbb{C}^2$-bundle $[k] \oplus \mathbb{P}^1 \times \mathbb{C}$ over $\mathbb{P}^1$ ($[k]$ is the $\mathbb{C}^1$-bundle over $\mathbb{P}^1$ corresponding to the divisor $k \cdot S$, S any point of $\mathbb{P}^1$).

Then Σ_k is a $\mathbb{P}^1$-bundle over $\mathbb{P}^1$ and denoting its obvious zero-section by B and typical fiber by F we note that $B \cdot B = k$, $B \cdot F = 1$ and $F^2 = 0$.

Now let E be the projectivization of the $\mathbb{C}^2$-bundle $[2B] \oplus \Sigma_k \times \mathbb{C}$ over Σ_k. Let Σ denote the obvious zero-section of E, and let $\pi : E \to \Sigma_k$ be the projection map. Denote $\pi^{-1}(F)$ by A and set $L = |2\Sigma + A|$.

Then it can be verified that L has W_k as generic element and has a non-generic fiber consisting of W_{k-1} and A intersecting transversely in a fiber of W_{k-1}. Upon following the blowing-up procedure described in Sect. 4 we can easily construct the desired family $Z \to \Delta$. We leave this construction as an exercise to the reader.

Having completed our discussion of simply-connected elliptic surfaces we now consider surfaces of general type.

VII. SURFACES OF GENERAL TYPE

In order to verify the truth of Conjecture $1'$ it is now enough to prove it holds for surfaces of general type. Unfortunately we are far from such a verification. The best general results known are contained in the following theorem of Moishezon [27].

THEOREM 7.1. [*Moishezon*]. *Let V be a compact complex surface of general type. Let* $\sigma_+ = \frac{1}{2}(b_2(V) + \sigma(V))$ *and* $\sigma_- = \frac{1}{2}(b_2(V) - \sigma(V))$. *Let* $K(X)$, $L(X)$ *be the cubic polynomials*

$$K(X) = 30{,}375X^3 + 68{,}850X^2 + 52{,}004X + 13{,}092$$

$$L(X) = 60{,}750X^3 + 141{,}750X^2 + 110{,}265X + 28{,}595.$$

Set $k = K[\sigma_+]$, $\ell = \max(0, L[\sigma_+] - \sigma_-)$. *Then*

$V \# kP \# \ell Q$ *is completely decomposable*

To prove this theorem one again uses a deformation theorem. In the past however, to study V we embedded it as the non-singular fiber of a 'nicely degenerating' family and examined what simply pieces V degenerated into. In this case one reverses the procedure and realizes V as the *singular fiber* of a family whose non-singular fiber has known topology.

We must make use of the following embedding Theorem of Bombieri [2].

THEOREM 7.2. Let V be a minimal non-singular simply-connected surface of general type. Let $N = P_5(V)-1$. Then there exists a holomorphic map $\Phi : V \to CP^N$ with image $W = \Phi(V)$ and points $p_1, \ldots p_m \in W$ such that if $L_i = \Phi^{-1}(p_i)$ and $P = \{p_i\}$, $L = \cup L_i$ then $V - L \xrightarrow{\Phi} W - P$ is a biholomorphic equivalence and each p_i is a rational double point of W.

Furthermore deg $W \leq 45p + 36$ *where* $p = \frac{1}{2}(b_2(V) + \sigma(V))$.

[We recall that a point $p \in V$ is a rational double point of V if it is isolated and for all sufficiently small open sets $U \subset V$ containing P we have $\pi_1(U - p)$ is finite but not trivial. (see [5] for thirteen other equivalent characterizations).]

Now suppose $W \subset CP^N$ is an irreducible algebraic surface in CP^N. We can always define a generic projection $\pi : CP^N \to CP^3$ and analyze the new singularities introduced in $\pi(W)$. One finds that generically there are only three types of new singularities introduced.

They are defined locally by equations of the following sort:

1) $z_1z_2 = 0$ (double lines)

2) $z_1z_2z_3 = 0$ (triplanar points)

3) $z_1^2 - z_2z_3^2 = 0$ (pinch point).

We call any singularity of the above type an ordinary singularity. If X is a complex 3-manifold and $V \subset X$ a complex 2 dimensional subvariety we say that V has canonical singularities if any singular point of V is either a rational double point or ordinary. Generalizing a theorem of Segre, Moishezon combines this with Bomberi's embedding theorem to show that every minimal non-singular simply connected surface of general type V admits a surjective holomorphic map $\psi : V \to W \hookrightarrow CP^3$ such that (1) W has only canonical singularities and $V - \Phi^{-1}(s) \overset{\Phi}{\to} W - S$ is biholomorphic, where $S \subset W$ is the singular locus of W and (2) the locus $S^* \subset S$ of ordinary singularities of W is an irreducible cone.

Since ψ is obtained by a combination of a projection and the map Φ of Theorem 7.2 we have that deg W is still $\leq 45p + 36$.

The following type of deformation Theorem is then used.

PROPOSITION 7.3. Let $W^3 \subset CP^N$ be a complex manifold and suppose F, G are degree r hypersurfaces of CP^N. Let $J_t = (t_0F + t_1G) \cap W$ be a pencil of hypersurfaces on W and suppose $V_1 = F \cap W$ is non-singular and $V_0 = G \cap W$ has canonical singularities such that the locus S^* of all ordinary singularities is an irreducible complex curves of genus g. Suppose also that V_0 and V_1 intersect normally in W.

Then if $\overline{V}$ is the minimal resolution of V_0 we have that if $\pi_1(\overline{V}) = \pi_1(V) = 0$, $S^* \cdot V_1 = b > 0$ in W and $\rho > 0$ is the number of pinch-points of V_0 and $v > 0$ is the number of triplanar points Then: $V_1 \# P$ is diffeomorphic to

$$V \# (2v + \rho + 2g - 1) P \# (v + 2\rho + b + 2g - 2) Q$$

(Note 1. We recall that $\Phi : V \to W$ is called a minimal resolution if (1) V is non-singular; (2) if S = singular locus of W then $V - \Phi^{-1}(S) \overset{\Phi}{\to} W - S$ is biholomorphic and (3) if $V' \overset{\Phi'}{\to} W$ is any other diagram having properties (1) and (2) then Φ' factors through V uniquely: (In fact there exists a 'blowing down map' $V' \overset{\rho}{\to} V$ such that ρ is the composition of a finite number n of σ processes and topologically $V' = V \# nQ$.)

Note 2: By the normal intersection of V_0 and V_1 we mean in this context that for all $X \in V_1 \cap V_0$ there exists a local complex coordinate system (z_1, z_2, z_3) on W with center X such that in some neighborhood U_* of X in W, V_1 is defined by the equation $z_3 = 0$ and V_0 is defined either by the equation $z_1 = 0$ (if V_0 is non-singular at X) or by $z_1 z_2 = 0$ (otherwise).)

The proof of Theorem 7.1 now proceeds as follows:

Let V be a simply connected surface of general type. We can without loss of generality suppose that V is minimal. Then we can find a hypersurface W of CP^3 of degree $r \leq 45p + 36$, $(p = \frac{1}{2}(b_2(V) + \sigma(V))$ such that W has canonical singularities and its locus of ordinary singularities is an irreducible curve and V is its minimal resolution. Furthermore we can then always find a non-singular hypersurface V_r intersecting W normally. Thus we find that by Proposition 7.3,

$$V \# k'P \# \ell' Q \text{ is diffeomorphic to } V_r \# P$$

where

$$k' = 2v + \rho + 2g - 1 = b^+(V_r) - b^+(V) + 1$$

$$\ell' = v + 2\rho + b + 2g - 2 = b^-(V_r) - b^-(V)$$

where $b^+(M)$ resp. $b^-(M)$ equals $\frac{1}{2}(b_2(M) + \sigma(M))$ resp. $\frac{1}{2}(b_2(M) - \sigma(M))$.

But we have shown in Sect. 3 that V_r is ACD and so

$$V_r \# P = k_r P \# \ell_r Q; \; k_r = \frac{r}{3}(r^2 - 6r + 11) \; \ell_r = \frac{r-1}{3}(2r^2 - 4^r + 3)$$

Thus $K' = -k_r - b^+(V)$ and $\ell' = \ell_r - b^-(V)$. Then in particular $k' \leq k\ (45p + 36)$ and $\ell' \leq \ell(45p + 36)$ and so the theorem follows.

Although these estimates gives us a tractable bound on resolving numbers of Complex Surfaces we are still a long way from showing that all such surfaces are *ACD*. All progress in resolving Conjecture I' has been made by considering families of surfaces which could be explicitly constructed. To quote Moishezon [ ]. "the "theoretical" Theorem 7.1 gives much weaker results than our "empirical knowledge". The interesting question is, how far can we move with such 'empirical achievements' in more general classes of simply-connected algebraic surfaces.

VIII. AFTERWORD

We have surveyed most of the currently known result on decomposing analytic surfaces. Clearly much remains to be done. The most promising path to showing that empirically defined classes of surfaces of general type are almost completely decomposable seems to be via the study of surfaces admitting fibrations of generic genus g with $g \geq 1$. In particular in [21] it is shown that certain classes of surfaces admitting hyperelliptic fibrations are *ACD*.

One can also use the fibration theorem to obtain results on handlebody decompositions of analytic surfaces. For example in [20] it is shown that any complete intersection surface admits a handlebody decomposition consisting only of a 0-handle, a 4-handle and some 2-handles! This is of use in constructing framed link representations of the surfaces via the Kirby Calculus [10].

Although surfaces with intersection forms of type II can't be completely decomposed it is possible that Type I surfaces can be so decomposed. Our own conviction is that this can't happen for non-rational surfaces. More precisely we conclude by restating two conjectures originally made in [16].

CONJECTURE A. *Suppose M is diffeomorphic to $kP \# \ell Q$ with $k \geq 2$. Then M does not admit a complex structure.*

A corollary of A is that any simply-connected non-rational complex analytic surface is not completely decomposable. Actually we go further than this. In general we shall say that a compact simply-connected 4-manifold W is decomposable if and only if there exist manifolds W_1, W_2 with $b_2(W_i) \geq 1$, $i = 1,2$ such that $W = W_1 \# W_2$. If W is not decomposable then we shall call it indecomposable. We then conjecture.

CONJECTURE B. *Suppose V is a minimal compact simply-connected complex analytic surface. Then V is indecomposable.*

Conjecture B implies, among other things, that for complex surfaces analytic minimality implies topological minimality. We note in [22] we can show that if m,n are relatively prime positive integers with $m+n$ even then

$$W_{2k+1}(m,n) \# S^2 \times S^2 = (k+1)\ [W_1 \# (S^2 \times S^2)]$$

(Recall that $W_1 = V_4$). Thus even type II surfaces can be rather thoroughly decomposed (though not, of course, completely decomposed) by the addition of only a single $S^2 \times S^2$.

I would like to thank the University of Georgia and the University of California, Berkely for their hospitality during by visits there, and especially thank J. Cantrell, J. Hollingsworth, Harsh Pittie and R. Kirby for their hospitality during my visits to their respective institutions. I would also like to thank B. Moishezon for arousing my interest in this area and again acknowledge the joint nature of many of the results presented.

REFERENCES

1. Reid, M., On Bogomolov's theorem, Mimeo'd Notes.
2. Bombieri, E., Canonical Models of Surfaces of General Type Publ. Math. I.H.E.S. 42 (1973) pp. 447-495.

3. Browder, W., Surgery on Simply-Connected Manifolds, Springer-Verlag, New York, 1972.
4. Dolgacev, I.V., On Severi's Conjecture concerning simply-connected algebraic surfaces, Soviet Math, Doklady, Vol. 7 (1966) 5; pp. 1169-72.
5. Durfee, A., Fourteen characterizations of rational double points (Preprint)
6. Gunning, R. C., Lectures on Vector Bundles on Riemann Surfaces, Math Notes, Princeton Univ. Press. (1967)
7. Hartshorne, R. Ample Subvarieties of Algebraic Varieties, Lecture Notes in Math 156, Springer-Verlag. Chapt. I.
8. Hirzebruch, F., The signature of Ramified Coverings., Global Analysis, Princeton Univ. Press. (1969), pp. 263-265.
9. Kas, A., On the deformation types of regular elliptic surfaces, Complex Analysis and Algebraic Geormetry., Cambridge University Press, Cambridge (1977), pp. 107-111.
10. Kirby, R., A calculus for framed links in S^3 (to appear).
11. Kleiman, S. L., A note on the Nakai-Moishezon test for ampleness of a divisor, Am. J. Math., 87 (1965) pp. 221-226.
12. Kodaira, K., On compact analytic surfaces I, II, III. Ann. of Math. 71 (1960) pp. 111-152; 77 (1963) pp. 563-626; 78 (1963) pp. 1-40.
13. Kodaira, K., On the structure of compact complex analytic surfaces I, II, IV. Am. J. of Math. 86 (1964) pp. 751-798; 88 (1966) pp. 682-72; 90 (1968) pp. 1048-1066.
14. Kodaira, K., On homotopy K^3 surfaces. Essays on Topology, Springer-Verlag (1970) pp. 58-69.
15. Kodaira, K. and Morrow, J., Complex Manifolds. Holt-Rinehart and Winston., N. Y. (1971).
16. Mandelbaum, R., and Moishezon, B., On the topological structure of non-singular algebraic surfaces in CP^3. Topology, Vol. 15 (1976) pp. 23-40.
17. Mandelbaum, R and Moishezon, B., On the topology of algebraic surfaces (to appear).
18. Mandelbaum, R., On Irrational connected sums. (to appear in Trans. Am. Math. Soc.)
19. Mandlebaum, R., On the topology of Elliptic Surfaces. (to appear in Advances in Math)
20. Mandlebaum, R., On special handlebody decompositions. (to appear).
21. Mandlebaum, R., (to appear)
22. Mandlebaum, R., (to appear)
23. Milnor, J., On simply-connected 4-manifolds. Symp. Internat. de. Top. Alg. (Mexico 1956) Mexico 1958. pp. 122-128.
24. Milnor, J. Morse Theory, Annals of Math Studies, Vol. 51 (1963).
25. Milnor, J., and Husemoller J., Symmetric Bilinear Forms, Springer-Verlag (1973)
26. Miyaoka, Y., On the Chern numbers of surfaces of general Type. Inventiones Math. 42 (1977) pp. 225-23 .

27. Moishezon, B., Complex surfaces and connected sums of complex projective planes. Springer-Verlag, Lecture Notes in Math., vol. 603
28. Moishezon. B., Surfaces of Fundamental type in Algebraic Surfaces "Proceeding of the Steklov Institute" No. 75 (1968) Am. Math. Soc. Providence (1967).
29. Moishezon, B., A criterion for projectivity of complete algebraic varieties. AMS translations (2) 63 (1967) pp. 1-50.
30. Mumford, D., Algebraic Geometry I: Complex Projective Varieties. Springer-Verlag (1976)
31. Nakai, Y., A criterion of an ample sheaf on a projective scheme, Am. J. Math 85 (1963) pp. 14-26.
32. Novikov, S. P., Homotopically equivalent smooth manifolds I, Translations Am. Math. Society (2) 48 (1965), pp. 271-396.
33. Pittie, H., Complex and almost complex manifolds in: Complex Analysis and its applications, Vol. III; International Atomic Energy Agency, Vienna, 1976. pp. 121-131.
34. Rohlin, V. A., A new result in the theory of 4-dimensional manifolds. Soviet. Math. Doklady 8 (1952). pp. 221-224. (in Russian)
35. Serre, J. P., Cours D'Arithmetique. Presses Universitaires de France (1970).
36. Shafarevitch, Basic Algebraic Geométry. Springer-Verlag (1975).
37. Sullivan, D., Infinitesimal Computations in Topology, Publ. Math. I.H.E.S. 47 (1977)
38. Wall, C.T.C., Surgery on Compact Manifolds, Academic Press, London and New York (1970).
39. Wall, C.T.C., Diffeomorphisms of 4-manifolds. J. London Math Soc. 39 (1964), 131-140.
40. Wall, C.T.C, On Simply-connected 4-manifolds, J. London Math Soc. 39 (1964) pp. 141-419.
41. Wavrik, J. J., Deformations of Banach coverings of complex manifolds. Amer. J. Math. 90 (1968) pp. 926-960.
42. Whitehead, J. H. C. On simply-connected 4-dimensional polyhedra. Comm. Math. Helv. 22 (1949) 48-92.
43. Whitney, H., Complex Analytic Varieties. Addison-Wesley (1972).
44. Yau, S. T., (to appear)
45. Hironaka, H., Modifications etc. Complex Analysis Conference, Minneapolis, 1964.
46. Fintushel R. A., Pao, S. P., Identification of certain 4-manifolds with group actions. (preprint).

HEEGAARD DIAGRAMS FOR CLOSED 4-MANIFOLDS

José María Montesinos

School of Mathematics
The Institute for Advanced Study
Princeton, N. J.

Deep results of Laudenbach and Poenaru are used to get an analogue of the 3-dimensional Heegaard diagrams for PL, closed, orientable 4-manifolds. The way to passing from one diagram to another representing the same 4-manifold is obtained, but in contrast with the 3-dimensional analogue, the problem of deciding whether a given diagram corresponds to a closed, 4-manifold is left open.

I. INTRODUCTION

Each closed, orientable, PL 4-manifold W^4 admits a handle presentation $W^4 = H^0 \cup \lambda H^1 \cup \mu H^2 \cup \gamma H^3 \cup H^4$ and we show that W^4 is uniquely determined by the cobordism between $\partial(H^0 \cup \lambda H^1)$ and $\partial(H^0 \cup \lambda H^1 \cup \mu H^2)$, defined by the 2-handles μH^2. This provides an analogue of the 3-dimensional Heegaard diagrams for closed, orientable, PL 4-manifolds.

We solve the equivalence relation problem for Heegaard diagrams (i.e. the way to pass from one diagram to another representing the same 4-manifold), but the problem of deciding whether a given diagram corresponds to a closed 4-manifold remains open. This is in contrast with the 3-dimensional analogue.

ISBN 0-12-158860-2

We illustrate the equivalence relation for Heegaard diagrams, giving a proof that $2V^4 \cong S^4$ for Mazur manifolds V^4, without appealing to the structure of $V^4 \times I$.

We finish with some examples of Heegaard diagrams for closed 4-manifolds and we suggest some applications.

I am indebted to Robert Edwards, Charles Giffen, Cameron Gordon and Laurence Siebenmann for helpful conversations.

II. THE HOMEOTOPY GROUP OF $\lambda \# S^1 \times S^2$

This is calculated, in an implicit way, in [3; p. 342] and we include the proof for completeness. We define $\lambda \# S^1 \times S^2$ as a connected sum of λ copies of $S^1 \times S^2$ if $\lambda > 0$, and as S^3 if $\lambda = 0$.

THEOREM 1. The homeotopy group of $\lambda \# S^1 \times S^2$ (i.e. the group of orientation-preserving autohomeomorphisms of $\lambda \# S^1 \times S^2$, quotient the subgroup of those isotopic to identity) is generated by sliding 1-handles, twisting 1-handles, permuting 1-handles and rotations.

Remark. The definitions of the first three kinds of generators are in [3] . For the fourth one, see [2], (see also Section 6).

Proof. Let φ be an orientation-preserving autohomeomorphism of $\lambda \# S^1 \times S^2$. Because the first three kinds of generators induce a system of generators for the automorphism group of $\pi_1(\lambda \# S^1 \times S^2)$, we can suppose that φ induces the identity on the fundamental group of $\lambda \# S^1 \times S^2$. But then (Lemma 3 of [3]) φ induces also the identity on $\pi_2(\lambda \# S^1 \times S^2)$; in other words, φa_i is homotopic to a_i, for each i, where $A = \{a_1, \ldots, a_\lambda\}$ is a set of transverse spheres to the handles of $\lambda \# S^1 \times S^2$. According to [2; Theorem 1 and proof of Lemma 5.1], up to isotopy, we can suppose that φ is the identity in A. Hence, using [2; 5, 4], we can compose φ with a sequence of rotations to get a map isotopic to the identity.

We remark, as in [3], that all of these generators extend to $\lambda \# S^1 \times B^3$. Note also, that there is an autohomeomorphism of $\lambda \# S^1 \times B^3$ which is orientation-reversing in the boundary. So we have;

THEOREM 2. Given a 4-manifold M^4, with $\partial M^4 = \lambda \# S^1 \times S^2$, the manifold $M^4 \cup \lambda \# S^1 \times B^3$ is independent of the way of pasting the boundaries together.

Remark. The results in [3] are a consequence of Theorem 2.

COROLLARY 1. Each closed, orientable 4-manifold with handle presentation $W^4 = H^0 \cup \lambda H^1 \cup \mu H^2 \cup \gamma H^3 \cup H^4$ is completely determined by $H^0 \cup \lambda H^1 \cup \mu H^2$.

Conversely (except for S^4, where the analogue of Waldhausen's result for Heegaard splittings of S^3 is true[1]), we do not know how many embeddings of a bouquet of γ 1-spheres there are in a closed W^4, such that the complementary space of the thickened bouquet has a representation with *0-*, *1-*, and *2*-handles. This seems to be an interesting question.

III. HEEGAARD SPLITTINGS

Let $W^4 = H^0 \cup \lambda H^1 \cup \mu H^2 \cup \gamma H^3 \cup H^4$ and let V^4 be a collar of $\lambda \# S^1 \times S^2$ in $H^0 \cup \lambda H^1$. Then, $V^4 \cup \mu H^2$ is a cobordism essentially from $\lambda \# S^1 \times S^2$ to $\gamma \# S^1 \times S^2$ that we call $C^4(\lambda, \gamma)$, and we have:

[1] *Robert Edwards and Cameron Gordon have pointed out that that analogue is true also for simply-connected 4-manifolds.*

COROLLARY 2. The cobordism $C^4(\lambda,\gamma)$ determines W^4.

We call $(W^4, C^4(\lambda,\gamma))$ a *Heegaard-splitting* of W^4. This is an analogue of a Heegaard splitting in M^3. Two Heegaard-splittings $(W^4, C^4(\lambda,\gamma))$ and $(W'^4, C'^4(\lambda',\gamma'))$ are *equivalent* if there is a homeomorphism of pairs between them. We have just proved that each W^4 has at least one Heegaard-splitting.

IV. HEEGAARD-DIAGRAMS FOR 4-MANIFOLDS

Consider the pair $(\lambda \# S^1 \times S^2, w)$, where w is a framed link in $\lambda \# S^1 \times S^2$. We can associate to this pair a cobordism from $\lambda \# S^1 \times S^2$ to the 3-manifold M^3 obtained by surgery on the framed link w, as follows. Form $(\lambda \# S^1 \times S^2) \times I$ and attach a number of 2-handles along the framed curves w, which we suppose are living in the $\times\{1\}$ boundary component of $(\gamma \# S^1 \times S^2) \times I$. If the new end M^3 of this cobordism is $\gamma \# S^1 \times S^2$, for some γ, we call $(\lambda \# S^1 \times S^2, w)$ a *Heegaard-diagram*.

The splitting $(W^4, C^4(\lambda,\gamma))$, in Section 3, can be represented by Heegaard-diagram $(\lambda \# S^1 \times S^2, w)$, which turns out to be, at the same time, a representative for the closed, orientable 4-manifold W^4.[1]

Two pairs $(\lambda \# S^1 \times S^2, w)$ and $(\lambda' \# S^1 \times S^2, w')$ are *equivalent* if there is a homeomorphism of pairs preserving the framings. Because such a homeomorphism extends to $C(\lambda,\mu)$, we see that equivalent Heegaard-diagrams provide equivalent Heegaard-splittings.

[1]*The term diagram stands for a pair $(\lambda \# S^1 \times S^2; w)$ even when the other end M^3 of the cobordism is not $\gamma \# S^1 \times S^2$, for any γ. In this case, the diagram represents the* <u>*bounded*</u> *4-manifold which is obtained by pasting $\lambda \# S^1 \times B^3$ with the cobordism associated to the diagram, along their boundaries. We emphasize that a Heegaard diagram represents always a* <u>*closed,*</u> *orientable 4-manifold.*

The problem of enumerating of all possible Heegaard diagrams corresponds to the following

PROBLEM. When is a pair $(\lambda \# S^1 \times S^2, w)$ *a Heegaard-diagram?*[1]

V. MOVES ON A HEEGAARD-DIAGRAM

We want to know when two Heegaard-diagrams are representatives of the same 4-manifold. It turns out to be (Theorem 3, below) that this happens if and only if the diagrams become equivalent after applying some moves, which we describe in the following.

Move i) (stabilization). The diagram $(\lambda \# S^1 \times S^2, w)$ changes to $(\lambda \# S^1 \times S^2 \# S^1 \times S^2, w \cup \alpha)$, where α is $S^1 \times \{*\}$ with framing $S^1 \times \{**\}$, $\{*\}$ and $\{**\}$ being two points of S^2, in the last connected summand of course.

Move ii) (stabilization). The diagram changes to $(\lambda \# S^1 \times S^2, w \cup \alpha)$, where α is the boundary of a disc in $\lambda \# S^1 \times S^2$ which does not cut w. The framing of α is a concentric curve in that disc.

Move iii) (band move, see [1] *and Fig. 1.)* The curve $\alpha \in w$, with framing a, changes to (α', a') by using another curve $\beta \in w$ with framing b, as follows. Let A (resp. A') be the framing annulus with $\partial A = \alpha \cup a$ (resp. $\partial A' = \beta \cup b$), and let $\hat{A}$ be a proper collar of b in A'. Let R be a ribbon connecting a with $\partial\hat{A} - b$, as in Figure 1, and let R' be a ribbon concentric to R. Then $(A \cup \hat{A} - R) \cup (R - R')$ is an annulus; the boundary containing a part of α (resp. a) is α' (resp. a').

[1] *Kirby calculus* [1] *is very useful for deciding this in concrete examples, (see example in section 7).*

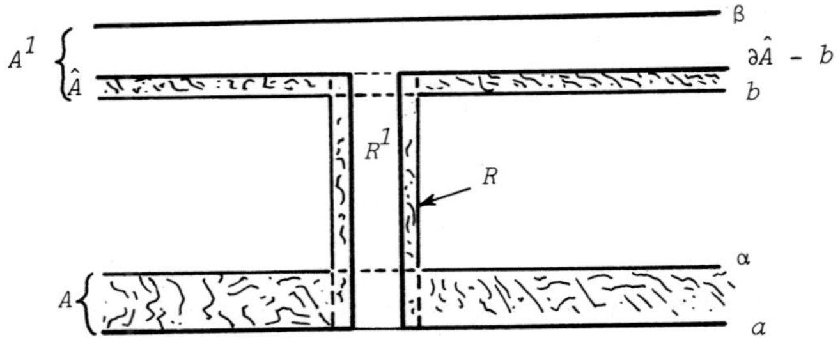

Fig. 1.

Now we state:

THEOREM 3. The diagrams $(\lambda \# S^1 \times S^2, w)$ *and* $(\lambda' \# S^1 \times S^2, w')$ *represent the same 4-manifold if and only if they become equivalent after applying a sequence of moves i), ii), iii), or their inverses.*

Proof. If $C^4(\lambda,\gamma)$ is the cobordism associated with the diagram $(\lambda \# S^1 \times S^2, w)$ we see that move iii), corresponding to a sliding of 2-handles [1], does not change the type of $C^4(\lambda,\gamma)$. Moves i) and ii) do change this type but do not change the 4-manifold W^4 associated with the diagram. To see this, note first that move ii) is the dual of a move i), and that (after filling the cobordism up with $\lambda \# S^1 \times B^3$) move i) correspond to a birth of a pair of cancelling handles in a handle presentation of W^4.

Now, we prove the other implication. We fill up $C^4(\lambda,\gamma)$ with $\lambda \# S^1 \times B^3$ and $\gamma \# S^1 \times B^3$ to get the closed 4-manifold W^4. We choose an arbitrary handle structure $H^0 \cup \lambda H^1$ (resp. $H^0 \cup \gamma H^1$) for $\lambda \# S^1 \times B^3$ (resp. $\gamma \# S^1 \times B^3$) to get a handle structure $H^0 \cup \lambda H^1 \cup \mu H^2 \cup \gamma H^3 \cup H^4$ for W^4, where μ is the number of components of the link w.

Doing the same with the other diagram $(\lambda' \# S^1 \times S^2, w')$ we get another handle presentation for W^4.

As in [1], we see that these two handle presentations are related by a sequence of the following moves:

1) Births and deaths of complementary (i.e. $(i, i+1)$-) handle pairs.

2) Handle slidings.

Also, as in [1], we can suppose that, in such a sequence of moves, there are no births or deaths of $(0,1)$-handle pairs or $(3,4)$-handle pairs.

Now a birth of a $(1,2)$-handle pair changes the diagram associated with the handle presentation by a move i). The 1-handle slidings do not change the equivalence class of the diagram. The 2-handle slidings are moves iii) [1]. The 3-handle slidings do not change the diagram. A birth of a $(2,3)$-handle pair modifies the cobordism $C^4(\lambda,\gamma)$ by the addition of a 2-handle along the boundary of a disc which is contained in $\gamma \# S^1 \times S^2$. We can push this disc off the 2-handles of the cobordism using moves iii): this time the diagram is changed by moves iii) and ii).

VI. A PARTICULAR MODEL

If we fix a particular model M_λ for $\lambda \# S^1 \times S^2$ we can define a Heegaard diagram (M_λ, w), where now the framed curves w live in M_λ.

For example, take $R^3 + \infty$ minus 2λ open balls $B = \{B_1, B_1', \ldots, B_\lambda, B_\lambda'\}$ of radius 1/4, B_i having its center at $(2i-1,0,0)$ and B_i' at $(2i,0,0)$. Then M_λ is the result of identifying the boundaries of B_i and B_i' by reflection in the plane $x = 2i - 1/2$.

We define some moves in this model which suffice to generate the group of autohomeomorphisms of M_λ, up to isotopy.

Move iv) (twisting 1-handles). This move takes place in a ball containing B_i and B_i' but meeting no others: outside of this ball, the move is the identity. The move is a permutation of the pair $(\partial B_i, \partial B_i')$ accomplished by rotation of 180° around the axis $\{x = 2i - 1/2,\ y = 0\}$.

Move v) (permuting 1-handles). The support of the move is a ball containing B_i, B_i', B_{i+1} and B_{i+1}', but meeting no others. The pair $(\partial B_i, \partial B_i')$ goes to $(\partial B_{i+1}', \partial B_{i+1})$ by rotation around the axis $\{x = 2i + 1/2,\ y = 0\}$.

Move vi) (sliding 1-handles). Take a ball C in $R^3 + \infty$ containing in its interior only a single ball B_i (or B_i') and such that ∂C cuts B only in two discs indentified by reflection (in $B_j \cup B_j'$, for instance). The support of the move is $C - B_i$ (which is a punctured solid torus in M_λ). We perform the move by pushing B_i along C, going through $\partial B_j \equiv \partial B_j'$ and coming back to the original position.

Move vii) (rotation). The support is a regular neighborhood of $\partial B_i \equiv \partial B_i'$ in M_λ, and consists in a complete rotation of ∂B_i around an axis. (i.e. this is the "suspension" of a Dehn-Lickorish twist).

Move viii) (symmetry). This is a reflection in the (x,y)-plane

Because these moves generate the group of autohomeomorphisms of $\lambda \# S^1 \times S^2$, up to isotopy, we have:

THEOREM 3'. *The diagrams (M_λ, w), $(M_{\lambda'}, w')$ represent the same 4-manifold if and only if they become isotopic after applying a sequence of moves i), ii), ..., viii).*

VII. THE DUAL DIAGRAM

Let $H^0 \cup \lambda H^1 \cup \mu H^2 \cup \gamma H^3 \cup H^4$ be a handle presentation for a closed, orientable, PL, 4-manifold W^4. There is a dual presentation with γ 1-handles, μ 2-handles and λ 3-handles. These two presentations define two Heegaard diagrams which are mutually *dual*.

Let $(\lambda \# S^1 \times S^2; w)$ be a diagram. A *framed meridian* for the component w_i of w is a meridian of $\partial U(w_i)$, framed by a parallel meridian.

THEOREM 4. If a Heegaard diagram $(\lambda \# S^1 \times S^2; w)$ is given, the associated dual diagram is $(M^3; n)$, where M^3 is obtained from $\lambda \# S^1 \times S^2$ by surgery in the framed w, and n is a system of framed meridians for w.

Proof. M^3 is $\partial(H^0 \cup \lambda H^1 \cup \mu H^2)$, which is obtained by surgery in the framed w. Let H_i^2 be one of the components of μH^2. Then H_i^2 is attached to M^3 by the cocore which is framed by the product structure of H_i^2. Isotoping this framed cocore out of H_i^2 it becomes a framed meridian for the attaching sphere of H_i^2, which is a component of w.

Example. The dual diagram of the Heegaard diagram of Figure 2a is obtained in Figures 2b to 2e. In the proof we change a 1-handle by $\bigcirc^0$ [1] because this also represents $S^1 \times S^2$. Note that we are also proving that the original diagram is in fact a Heegaard diagram (compare footnote 3).

[1] *The notation $\bigcirc^n$ means that the framing has linking number n with the curve, in a regular neighbourhood of a disc bounding the curve.*

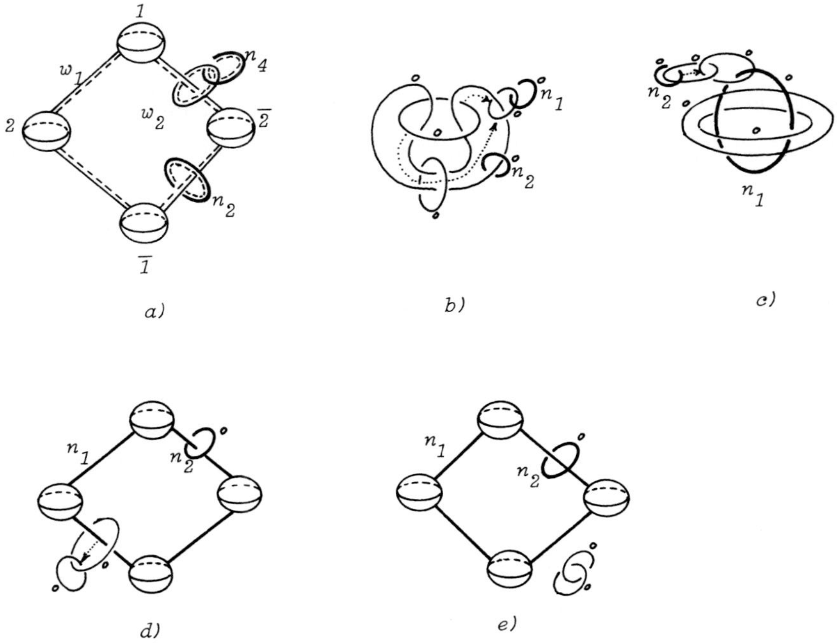

Fig. 2.

Remark. In the last example we illustrate the fundamental feature of Kirby's calculus [1], i.e. given a Heegaard diagram $(S^3;w)$ of a manifold W^4 (which necessarily is simply connected), by adition of k copies of $\bigcirc^{\pm 1}$ (see footnote 4) and moves we get $(S^3; \bigcirc^{\pm 1}, \overset{m}{\dots}, \bigcirc^{\pm 1})$. This means that $W^4 \# k\ (\pm\mathbb{C}P^2) \cong m(\pm\mathbb{C}P^2)$. Thus, Kirby's calculus cannot work with framed links in $\lambda \# S^1 \times S^2$, if $\lambda > 0$.

VIII. EXAMPLE: HEEGAARD DIAGRAMS FOR $2W^4$

Let $(\lambda \# S^1 \times S^2;w)$ be a pair where w is a framed link of μ components. Let W^4 be the orientable 4-manifold with handle presentation $H^0 \cup \lambda H^1 \cup \mu H^2$, where μH^2 is attached along the framed w.

LEMMA 1. Let n be a system of framed meridians for w, then $(\lambda \# S^1 \times S^2; w \cup n)$ is a Heegaard diagram for $2W^4$.

Proof. $2W^4$ has a handle presentation $H^0 \cup \lambda H^1 \cup \mu H^2 \cup \mu H'^2 \cup \lambda H^3 \cup H^4$, obtained by doubling the presentation $W^4 = H^0 \cup \lambda H^1 \cup \mu H^2$. The cocore of H_i^2 is the attaching sphere of the corresponding $H_i'^2$, and is framed by the product structure of H_i^2. Isotoping this framed cocore out of H_i^2 it becomes the framed meridian n_i.

Corollary 3. $(S^3;$ 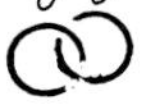

$)$ *represents* $2(S^2 \times D^2) \cong S^2 \times S^2$, *and* $(S^3;$ 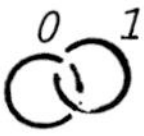

$)$ *represents* $2(\mathbb{C}P^2\text{-ball}) \cong \mathbb{C}P^2 \# -\mathbb{C}P^2 \cong S^2 \tilde{\times} S^2$. *(see footnote 4).*

The main idea for studying $2W^4$ in a purely four dimensional setting is illustrated by the first two examples.

Examples:

a) *Mazur manifolds* - $S^1 \times S^2 \cong \partial(S^1 \times B^3)$ is illustrated in Figure 3 by $(R^3 + \infty)$ - Int $B_1 \cup B_2$, identifying ∂B_1 with ∂B_2 by reflection in the mediatrix plane of the centers of B_1, B_2. The curve w is Mazur curve [6] (with arbitrary framing), and n is a framed meridian for w. By Lemma 1 $(S^1 \times S^2; w, n)$ is a Heegaard diagram for $2W^4$. Mazur proved $2W^4$ to be S^4 by looking at $2W^4$ as the boundary of $W^4 \times I$. We prove this in Figure 3 by using Heegaard moves, because the Heegaard diagrams $(S^1 \times S^2; w, n)$ and $(S^3; \phi)$, being equivalent, represent the same closed 4-manifold S^4.

b) In the example a) we have avoided Mazur phenomenon without passing to $W^4 \times I$. In this example we see that the framing

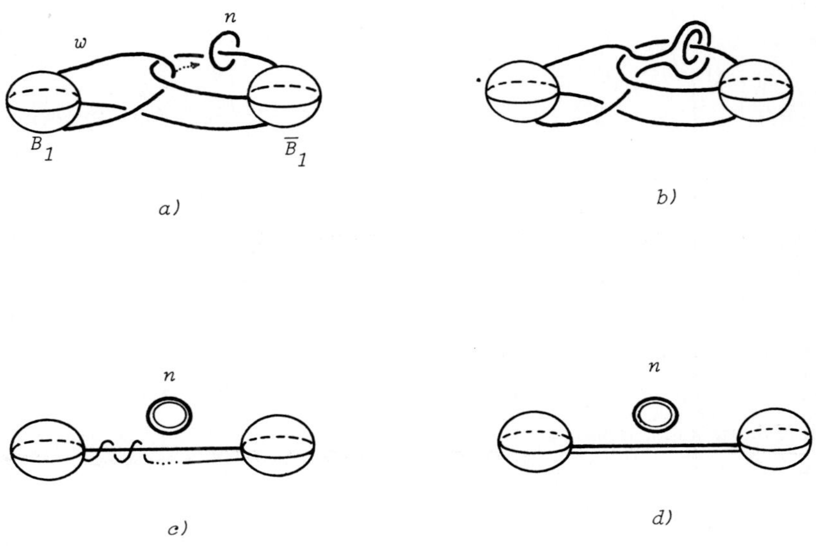

Fig. 3.

of a curve w with associated meridian n can be modified mod 2 (see Fig. 4.) This illustrates why framings in $\partial\ (\#\ S^1 \times B^4) \cong \partial(S^1 \times B^3 \times I)$ are in $\mathbb{Z}_2$.

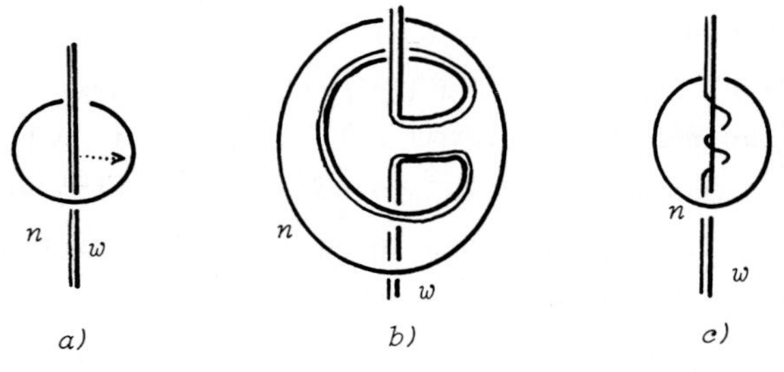

Fig. 4.

c) *Heegaard diagrams for* $F_g \times S^2$. Let F_g be a closed, orientable surface of genus g. Since $F_g \times S^2 = 2(F_g \times D^2)$ we can apply Lemma 1, after getting a diagram (see footnote 2) for $F_g \times D^2$.

A handle presentation $F_g = H^0 \cup 2g\, H^1 \cup H^2$ is shown in Figure 5a (for $g=2$). Hence Figure 5b is a diagram $(2g \# S^1 \times S^2; w)$ for $F_g \times D^2$. Thus, according to Lemma 1, $(2g \# S^1 \times S^2; w, n)$ is a Heegaard diagram for $F_g \times S^2$.

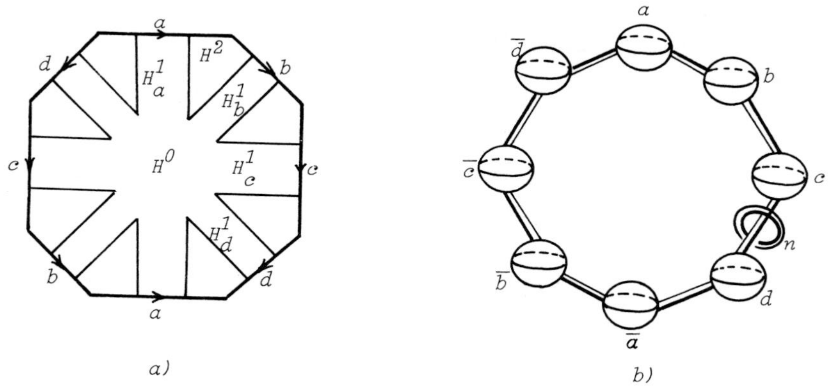

Fig. 5.

IX. HEEGAARD DIAGRAMS FOR MAPPING TORUS.

This construction is a generalization of [4].

A Heegaard diagram $(F_\lambda; v, w)$ for a closed, orientable 3-manifold M^3 is a scheme for a handlebody presentation $M^3 = H^0 \cup \lambda H^1 \cup \lambda H^2 \cup H^3$, where $\partial(H^0 \cup \lambda H^1) = F_\lambda$ and the system of belt-spheres (resp. attaching spheres) of λH^1 (resp. λH^2) is v (resp. w).

We can represent F_λ as the intersection of the (x,y)-plane with the model M_λ, for $\lambda \# S^1 \times S^2$, described in section 6.

The system v is represented by $\partial B \cap (x,y)$-plane, and w lies on F_λ. Figure 6 illustrates the case $M^3 = L(3,1)$.

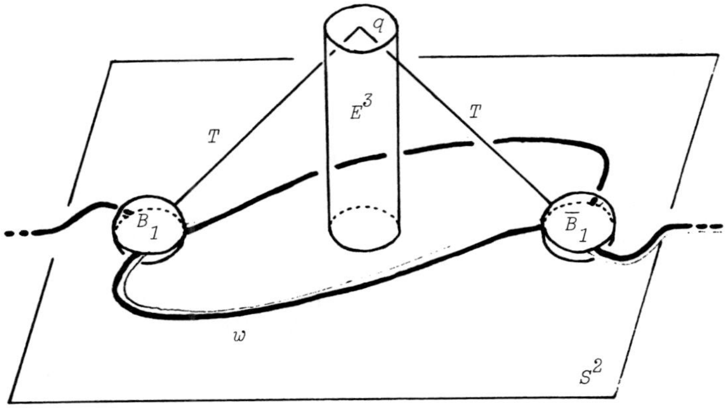

Fig. 6.

Let B^3 be a 3-ball in M^3. The punctured manifold $M_0^3 = M^3 - \text{int } B^3$ has the handlebody presentation $M_0^3 = H^0 \cup \lambda H^1 \cup \lambda H^2$. The following Lemma is then clear.

LEMMA 2. *$(\lambda \# S^1 \times S^2; w')$, where w' is the system w of $(F_\lambda; v, w)$ framed by parallel curves lying on F_λ, is a diagram (see footnote 2) for $M_0^3 \times [0,1]$.*

The boundary of $M_0 \times [0,1]$ is $M\#{-M}$, hence $(\lambda \# S^1 \times S^2; w)$ is also a surgery presentation for $M^3\#{-M^3}$. The operation # is performed along a 2-sphere S^2 which is F_λ, cut along w' and closed with 2λ discs going by the λ 2-handles. The natural homeomorphism $k : M_0^3 \times \{0\} \to M_0^3 \times \{1\}$ is generated by reflection in the (x,y)-plane. Clearly (see Diagram 1) $M^3 \times [0,1]$ is obtained by attaching a 3-handle to $M_0^3 \times [0,1]$ along S^2.

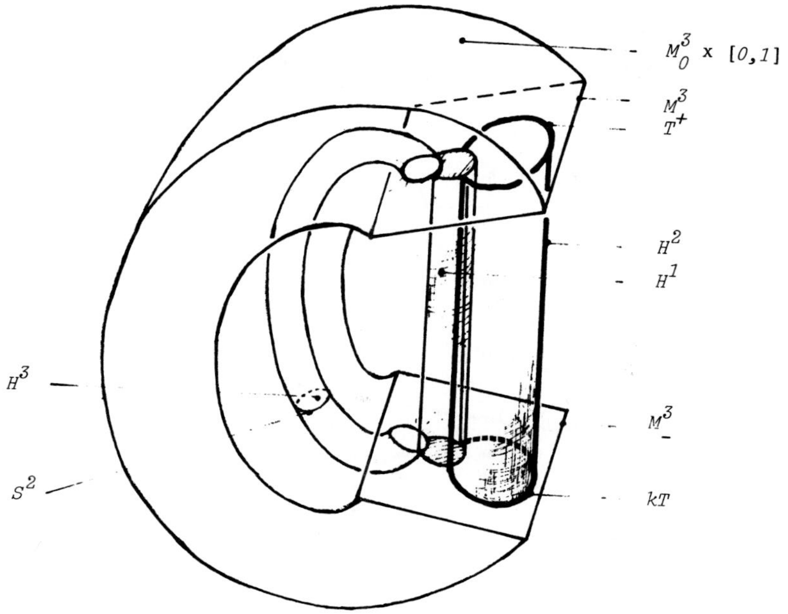

Diagram 1.

Let E^3 be a regular neighbourhood in $\{(x,y,z) | z \geq 0\}$ of a point in $F_\lambda - w$. Let T be the joint of $Q \in \text{Int } E^3$ with the 2λ north poles of the system B in the model M_λ (see Figure 6). Thus T is a 1-spine of the handlebody $M_\lambda \cap \{(x,y,z) | z \geq 0\}$.

If an orientation preserving autohomeomorphism of M^3 is given, we can suppose that, up to isotopic deformation, h is an autohomeomorphism of M_0^3 which is the identity in E^3 and also that $hT \subset M_\lambda \cap \{(x,y,z) | z \geq 0\}$.

We begin to build the mapping torus of h, $(M^3 \times [0,1])/(x = khx \ x \in M^3 \times \{0\})$, by taking a 1-handle in $M^3 \times [0,1]$ with attaching sphere $Q \cup kQ$, so that if $\alpha : \partial B^1 \times B^3 \to M^3 \times \{0,1\}$ is the attaching map, the composition

$$\alpha(\{0\} \times B^3) \xrightarrow{\alpha^{-1}} \{0\} \times B^3 \to \{1\} \times B^3 \xrightarrow{\alpha} \alpha(\{1\} \times B^3)$$

equals $k|\alpha(\{0\}\times B^3)$. The effect of this in the diagram is to take off balls C^3 and kC^3, centered at Q and kQ, respectively, and to identify their boundaries by k (see Figure 7).

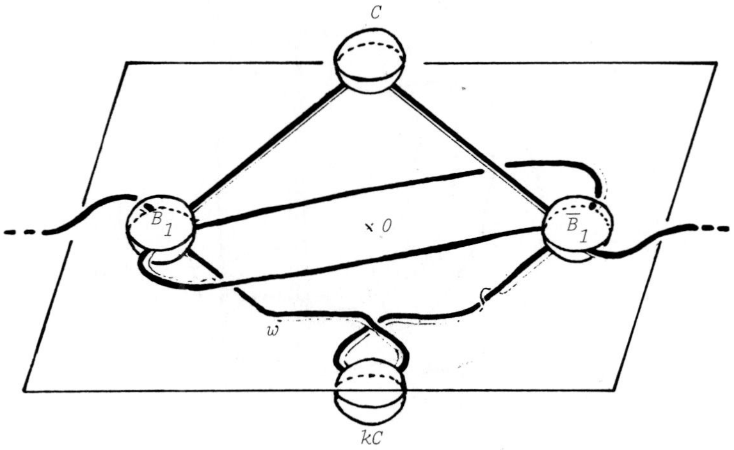

Fig. 7.

We continue the construction of the mapping torus by connecting T with khT, out of $C^3 \cup kC^3$, as suggested by Diagram 1, i.e. by adding λ 2-handles along the curves $(T \cup khT) \cap M_\lambda$, with framings invariants by kh. The effect in the diagram is illustrated in Figure 7.

Thus, the Heegaard diagram of the mapping torus is completed, because the result of connecting the rest of $M^3 \times \{0\}$ with the rest of $M^3 \times \{1\}$ is equivalent to take λ 3-handles and one 4-handle.

Example. Figure 7 represents a Heegaard diagram for the mapping torus of $h : L(3,1) \to L(3,1)$, h being the involution inducing the 2-fold cyclic covering $L(3,1) \to S^3$, branched over the trefoil knot (h is generated by a central reflection on the (x,y)-plane).

X. HEEGAARD DIAGRAMS FOR OPEN BOOKS WITH BINDING S^2

Let h be an orientation preserving autohomeomorphism of M_0^3 which is the identity in the boundary. Let V^4 be the mapping torus of h. Now, ∂V^4 has a natural product structure $\partial V^4 \cong S^1 \times S^2$, and we consider the closed 4-manifold W^4 (resp. $\tilde{W}^4$) which is obtained by pasting the boundaries of V^4 and $B^2 \times S^2$ together so that the product structures agree (resp. does not agree essentially). From the point of view of handle addition this can be done in two steps. In the first step, we add a 2-handle along the curve S^1 x point $\subset S^1 \times S^2 \cong \partial V^4$, and afterwards we add a 4-handle to close the manifold. The difference between W^4 and $\tilde{W}^4$ depends on the framing of the handle addition.

Now the naturally framed S^1 x point $\subset \partial V^4$ appears in the diagram for the mapping torus of h as the joint of two points in ∂C^3 with their images by k (See Diagram 1). The inclusion of this framed curve in the diagram for the mapping torus of h gives a Heegaard diagram for W^4. The diagram for $\tilde{W}^4$ is obtained by changing the framing of S^1 x point by a complete twist.

Example. The diagram of Figure 8a corresponds to an open book with leave $L(3,1)_0$ and monodromy h as in the example of section 9. This manifold is S^4 [5] because the binding is the 2-twist spun knot of the trefoil knot. A different proof is given in Figures 8b to 8d, using Heegaard moves.

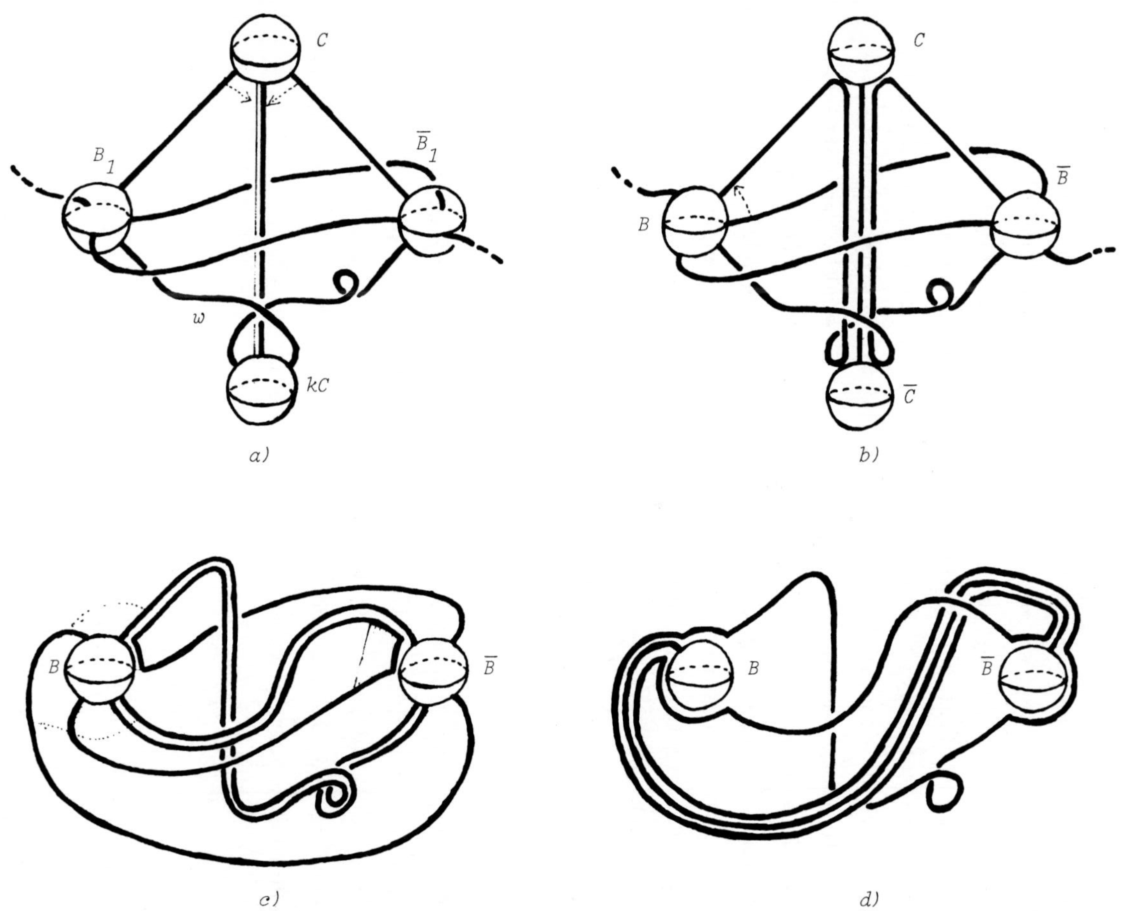
C
B_1
$\overline{B}_1$
w
kC
a)
C
B
$\overline{B}$
$\overline{C}$
b)
B
$\overline{B}$
c)
B
$\overline{B}$
d)

REFERENCES

1. Kirby, R., A calculus for framed links in S^3. (to appear).
2. Laudenbach, F., Sur les 2-spheres d'une variete de dimension 3. Annals of Math., 97 (1973), 57-81.
3. Laudenbach, F., and Poenaru, V., A note on 4-dimensional handlebodies, Bull. Soc. Math., France, 100 (1972), 337-347.
4. Akbulut, S., and Kirby, R., An exotic involution of S^4. (preprint)
5. Zeeman, E. C., Twisting spun knots. Trans. Amer. Soc. 115 (1965) 471-495.
6. Mazur, B., A note on some contractible 4-manifolds. Ann. of Math. 73 (1961) 221-228.

DECOMPOSITIONS OF E^3 WITH COUNTABLY MANY NON-DEGENERATE ELEMENTS

Michael Starbird[1]

Department of Mathematics
The University of Texas at Austin
Austin, Texas

Edythe P. Woodruff[2]

Department of Mathematics
Trenton State College
Trenton, Jew Jersey

Let G be an u.s.c. decomposition of E^3 into points and countably many tame cellular polyhedra. It is proved in this paper that E^3/G is homeomorphic to E^3. This theorem extends Bing's result [Ann. of Math. 65 (1957) 363-374] that E^3/G is homeomorphic to E^3 if G is an u.s.c. decomposition of E^3 into points and countably many tame arcs. The proof uses Woodruff's 2-sphere Theorem [Trans. Am. Math. Soc. 232 (1977), 195-204] which states that if an u.s.c. decomposition of E^3 satisfies the hypothesis that, given any non-degenerate element P_0 of G and open set U containing P_0, there is an open set V such that $P_0 \subset V \subset U$ and Bd V is a 2-sphere which misses the non-degenerate elements of G; then it follows that E^3/G is homeomorphic to E^3.

[1]*This research was supported in part by NSF Grant MCS 76-07242-A01.*

[2]*This reserach was supported in part by a Trenton State College Faculty and Institutional Research Award.*

ISBN 0-12-158860-2

Let G be an upper semi-continuous decomposition of E^3 into points and countably many tame cellular polyhedra. It is shown here that E^3/G is homeomorphic to E^3. This result extends Bing's theorem [2, Theorem 3] which states that E^3/G is homeomorphic to E^3 if G is an upper semi-continuous decomposition of E^3 into points and countably many tame arcs. (See [7] in these *Proceedings* for a brief survey of related results.)

Many techniques in this paper were first used by Woodruff in "Decompositions having a quasi-null collection of locally connected cellular nondegenerate elements", which was a preliminary report presented at the Louisiana State University Topology Conference in March, 1977.

The proof below makes use of Woodruff's 2-sphere Theorem [6]. An alternative proof of Theorem 1 below which does not make use of Woodruff's 2-sphere theorem can be found in [5, Theorem 4.1].

Definition. A subset X of E^3 is a *tame polyhedron* if and only if there is a homeomorphism h of E^3 to itself so that $h(X)$ is a rectilinear simplicial complex.

THEOREM 1. *Let G be an upper semi-continuous decomposition of E^3 into points and countably many tame, cellular polyhedra $\{P_i\}_{i\in\omega}$. Then E^3/G is homeomorphic to E^3.*

Proof. Using [3, Corollary to Theorem 5] repeatedly, one can find a homeomorphism of E^3 onto itself which makes each P_i *PL*, so we assume that each P_i is *PL*.

Let P_0 be an arbitrary non-degenerate element of G, and let U be a saturated open set containing P_0. The strategy of the proof of Theorem 1 is to find a sequence of maps $f_0, f_1, f_2, \ldots$ of the 2-sphere S into U so that $\lim_{i\to\infty} f_i$ is a monotone map f of S into $U - \bigcup_{i\in\omega} P_i$, and P_0 is contained in a bounded component of $E^3 - f(S)$. Using R.L. Moore's characterization of the 2-sphere

[4], one can find a subset of $f(S)$ which is homeomorphic to a 2-sphere and which contains P_0 in its bounded complementary domain. Then by Woodruff's 2-sphere Theorem, stated below, we deduce that E^3/G is homeomorphic to E^3.

WOODRUFF's 2-SPHERE THEOREM. [6] *Let G be an upper semi-continuous decomposition of* E^3. *If, given any non-degenerate element* P_0 *of G and open set U containing* P_0, *there is an open set V such that* $P_0 \subset V \subset U$ *and Bd V is a 2-sphere which misses the non-degenerate elements of G, then* E^3/G *is homeomorphic to* E^3.

The map f_0 is a *PL* embedding of the 2-sphere S into U so that $f_0(S) = S_0$ is a *PL* 2-sphere in general position with respect to each P_i and containing P_0 in the bounded component of $E^3 - S_0$. In general, f_{i+1} will be a map of S into U such that f_{i+1} agrees with f_i on all of S except for a finite number of disjoint disks on S. The maps f_i for $i > 0$ will not necessarily be homeomorphisms; however, for each i, $f_i(S)$ is a 2-sphere whose image contains P_0 in its interior. Sufficient control will be exercised in defining the f_i's so that $\lim_{i\to\infty} f_i$ is a monotone map f whose image is contained in U and so that P_0 is contained in a bounded component of $E^3 - f(S)$.

Finding nice neighborhoods for the P_i*'s.* It is possible to construct especially nice cellular neighborhoods for the P_i's. The following lemma specifies this niceness.

LEMMA 2. Let P be a cellular polyhedron in E^3 *and* Σ *be a PL 2-sphere in general position with respect to P. Then there is a sequence of PL 3-cells* $\{M_i\}_{i\in\omega}$ *each containing P in its interior such that*

(a) $M_{i+1} \subset \operatorname{Int} M_i$ *for each i,*

(b) $\bigcap_{i \in \omega} M_i = P$,

(c) $Bd\ M_0 \cap \Sigma$ *is a finite number of simple closed curves, and*

(d) *there is a homeomorphism* $h : (Bd\ M_0 \cap \Sigma) \times (0,1] \rightarrow (M_0 - P) \cap \Sigma$ *such that* $h((Bd\ M_0 \cap \Sigma) \times \{1/n\}) = h((Bd\ M_0 \cap \quad) \times (0,1 \quad) \cap Bd\ M_n$

Proof. Incorporate $P \cup \Sigma$ into a triangulation T of E^3. Let M_0 be the simplicial neighborhood of P with respect to $T^{(2)}$, the second barycentric subdivision of T. Let M_1 be the simplicial neighborhood of P in $T^{(4)}$. In general M_i is the simplicial neighborhood of P in $T^{(2i+2)}$. One can easily obtain the conclusions of this lemma from these neighborhoods of P.

In modifying the map f_i to obtain f_{i+1}, we will make use of nice neighborhoods of P_i. Suppose $f_i(S) = S_i$ is a *PL* 2-sphere in general position with respect to all the P_i's. We record these nice properties in the following lemma.

LEMMA 3. *There is a sequence of PL 3-cells* $\{M_j(P_i)\}_{j \in \omega}$ *containing* P_i *satisfying the conclusions of Lemma 2 with respect to the 2-sphere* S_i *and in addition satisfying the following conditions:*

(e) *for every* $j \in \omega$, $M_j(P_i)$ *is in general position with respect to each* P_k, $(k \neq i)$;

(f) *if for some* j *and* k, $P_k \cap Bd\ M_j(P_i) \neq \emptyset$, *then* $P_k \cap M_{j+1}(P_i) = \emptyset$; *and*

(g) *for no* k *does one component of* $P_k \cap S_i$ *meet two components of* $M_0(P_i) \cap S_i$.

Proof. Use the upper semi-continuity of the decomposition G to select, from a sequence of 3-cells which satisfy the conclusions of Lemma 2 and (e) above, a subsequence satisfying (f) as well. Note that (f) makes (g) automatically true for any P_k which intersects $M_1(P_i)$. Using part (d) of Lemma 2, we can renumber the subscripts and obtain conclusion (g) for $M_0(P_i)$. The homeomorphism h of Lemma 2 (d) must be redefined for the new numbering.

Obtaining f_{i+1} from f_i. Suppose the function $f_i : S \to E^3$ has been produced so that $f_i(S) = S_i$ is a *PL* 2-sphere in general position with respect to each P_j and so that f_i has only a finite number of non-degenerate point inverses each of which is a disk. Let $\{M_j(P_{i+1})\}_{j\in\omega}$ be a sequence of 3-cell neighborhoods of P_{i+1} satisfying the conclusions of Lemma 3. We choose $M_0(P_{i+1})$ to lie in such a small neighborhood of P_{i+1} that $P_0 \cap M_0(P_{i+1}) = \emptyset$ and that a further condition on the size of $M_0(P_{i+1})$, which will be specified in the next section, is also satisfied. In this section $M_j(P_{i+1})$ will be abbrieviated M_j.

Let E be a disk-with-holes component of S_i - Int M_0 such that $E \cup M_0$ separates P_0 from $E^3 - U$. Let $\{J_i\}_{i=1}^n$ be the simple closed curve components of $Bd\ E = M_0 \cap E$. By renumbering, if necessary, we suppose that J_1 bounds a disk D_1 on $Bd\ M_0$ such that $D_1 \cap (\bigcup_{i=2}^n J_i) = \emptyset$. Let J_1' and D_1' be the simple closed curve and disk, respectively, on $Bd\ M_1$ which correspond to J_1 and D_1 via the homeomorphism of Lemma 2 (d). We assume that J_1' does not meet a point with a non-degenerate inverse under f_i. Let $\tilde{D}_1$ be the disk on S bounded by $f_i^{-1}(J_1')$ which misses $f_i^{-1}(E^3 - M_0)$. Let C be a bicollar on $Bd\ \tilde{D}_1$ so that $f_i(C) \subset M_0$ and C misses the non-degenerate point inverses of f_i. Let $f_i^1 : S \to E^3$ be a map such that

$f_i^1|S - \tilde{D}_1 = f_i|S - \tilde{D}_1$,

$f_i^1(\tilde{D}_1) = D_1'$, and

$f_i^1|\tilde{D}_1$ has one non-degenerate point inverse, namely $Cl(\tilde{D}_1 - C)$.

The map f_i^1 is modified to obtain f_i^2. In general, if f_i^j has been defined, f_i^{j+1} is obtained as follows. Let J_{j+1} be a simple closed curve in $Bd\ M_0 \cap S_i$ so that J_{j+1} bounds a disk D_{j+1} on

$Bd\ M_0$ so that $D_{j+1} \cap (\bigcup_{i=j+2}^{n} J_i) = \emptyset$. Find J'_{j+1} and D'_{j+1} on $Bd\ M_{j+1}$ as before. Obtain $\tilde{D}_{j+1}$ on S as before and redefine f_i^j on $\tilde{D}_{j+1}$ to obtain f_i^{j+1}.

Let f_i^n be called f_{i+1}. While the construction of f_{i+1} is fresh in mind let us make several observations about it.

1) $f_{i+1}(S) = S_{i+1}$ is a *PL* 2-sphere which misses P_{i+1};

2) f_{i+1} has a finite number of non-degenerate point inverses each of which is a disk;

3) if $A_1 = f_i^{-1}(y)$ for some $y \in S_i$ and $A_2 = f_{i+1}^{-1}(z)$ for some $z \in S_{i+1}$, then either $A_1 \cap A_2 = \emptyset$ or $A_1 \subset \text{Int } A_2$;

4) if for some $x \in S$, $f_{i+1}(x) \neq f_i(x)$, then $f_{i+1}(x) \in M_1(P_{i+1})$;

5) if $x \notin \text{Int } f_{i+1}^{-1}(y)$ for any $y \in S_{i+1}$, then either $f_{i+1}(x) = f_i(x)$ or both $f_i(x)$ and $f_{i+1}(x)$ are in $M_0(P_{i+1})$; and

6) if C_1 and C_2 are two components of $S_{i+1} \cap M_0(P_{i+1})$ and P_j intersects both C_1 and C_2, then P_j intersects the corresponding components $\hat{C}_1$ and $\hat{C}_2$ of $S_i \cap M_0(P_{i+1})$. (By corresponding we mean that the boundary of the disk C_i $(i = 1,2)$ is a boundary component of $\hat{C}_i$ $(i = 1,2)$.)

All of these observations should be clear except perhaps (6). So we prove below that (6) is true.

Proof of (6). Suppose P_j meets C_1 and C_2 in points x and y, respectively. If $x \in \hat{C}_1$ and $y \in \hat{C}_2$, then P_j does meet $\hat{C}_1$ and $\hat{C}_2$ so there is no contradiction. If $x \notin \hat{C}_1$, then $x \in D'_k$ for some k. Recall that $D'_k \subset Bd\ M_k(P_{i+1})$. Now $y \in \hat{C}_2$, because if not, then $y \in D'_m$ for some $m \neq k$ in which case $y \in D'_m \subset Bd\ M_m(P_{i+1})$. But Conclusion (f) of Lemma 3 guarantees that no P_j intersects the boundaries of two M_k's . So $x \in D'_k$ and $y \in \hat{C}_2$. The choice of D'_k to be innermost among the curves that are left at that stage implies that $\hat{C}_1 \cap (M_{k-1} - M_{k+1})$ separates Int D'_k from

$\hat{C}_2 \cap (M_{k-1} - M_{k+1})$ in $M_{k-1} - M_{k+1}$. Hence the connected set P_j must intersect $\hat{C}_1$, proving (6). (See Figure 1.)

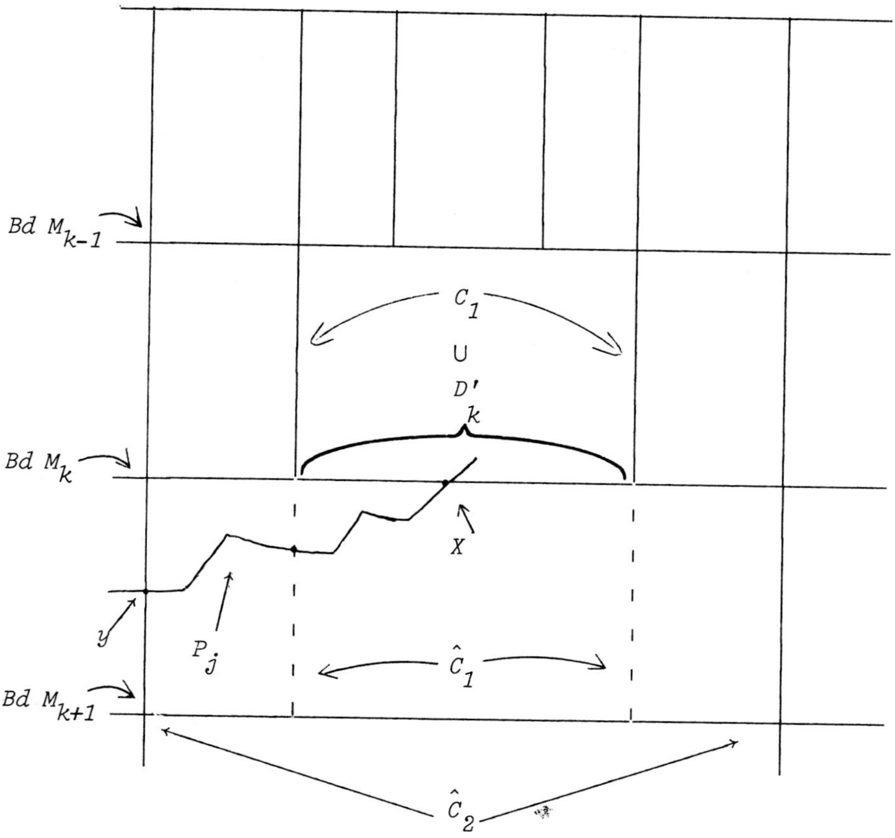

Figure 1

Restraining the size of $M_0(P_{i+1})$. For each i in ω, we describe a collection V_i of saturated open sets and then require $M_0(P_{i+1})$ to be in one of them, thus restraining its size.

Let $V_0 = \{U\}$. The ith collection V_i will be chosen after f_i has been defined. Recall that $f_i(S) = S_i$ is a *PL* 2-sphere.

There is a collection $V_i = \{V_{i,j}\}_{j \in \omega}$ of disjoint, saturated, connected open sets such that

(i) for each $j \in \omega$, there is a (unique) $k \in \omega$ such that $Cl(V_{i,j}) \subset V_{i-1,k}$;

(ii) if for some j, $S_i \cap P_j \neq \emptyset$, then $P_j \subset V_{i,k}$ for some k;

(iii) if $j \neq k$, then $Cl(V_{i,j}) \cap Cl(V_{i,k}) = \emptyset$;

(iv) if the diameter of $V_{i,j}$ is d, then there is a $P_k \subset V_{i,j}$ such that $S_i \cap P_k \neq \emptyset$ and the diameter of P_k is greater than $d/2$;

(v) $Cl(\bigcup_{j \in \omega} V_{i,j}) \cap (\bigcup_{j \leq i} P_j) = \emptyset$;

(vi) no two components of $S_i \cap M_0(P_i)$ meet the same component of $S_i \cap Cl(V_{i,j})$ for any j (Conclusion (g) of Lemma 3 was designed to make this property obtainable.);

(vii) if for some j and k, $Cl(V_{i,j}) \cap Bd\ M_k(P_i) \neq \emptyset$, then $Cl(V_{i,j}) \cap M_{k+1}(P_i) = \emptyset$ (Condition (f) of Lemma 3 makes this property obtainable.);

(viii) if no P_j intersects both of two components E_1 and E_2 of $M_0(P_i) \cap S_i$, then for no k does $Cl(V_{i,k})$ intersect both E_1 and E_2.

To produce V_i first consider the compact set $\bigcup\{P_k | k > i, P_k \cap S_i \neq \emptyset$, and $\text{diam}\, P_k \geq 1\}$. It has a cover by open sets satisfying (iii), (v), (vi), (vii), and (viii). The intersections of each of these with each element in V_{i-1} is a cover which also satisfies (i). Using methods in [1], we find a subcover by open sets which also satisfies (iv). To get saturation, we subtract from each component K of this cover the set $\bigcup\{P_k | P_k \cap Bd\ K \neq \emptyset\}$. Thus we have constructed a finite subcollection of V_i which satisfies all conditions except possibly (ii). By similar methods, we add elements so the cover is of $\bigcup\{P_k | k > i, P_k \cap S_i \neq \emptyset$, and $\text{diam}\ P_k \geq 1/2\}$. Continuing in this manner, we can produce the cover of all $\bigcup\{P_k | k > i$ and $P_k \cap S_i \neq \emptyset\}$.

The condition on the size of $M_0(P_{i+1})$ mentioned in "obtaining f_{i+1}" is simply that it lie wholly in a $V_{i,k}$ for some k. (Of course, if S_i misses P_{i+1} altogether we do not construct $M_0(P_{i+1})$ at all, but let $f_{i+1} = f_i$.)

This much completes the definitions of the maps $f_i : S \to E^3$. It remains to prove that $\lim_{i\to\infty} f_i = f$ exists and is continuous, that $f(S) \subset U$ and $f(S) \cap (\bigcup_{i\in\omega} P_i) \neq \emptyset$, that f is monotone, and that P_0 is contained in a bounded component of $E^3 - f(S)$.

Proof that $\lim_{i\to\infty} f_i = f$ *is a desired monotone map.* The first lemmas in this section are used in the proof of Proposition 9 below which states that $\lim_{i\to\infty} f_i$ exists and is continuous. The first two of these lemmas deal with the V_i's.

LEMMA 4. For each $\varepsilon > 0$ *and* $i \in \omega$, *there are only finitely many sets in the collection* V_i *with diameter greater than* ε.

Proof. The union of all polyhedra P_j which meet S_i and have diameter greater than or equal to $\varepsilon/2$ is a compact set. Hence a finite number of the $V_{i,k}$'s cover that set. By Property (iv) of V_i, every other $V_{i,k}$ has diameter less than ε.

LEMMA 5. For each $\varepsilon > 0$ *there are only finitely many* V_i*'s which contain a set with diameter greater than* ε.

Proof. The union of all polyhedra P_j with diameters greater than or equal to $\varepsilon/2$ is compact. Using Property (v), we see that $\{E^3 - Cl(\bigcup_{k\in\omega} V_{i,k})\}_{i\in\omega}$ is an open cover of this compact set. Find a finite subcover. Then for any i

where $E^3 - Cl(\bigcup_{k\in\omega} V_{i,k})$ is not in that finite subcover, V_i contains no set of diameter greater than ε by Property (iv) of V_i.

Definition. Let F be the set of points on S which are not interior to a point inverse for any f_i.

LEMMA 6. *Let $x \in F$. Then $f(x) = \lim_{i\to\infty} f_i(x)$ exists and $f_i(x) \in V_{i,j}$ for some j, if and only if $f(x) \in V_{i,j}$.*

Proof. If $f_i(x) \in V_{i,j}$ then by Observation (4) about f_{i+1} and the fact that $M_0(P_{i+1})$ was contained in some $V_{i,j}$, we know that $f_{i+1}(x) \in V_{i,j}$. Property (i) of the V_i's implies that if the maps $f_j(x)(j > i+1)$ are ever different from $f_{i+1}(x)$, each $f_j(x)$ lies in a single $V_{i+1,k}$. By Property (i) $Cl(V_{i+1,k}) \subset V_{i,j}$. So if $\lim_{i\to\infty} f_i(x)$ exists it lies in $V_{i,j}$. However, Lemma 5 guarantees that the limit does exist.

If $f_i(x) \notin V_{i,j}$, then for no $k > i$ is $f_k(x) \in V_{i,j}$. So $f(x) \notin V_{i,j}$. Thus Lemma 6 is proved.

LEMMA 7. *The map f equal to $\lim_{i\to\infty} f_i$ exists and is continuous on F.*

Proof. To see that f is continuous on F, let W be an open set in E^3 and $y \in W$. Then there are an ε so that $N_{2\varepsilon}(y) \subset W$ and an n, by Lemma 5, so that each $V_{n,j}$ has diameter less than ε. So the open set $A = N_\varepsilon(y) \cup \bigcup\{V_{n,j} | V_{n,j_1} \cap N_\varepsilon(y) \neq \emptyset\}$ is contained in W. But by Lemma 6, $f^{-1}(A) \cap F = f_n^{-1}(A) \cap F$. Since f_n is continuous, f is continuous on F.

LEMMA 8. *Let G be a component of $S - F$. Then $\lim_{i\to\infty} f_i(cl G)$ is a single point.*

Proof. Recall that for each i, f_i is a map with a finite number of nondegenerate point inverses each of which is a disk. Observation (3) about f_{i+1} says that if E_i is a point inverse of f_i and E_{i+1} is a point inverse of f_{i+1}, either $E_i \cap E_{i+1} = \emptyset$ or $E_i \subset \text{Int } E_{i+1}$. Therefore $G = \bigcup_{i\in\omega} \text{Int } E_i$ where, for each i, $E_i = f_i^{-1}(x)$ for some x and $\text{Int } E_i \subset \text{Int } E_{i+1}$.

Note that $Bd\ G$, being the lim sup of the $Bd\ E_i$'s, is a non-degenerate continuum. If only a finite number of the E_i's are different, the result is obvious, so we assume that there are an infinite number of different E_i's. Note that for each i and j, $f_i(Bd\ E_j)$ lies in one element of V_k or else belongs to $E^3 - \bigcup_{k\in\omega} V_{i,k}$ and is a single point. Observe that for each i, $f_i(Bd\ G)$ is non-degenerate. Therefore diam $f_i(Bd\ G) > \varepsilon$ for some $\varepsilon > 0$. Hence, there is an integer n so that $f_i(Bd\ E_j) > \varepsilon$ for all $j > n$, since f_i is continuous. Using the facts that each $f_i(Bd\ E_j)$ is wholly in one $V_{i,k}$, Lemma 4, and Property (iii) of V_i, we conclude that there is an integer m and set $V_{i,k}$ so that $f_i(Bd\ E_j) \subset V_{i,k}$ for each $j > m$. Therefore $f_i(Bd\ G) \subset Cl(V_{i,k})$, but since $f_{i+1}(Bd\ G) \subset V_{i+1,k}$ for some k, then $f_i(Bd\ G) \subset V_{i,k}$. Also note that for $s > m$, $f_s(Cl\ G) \subset V_{i,k}$. Since by Lemma 5 the diameters of the $V_{i,k}$'s go to 0 as i goes to infinity we conclude that $\lim_{i\to\infty} f_i(Cl\ G)$ is a single point.

PROPOSITION 9. *The map* $\lim_{i\to\infty} f_i = f$ *exists and is continuous.*

Proof. Let W be an open set in E^3 and $x \in S$ such that $f(x) \in W$. If $x \in S - F$, the result follows from Lemma 8 since the component of $S - F$ containing x is mapped to a single point. If $x \in F$, by Lemma 7 there is an $\varepsilon > 0$ such that $f(N_\varepsilon(x) \cap F) \subset W$. But Lemma 8 implies that $f(N_\varepsilon(x)) = f(N_\varepsilon(x) \cap F)$. So the Proposition is proved.

PROPOSITION 10. The set $f(S)$ does not intersect $\bigcup_{i \in \omega} P_i$.

Proof. Property (v) of the V_i's guarantees this fact.

The next lemmas will be used in proving that f is monotone. Recall that F is the set of all points of S which are not interior to a point inverse for any f_i.

LEMMA 11. If $x, y \in F$ and $f(x) = f(y)$, then for each i for which $f_{i+1}(x) \neq f_i(x)$, $f_{i+1}(x)$ and $f_{i+1}(y)$ lie on the same component of $M_0(P_{i+1}) \cap S_{i+1}$.

Proof. By Observation (4) about f_{i+1}, we know that $f_{i+1}(x)$ lies in $M_1(P_{i+1})$. Let C_1 be the component of $M_0(P_{i+1}) \cap S_{i+1}$ on which $f_{i+1}(x)$ lies.

Case 1. Suppose $f_{i+1}(y) \notin M_0(P_{i+1})$. Then Property (vii) of V_{i+1} would guarantee that $f_{i+1}(x)$ and $f_{i+1}(y)$ are not in the same $V_{i+1,j}$. (Recall that V_{i+1} is defined after f_{i+1} has been defined.) But then $f(x) \neq f(y)$ by Lemma 6.

Case 2. Suppose $f_{i+1}(y)$ belongs to a component C_2, not equal to C_1, of $M_0(P_{i+1}) \cap S_{i+1}$. There are now two possibilities.

Possibility 1. Suppose $f_{i+1}(x)$ and $f_{i+1}(y)$ do not lie in the same $V_{i+1,k}$. Then $f(x) \neq f(y)$ by Lemma 6.

Possibility 2. Both $f_{i+1}(x)$ and $f_{i+1}(y)$ belong to $V_{i+1,j}$ for some j. Property (vi) of V_{i+1} guarantees that $f_{i+1}(x)$ and $f_{i+1}(y)$ lie on different components, D_1 and D_2 respectively, of $Cl(V_{i+1,j}) \cap S_{i+1}$. Let A be the set of all P_k's which intersect both D_1 and D_2. Observation (6) about the f_i's can be used to show that for any $m > i + 1$, the set of P_k's which intersect both $f_m \circ f_{i+1}^{-1}(D_1)$ and $f_m \circ f_{i+1}^{-1}(D_2)$ is a subset of A.

The union of the P_k's in A is a compact set; therefore, there is an integer n such that $f_n \circ f_{i+1}^{-1}(D_1)$ and $f_n \circ f_{i+1}^{-1}(D_2)$ do not intersect the same P_k. Since we are supposing that $f(x) = f(y)$, either $\{f_i(x)\}_{i\in\omega}$ or $\{f_i(y)\}_{i\in\omega}$ is infinite. Suppose $\{f_i(x)\}_{i\in\omega}$ is infinite. Let p be an integer greater than n such that $f_{p+1}(x) \neq f_p(x)$. Then $f_{p+1}(x) \in M_1(P_{p+1})$. If $f_{p+1}(y) \notin M_0(P_{p+1})$, then $f_{p+1}(x)$ and $f_{p+1}(y)$ do not lie in the same $V_{p+1,j}$ and hence $f(x) \neq f(y)$. If $f_{p+1}(y) \in M_0(P_{p+1})$, then $f_{p+1}(y)$ belongs to a component E_2 of $M_0(P_{p+1}) \cap S_{p+1}$ while $f_{p+1}(x)$ belongs to a component E_1 of $M_0(P_{p+1}) \cap S_{p+1}$. Note that $E_1 \subset f_{p+1} \circ f_{i+1}^{-1}(D_1)$ and $E_2 \subset f_{p+1} \circ f_{i+1}^{-1}(D_2)$. Since $p > n$, no P_k intersects both E_1 and E_2. But then by Property (viii) of V_{p+1}, $f_{p+1}(x)$ does not lie in the same $V_{p+1,j}$ as $f_{p+1}(y)$. Hence by Lemma 6, $f(x) \neq f(y)$ and the lemma is proved.

PROPOSITION 12. The map f is monotone.

Proof. Proposition 12 is equivalent to the statement that if $f(x) = f(y)$, then there is a continuum C in S so that $x, y \in C$ and $f(C) = f(x) = f(y)$. First note that if Proposition 12 were false there would be points x and y in F which witness its falseness. This assertion is true since if x and y were any points witnessing the falseness and x were in a component G of $S - F$, then any point x' on $Bd\, G$ with y would equally witness the falseness since $Cl(G)$ is connected and $f(Cl(G)) = f(x)$.

Suppose then that $x, y \in F$ and $f(x) = f(y)$. Suppose $\{f_i(x)\}_{i\in\omega}$ is infinite. By Lemma 11, for each i such that $f_i(x) \neq f_{i+1}(x)$, $f_{i+1}(x)$ and $f_{i+1}(y)$ belong to the same component B_{i+1} of $M_0(P_{i+1}) \cap S_{i+1}$. Let B be the lim sup of $\{\tilde{B}_{i+1} = f_{i+1}^{-1}(B_{i+1}) \mid f_i(x) \neq f_{i+1}(x)\}$. Then each $\tilde{B}_{i+1}$ is a continuum containing x and y, so B is a continuum containing x and

y. Since as $i \to \infty$, diam $B_{i+1} \to 0$, $f(B) = f(x) = f(y)$ and the proposition is proved.

PROPOSITION 13. The set P_0 is contained in a bounded component of $E^3 - f(S)$.

Proof. This proposition is true since for each $i \in \omega$, P_0 is contained in the bounded component of $E^3 - f_i(S)$.

These Propositions prove that $f(S)$ is a cactoid which contains P_0 in a bounded component of its complement. Moore's Theorem [4] implies that a subset of it is a 2-sphere and Woodruff's Theorem [6] implies that E^3/G is homeomorphic to E^3 as claimed.

REFERENCES

1. S. Armentrout, On embedding decompositions spaces of E^n in E^{n+1}, Fund. Math. 61 (1967), 1-21.
2. R.H. Bing, Upper semi-continuous decompositions of E^3, Ann. of Math. 65(1957), 363-374.
3. R.H. Bing, An alternative proof that 3-manifolds can be triangulated, Ann. of Math. 69 (1959), 37-65.
4. R.L. Moore, Concerning upper semi-continuous collections of continua, Trans. Amer. Math. Soc. 27 (1925), 416-428.
5. M. Starbird, Cell-like, 0-dimensional decompositions of E^3, (preprint).
6. E.P. Woodruff, Decomposition spaces having arbitrarily small neighborhoods with 2-sphere boundaries, Trans. Amer. Math. Soc. 232 (1977), 195-204.
7. E. P. Woodruff, Cellular Decompositions of E^3, (These *Proceedings*).

DECOMPOSITIONS OF E^3 INTO CELLULAR SETS

Edythe P. Woodruff[1]

Department of Mathematics
Trenton State College
Trenton, New Jersey

This paper is a brief survey of results for countable, cellular u.s.c. decompositions of E^3 which are topologically E^3.

It is well known that cellularity of each nondegenerate element in a decomposition G of E^2 is a sufficient condition for $E^2/G \approx E^2$ (See Moore [12].), but that this condition on a decomposition G of E^3 is not sufficient for $E^3/G \approx E^3$ (See Bing [5] for the first such example.) Since countability is a useful hypothesis for many arguments, it is a natural one to add. All results discussed in this paper assume the decomposition has a countable number of nondegenerate elements.

Bing has put some limits on results for countable, cellular decompositions of E^3. Among the added hypotheses which he showed in [4] do give the homeomorphism are the conditions that each nondegenerate element be a tame arc or a starlike set. On the other hand, in [6] he produced a counterexample to $E^3/G \approx E^3$ in which each nondegenerate element in G, although not an arc, is a cellular set and lies in one of two given planes. In unpublished work announced in the Notices [10] , Gillman and

[1]*This research was supported in part by a Trenton State College Faculty and Institutional Research Award.*

ISBN 0-12-158860-2

Martin sharpened the results for arcs. They state that if each arc in the countable set of cellular arcs is locally tame except at perhaps a finite number of points, then the decomposition space is topologically E^3; but that there is an example of a decomposition G into points and countably many wild cellular arcs such that $E^3/G \not\approx E^3$. A new similar counterexample by Bing and Starbird appears in these *Proceedings* [7] .

Following Gillman and Martin's announcement, Armentrout [1, 2] suggested that one consider separately the three problems concerning decompositions into points and a countable number of tame cellular sets each of which is a (1) 3-cell; (2) 2-cell; or (3) one of certain trees, or more generally dendrites.

Schurle and Snay attacked the third question. Schurle [13] proved the homeomorphism for a decomposition into points and a countable set of trees, each consisting of finitely many tame arcs. (Recall that a tree is a continuum which is the union of a finite number of arcs and contains no simple closed curve. A dendrite is a locally connected continuum which contains no simple closed curve.) Snay [15] was able to generalize the theorem to a countable number of dendrites, which were required to be "flexible". Both Schurle's and Snay's hypotheses allow certain examples involving wild sets.

When the nondegenerate elements are not 1-dimensional, it has been helpful to assume that they form a null collection. (A null collection is one such that, given $\varepsilon > 0$, there are only a finite number of elements with diameter greater than ε.) Meyer [11] proved that the decomposition G yields E^3 if G consists of one point sets and a null collection of tame 3-cells, and Bean [3] generalized this to a null sequence of tame sets which are star-like-equivalent.

A useful condition to impose on a cellular set X is that there exist a map $f : S^2 \times I \to E^3$ such that f is a homeomorphism of $S^2 \times [0, 1)$ into $E^3 - X$, $f(S^2 \times 1) \subset X$, and $f(S^2 \times I) \cup X$ contains an open neighborhood of X. Gerlach [9] calls such a set X a "fibered cellular set", and proves that a decomposition G into points and a null collection of fibered cellular sets gives the results that $E^3/G \approx E^3$. Schurle [14] calls such sets "strongly cellular" and proves that the decomposition space for points and a countable number of strongly cellular sets is the same as the decomposition space for points and a countable number of tame 3-cells. Below we refer to such sets as "cellular sets which have a mapping cylinder neighborhood".

The following theorem was proved independently and simultaneously by Starbird and Woodruff [17], and by Everett [8]. These proofs are the next two papers in these *Proceedings*.

THEOREM 1. Suppose that G is an u.s.c. decomposition of E^3, the collection of nondegenerate elements is countable, and each one is cellular and polyhedral. Then $E^3/G \approx E^3$.

The impetus for the simultaneous proofs is Woodruff's 2-sphere Theorem [18].

More recently, Starbird [16] found an alternate proof; and John Walsh, using Starbird's methods, proved the following result. Theorem 2 also follows from Theorem 1 and Schurle's work cited above.

THEOREM 2. Let G be an u.s.c. cellular decomposition of E^3 with countably many nondegenerate elements. If each element of G has a mapping cylinder neighborhood in E^3, then $E^3/G \approx E^3$.

Armentrout's three questions are now fully answered, since Theorem 1 answers the first two and Theorem 2 and Snay's work both answer the third question.

Some appropriate questions now are the following.

Question 1. Can Theorem 1 or 2 be extended to higher dimensions?

Question 2. In Theorem 2 how can the mapping cylinder neighborhood hypothesis be weakened and still yield $E^3/G \approx E^3$?

Question 3. Is there a counterexample to $E^3/G \approx E^3$ in which the decomposition G consists of points and countably many cellularly embedded 2-cells? 3-cells?

REFERENCES

1. Steve Armentrout, Monotone decompositions of E^3, Ann. of Math. Studies 60 (1966), 1-25.
2. Steve Armentrout, A survey of results on decompositions, Proc. Univ. of Okla. Topology Conf. 1972, Dept. of Math., Univ. of Okla., Norman, Okla..
3. Ralph J. Bean, Decompositions of E^3 with a null sequence of starlike equivalent non-degenerate elements are E^3, Ill. J. Math. 11, (1968), 21-23.
4. R. H. Bing, Upper semicontinuous decompositions of E^3, Ann. of Math. 65 (1957), 363-374.
5. R. H. Bing, A decomposition of E^3 into points and tame arcs such that the decomposition space is topologically different from E^3, Ann. of Math. 65 (1957), 484-500.
6. R. H. Bing, Point-like decompositions of E^3, Fund. Math. 50 (1962), 431-453.
7. R. H. Bing and Michael Starbird, A decomposition of S^3 with a null sequence of cellular arcs, (These *Proceedings*).
8. Dan Everett, Shrinking countable decompositions of E^3 into points and tame cells, (These *Proceedings*).
9. Mary Ann Davis Gerlach, Some fibered cellular decompositions of E^3 give E^3, Doctoral Dissertation, Univ. of Wisc., 1972.
10. David S. Gillman and James M. Martin, Countable decompositions of E^3 into points and point-like arcs, Notices Amer. Math. Soc. 10 (1963), 74.
11. Donald V. Meyer, A decomposition of E^3 into points and a null family of tame 3-cells is E^3, Ann. of Math. 78 (1963), 600-604.
12. R. L. Moore, Concerning upper semi-continuous collections of continua, Trans. Amer. Math. Soc. 27 (1925), 416-428.
13. Arlo W. Schurle, Decompositions of E^3 into points and countably many trees, Bull. Amer. Math. Soc. 75 (1969), 422-425.
14. Arlo W. Schurle, Decompositions of E^3 into strongly cellular sets, Ill. J. of Math. 18 (1974), 160-164.
15. Richard A. Snay, Decompositions of E^3 into points and countably many flexible dendrities, Pac. J. of Math. 48 (1973), 503-509.

16. Michael Starbird, Cell-like 0-dimensional decompositions of E^3, (to appear).
17. Michael Starbird and Edythe P. Woodruff, Decompositions of E^3 with countably many nondegenerate elements (These *Proceedings*).
18. Edythe P. Woodruff, Decomposition spaces having arbitrarily small neighborhoods with 2-sphere boundaries, Trans. Amer. Math. Soc. 232 (1977), 195-204.

TOPOLOGY OF MANIFOLDS

THE STRUCTURE OF GENERALIZED MANIFOLDS HAVING NONMANIFOLD SET OF TRIVIAL DIMENSION

J. W. Cannon[1]

Department of Mathematics
University of Wisconsin
Madison, Wisconsin

J. L. Bryant
R. C. Lacher

Department of Mathematics
Florida State University
Tallahassee, Florida

Dedicated to R. L. Wilder in honor of his seventieth birthday.

Let Y denote a generalized manifold of dimension $n \geq 5$, and suppose that the nonmanifold set $N(Y)$ of Y has dimension k in the trivial range $2k + 2 \leq n$. Then we prove that Y is a cell-like decomposition of a manifold and that Y is a manifold if and only if it has the DISJOINT DISK PROPERTY (≡ singular 2-disks f, $g : B^2 \to Y$ can be approximated by disjoint singular 2-disks f',

[1]*Bryant spoke on these topics (as joint work with Lacher) on the first day of the conference, after which Cannon showed the other two authors his manuscript containing their results. This paper is essentially that manuscript of Cannon. The techniques used by Bryant and Lacher were surprisingly similar to those herein.*

Each of the authors was supported in part by the National Science Foundation.

ISBN 0-12-158860-2

$g' : B^2 \to Y$ ($Im\ f' \cap Im\ g' = \emptyset$)). If a certain 1-LCC conjecture is true, then our results can be extended to cover the codimension three case $n - k \geq 3$. Our techniques supply an alternative to surgery for proving that homology n-spheres, $n \geq 4$, bound contractible $(n + 1)$ - manifolds; and they suggest an approach to the problem of determining which homology 3-spheres bound contractible 4-manifolds.

I. INTRODUCTION

Let Y denote a generalized manifold of dimension $n \geq 5$, and suppose the nonmanifold set $N(Y)$ of Y had dimension $k \leq n - 3$ (definitions appear in Section 2). Then we show how to detect homologically a sequence of (pinched) crumpled n-cells in Y that capture the homotopic nontriviality of the embedding $N(Y) \subset Y$. If we replace each of the crumpled n-cells by a (pinched) real n-cell, we obtain a generalized n-manifold X, such that the embedding $N(X) \subset X$ is 1-LCC, and we obtain a proper cell-like map $h : X \to Y$ from X onto Y which maps $N(X)$ injectively into $N(Y)$.

CONJECTURE. Let X denote a generalized manifold of dimension $n \geq 5$, and suppose the nonmanifold set $N(X)$ of X has dimension $k \leq n - 3$. If the embedding $N(X) \subset X$ is 1-LCC, then X is a topological manifold.

We shall prove the conjecture in the trivial range $2k + 2 \leq n$. As a corollary we obtain the following theorem which generalizes the theorem of Bryant and Lacher for $k = 0$ [6].

THEOREM. Let Y denote a generalized manifold of dimension $n \geq 5$, and suppose the nonmanifold set $N(Y)$ of Y has dimension k in the trivial range $2k + 2 \leq n$. They Y is a cell-like decomposition space of an n-manifold-without-boundary.

Applying the results of [11], we obtain the following corollary.

COROLLARY. Let Y be as in the theorem. Then Y is a manifold if and only if Y satisfies the disjoint disk property.

Definition. A space Y satisfies the disjoint disk property provided maps $f, g : B^2 \to Y$ can be approximated by maps f', $g' : B^2 \to Y$ satisfying $Im\ f' \cap Im\ g' = \emptyset$.

Our arguments prove the following related theorem.

THEOREM. Suppose Y is a generalized n-manifold such that the nonmanifold part N(Y) of Y lies in a closed subset C of Y that is a generalized (n - 1)-manifold. If $n \geq 4$ and if $C \times E^m$ is a manifold ($m \geq 1$), then $Y \times E^m$ is a manifold. If $n \geq 5$ and C is an (n - 1)-manifold, then Y is a cell-like decomposition of a manifold-without-boundary.

We remark that our proofs supply an alternative to surgery for proving that homology n-spheres, $n \geq 4$, bound contractible $(n + 1)$-manifolds. We suggest as an interesting problem that of using our techniques to determine, if possible, which homology 3-spheres bound contractible 4-manifolds.

Our methods also give new proofs of some of F. Tinsley's *1-LCC* taming theorems, the F. D. Ancel-J. W. Cannon locally flat approximation theorem for embedded codimension-one manifolds [13], and R. J. Daverman's high dimensional analogue [15] of the Hosay-Lininger Theorem [21, 26]. We shall discuss some of these applications in another paper.

The longest section of the paper is Section 6 and its length is due to the contortions we go through to improve our results from $2k + 3 \leq n$ to $2k + 2 \leq n$. Much of Section 6 can be omitted on a first reading.

Cannon intends to write another paper exposing the ideas of the proof in simple cases. The reader may well wish to study that paper first.

The sources of Cannon's ideas for this paper were primarily the proofs in [11]. Those ideas in turn were stimulated by R. D. Edwards' study of the double suspension problem [17].

We insert here some historical notes.

Generalized *homology* manifolds seem to have been first introduced and studied extensively by R. L. Wilder (see [32]) in his program of extending planar results such as the Jordan Curve Theorem and Schoenflies Theorem to higher dimensions. Wilder mentions that he often discussed the desirability of such a program with his teacher R. L. Moore. However, Wilder also mentions that Moore always seemed somewhat disappointed at the highly algebraic turn taken by Wilder's study of manifolds. Our study of generalized manifolds, while using algebraic tools pioneered by Wilder, returns to very geometric results. In order to obtain geometric results we have been forced to use a more restrictive definition of generalized manifold than that employed by Wilder. The additional restrictions are that the homology manifolds discussed be homology manifolds over the integers, locally contractible, separable metric, and finite dimensional.

Generalized manifolds as geometric objects have arisen in at least three settings in recent years:

1) In taming theory [9, 10, 31].
2) In the double suspension problem [17, 18, 11].
3) In the resolution problem [5, 6, 10].

A good survey of related results and questions is R. C. Lacher's paper [23].

II. PREREQUISITES ON GENERALIZED MANIFOLDS

A generalized n-manifold is an ENR (Euclidean neighborhood retract ≡ retract of an open subset of some Euclidean space E^k) such that for each $x \in Y$, $H_*(Y, Y - \{x\}; \mathbb{Z}) \approx H_*(E^n, E^n - \{0\}; \mathbb{Z})$. The most important property of a generalized n-manifold Y is that duality is valid in Y:

THEOREM [1, 4, 32]. *A generalized n-manifold Y is locally orientable. If Y is orientable and (A,B) is a closed pair in Y, then there is a duality isomorphism*

$$H^*_C(A,B;\ \mathbb{Z}) \xrightarrow{D} H_{n-*}(Y-B,\ Y-A;\ \mathbb{Z})\ .$$

This duality is natural in the sense that if $(A,B) \subset (A', B')$, *then*

$$\begin{array}{ccc} H^*_C(A,B;\ \mathbb{Z}) & \xrightarrow{D} & H_{n-*}(Y-B,\ Y-A;\ \mathbb{Z}) \\ \uparrow & & \uparrow \\ H^*_C(A',B';\mathbb{Z}) & \xrightarrow{D} & H_{n-*}(Y-B',\ Y-A';\ \mathbb{Z}) \end{array}$$

commutes (vertical arrows induced by inclusions).

Generalized manifolds have very strong geometric properties as well. One may wish to consult, for example [10].

The set of those points of a generalized manifold M having locally euclidean neighborhoods is obviously open in M. The complement in M is a closed subset of M called the nonmanifold set of M.

Much of the argumentation in this paper relies on [11]. We recommend that the reader consult that paper at least generally before attempting to read this paper for further references, definitions, theorems, and conjectures.

III. DETECTING CRUMPLED n-CELLS

The crumpled n-cells we use are homeomorphic with what R. D. Daverman has called *inflations* [14] of certain standard 3-dimensional crumpled cubes. These crumpled n-cells arise as the natural compactifications of regular neighborhoods of certain infinite 2-complexes. We describe the 2-complexes as follows.

The basic unit is the disk-with-handles. A disk-with-handles is a compact, connected, orientable 2-manifold with a single boundary component $Bd\ D$. If D is, in fact, not a disk, then there exist simple closed curves A_1 and B_1 in Int D meeting transversely in a single point; and the identification space $D/(A_1 \cup B_1)$ is a disk-with-handles having one handle fewer than D. An iteration yields disjoint pairs (A_1,B_1), ..., (A_k,B_k) of simple closed curves in Int D such that A_i meets B_i transversely in a single point and $D/\ A_1 \cup B_1, \ldots, A_k \cup B_k$ is a disk. The union $J = A_1 \cup B_1 \cup \ldots \cup A_k \cup B_k$ is called a *complete handle curve* for D.

Infinitely many disks-with-handles are sewn together in the following way to form what we shall call a (geometric) *infinite commutator*. Let D_0 be a disk-with-handles having prescribed complete handle curve $J_0 = A_1 \cup B_1 \cup \ldots \cup A_k \cup B_k$ as in the preceding paragraph. Let D_1 be the disjoint union of $2k$ disks-with-handles $D(A_1)$, $D(B_1)$, ..., $D(A_k)$, $D(B_k)$, one for each of the simple closed curves of J_0. Identify $Bd\ D(A_i)$ with A_i and $Bd\ D(B_i)$ with B_i, $i = 1, \ldots, k$. Iterate the procedure by choosing a prescribed complete handle curve J_1 for D_1, component by component, by letting D_2 be a disjoint union of disks-with-handles, one for each simple closed curve of J_1, and by identifying each boundary in D_2 with its correspoinding simple closed curve in J_1. After infinitely many iterations, one obtains an infinite 2-dimensional complex $D = D_0 \cup D_1 \cup D_2 \cup D_3 \cup \ldots$, $D_{i+1} \cap D_i = J_i = Bd\ D_{i+1}$, which we call an infinite commutator, with stages D_i and boundary $Bd\ D \equiv Bd\ D_0$. In [11] we defined a *generalized 2-disk* as the Freudenthal compactification D^+ of an infinite commutator D.

THEOREM 3.1. Let D be an infinite commutator, M a PL n-manifold-without-boundary, $n \geq 5$, $f : D \to M$ a closed PL embedding, and N a regular neighborhood of f(D) in M. Then the PL homeomorphism type of the pair (N,f(D)) is completely determined by the space D and the dimension n.

Proof. According to [25, Corollary 2, p. 220], the PL homeomorphism types of thickenings (regular neighborhoods) of D in an n-manifold, $n > 2 \cdot \dim D$, correspond bijectively to the elements of the homotopy set $[D, BPL]$, where BPL is the classifying space for stable PL microbundles. Thus is suffices to show that any two maps from D into BPL are homotopic.

The homotopy groups of BPL coincide with those of BO in low dimensions [24, Theorem (Hirsch-Maxur), p. 384]. Since BO is connected and $\pi_1(B)) \approx \mathbb{Z}_2$ and $\pi_2(B)) \approx \mathbb{Z}_2$ [27, pp. 214-216], it follows that BPL is also connected with $\pi_1(BPL) \approx \mathbb{Z}_2$ and $\pi_2(BPL) \approx \mathbb{Z}_2$. We complete the proof of our theorem by using these facts about the homotopy of BPL to contract any map $g : D \to BPL$ to a prescribed point $*$ of BPL.

Pick a base point $x \in Bd\ D_0$. Each disk-with-handles in each D_i, $i \geq 1$, has a natural base point, namely the single point where its boundary intersects another disk-with handles of D_i. Thus in D_i, $i \geq 1$, disks-with-handles share base points in pairs. Let the collection of all base points serve as the vertex set of a graph G in which vertices are joined by an edge if and only if they are distinct and the corresponding disks-with-handles intersect. We may realize G geometrically in D so that $G \cap (Bd\ D_0 \cup J_0 \cup J_1 \cup \ldots)$ is the set of base points. We form two subsidiary identification spaces $D' = D/G$ and $D'' = D/(G \cup J_0 \cup J_1 \cup \ldots)$ with identification maps $p : D \to D'$ and $q : D' \to D''$. We shall homotop g first so that it factors through p, then so that it factors through $q \circ p$. Finally, we show that if the adjustment of g has been carefully made, then the map $g \circ p^{-1} \circ q^{-1} : D' \to BPL$ is contractible.

Step 1. Factoring g through p. Since G is a contractible space, $g|G$ is a contractible map. Using the homotopy extension property, we may therefore homotop g to a map which takes G to $*$. We assume that the original g had this property. Then $g' = g \circ p^{-1}$: $(D', p(x)) \to (BPL, *)$ is a continuous function.

Step 2. Factoring $g' = g \circ p^{-1}$ through q. The sets $B = p(J_0 \cup J_1 \cup \ldots)$, $B_0 = pJ_0$, $B_1 = pJ_1, \ldots$ are bouquets of loops wedged at $p(x)$. Let A denote a simple closed curve in B_i, $i \geq 0$. Then A is nullhomologous in $p(D_{i+1})$. Thus $g'|A$ is nullhomogous in BPL. But $\pi_1(BPL) \approx \mathbb{Z}_2$ is abelian so that $H_1(BPL) = \pi_1(BPL)$. Thus $g'|A$ is nullhomotopic in BPL. By a technique to be expounded in the next step of this proof, we choose a particular contraction for $g'|A$ for each A in each B_i. Using the homotopy extension property, we use these contractions to homotop g' so that g' takes all of B to $*$ in BPL. We may assume that the original g' had this property so that $g'' = g' \circ q^{-1}$: $(D'', q \circ p(x)) \to (BPL, *)$ is a continuous function.

Step 3. Contracting g''. The space D'' is the wedge of a single disk $q \circ p(D_0)$ and infinitely many 2-spheres, one for each disk-with-handles of each D_i, $i \geq 1$. The map g'' is contractible if and only if, for each 2-sphere S of D'', $g''|S$ is contractible. We now indicate how the choices in Step 2 can be made so that each $g''|S$ is contractible.

Let A denote a simple closed curve in B_i, $i \geq 0$. Since $\pi_2(BPL) \approx \mathbb{Z}_2$, there are up to homotopy precisely two contractions of $g'|A$ in BPL. For each $i \geq 1$, let n_i denote the number of simple closed curves in B_i. Then each n_i-tuple of 0's and 1's represents a choice for each $A \subset B_i$ of one of the two possible contractions of A. Let C_i denote the set of all possible n_i-tuples of 0's and 1's. There exist natural functions

$C_0 \xleftarrow{f_1} C_1 \xleftarrow{f_2} C_2 \longleftarrow \ldots$ forming an inverse system of finite sets defined as follows: let α_i be an n_i-tuple of 0's and 1's, $i \geq 1$; contract the curves of B_i according to α_i; if A is a simple closed curve of B_{i-1}, then A bounds a disk-with-handles H of $p(D_i)$; the curves of $p(J_i) \cap H$ are curves of B_i contracted according to α_i; if this contraction is extended to H rel A, one obtains a new map $g'|H$ which factors through the identification $H \xrightarrow{\pi} H/(p(J_i) \cap H)$; but $H/(p(J_i) \cap H)$ is a disk bounded by A so that $g' \circ \pi^{-1} : H/(p(J_i) \cap H) \to BPL$ defines a contraction of A and hence the 0 or 1 corresponding to that contraction in the n_{i-1}-tuple $f_i(\alpha_i) \in C_{i-1}$. We choose an element $\alpha = (\alpha_0, \alpha_1, \ldots)$ of the (nonempty [29, Lemma, p. 13]) inverse limit of this system. If Step 2 is carried out using the contractions indicated by α, then each $g''|S$, hence g'', is contractible.

COROLLARY 3.2. *Let D be an infinite commutator, M a PL n-manifold, $n \geq 5$, $f : D \to M$ a closed PL embedding, and N a regular neighborhood of $f(D)$ in M. Then the Freudenthal compactification $(N^+, f(D)^+)$ of the pair (N,D) consists of a crumpled n-cell N^+, with wild set in $N^+ - N = f(D^+) - f(D)$, and a generalized 2-disk [11] $f(D^+)$, which is a strong deformation retraction of N^+.*

Proof. The argument of [11, Sections 5 and 6] may be reinterpreted as showing that the corollary is true for some embedding $f : D \to (E^n$ - Cantor Set$)$. Thus, by the theorem, the corollary is true for all closed embeddings. If we had recognized more clearly at the time what was happening, we might have quoted an argument of R. J. Daverman [14] in place of much of [11, Sections 5 and 6].

Definition. If D is an infinite commutator and $n \geq 5$, we use the symbol $C_n(D)$ to denote the crumpled n-cell associated with D by the corollary.

COROLLARY 3.3. *Let $C_n(D)$ be as in the preceding definition. Then there is a map $f : B^n \to C_n(D)$ from the standard n-cell B^n onto*

$C_n(D)$ such that the nondegenerate point preimages of f are the preimages of points of $D^+ - D \subset Bd\ C_n(D)$ and such that these preimages form a Cantor set of arcs from a (possibly wild) interior collar on $S^{n-1} = Bd\ B^n$.

Proof. Again this is a restatement of facts proved in [11, Sections 5 and 6].

THEOREM 3.4. *Suppose D is an infinite commutator, M is an n-manifold-without-boundary, $n \geq 5$, and $f : D \to M$ is a closed map. Then f can be approximated by a closed 1-LCC embedding.*

Proof. Since M satisfies the disjoint disk property locally (see [11]), it is easy to deduce Theorem 3.4 by the methods of [11, Section 3].

THEOREM 3.5. *Suppose D is an infinite commutator, M is an n-manifold-without-boundary, $n \geq 5$, and $f : D \to M$ is a closed 1-LCC embedding. Then $f(D)$ has a neighborhood N in M that admits a PL triangulation with $f(D)$ a subpolyhedron.*

Proof. By [7,8], $f(D)$ is locally tame. Thus it is easy to construct an open neighborhood N of $f(D)$ of which $f(D)$ is a strong deformation retraction. The obstruction to imposing a PL structure on N lies in $H^4(N; \mathbb{Z}_2) \approx H^4(D; \mathbb{Z}_2) = 0$ [22]. By [7,8], $f(D)$ can be ambiently isotoped to a PL embedding in N. The reverse of the isotopy deforms the PL structure on N so that $f(D)$ is a subpolyhedron.

THEOREM 3.6. *Suppose Y is a generalized n-manifold, and Z is a closed subset of Y of dimension $\leq n-3$. Then if $f : B^2 \to Y$ is a map and $\varepsilon > 0$, there is a generalized 2-disk D^+ and a map $g : D^+ \to N_\varepsilon(Im\ f)$ satisfying the following conditions:*

i) $g|BD\ D^+$ *is a loop within* ε *of* $f|BD\ B^2$, *and*
ii) $g^{-1}(Z) = D^+ - D$.

Proof. The proof is exactly that of [11, Theorem 3] except that, without the disjoint disk property, one cannot force g to be a 1-*LCC* embedding.

THEOREM 3.7. Suppose Y *is a generalized n-manifold, and* Z *is a closed subset of* Y *that is a generalized (n-1)-manifold which separates* Y *into disjoint open sets* U *and* V *of which* Z *is the common boundary. Then if* $f : B^2 \to Cl\ U$ *is a map and* $\varepsilon > 0$, *there is a generalized 2-disk* D^+ *and a map* $g : D^+ \to N_\varepsilon(\mathrm{Im}\ f) \cap Cl\ U$ *satisfying the following conditions:*
i) $g|BD\ D^+$ *is a loop within* ε *of* $f|BD\ B^2$, *and*
ii) $g^{-1}(Z) = D^+ - D$.

Proof. The pair $(Cl\ U, Z)$ satisfy the same homotopy and homology properties of the pair (Y,Z) of Theorem 3.6 used in the proof of Theorem 3.6. Thus the same argument proves Theorem 3.7.

IV. REPLACING CRUMPLED CELLS BY REAL CELLS

We consider a generalized n-manifold Y containing a null sequence of (pinched) crumpled n-cells and give a careful description of the process of replacing the crumpled n-cells by (pinched) real n-cells. Possibly the most subtle point is that of determining the appropriate topology for the newly constructed space. To make the choice of topology obvious we embed Y as a closed subset of some high dimensional Euclidean space E^k.

Setting:

Y, a generalized n-manifold embedded as a closed subspace of E^{2n+1};

$f_i : B_i \to C_i$ $(i = 1,2,3, \ldots)$, continuous surjective functions from n-cells B_i to crumpled n-cells C_i such that the nondegenerate point preimages of f_i are collar fibers of some (possibly wild) interior collar on $S_i = Bd\ B_i$;

$g_i : C_i \to Y$ $(i = 1,2,3, \ldots)$, continuous functions that embed Int C_1, Int C_2, Int $C_3, \ldots$ disjointly in Y as a null sequence of sets (necessarily open in Y).

Construction:

Consider $Y \subset E^{2n+1} = E^{2n+1} \times \{0\} \subset E^{2n+1} \times E^2 = E^{2n+3}$. Let $p_1, p_2, p_3, \ldots$ denote distinct points in the unit circle of E^2. Then $H_1 = E^{2n+1} \times [0,\infty) \cdot p_1$, $H_2 = E^{2n+1} \times [0,\infty) \cdot p_2$, ... are half $(2n + 2)$-spaces in E^{2n+3} intersecting only in their common boundary E^{2n+1}. Fix i. Then $g_i \circ f_i : B_i \to Y$ takes B_i into $E^{2n+1} = Bd\ H_i$. There is clearly a map $h_i : B_i \to H_i$ such that $h_i|$Int B_i is an embedding into Int $H_i = E^{2n+1} \times (0,\infty) \cdot p_i$, $h_i|Bd\ B_i = g_i \circ f_i$, and Diam $(Im\ h_i) < 2 \cdot$ Diam $(Im\ g_i \circ f_i) = 2 \cdot$ Diam $(Im\ g_i)$. We define X to be the subspace of E^{2n+3} given by $X = [Y - \cup g_i\ (\text{Int}\ C_i)] \cup [\cup h_i\ (\text{Int}\ B_i)]$.

THEOREM 4.1. The space X constructed above is a generalized n-manifold embedded as a closed subspace of E^{2n+3}.

Proof. X is closed in E^{2n+3}: This fact is obvious since $Im\ h_1$, $Im\ h_2$, ... form a null sequence of compact sets.

It remains to show that X is locally contractible and has the proper local homology. We first show that X and Y are nicely homotopy equivalent:

X and Y are homotopy equivalent rel $X \cap Y$:

Since f_i takes $Bd\ B_i$ homeomorphically onto $Bd\ C_i$, there is a map $e_i : C_i \to B_i$ with $e_i \circ f_i \mid Bd\ B_i = id$. Then $e_i \circ f_i \sim id$ rel $Bd\ B_i$ and $f_i \circ e_i \sim id$ rel $Bd\ C_i$ ($\sim$ denotes "homotopic to").

Let $B_i' = h_i B_i$, $C_i' = g_i C_i$, $f_i' = g_i \circ f_i \circ h_i^{-1}$, and $e_i' = h_i \circ e_i \circ g_i^{-1}$. Then $f_i' : B_i' \to C_i'$ and $e_i' : C_i' \to B_i'$ are continuous functions such that $e_i' \circ f_i' \simeq id$ rel $h_i\, Bd\, B_i$ and $f_i' \circ e_i' \sim id$ rel $g_i\, Bd\, C_j$.

Piecing the e_i' and f_i' together, we obtain homotopy equivalences $f' : X \to Y$ and $e' : Y \to X$ defined by the formulae

$$f'(x) = \begin{cases} x & \text{if } x \in X \cap Y \\ f_i'(x) & \text{if } x \in B_i' \end{cases}$$

$$e'(y) = \begin{cases} y & \text{if } y \in X \cap Y \\ e_i'(y) & \text{if } y \in C_i' \end{cases} .$$

Then $e' \circ f' \sim id$ rel $X \cap Y$, the homotopy taking each B_i' into itself. Also $f' \circ e' \sim id$ rel $X \cap Y$, C_i' being carried into C_i'.

X is locally contractible:

Suppose $x \in X$ and $\alpha > 0$ given. We must find an $\varepsilon > 0$ such that the ε-neighborhood $N_\varepsilon(x,X)$ of x in X contracts in the α-neighborhood $N_\alpha(x,X)$ of x in X. Since the individual sets $h_i(\text{Int } B_i)$ are open in X and locally contractible, it suffices to consider the case $x \in X \cap Y$.

Choose $\beta > 0$ such that $e'[N_\beta(x,Y)] \subset N_\alpha(x,X)$. Using the local contractibility of Y, choose $\gamma > 0$ such that there is a contraction $r : N_\gamma(x,Y) \times I \to N_\beta(x,Y)$, $r_0 = id$, $r_1 =$ constant. Choose $\delta > 0$ such that $f'[N_\delta(x,X)] \subset N_\gamma(x,Y)$. Since $id \sim e' \circ f'$ rel $X \cap Y$, there is a neighborhood $N_\varepsilon(x,X)$ in $N_\delta(x,X)$ such that $id \mid N_\varepsilon(x,X) \sim e' \circ f' | N_\varepsilon(x,X)$ in $N_\alpha(x,X)$. We may assume $\alpha > \beta > \gamma > \delta > \varepsilon > 0$.

Then, with all maps restricted to $N_\varepsilon(x,X)$, we find that $id \sim e' \circ f' = e' \circ r_0 \circ f' \sim e' \circ r_1 \circ f' =$ constant in $N_\alpha(x,X)$. This completes the proof that X is locally contractible.

X is an ENR (Euclidean neighborhood retract):

This is a well-known consequence of the fact that X is a locally contractible closed subset of E^{2n+3}.

Since X is an *ENR*, it makes sense to talk about cell-like sets in X. And any two homology theories we choose to consider will coincide for X.

The map $f' : X \to Y$ is a closed, surjective map whose point preimages are cell-like in X (in fact, contractible):

The nondegenerate point preimages of f' are unions of arcs arising as follows. Let i be a positive integer and examine the map $f_i : B_i \to C_i$. The nondegenerate point preimages of f_i are by hypothesis all fibers of an interior collar $Bd\ B_i \times [0,1]$ on $Bd\ B_i = Bd\ B_i \times \{0\}$. If $\{x\} \times [0,1]$ is one of the fibers collapsed by f_i, then $h_i(\{x\} \times [0,1])$ is an arc collapsed by f'.

There is a deformation of $\overset{\infty}{\cup} h_i(Bd\ B_i \times [0,1])$ which starts at the identity and contracts each fiber $h_i(\{x\} \times [0,1])$ uniformly toward $h_i(x,0)$. This deformation contracts simultaneously each of the point preimages of f'.

$H_*(X, X-\{x\}; \mathbb{Z}) \cong H_*(E^n, E^n - \{0\}; \mathbb{Z})$ for each $x \in X$:

We prove this homology fact in two steps. We clearly need check only the case $x \in X \cap Y$.

Step 1. Examine the commutative diagram ($\mathbb{Z}$ coefficients understood)

$$\begin{array}{ccccccccc}
\cdots\to H_k(X-(f')^{-1}f'(x)) & \to & H_k(X) & \to & H_k(X,X-(f')^{-1}f'(x)) & \to & H_{k-1}(X-(f')^{-1}f'(x)) & \to & H_{k-1}(X)\to\cdots \\
f'_*\downarrow & & \downarrow f'_* & & \downarrow f'_* & & \downarrow f'_* & & \downarrow f'_* \\
\cdots\to H_k(Y-f'(x)) & \to & H_k(Y) & \to & H_k(Y,Y-f'(x)) & \to & H_{k-1}(Y-f'(x)) & \to & H_{k-1}(Y)\to\cdots
\end{array}$$

By the Vietoris-Begle mapping theorem (easily provable in our case by means of the Approximation Theorem of [9, 14. Approximation Theorem, p. 70]), all except the center vertical arrow are isomorphisms since f' is cell-like. Thus the center arrow is an isomorphism by the five lemma.

Step 2. Examine the commutative diagram ($\mathbb{Z}$ coefficients understood)

$$\begin{array}{ccccccccc}
\cdots \tilde{H}_k(X-\{x\}) & \to & \tilde{H}_k(X) & \to & H_k(X,X-\{x\}) & \to & \tilde{H}_{k-1}(X-\{x\}) & \to & \tilde{H}_{k-1}(X)\to\cdots \\
\uparrow\alpha & & \uparrow\beta & & \uparrow\gamma & & \uparrow\delta & & \uparrow\varepsilon \\
\cdots \tilde{H}_k(X-(f')^{-1}f'(x)) & \to & \tilde{H}_k(X) & \to & H_k(X,X-(f')^{-1}f'(x)) & \to & \tilde{H}_{k-1}(X-(f')^{-1}f'(x)) & \to & \tilde{H}_{k-1}(X)\to\cdots
\end{array}$$

We wish to prove the central vertical arrow γ is an isomorphism by proving that the first and fourth vertical arrows α and δ are isomorphisms. We consider only α.

(1) α is monomorphic: Let Z denote a reduced k-cycle in $X - (f')^{-1}f'(x)$ which bounds a $k + 1$ chain B in $X - \{x\}$. Only finitely many of the sets $Im\ h_j$ can hit both x and B. Let $Im\ h_i$ be one of them. Let $Bd\ B_i \times [0,1]$ be an interior collar on $Bd\ B_i$ such that the nondegenerate point preimages of $f_i : B_i \to C_i$ are fibers of this collar. Let X_i denote the compact subset of $Bd\ B_i$ mapped by h_i to x. Without changing Z and without changing $B - h_i(\text{Int } B_i)$, we shall change B in $h_i(\text{Int } B_i)$ so that the new B misses $h_i(X_i \times [0,1])$. A finite iteration will then show that Z bounds in $X - (f')^{-1}f'(x)$.

Subdividing B if necessary, we find that B is the sum of two chains B' and B'', where B' is disjoint from Z and lies in $h_i(\text{Int } B_i)$ while B'' misses $h_i(X_i \times [0,1])$. Thus $\partial B'$ is a reduced k-cycle in $h_i(\text{Int } B_i - X_i \times (0,1])$. Assume that B_i is a standard n-cell in S^n. Then

$$\tilde{H}_k[h_i(\text{Int } B_i - X_i \times (0,1])]$$

$$\approx \tilde{H}_k \{S^n - [(S^n - \text{Int } B_i) \cup (X_i \times (0,1]))\}$$

$$= 0$$

since $(S^n - \text{Int } B_i) \cup (X_i \times [0,1])$ is contractible. We conclude that $\partial B' = \partial B'''$ for some $(k+1)$-chain in h_i (Int $B_i - X_i \times (0,1]$). Thus $B' + B'''$ is the desired alteration of B. We conclude that Z bounds in $X - (f')^{-1}f'(x)$ and that α is monomorphic.

(2) α is epimorphic: Let Z denote a reduced k-cycle in $X - \{x\}$. Suppose $Im\ h_i$ hits both x and Z. As in (1), $Z = B' + B''$, B' in h_i (Int B_i) and B'' missing h_i $(X_i \times [0,1])$. As in (1), $\partial B'$ bounds B''' in h_i (Int $B_i - X_i \times (0,1]$). Since H_k $(h_i(\text{Int } B_i)) = 0$, the cycle $B' - B'''$ is the boundary of a $(k + 1)$-chain K in h_i (Int $B_i) \subset X - \{x\}$. Thus $Z - (B''' + B'') =$ $(B' + B'') = B' - B''' = \partial K$ so that Z is homologous in $X - \{x\}$ to a cycle which misses h_i $(X_i \times [0,1])$. A finite iteration moves Z off all of $(f')^{-1}f'(x)$ so that α is epimorphic.

Arguments like (1) and (2) prove that

$$\gamma : H_k(X,X-(f')^{-1}f'(x)) \to H_k(X,X-\{x\})$$

is isomorphic. This completes Step 2. Steps 1 and 2 together show that $H_*(X,X-\{x\}) \cong H_*(Y,Y-\{x\})$. Since Y is a generalized n-manifold, $H_*(Y,Y-\{x\}) \cong H_*(E^n,E^n-\{0\})$. This fact completes the proof that X is a generalized n-manifold.

CONJECTURE 4.2. *If Y is an n-manifold-without-boundary and X is constructed from Y as above, then X is an n-manifold.*

V. IMPROVING GENERALIZED MANIFOLDS

We now apply the tools of Sections 3 and 4 to prove the result explained in the first paragraph of the introduction (Section 1).

Setting:

Y, a generalized n-manifold embedded as a closed subset of E^{2n+1}, $n \geq 5$;

$N(Y)$, the nonmanifold set of Y with dim $N(Y) \leq n - 3$;

Z, a closed subset of Y having dimension $\leq n-3$ with $N(Y) \subset Z$.

THEOREM 5.1. *Assume the setting. Then there exist a generalized n-manifold X, a 1-LCC closed embedding $f : Z \to X$ with the nonmanifold set $N(X)$ of X contained in $f(Z)$, and a proper cell-like surjection $f' : X \to Y$ such that $f' \circ f = id_Z$.*

Proof. Let $\alpha_1, \alpha_2, \alpha_3, \ldots : B^2 \to Y$ be dense in the space of singular disks in Y. By Theorem 3.6, there exist generalized 2-disks ${}^1D^+$, ${}^2D^+$, ${}^3D^+$, ... and maps $\beta_i : {}^iD^+ \to N_{1/i}$ $(Im\ \alpha_i, Y)$ satisfying the following conditions:

(i) $\beta_i \mid Bd\ {}^iD^+$ is a loop within $1/i$ of $\alpha_i \mid Bd\ B^2$ and

(ii) $\beta_i^{-1}(Z) = {}^iD^+ - {}^iD$.

Since the nonmanifold part $N(Y)$ of Y lies in the set Z, it follows from (ii) that, for every i, $\beta_i({}^iD)$ lies as a closed set in the *manifold* $Y - Z$. Hence, we may assume by Theorem 3.4 and the techniques used in its proof that the maps $\beta_i \mid {}^iD : {}^iD \to Y - Z$ are closed 1-LCC embedding with disjoint images.

By Theorem 3.5, each of the sets $\beta_i({}^iD)$ has a neighborhood N_i in $Y - Z$ that admits a PL triangulation with $\beta_i({}^iD)$ a subpolyhedron. We may thus use Theorem 3.1 and Corollary 3.2 to choose (pinched) crumpled n-cells in the $C\ell\ N_i$ containing the $\beta_i({}^iD^+)$. However, since we want these crumpled n-cells to have disjoint interiors and to form a null sequence we must deviate a bit from the obvious as follows:

Stage 1. Choose a stage ${}^1D_k$ of ${}^1D$ such that the image of each component of ${}^1D - ({}^1D_1 \cup \ldots \cup {}^1D_k)$ under β_1 has diameter less than 1. Let U_1 denote the interior of a regular neighborhood of

${}^1D_1 \cup \dots \cup {}^1D_k$ in ${}^1D$. Then ${}^1D - U_1$ is a finite disjoint union of infinite commutators mapped by β_1 to sets of diameter less than 1. It is these infinite commutators that we shall thicken in N_1 rather than $\beta_1({}^1D - U_1)$ itself.

Pick a PL triangulation of N_1 such that $\beta_1({}^1D - U_1)$ covers a subcomplex. Choose a second derived subdivision of N_1 such that, if M_1 is the simplicial neighborhood of $\beta_1({}^1D - U_1)$ in the second derived, then the following two conditions are satisfied:

(1) Each component of M_1 has diameter less than one.

(2) Each simplex of M_1 hitting $\beta_1({}^1D_j \cup {}^1D_{j+1} \cup \dots)$ has diameter less than $1/j$ and misses $Im\ \beta_j$.

If $M_{11}, \dots, M_{1k_1}$ are the components of M_1 whose closures hit Z, then the Freudenthal compactifications $C_{11}, \dots, C_{1k_1}$ are crumpled n-cells by Theorem 3.1 and Corollary 3.2. There is a unique map $g_{1j} : C_{1j} \to C\ell\ M_{1j} \subset Y$ that is the identity on M_{1j}. By Corollary 3.3, there exist n-cells $B_{11}, \dots, B_{1k_1}$ and maps $f_{1j} : B_{1j} \to C_{1j}$ having the simple properties of Corollary 3.3 and the setting of Section 4.

Stage m. Assume all $f_{ij} : B_{ij} \to C_{ij}$ and $g_{ij} : C_{ij} \to C\ell\ M_{ij} \subset Y$ chosen for $i < m$. Proceed as in Stage 1 with the following minor adjustments. Choose stage ${}^mD_\ell$ of ${}^mD$ such that the image of each component of ${}^mD - ({}^mD_1 \cup \dots \cup {}^mD_\ell)$ under β_m has diameter less than $1/m$ and misses all of the M_{ij}, $i < m$. Pick U_m as U_1 was picked. Triangulate U_m as U_1 was triangulated so that $\beta_m({}^mD - U_m)$ covers a subcomplex. Choose the second derived so that simplicial neighborhood M_m of $\beta_m({}^mD - U_m)$ satisfies

(1_m) Each component of M_m has diameter less than $1/m$ and misses all previous M_{ij}'s

(2_m) Each simplex of M_m hitting $\beta_m({}^mD_j \cup {}^mD_{j+1} \cup \dots)$ diameter less than $1/j$ and misses $Im\ \beta_j$.

Take $M_{m1}, \dots, M_{mk_m}$ to be the components of M_m whose closures hit Z. Let $C_{m1}, \dots, C_{mk_m}$ denote the corresponding Freudenthal compactification and $g_{mj} : C_{mj} \to Cl\, M_{mj} \subset Y$ and $f_{mj} : B_{mj} \to C_{mj}$ the associated maps.

One finally obtains by the above method countably many maps $f_{ij} : B_{ij} \to C_{ij}$ and $g_{ij} : C_{ij} \to Y$ satisfying the setting for Section 4. Let X Be the generalized n-manifold obtained by the construction of Section 4, $f' : X \to Y$ the map defined in that section and $f : Z \to X$ the identity embedding. Clearly f is a closed embedding with $f' \circ f = id_Z$ since $f'|Z = f|Z = id$. The map f' is a proper cell-like surjection by Section 4. Since each of the sets $g_{ij}(C_{ij})$ intersects Z and the $g_{ij}(C_{ij})$'s form a null sequence, only finitely many changes are made near any point of $Y - Z$. Thus $X - Z$ is a manifold. It remains only to check that the embedding $Z \subset X$ is *1-LCC*.

The inclusion $i : Z \to X$ is *1-LCC*:

The proof requires a good deal of unpleasant detail, which we divide into a number of parts: (i), (ii), (iii), and (iv). As basic setting we assume given $z \in Z$ and a neighborhood $\varepsilon : Z \to X$ of i in $Z \times X$. We seek a neighborhood ε_0 of i such that loops in $\varepsilon_0(z) - Z$ shrink in $\varepsilon(z) - Z$.

(i) Estimates. Define $f'' : X \to X$ by the formula

$$f''(x) = \begin{cases} f'(x) & \text{if} \quad x \in f'^{-1}f'(Z) \\ \emptyset & \text{otherwise.} \end{cases}$$

Consider the composite $i = f'' \circ id_X \circ f'^{-1} \circ id_Y \circ f' \circ i : Z \to X$. By the Composition Theorem [9, Appendix I], there exist neighborhoods $\varepsilon_1 : Z \to X$ of i in $Z \times X$, $\varepsilon_2 : Y \to Y$ of $id\ Y$ in $Y \times Y$, ε_3 of id_X in $X \times X$, and $L : X \to X$ of f'' in $X \times X$ such that $L \circ \varepsilon_3 \circ f'^{-1} \circ \varepsilon_2 \circ f' \circ \varepsilon_1 \subset \varepsilon$.

(ii) Retracting $f'^{-1} f(Z)$ close to Z. We claim that there exist a neighborhood M of $f'^{-1} f'(Z)$ in X and a map $r : M \to X$ in L such that r fixes a neighborhood N of Z and takes $M - Z$ into $X - Z$.

Proof of claim. Let $h : B_h \to X$ denote a map from an n-cell B_h into X defining one of the pinched n-cells in X used in changing Y to X. Let $S_h = Bd\, B_h$, and let $S_h \times [0,1]$ $(S_h = S_h \times \{0\})$ denote the (possibily wild) interior collar on S_h used in defining the nondegenerate point preimages of $f' \circ h : B_h \to g(C_h) = C\ell(M_h) \subset Y$. Let p_h be a point of $B_h - (S_h \times [0,1])$, and let $r_h : B_h - \{p_h\} \to S_h \times [0,1]$ denote a retraction with $r_h (B_h - S_h \times [0,1]) = S_h \times \{1\}$. (If the reader does not know how to construct r_h, simply retract a neighborhood of $(S_h \times [0,1])$.) Pick $t_h \in (0,1]$, and let $s_h : S_h \times [0,1] \to S_h \times [0,t_h]$ be the retraction that, for each $x \in S_h$, takes $\{x\} \times [t_h,1]$ to (x,t_h). It is clear that, for some $t_h \in (0,1]$ and some neighborhood M_h of $f'^{-1}f'(Z) \cap Im\, h$ in $h(B_h - \{p_h\})$, $h \circ s_h \circ r_h \circ h^{-1}|M_h$: $M_h \to X$ lies in L.

There is a neighborhood M_1 of Z in X such that the inclusion $M_1 \subset X$ lies in L. Let H be the set of defining maps $h : B_h \to X$ such that $Im\, h \not\subset M_1$.

$$M = [M_1 - \cup\{Im\, h | h \in H\}] \cup [\cup\{M_h | h \in H\}],$$

and define $r : M \to X$ by the formula

$$r(x) = \begin{cases} x & \text{if } x \in M_1 - \cup \{Im\, h | h \in H\} \\ h \circ s_h \circ r_h \circ h^{-1}(x) & \text{if } x \in M_h. \end{cases}$$

Then r satisfies the claim.

(iii) Choosing $\varepsilon_0 : Z \to X$. By reducing the neighborhoods ε_1, ε_2, and ε_3, we may require that $\varepsilon_3 \circ f'^{-1} \circ \varepsilon_2 \circ f' \circ \varepsilon_1(Z) \subset M$ and that $\varepsilon_1(Z) \subset N$ (Composition Theorem [9, Appendix I]). We choose a neighborhood $\varepsilon_0 : Z \to X$ of $i : Z \to X$ such that $\varepsilon_0(Z)$ is contractible in $\varepsilon_1(Z)$.

(iv) Shrinking loops missing Z. Let $\alpha : S^1 \to \varepsilon_0(z) - Z$ be a continuous function. We shall contract α in $\varepsilon(z) - Z$.

Since dim $[f'^{-1} \circ f'(Z) - Z] \leq 1$, we may homotop α in $\varepsilon_0(z) - Z$ so that it misses $f'^{-1} f'(Z)$. Thus $f'^{-1} \circ f' \circ \alpha = \alpha$. By the Composition Theorem [9, Appendix I] we may therefore reduce $\varepsilon_2 : Y \to Y$ so that any loop in the neighborhood $f'^{-1} \circ \varepsilon_2 \circ f' \circ \alpha$ of α is homotopic to α in $\varepsilon_0(z) - Z$.

Since $\varepsilon_0(z)$ contracts in $\varepsilon_1(z)$, we may extend α to a map $\alpha^* : B^2 \to \varepsilon_1(z)$. The composite $f' \circ \alpha^* : B^2 \to Y$ is a singular disk in Y, hence is the uniform limit of some subsequence of $\alpha_1, \alpha_2, \ldots : B^2 \to Y$. Choose m very large with $2/m < \varepsilon_3$, $Im\ (\beta_m : {}^mD^+ \to Y) \subset \varepsilon_2(Im\ \alpha^*)$, and $(\beta_m \mid Bd\ {}^mD^+) \subset \varepsilon_2 \circ f' \circ \alpha$. It is easy to check that $f'^{-1} \circ \beta_m \mid Bd\,{}^mD^+$ lies in $f' \circ \varepsilon_2 \circ f' \circ \alpha$ so that α is homotopic to $f'^{-1} \circ \beta_m \mid Bd\,{}^mD^+$ in $\varepsilon_0(z) - Z$. Thus it suffices to shrink $f'^{-1} \circ \beta_m \mid Bd\,{}^mD^+$ in $\varepsilon(z) - Z$.

Recall the manner in which portions of $\beta_m({}^mD^+)$ were thickened in $Y - Z$ to form M_m. Let $M_{m1}, \ldots, M_{mk}$ denote the components of M_m whose closures hit Z. Note that each M_{mj} is a pinched crumpled n-cell of diameter less than $1/m$. Recall the spaces and maps

$$\begin{array}{ccccc} B_{mj} & \xrightarrow{h_{mj}} & B'_{mj} & & \subset X \\ \downarrow f_{mj} & & \downarrow f'_{mj} & & \downarrow f' \\ C_{mj} & \xrightarrow{g_{mj}} & C\ell\ M_{mj} & & \subset Y \end{array}$$

associated with the sets $Cl\ M_{mj}$ in the construction. It is clear that $f'^{-1} \circ \beta_m \mid Bd\ {}^mD^+$ contacts in $[f'^{-1} \circ \beta_m({}^mD)] \cup \cup h_{mj}(\text{Int } B_{mj}) \subset (X - Z) \cap 2/m \circ f'^{-1} \circ \varepsilon_2 \circ (Im\ \alpha^*) \subset [\varepsilon_3 \circ f'^{-1} \circ \varepsilon_2 \circ f' \circ \varepsilon_1(z)] - Z \subset M - Z$. Note that $r : M - Z \to X - Z$ fixes $f'^{-1} \circ \beta_m\ (Bd\ {}^mD^+)$, hence proving that $f'^{-1} \circ \beta_m \mid Bd\ {}^mD^+$ shrinks in $L \circ \varepsilon_3 \circ f'^{-1} \circ \varepsilon_2 \circ f' \circ \varepsilon_1(z) - Z \subset \varepsilon(z) - Z$. This completes the proof that $i : Z \to X$ is 1-*LCC*.

VI. RECOGNIZING MANIFOLDS

In this section we prove the conjecture of the introduction (section 1) for nonmanifold sets of trivial dimension. The tool we use is a certain codimension-one version of the conjecture; this version has been proved by C. L. Seebeck III [28, Theorem 6.3]. The Seebeck theorem was first stated [12], and presumably also proved, by A. V. Cernavskii; we have been unable to determine wheither Cernavskii's proof has appeared in any version.

[Seebeck informs us (September, 1977) that he does not intend at present to publish his proof because S. Ferry [20] has proved a more general theorem. To make it possible therefore for the expert reader to complete the proof of Theorems 6.1 and 7.1 in the absence of a printed proof of Cernavskii-Seebeck-(Ferry) theorem, we make the following comments on the proof. For the proof of Theorem 6.1 it suffices to prove Cernavskii-Seebeck in the case where M is connected and M is locally collared in $U \cup M$ at some point of M. The homotopy hypotheses allow one to homotopically slide the collar about in $U \cup M$. These stretchings form the basis of an engulfing argument that allows one to geometrically slide the collar along M in $M \cup U$ to prove M locally collared at every point of M from U. The more general case requires the Kirby torus trick (see [22]). The reader will therefore presumably have to await Ferry's paper for the general case used in Theorem 7.2.]

THEOREM (Cernavskii-Seebeck). Let U be an open n-manifold, M an open (n - 1)-manifold, and suppose $X' = U \cup M$ is a metric space such that $U \cap M = \emptyset$, U is locally homotopically k-connected at each point of M for $k = -1, 0, 1, \ldots, n$, and $n \geq 5$. Then X' is an n-manifold with boundary M.

From the Cernavskii-Seebeck theorem we deduce the following theorem.

THEOREM 6.1. Let X denote a generalized manifold of dimension $n \geq 5$, and suppose the nonmanifold set N(X) of X has dimension k, $2k + 2 \leq n$. If the embedding $N(X) \subset X$ is 1-LCC, then X is a topological manifold.

Proof of the case $k = 0$, N(X) compact. (We present a proof of the simplest interesting special case in order to motivate the hard work to come later.)

If K_∞ is a Cantor set in a connected n-manifold M, $n \geq 2$, then, by a standard technique, K_∞ lies in the boundary of an n-cell in M: one simply starts with a small n-cell and then pushes out branching feelers nearer and nearer to K_∞ (Fig. 1).

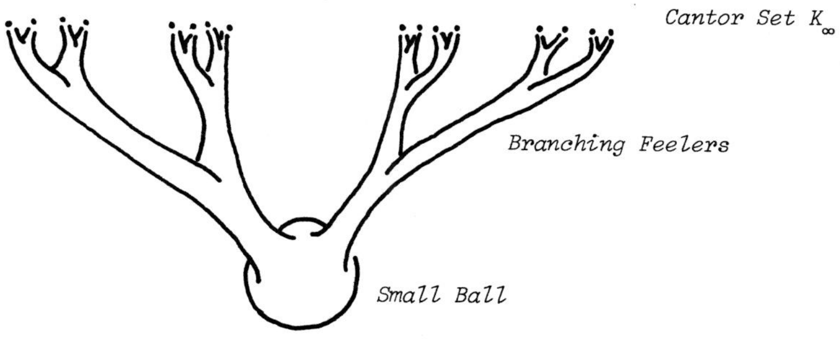

Fig. 1.

Precisely the same construction works in X to show that $N(X)$ (dim $N(X) = 0$) lies in the surface of an n-ball B in X. If the most simple-minded construction is followed, then it will also be the case that the inclusions $N(X) \subset Bd\ B$ and $(Bd\ B) - N(X) \subset X - N(X)$ are 1-LCC; we assume that this is the case. Then, with $U = X - B$, $M = Bd\ B$, and $X' = U \cup M$, the hypotheses of the Cernavskii-Seebeck theorem are satisfied. Hence X' is an n-manifold with boundary $Bd\ B$. We conclude that $X = X' \cup_{Bd\ B} B$ is an n-manifold-without-boundary.

Initial comments on the general case: We shall generalize the process of pushing feelers out toward a Cantor set. To motivate the generalization, we reconsider the pushing process as that of pushing a ball out along an infinite graph compactified (Figure 2).

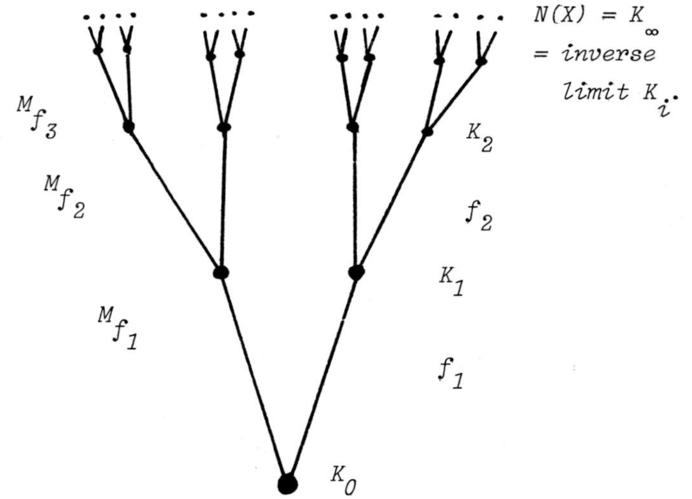

Fig. 2.

The graph itself, before compactification, is, in a natural way, the union of infinitely many mapping cylinders. The nonmanifold set $N(X)$ is the inverse limit associated with the maps defining the mapping cylinders. Our generalization of the procedure for $K = 0$ requires that we complete three steps:

I. Define the appropriate analogue of the compactified graph from Figure 2.

II. Embed the space thus defined in such a manner that it can be used to push a ball toward $N(X)$.

III. Describe the way in which a ball is pushed toward $N(X)$ along the object embedded in Step II.

These steps are relatively straightforward except in the single case of II when $2k + 2 = n$.

I. An approximate cone over a finite dimensional compactum.

Let each of $B_0, B_1, \ldots$ denote a standard unit ball centered at the origin in some Euclidean space, and examine the product $B = B_0 \times B_1 \times \ldots$. Define a continuous multiplication $(\cdot) : [0,\infty] \times B \to B$ by the formulae

$$s \cdot (b_0, b_1, \ldots) = \begin{cases} (1\cdot b_0, \ldots, 1\cdot b_{[s]-1}, (s-[s])\cdot b_{[s]}, 0, 0, \ldots) & \text{if } s \in [0,\infty) \\ (b_0, b_1, \ldots) & \text{if } s = \infty. \end{cases}$$

Then $(\cdot)$ defines a contraction of B since $\infty \cdot b = b$ and $0 \cdot b = 0$ $(\equiv (0,0,\ldots))$. (The symbol $[\cdot]$ is the greatest integer symbol.)

Let $K_0 \xleftarrow{f_1} K_1 \xleftarrow{f_2} K_2 \longleftarrow \ldots$ denote an (inverse) system of nonempty compact spaces and continuous maps, $K_i \subset B_i - \{0\}$, K_i joinable to 0 in B_i. Define an embedding $g_i : K_i \to B$ by the formula $g_i(x) = (x_0, \ldots, x_j, \ldots x_i, 0, 0, \ldots)$ where $x_i = x$ and $x_j = f_{j+1} \circ \ldots \circ f_i(x)$ for $j < i$. Let $L_i = g_i(K_i)$. Note that the following diagram commutes

$$\begin{array}{ccc} K_{i-1} & \xleftarrow{\quad f_i \quad} & K_i \\ \Big\downarrow g_{i-1} & & \Big\downarrow g_i \\ L_{i-1} & \xleftarrow{\quad i \quad} & L_i \end{array}$$

where $L_{i-1} \xleftarrow{i} L_i$ denotes the multiplication by the integer $i \in [0,\infty]$ defined above, $y \mapsto i \cdot y$.

Consider the union $L = [0,1] \cdot L_0 \cup [1,2] \cdot L_i \cup \ldots$. Then L is called the infinite mapping cylinder associated with the system $L_0 \xleftarrow{1} L_1 \xleftarrow{2} L_2 \longleftarrow \ldots$. Let L^+ denote the closure of L in B. Then L^+ is called an approximate cone over the inverse limit $L_\infty = L^+ - L$.

Since the systems $K_0 \xleftarrow{f_1} K_1 \xleftarrow{f_2} K_2 \longleftarrow \ldots$ and $L_0 \xleftarrow{1} L_1 \xleftarrow{2} L_2 \longleftarrow \ldots$ are naturally isomorphic, we may define K, K^+, and K_∞ in an obvious way up to homeomorphism.

THEOREM. Suppose $L_0 \xleftarrow{1} L_1 \xleftarrow{2} L_2 \longleftarrow \ldots$ *is an inverse system as above with infinite mapping cylinder* $L = [0,1] \cdot L_0 \cup [1,2] \cdot L_1 \cup \ldots$, *approximate cone* L^+, *and inverse limit* L_∞. *Then* L_∞ *is nonempty and compact, and* L^+ *is contractible.*

Proof. Let M_i be the inverse image under the projection map $i + 1 : B \to B_0 \times \ldots \times B_i \times 0 \times 0 \times \ldots$ of the compact space $L_i \subset B_0 \times \ldots \times B_i \times 0 \times 0 \ldots$. Then M_i is nonempty and compact, $M_0 \supset M_1 \supset \ldots$, and $L_\infty = \bigcap_{i=0}^{\infty} M_i$; thus L_∞ is nonempty and compact. Since L_∞ can also be expressed as the set $\{x = (x_0, x_1, \ldots) \in B \mid x_i \in L_i,\ i - x_i = x_{i-1}\}$, it follows easily that $(\cdot) \mid ([0,\infty] \times L^+) : [0,\infty] \times L^+ \to B$ takes its values in L^+, hence contracts L^+.

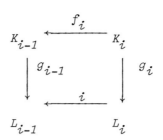

Remark. It is also worthwhile to note that L has the identification topology $L'/\sim$ where L' is the disjoint union $L' = ([0,1] \times L_0) \cup ([1,2] \times L_1) \cup \ldots$ and $\sim$ is the equivalence relation which identifies $\{0\} \times L_0$ to the single point $0 \cdot L_0$ and hich identifies $\{i\} \times L_i$ with its image in $\{i\} \times L_{i-1}$ under the map $L_{i-1} \xleftarrow{i} L_i$. Thus L is indeed an infinite mapping cylinder. If each L_i is a topological polyhedron and each map $L_{i-1} \xleftarrow{i} L_i$ is PL with respect to the given PL structures on L_i and L_{i-1}, then L may be triangulated as a locally finite but globally infinite polyhedron, each set $L_{i-1} \cup_i [i,i+1] \cdot L_i$ having the structure of a PL mapping cylinder.

COROLLARY. Suppose L_∞ is a compactum (compact metric space) of dimension $k < \infty$. Then there is an inverse system $L_0 \xleftarrow{1} L_1 \xleftarrow{2} L_2 \longleftarrow \ldots$ of finite k-dimensional topological polyhedra with PL bonding maps 1, 2, ... having inverse limit L_∞ and approximate cone L^+, with $L = L^+ - L_\infty$ PL as in the remark.

Proof. That some inverse system of finite k-dimensional polyhedra with PL bonding maps has L_∞ as inverse limit is well-known. The existence of the approximate cone L^+ and PL structure on L is a consequence of our construction and remark.

II. Embedding an approximate cone.

Let us return to the setting the Theorem 6.1, with X, $N(X)$, $n \geq 5$, $N(X) \subset X$ 1-LCC, $k \geq \dim N(K)$, $2k+2 \leq n$ forming the data. Let Z denote a closed subset of X, being 1-LCC embedded in X, having dimension $\leq k$, and containing $N(X)$.

THEOREM. Let p be a point of Z. Then p has a compact neighborhood L_∞ in Z, and L_∞ has an approximate cone $L^+ = L \cup L_\infty$, L PL of dimension k + 1, such that the natural identification space $L^+ \cup_{L^\infty} Z$ admits a 1-LCC embedding in X that is the identity on Z.

Remark. If $2k + 3 \leq n$, then Step II with its Theorem is easy by general position arguments applied chart by chart. The case $2k + 2 = n$ is really quite difficult and necessitates the lengthy complications which we now describe. The reader may choose to ignore the following proof at a first reading since it improves the result only by one dimension.

Proof of the theorem for Step II. A generalized n-manifold is locally orientable [3]. Hence p has an orientable neighborhood U, and duality arguments are valid in U [1, 4, 32].

Since a generalized manifold is locally contractible, there exist open neighborhoods $U \supset V_2 \supset V_1 \supset V_0$ of p in U, each contractible in the previous (see Figure 3).

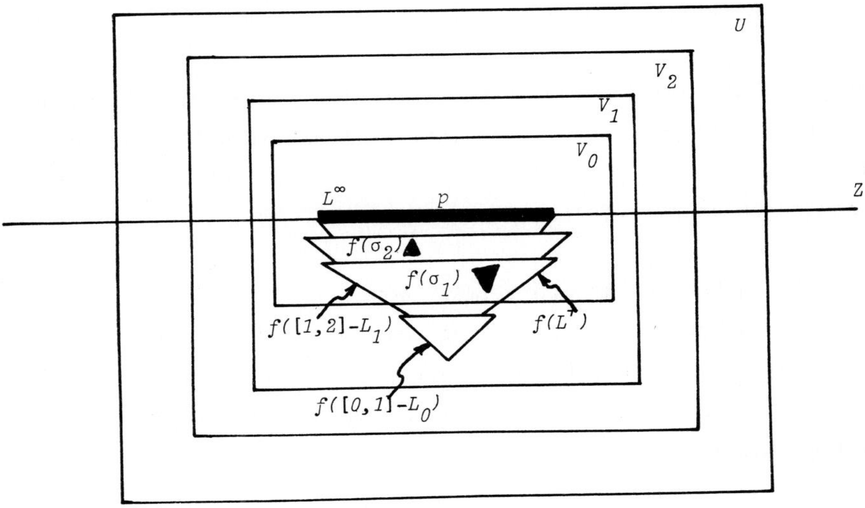

Fig. 3. Embedding the approximate cone L^+ in U. (The intersection $f(\sigma_1) \cap f(\sigma_2)$ is not indicated in the figure.)

Choose a compact neighborhood L_∞ of p in $Z \cap V_0$. By the Corollary in Step I, there is a PL inverse system $L_0 \longleftarrow L_1 \longleftarrow \ldots$ with inverse limit L_∞, approximate cone L^+, and infinite PL mapping cylinder $L = L^+ - L_\infty = [0,1]$. $L_0 \cup [1,2] \cdot L_1 \cup \ldots$ of dimension $k + 1$. We assume that $2(k + 1)$ is exactly n. All other cases are easier. Since V_0 is an ANR and contacts in V_1, it is an easy matter to construct a map $f : L^+ \to V_1$ that is the inclusion map on L_∞.

Since duality is valid in V_1, $V_1 - Z$ is ℓc^j rel $(Z \cap V_1)$ for $j = n - (k + 2) = (2k + 2) - (k + 2) = k$ (see Section 2 above). Since $V_1 - Z$ is LC^1 rel $Z \cap V_1$ by hypothesis, it follows from the Local Hurewicz Isomorphism Theorem (see, for example, [9, Theorem 37]) that $V_1 - Z$ is LC^k rel $(Z \cap V_1)$. Thus [19, Theorem 2] implies that f may be adjusted rel L_∞ so that $f(L) \cap Z = \emptyset$.

Working coordinate chart by coordinate chart in $V_0 - Z$, using the Bryant-Seebeck codimension three *1-LCC* taming theorem for polyhedra [7, 8] and general position in each chart in turn, it is possible to adjust $f|L$ so that it is a *1-LCC* immersion with isolated double point singularities where two $(k+1)$-simplexes of L have images under f meeting transversely at a single point. Interpolating more stages in $L = [0,1] \cdot L_0 \cup [1,2] \cdot L_1 \cup \ldots$ if necessary and adjusting the product structure locally if necessary, we may assume that at most one point of the singular set $S(f)$ lies in any one of the sets $(i, i + 1) \cdot L_i$, and none in any of the sets L_i. We shall show how to remove a single double point, and simply indicate why the obvious iteration can be so carefully performed that it converges to the desired embedding of $(Z \cap U) \cup L_\infty$ in U.

Removing a single double point:

Let σ_0 and σ_1 denote $(k+1)$-simplexes of $L - (L_0 \cup L_1 \cup L_2 \cup \ldots)$ that are individually *1-LCC* embedded by f but whose images meet transversely in a single point in $V_1 \subset U$. We assume that $\sigma_0 \subset (0,1) \cdot L^+$, that $\sigma_1 \subset (L^+ - [0,i] \cdot L^+)$, and that

$\sigma_0 \cap f^{-1}(\sigma_1)$ is the only singularity of f in $[0,i] \cdot L^+$. We shall alter f on σ_0 only so as to push the singularity $f(\sigma_0) \cap f(\sigma_1)$ toward and beyond $L_\infty \equiv f(L_\infty)$ so that the singularity disappears. The argument proceeds in four steps, (1) - (4). The second step removes the singularity homologically, the third removes it homotopically, the forth geometrically.

(1) Neighborhood Control. Let $p_i = \sigma_i \cap f^{-1}(\sigma_{1-i})$, $i = 0,1$. Let $t \in [0,\infty]$ be the unique scalar such that $p_1 \in t \cdot L_\infty \subset L^+$. Let $P_1 = t^{-1}(p_1)$ (see Fig. 4). Choose an open neighborhood W of $f(P_1)$ in V_1 such that $C\ell W$ is a compact subset of V_1 hitting $f(L)$ only in $(L^+ - [0,1] \cdot L^+) \cup f(\text{Int}\ \sigma_0)$.

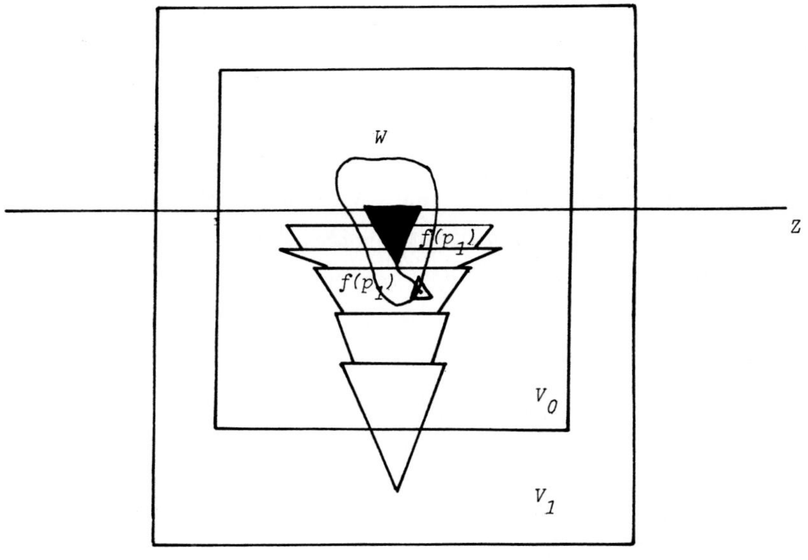

Fig. 4. Neighborhood Control.

(2) Duality. Let τ_0 denote a $(k+1)$-simplex in Int σ_0 containing p_0 in its interior. We claim that the singular k-cycle $f|Bd\ \tau_0$ bounds homologically in $W - f(L^+ - \sigma_0)$. The technical tool we shall use in demonstrating this fact is a duality lemma proved below. Its application requires that we first show the existence of a certain homotopy of $Z \cup f(L^+)$ in U.

LEMMA (2)1 (Homotopy). There exists a proper homotopy $F : \{[Z \cup f(L^+ - [0,i) \cdot L^+)] \cap U\} \times I \to U$ *such that only tracks of points in* W *hit* $f(\tau_0)$, *no track hits* $f(Bd\ \tau_0)$, *and* $Im\ F_1$ *misses* $f(\tau_0)$ *entirely.*

Proof of Lemma (2)1. If there were no singularities in $f|L^+ - [0,i) \cdot L^+$, then we could use the multiplication $(\cdot) : [i,\infty] \times (L^+ - [0,i) \cdot L^+) \to (L^+ - [0,i) \cdot L^+)$, together with the homotopy extension property for $(Z \cap U) - L_\infty$, to obtain the desired homotopy. Even with singularities, however, we obtain by the method described a multivalued homotopy F'' of the desired sort whose only drawback is that it splits a double point α into two points α', α'' in the course of the homotopy. We first repair the splitting in a cell-like way as follows. Let α denote a double point with split images α' and α''. Then F'' describes an actual (single valued) homotopy of both α' and α''. Think of α' and α'' as endpoints of an interval $I_\alpha = [0_\alpha, 1_\alpha]$, $\alpha' = 0_\alpha$, $\alpha'' = 1_\alpha$. Extend F'' to all of $I_\alpha \times I$ so as to take $I_\alpha \times \{0\}$ to α. By general position we require that $F''(I_\alpha \times I) \cap f(\tau_0) = \emptyset$. If, for every $t \in [0,1]$, $F''((\alpha', t))$ and $F''((\alpha'',t))$ are very close pointwise, as they must be for a double point very near L_∞, we may also require that each $F''(I_\alpha \times \{t\})$ be very small. With the controls just described, the multivalued homotopy F'' is actually the composite of a cell-like embedding relation $(Z \cap U) \cup f(L^+ - [0,i) \cdot L^+) \to (Z \cap U) \cup (L^+ - [0,i) \cdot L^+) \cup (\bigcup_\alpha I_\alpha)$, (where the latter space is simply $(Z \cap U) \cup (L^+ - [0,1) \cdot L^+)$ with a null sequence of arcs joining singularities) and an an actual homotopy

$$[(Z \cap U) \cup (L^+ - [0,i) \cdot L^+) \cup (\bigcup_\alpha I_\alpha)] \times I \to U.$$

If we think of $(Z \cap U) \cup (L^+ - [0,i) \cdot L^+) \cup (\bigcup_\alpha I_\alpha)$ as embedded as a closed set in a high dimensional Euclidean space E^m, the actual homotopy may be extended to a neighborhood in E^m. The cell-like relation may be approximated by an actual continuous function into that neighborhood [9, Approximation Theorem], and the composite of the actual homotopy and approximating continuous map essentially provide the homotopy claimed by Lemma (2)1.

LEMMA (2)2. Suppose U is an orientable generalized n-manifold, $(A,B) \subset (A',B')$ are closed pairs in U, with $Cl(A - B)$ and $Cl(A' - B')$ compact subsets of U, and x is an element of $H_*(U - A', U - B')$. If there is a homotopy $F: (A,B) \times I \to (A',B')$ such that the inclusion induced homomorphism $\alpha : H_*(U - B', U - A') \to H_*(U - F_1(B), U - F_1(A))$ takes x to 0, then the homomorphism $\beta : H_*(U - B', U - A') \to H_*(U - B, U - A)$ takes x to 0.

Proof. We consider the commutative diagram,

$$\begin{array}{ccccc} \check{H}^{n-*}(F_1(A),F_1(B)) & \xleftarrow{\alpha'} & \check{H}^{n-*}(A',B') & \xrightarrow{\beta'} & \check{H}^{n-*}(A,B) \\ D_- \downarrow \cong & & D \downarrow \cong & & D_+ \downarrow \cong \\ H_*(U - F_1(A), U - F_1(B)) & \xleftarrow{\alpha} & H_*(U - A', U - B') & \xrightarrow{\beta} & H_*(U - A, U - B), \end{array}$$

where α, α', β, and β' are induced by inclusions and D_-, D, and D_+ are duality isomorphisms.

Recalling the definition of Cech cohomology, we have $\check{H}^{n-*}(A',B') = \varinjlim H^{n-*}(R',S')$, where $H^{n-*}(R',S')$ denotes singular cohomology of an open pair (R',S') in U, $(A',B') \subset (R',S')$ and the direct limit is taken over all such pairs (R',S'). The isomorphism

D is, in turn, the direct limit of homomorphisms $H^{n-*}(R',S') \to H_*(U-B',U-A')$ (induced by cap product and excision); the point of this latter observation is that $Im\ D = \cup Im\ H^{n-*}(R'S')$ so that we may choose (R',S') with $x \in Im\ H^{n-*}(R',S')$. Let ξ be a singular cocycle, $\xi \in Z^{n-*}(R',S')$, representing x $(x = Im\ [\xi])$.

Since $\alpha'[\xi \mid (A',B')] = D_-^{-1} \circ \alpha \circ D[\xi \mid (A',B')] = D_-^{-1} \circ \alpha(x) = 0$ by hypothesis, $[\xi \mid (F_1(A),F_1(B)]$ is the zero element of $\check{H}^{n-*}(F_1(A),F_1(B)) = \varinjlim H^{n-*}(R'',S'')$, where (R'',S'') ranges over the set of open pairs in U containing $(F_1(A),F_1(B))$. We may therefore choose (R'',S'') so small that $(R'',S'') \subset (R',S')$ and $\xi \mid (R'',S'')$ is a coboundary, $\xi \mid (R'',S'') = \delta\zeta$, $\zeta \in C^{n-*-1}(R'',S'')$.

We may extend $F : (A,B) \times I \to (A',B')$ to a homotopy $F' : (R,S) \times I \to (R',S')$ where (R,S) is an open pair in U, $(A,B) \subset (R,S)$, and $F_1'(R,S) \subset (R'',S'')$.

The homotopy F' allows us to define a chain homotopy $d : C_*(R',S') \to C_{*+1}(R',S')$ with $id = F'_1 + d\partial + \partial d$. Then $\xi \mid (R,S) = \delta(\zeta F_1' + \xi d) \mid (R,S)$. Thus $\zeta \mid (R,S)$ is a coboundary its image in $H_*(U-B,U-A)$, which by commutativity of our first diagram is $\beta(x)$, is therefore 0.

Completion of Step (2). It follows from Lemmas (2)1 and (2)2, that $f|\tau_0$ represents the zero element in $H_{k+1}(U-B,U-A)$ $A = (Z \cap U) \cup f(L^+ - [0,i) \cdot L^+)$ and $B = A - W$. Let C denote a singular $(k+2)$-chain in $U-B$ such that $\partial C = (f|\tau_0) + C'$, with $C' \subset U-A$. Note that the image of C hits $(Z \cap U) \cup f(L^+ - [0,i) \cdot L^+)$ only in W. Subdivide C so finely that any singular simplex hitting $(Z \cap U) \cup f(L^+ - [0,i) \cdot L^+)$ lies in W and hits $f([0,i] \cdot L^+)$ only in $f(\text{Int}\ \tau_0)$. Let C'' denote the subchain of C consisting of those singular simplexes hitting $(Z \cap U) \cup f(L^+ - [0,i) \cdot L^+)$. Then $\partial(C-C'') = f|\tau_0 + C' - \partial C''$ so that $f|\tau_0 \sim C' \sim f \mid \tau_0 - \partial C''$ which is a relative cycle bounded by $f \mid Bd\ \tau_0$ in $W - f(L^+ - \sigma_0)$ as desired.

(3) The Hurewicz isomorphism. We shall show that $f \mid Bd\ \tau_0$ is contractible in V_2 hitting $(Z \cap U) \cup f(L^+)$ only in $f((\text{Int}\ \sigma_0) - \{p_0\})$. By step (2) there is a finite simplicial complex P containing $Bd\ \tau_0$ as a k-dimensional subcomplex in which $Bd\ \tau_0$ is nullhomologous and a continuous function $g : P \to W$ extending $f \mid Bd\ \tau_0$ such that $g(P)$ hits $(Z \cap U) \cup f(L^+)$ only in $f((\text{Int}\ \sigma_0) - \{p_0\})$. Let $P \cup C(P^{(k-1)})$ denote P together with the cone $C(P^{(k-1)})$ on the $(k-1)$-skeleton $P^{(k-1)}$ of P. Then $P \cup C(P^{k-1})$ is $(k-1)$-connected so that the Hurewicz isomorphism theorem implies that $\pi_k(P \cup C(P^{(k-1)})) \to H_k(P \cup C(P^{(k-1)}))$ is an isomorphism. Thus $Bd\ \tau_0$, being nullhomologous in P, is nullhomotopic in $P \cup C(P^{(k-1)})$. Since $W \subset V_1$ and V_1 contracts in V_2, we may extend g to a map $g^* : P \cup C(P^{k-1}) \to V_2$. Since $V_2 - Z$ is LC^k rel Z, we may choose g^* to miss Z and, by general position, we may choose g^* to hit $f(L)$ only in a subset of $f((\text{Int}\ \sigma_0) - \{p_0\})$. The desired result follows.

(4) The Whitney Trick. Steps (1), (2), and (3) taken together show that $f|\sigma_0$ can be altered so that $f(\text{Int}\ \sigma_0) \cap f(L^+ - \sigma_0) = \emptyset$. The altered f takes σ_0 into V_2. After a chart by chart general positon argument using the Bryant-Seebeck 1-LCC taming theorem, we may assume that $f|\sigma_0$ has only a finite number of singularities, each a double point at which two subsimplexes of σ_0 have image meeting transversely at a single interior point. The singular set $S(f|\sigma_0)$ lies in a 1-dimensional collapsible subcomplex K of Int σ_0. The cone $C(f(K))$ over the image $f(K)$ in V_2 1-LCC embeds in U since V_2 contracts in U. As before we may adjust $C(f(K))$ rel $(f(K))$ so that $C(f(K)) \cap f(L^+) = f(K)$. Then $C(f(K))$ is cellular in U and may be identified to a point in U. The result is a 1-LCC embedding of σ_0/K. We identify σ_0/K and σ_0 and alter f so that it is the old f followed by the cellular collapse. This completes the alteration of $f|\sigma_0$ so that $S(f) \cap \sigma_0 = \emptyset$ and removes one double point of f.

Removing all double points by iteration:

The only point to be noted is that, as double points approach $L_\infty \subset Z$, the sets W of Steps (1) and (2) can become smaller and smaller, hence the alterations of Steps (3) and (4) can become smaller and smaller. Thus the process described can be made to converge to the desired embedding $L^+ \cup Z \to X$. This completes Step II.

III. Pushing an n-ball along an embedded approximate cone.

THEOREM. Suppose L^+ is an approximate cone over a compactum L_∞ in a generalized n-manifold X, and Z is a closed subset of X containing the nonmanifold set of X such that $Z \cap L^+ = L_\infty$. Then the following are true.

(1) If $\dim L^+ \leq n - 3$, *then after slight adjustment one may assume the embedding $L \subset X - Z$ is 1-LCC.*

(2) If $L \subset X - Z$ is 1-LCC, $\dim L^+ \leq n - 3$, *and $n \geq 5$, then L has a neighborhood N in $X - Z$ which admits a PL triangulation in which L is a subcomplex.*

(3) If L has a neighborhood N in $X - Z$ which admits a PL triangulation with L as a subcomplex, then there exists an n-ball B in $(N - Z) \cup L_\infty$ containing L in its interior and L_∞ in its boundary; if $k =$ $\dim L^+ \leq n - 3$, *then $L_\infty \subset Bd\ B$ may be chosen to be 1-LCC.*

Proof of the theorem for Step III.

(1) : This conclusion is an immediate consequence of M. A. Stan'ko's *1-LCC* approximation Theorem [30].

(2) : The embedding $L \subset X - Z$ is locally tame by the Bryant-Seebeck *1-LCC* taming theorem [7, 8]. Thus a neighborhood N of L strong deformation retracts to L. But L is contractible. Thus N is contractible and hence, by [22], also PL triangulable. Finally, L is tame in the triangulated N by [7, 8]. The inverse of a taming homeomorphism triangulates N with L as a subcomplex.

(3) : Using Marshall Cohen's theory of relative regular neighborhoods [13] and the fact that each stage $[0,1] \cdot L_0 \cup [1,2] \cdot L_1 \cup \ldots \cup [i,i+1] \cdot L_i$ collapses to the previous stage $[0,1] \cdot L_0 \cup [1,2] \cdot L_1 \cup \ldots \cup [i-1,i] \cdot L_{i-1}$, it is easy to expand a ball neighborhood B_0 of $0 \cdot L_0$ out along the inverse of the collapse. Some care must be taken in order to ensure that the process converges to an embedding $f_\infty : B_0 \to X$ whose image $B = f_\infty(B_0)$ has the desired properties. The argument is probably best left to the reader. However we give some indications of the necessary precautions.

The construction:

Take B_0 to be a relative regular neighborhood of the cone $[0,1] \cdot L_0$ intersecting L^+ precisely in $[0,1] \cdot L_0$, $(Bd\ B_0) \cap L^+ = L$. Assume PL n-balls $B_0 \subset \ldots \subset B_{i-1} \subset Z - X$ and PL homeomorphisms $f_0 : B_0 \to B_1, \ldots, f_{i-2} : B_{i-2} \to B_{i-1}$ already chosen with $[0,1] \cdot L_0 \cup \ldots \cup [j,j+1) \cdot L_j = L^+ \cap \text{Int } B_j$ and $L_j = (j+1) - L_j = L^+ \cap Bd\ B_j$ $(j = 0, \ldots, i-1)$. Since $[i,i+1] \cdot L_i$ collapses to $B_{i-1} \cap ([i,i+1] \cdot L_i)$ in an arbitrarily close approximation to the map $i : [i,i+1] \to L_{i-1}$ it is easy to construct a relative regular neighborhood N_i of $[i,i+1] \cdot L_i$ in $X - \text{Int } B_{i-1}$ such that $B_i = B_{i-1} \cup N_i$ is an n-cell meeting L^+ in the desired way and to construct a PL homeomorphism $f_{i-1} : B_{i-1} \to B_i$ that moves points only very near the fiber lines $[i,i+1] \cdot x$, $x \in L_i$.

Forcing the compositions $f_{i-j} \circ \ldots \circ f_0 : B_0 \to B_i$ to converge to a limit map $f_\infty : B_0 \to Z$:

Points x moved by $f_{j+i} \circ \ldots \circ f_i : B_i \to B_{j+i+1}$ may be chosen to be moved very close to the sets $(i+1)^{-1}(x')$, x' being a point of L_i very near x. Since $\lim_{i\to\infty} \text{Diam } (i+1)^{-1}(x') = 0$, the distance $d(f_{j+1} \circ \ldots \circ f_i, id)$ may be kept arbitrarily small for i large. This makes the sequence of compositions Cauchy so that f_∞ exists.

Forcing f_∞ *to be 1-1:*

There is a distance $\delta_i > 0$ such that points of B_0 at distance $\geq 1/i$ from each other have images under $f_{i-1} \circ \cdots \circ f_0$ at distance $\geq \delta_i$. Let $R : L^+ \to L^+$ be the relation which takes each point $t \cdot x$, $(t \in i,i+1$, $x \in L_i)$ to the set of points y such that $x \in [0,\infty] \cdot y$. We may choose $\varepsilon_i > 0$ so small that, if $x,y \in L_i$ are at distance $\geq \delta_i/2$ apart, then $R(x)$ and $R(y)$ are at distance $\geq \varepsilon_i$ apart. Then by choosing f_j, $j \geq i$, to move points only very near $R \mid L^+ - [0,j+1) \cdot L^+$, f_∞ will be 1-1.

Forcing $B \cap Z = L_\infty$:

This is automatic provided again that f_j is very near $R \mid L^+ - [0,j+1) \cdot L^+$.

Forcing the embedding $L \subset Bd\ B$ *to be 1-LCC provided* $k = \dim L^+ \leq n-3$:

We choose a regular neighborhood N_i of L_i in $Bd\ B_i$ so that it admits a $1/i$-collapse to L_i. The homeomorphism $\lim_{j\to\infty} f_{j+1} \circ \cdots \circ f_i$ can be chosen to move only those points of $Bd\ B_i$ in Int N_j. It then follows that $\lim_{j\to\infty} f_{j+1} \circ \cdots \circ f_i\ (N_i)$ is a mapping cylinder neighborhood of L_∞ in $Bd\ B_i$ with fibers of small length and tame k-dimensional spine. Thus the geometric dimension (demension (sic)) of L_∞ in $Bd\ B$ is $\leq k$ (see [16]). Hence, since $k \leq (n-1) - 3$, the embedding $Z \subset Bd\ B$ is 1-LCC.

This completes III.

Proof of Theorem 6.1. By parts I, II, and III, each $p \in N(X)$ has a neighborhood L_∞ in $N(X)$ that lies in the surface of an n-ball B in X, $B \cap Z = L_\infty$, $L_\infty \subset Bd\ B$ 1-LCC, $(Bd\ B - Z) \subset X - Z$ 1-LCC. A standard argument then shows $B \subset X$ 1-LCC. Hence the Cernavskii-Seebeck theorem applies as before to show that X is a manifold at each point of the interior of L_∞ (relative to Z). Since p was arbitrary, X is a manifold.

VII. SUMMARY

THEOREM 7.1. Let Y denote a generalized manifold of dimension $n \geq 5$, *and suppose the nonmanifold set* $N(Y)$ *of Y has dimension k in the trivial range* $2k + 2 \leq n$. *Then Y is a cell-like decomposition of an n-manifold X without boundary.*

Proof. By Theorem 5.1 there exist a generalized n-manifold X, a 1-LCC closed embedding $f : Z \to X$ with the nonmanifold set $N(X)$ of X contained in $f(Z)$ and a proper cell-like surjection $f' : X \to Y$ such that $f' \circ f = id_Z$. By Theorem 6.1, X is a manifold.

COROLLARY. The space Y is a manifold if and only if it satisfies the disjoint disk property.

Proof. Since Y is a cell-like decomposition of a manifold by the theorem the result follows from [11].

THEOREM 7.2. Suppose Y is a generalized n-manifold such that the nonmanifold part N(Y) of Y lies in a closed subset C of Y that is a generalized $(n - 1)$*-manifold. If* $n \geq 4$ *and if* $C \times E^m$ *is a manifold* $(m \geq 1)$ *then* $Y \times E^m$ *is a manifold. If* $n \geq 5$ *and C is an* $(n - 1)$*-manifold, then Y is a cell-like decomposition of a manifold-without-boundary.*

Proof. Suppose $C \times E^m$ is a manifold. Then the infinite commutators constructed in the proof of Theorem 5.1 can be constructed in $(Y \times E^m) - (C \times E^m)$ so that the (crumpled cell-real cell) replacement coverts $Y \times E^m$ into a generalized $(n + m)$-manifold X with $C \times E^m$ 1-LCC embedded in X and containing the nonmanifold set of X. By the Cernavskii-Seebeck-(Ferry) Theorem X is a manifold. If $m \geq 1$, then $Y \times E^m$ satisfies the relative disjoint disk property discussed in [10] so that $Y \times E^m$ is a manifold by [11].

VIII. OPEN QUESTIONS

The most important open questions seem to be the following.

Question 1. Is the conjecture of Section 1 true?

Question 2. Is the conjecture of Section 1 true provided $N(X)$ lies in a topological polyhedron $P \subset X$ with $P \subset X$ *1-LCC*?

Question 3. How does one deal with generalized manifolds with $N(X) = X$?

Question 4. Is Conjecture 4.2 from the end of Section 4 true?

REFERENCES

1. Ancel, F. D., The locally flat approximation of cell-like embedding relations, Doctoral thesis, University of Wisconsin at Madison, 1976.
2. Ancel, F. D., and Cannon, J. W., The locally flat approximation of cell-like embedding relations, Ann. of Math., to appear.
3. Bredon, G. E., Wilder manifolds are locally orientable, Proc. Nat. Acad. Sci. U.S. 63 (1969), pp. 1079-1081.
4. Bredon, G. E., Sheaf Theory, McGraw Hill, New York, 1967.
5. Bryant, J. L., and Lacher, R. C., Blowing up homology manifolds, J. London Math. Soc. (2) 16 (1977).
6. ______, Resolving 0-dimensional singularities in generalized manifolds, Proc. Camb. Phil. Soc. (to appear).
7. Bryant, J. L., and Seebeck, C. L., III, Locally nice embeddings of polyhedra, Quart. J. Math. (Oxford) (2) 19 (1968), pp. 257-274.
8. ______, Locally nice embeddings in codimension three, Quart. J. Math. (Oxford) (2) 21 (1970), pp. 265-272.
9. Cannon, J. W., Taming cell-like embedding relations, in Geometric Topology, (L. C. Glaser and T. B. Rushing, editors), Lecture Notes in Mathematics #43, Springer-Verlag, New York, 1975, pp. 66-
10. ______, Taming codimension-one generalized submanifolds of S^n, Topology, 16 (1977), pp. 323-334.
11. ______, Shrinking cell-like decompositions of manifolds. Codimension Three., Ann. of Math., to appear.
12. Cernavskii, A. V., Coincidence of local flatness and local simple-connectedness for embeddings of $(n-1)$-dimensional manifolds in n-dimensional manifolds when $n > 4$, Mat. Sbornik 91 (133) (1973), pp. 297-304.

13. Cohen, M. M., A general theory of relative regular neighborhoods, Trans. Amer. Math. Soc. 136 (1969), 189-229.
14. Daverman, R. J., Sewings of closed n-cell complements, Trans. Amer. Math. Soc., to appear.
15. ______________, Every crumpled n-cube is a closed -cell-complement. Michigan Math. J., 24 (1977), pp. 225-241.
16. Edwards, R. D., Demension theory, I, in Geometric Topology (L. C. Glaser and T. B. Rushing, editors), Lecture Notes in Mathematics #438, Springer-Verlag, New York, 1975, pp. 195-211.
17. ______________, The double suspension of a certain homology 3-sphere is S^5, Notices Amer. Math. Soc., 22 (1975), A 334. Abstract #75T-G33.
18. ______________, Approximating certain cell-like maps by homeomorphisms, Ann. of Math, to appear.
19. Eilenberg, S., and Wilder, R. L., Uniform local connectedness and contractibility, Amer. J. Math. 64)1942), pp. 613-622.
20. Ferry. S.
21. Hosay, N., The sum of a real cube and a crumpled cube is S^3, Notices Amer. Math. Soc. 10 (1963), 666. Abstract #607-17.
22. Kirby, R. C., and Siebenmann, L. C., Foundational Essays on Topological Manifolds, Princeton Univ. Press, Princeton, N. J., 1977.
23. Lacher, R. C., Cell-like mappings and their generalizations, Bull. Amer. Math. Soc. 83 (1977), 495-552.
24. Lashof, R., and Rothenberg, M., Microbundles and smoothings, Topology 3 (1964), pp. 354-388.
25. Lickorish, W. B. R., and Siebenmann, L. C., Regular Neighbourhood and the stable range, Trans. Amer. Math. Soc., 139 (1969), 207-230.
26. Lininger, L. L., Some results on crumpled cubes, Trans. Amer. Math. Soc. 118 (1965), pp. 534-549.
27. Milnor, J. W., and Stasheff, J. D., Characteristic Classes, Princeton Univ. Press, Princeton, N. J., 1974.
28. Seebeck, C. L., III, Codimension one manifolds that are locally homotopically unknoted on one side I and II, preprint.
29. Serre, J. P., A Course in Arithmetic, Springer Verlag, New York, 1970.
30. Stan'ko, M. A., Approximation of imbeddings of compacta in codimensions greater than two, Dokl. Akad. Nauk SSSR 198 (1971), Soviet Math. Dokl. 12 (1971), pp. 906-909.
31. Tinsley, F., Doctoral Thesis, University of Wisconsin at Madison, 1977.
32. Wilder, R. L., Topology of Manifolds, Amer. Math. Soc., Colloq. Publ., Vol. 32, Amer. Math. Soc., Providence, R. I., 1963.

EMBEDDINGS AND IMMERSIONS OF FOUR DIMENSIONAL MANIFOLDS IN R^6

Sylvanin E. Cappell[1]

Courant Institute of Mathematical Sciences
New York University
New York, N.Y.

Julius L. Shaneson[1]

Department of Mathematics
Rutgers University
New Brunswick, N. J.

This announcement gives necessary and sufficient conditions for an orientable closed differentiable or P.L. four-dimensional manifold to have in R^6

1. A differentiable or P.L. locally flat embedding; or
2. A P.L. (but not necessarily locally flat) embedding; or
3. A differentiable or P.L. locally flat immersion; or
4. A P.L., but not necessarily locally flat immersion.

All of these results, except for (3) which is just a restatement of the Smale-Hirsch theory of smooth immersions in this context, seem to be new even in the easier simply connected case. In what follows, M is always a connected oriented four dimensional P.L. manifold. Recall, that every such four manifold has a unique differentiable structure.

[1]*Supported by NSF Grant*

ISBN 0-12-158860-2

THEOREM 1. The following are equivalent.

(i) *M is stably parallelizable (i.e. the tangent bundle of M plus a trivial line bundle is trivial)*

(ii) $\omega_2(M) = 0$ *and* $p_1(M) = 0$

(iii) *M is almost parallelizable and the index of M is zero*

(iv) *M embeds smoothly in* R^6

(v) *M has a locally flat P.L. embedding in* R^6.

Here, $\omega_2(M)$ (resp. $p_1(M)$) denotes the second Stiefel-Whitney (resp. first Pontrjagen) class. Almost parallelizable means that M - {a point} is parallelizable.

THEOREM 2. The following are equivalent.

(i) *M is almost parallelizable.*

(ii) $\omega_2(M) = 0$

(iii) *M has a P.L. (but not necessarily locally flat) embedding in* R^6

(iv) *M has a P.L. embedding in* R^6 *which is locally flat except possibly at one point of M.*

Embedding of four manifolds in R^7 were studied by Hirsch [8] in the P.L. case and by Boechat and Haefliger. [1] in the smooth case.

Theorem 3 is a formulation of the Smale-Hirsch [9][7] criteria for smooth immersion and the Haefliger-Poenaru [6] criteria for PL immersion in the present case.

THEOREM 3. The following are equivalent.

(i) $\omega_2(M)$ *is the reduction of an integral class* $x \in H^2(M;Z)$ *satisfying* $\langle x^2, [M]\rangle = -3I(M)$, *where* $I(M)$ *denotes the index of M.*

(ii) *M has a smooth immersion in* R^6

(iii) *M has a PL locally flat immersion in* R^6

THEOREM 4. The following are equivalent.

(i) The Euler characteristic of M is even.

(ii) M has a P.L. (but not necessarily locally flat) immersion in R^6.

(iii) M has a P.L. immersion in R^6 which is locally flat except possibly at one point.

For example, for any orientable 4-manifold, either M or M blown up at one point has an immersion in R^6.

REFERENCES

1. Boechat, J., and Haefliger, A., Plongements differentiables des verietes orientees de dimension 4 dans R^7, Essays on Topology and Related Topics, Memoirs dedies a Georges de Rham, Springer-Verlag (1970), 156-166.
2. Cappell, S., and Shaneson, J., The codimension two placement problem and homology equivalent manifolds, Ann. of Math. 99 (1974), 277-348.
3. __________, P. L. embeddings and their singularities, Ann. of Math. 103 (1976) 163-228.
4. __________, Immersion and singularities, Ann. of Math. 105 (1977), 539-552.
5. __________, Totally spineless manifolds, Illinois J. of Math. 83 (1977), 231-239.
6. Haeflinger, A., and Poenaru, V., Les classifications des immersions combinatoires, Institut des Hautes Etudes Scientifiques, Publications Mathématiques 23, 651-667.
7. Hirsch, M., Immersions of manifolds, Trans. Amer. Math. Soc. 93 (1959), 242-276.
8. __________, On embedding 4-manifolds in R^7, Proc. Cambridge Phil. Soc. 61 (1965), 657-658.
9. Smale, S., The classification of immersions of spheres in Euclidean spaces, Ann. of Math. 69 (1959), 327-344.

GENERAL POSITION MAPS FOR TOPOLOGICAL MANIFOLDS

Jerome Dancis

Department of Mathematics
University of Maryland
College Park, Maryland

For each proper map $f : M \to q$ of a topological m-manifold M into a topological q-manifold Q, $m \leq q - 3$, we build a "topological general-position" approximating map g such that:

(i) the set of singularities S of g is a locally-finite simplicial (2m - q)-complex which is locally-tamely embedded in M;

(ii) the set g(S) is another complex which is locally-tamely embedded in Q;

(iii) the map $g| : S \twoheadrightarrow g(S)$ is piecewise linear; and

(iv) $g \mid M - S$ is a locally-flat embedding.

The purpose of this paper is to announce a general position lemma for maps between topological manifolds.

Definition. Let $g : M \to Q$ be a continuous map of an m-manifold M into a q-manifold Q. Let S_g be the singular set of g. This map g is a *topological general position map* if

(a) S_g and $g(S_g)$ have triangulations as locally-finite $(2m-q)$-simplicial-complexes such that $g : S_g \twoheadrightarrow g(S_g)$ is piecewise linear.

(b) $g|M-S_g$ is a locally flat embedding

(c) For each point $p \in g(S_g)$:

ISBN 0-12-158860-2

(i) $g^{-1}(p)$ is a finite set $\{a_1, a_2, \ldots, a_r\}$;

(ii) there are disjoint euclidean neighborhoods E^q and $\{E_i^m\}$ such that $p \in E^q \subset Q$ and each $a_i \in E_i^m \subset M$ such that

$$g \mid : \cup E_i^m \longrightarrow E^q$$

is piecewise linear;

(iii) the triangulations of E^q and $g(S_g)$ are compatible, i.e., $g(S_g) \cap E^q$ is a subpolyhedron of both $g(S_g)$ and E^q. The triangulations of S_g and each E_i^m are also compatible.

Definition. A continuous map $f : X \to Y$ is *proper* means that the inverse of a compact set is always a compact set.

TOPOLOGICAL GENERAL POSITION LEMMA. Let $f : M \to Q$, be a proper map of a topological m-manifold into a topological q-manifold, $m \leq q - 3$. Then there is a general position map $g : M \to Q$ which approximates f. (If $\varepsilon > 0$ is given, then g may be constructed so that $d(f,g) < \varepsilon$).

COROLLARY. Let $f : (M, \partial M) \to (Q, \partial Q)$ be a proper map of a topological m-manifold M into a topological q-manifold Q, $m \leq q - 3$. Suppose $f \mid : \partial M \to \partial Q$ is a topological general position map. Then f may be approximated by a general position map $g : (M, \partial M) \to (Q, \partial Q)$ such that $g \mid \partial M = f \mid \partial M$ (and g is homotopic to f).

The only tools that are used in our proof are these Taming Lemmas:

TAMING LEMMA 1. (Bryant, Seebeck, Cernavskii, Homma and Miller). Let K be a k-complex in the interior of a combinatorial q-manifold Q, $k \leq q - 3$. Let the interior of each simplex of K be *locally flat in Q and let $\varepsilon > 0$. Then there is an ambient ε-isotopy*

$$\{H_t : Q \twoheadrightarrow Q, t \in [0,1] \quad \text{and} \quad H_0 = 1\}$$

such that

(i) $H_1|K$ *is piecewise linear*

(ii) $H_t(x) = (x)$, $d(x,K) > \varepsilon$ *and* $t \in [0,1]$.

Furthermore, if L is a subcomplex of both K and Q, then we may insist that $H_1|L = 1$.

Remark. This Taming Lemma 1 is a corollary of the theorems of [3], [8], [4] and [2]. Of course [3] uses the ideas of [7].

TAMING LEMMA 2. (Bryant, Seebeck, Cernavskii, Homma and Miller). Let $h : E^q \to Q$ *be a coordinate chart of a topological q-manifold Q. Let* $g : K \to h(E^q)$ *be a continuous map whose domain is a finite simplicial k-complex K,* $k \leq q-3$. *Suppose g(K) has a triangulation as a simplicial complex such that* $g : K \twoheadrightarrow g(K)$ *is a piecewise linear map. Suppose that the interior of each simplex of g(K) is a locally flat subset of Q. Then there is a "replacement" coordinate chart* $h' : E^q \to Q$ *such that*

$$h'^{-1} \circ g : K \to E^q$$

is piecewise linear (and such that $h'(E^q) = h(E^q)$ *and h and h' agree off some compact set).*

Furthermore if L is a subcomplex of K and if $h^{-1}g : L \to E^q$ *is already piecewise linear, then h' will agree with h on* $h^{-1}g(L)$.

Remark. Taming Lemma 2 is a corollary of Taming Lemma 1.

History. We established the Topological General Position Lemma for the dimension range $m \leq (2/3)q - 1/3$ in [6]. The case of the trivial range $2m+2 \leq q$, where g is a locally-flat embedding, was established in [5].

Our method of doing patch by patch arguments on a topological manifold is to do an induction on the sets $\{N_i\}$ of something we call a "standard representation", namely

A *standard representation* of a compact topological n-manifold M is a collection of compact subsets

$$\{N_i,\ I_i,\ M_i',\ M_i'',\quad i = 1, 2, \ldots, r\}$$

of M where

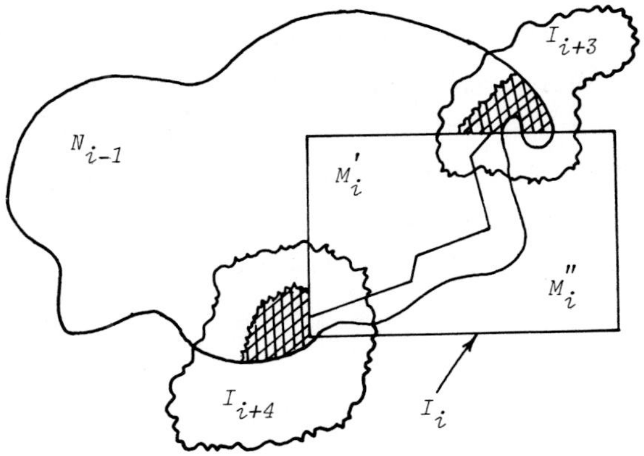

Fig. 1. Note: The shaded area is $N_{i-1} - N_i$ = *Buffer #i.*

(a) The I_i are locally flat n-cells whose interiors cover M

(b) (i) $N_1 = I_1$

(ii) $N_i = (N_{i-1} \cup I_i)$ - Buffer # i, where Buffer # i is the cross hatched area, in Figure 1, which separates M_i'' from $N_{i-1} - I_i$.

(iii) $N_r = M$

(c) $M_i' \approx I_i \cap N_{i-1}$, actually $M_i' = (I_i \cap N_{i-1}) - (Nbhd\ \partial N_{i-1})$.

(d) $M_i' \cup M_i'' = I_i$ and M_i' and M_i'' are combinatorial submanifolds of I_i, $M_i'' = C\ell(I_i - M_i')$.

Remark. The standard representation is formally presented in Section 2 of [6].

In this paper, we shall present only the following proof.

Outline of Proof of Topological General Position Lemma when M is compact and $\partial M = \phi = \partial Q$. The epsilonics are omitted.

Let $f : M \longrightarrow Q$ satisfy the hypotheses of this lemma. The proof will be a Big Induction (which includes two small inductions) over a standard representation $\{I_i, M_i', M_i'', N_i, i=1,2,\ldots, r\}$ of M, where each $f(I_i)$ is contained in some euclidean patch E_i^q of Q.

Big Induction. We shall construct approximations $g_i : M \longrightarrow Q$ of $f, i=1,2,\ldots,r$ such that each $g_i|N_i$ is a topological general position map. Also g_i will differ from g_{i-1}, on $N_{i-1} - M_i''$, by an ambient isotopy of Q, with support in E_i^q; except possibly on Buffer # i (the cross hatched section).

One begins by approximating $f| : I_1 \longrightarrow E_1^q$ by a general position PL map, and then extending this approximation to a map $g_1 : M \longrightarrow Q$.

The construction of each of the other maps $g_2, g_3, \ldots, g_r$ will consist of two parts:

Part (i). We will change the coordinate charts in order to make $g_{i-1}: |M_i' \longrightarrow E_i^q$ a piecewise linear map. Then we may define $g_i|: I_i \to E_i^q$ to be a piecewise linear, general-position map which agrees with g_{i-1} on the set M_i' of the standard representation.

Some details for these two parts will follow the basic description of Part (ii).

Part (ii). In this part, we will make the intersection

$$g_i(M_i'') \quad \cap \quad g_{i-1}[N_i - I_i - (\text{Buffer } \# \ i)]$$

into a polyhedron.

We will accomplish this by changing each of the following 3 intersections into polyhedra, in *this order:* (Note: Collar # i is a collar of $(\partial I_i) \cap N_{i-1}$, see Figure 2).

(a) $g_i(M_i'') \cap g_{i-1}$ (a *Nbhd* S_{i-1})

(b) $g_i(M_i'') \cap g_{i-1}(N_{i-1}$ - *Nbhd* S_{i-1} - Collar # i)

(c) $g_i(M_i'') \cap g_{i-1}$ (Collar # i - *Nbhd* S_{i-1} - Buffer # i)

These changes will be performed by an ambient isotopy (on Q Rel $g_i(M_i')$).

We will then paste the new map

$$g_{i-1}|N_{i-1} - (\text{Buffer } \# \ i) \text{ and the map } g_i|I_i$$

together to obtain $g_i|N_i$. The pasting is all right because g_{i-1} and g_i agree on M_i' and because the map on the Buffer zone will be corrected at later stages. Furthermore, since the set $g_{i-1}(N_i - I_i)$ is changed by an ambient isotopy, the changes will *not* affect the already-created general-position set of singularities of

$$g_{i-1} \mid N_i - I_i - (\text{Buffer } \# \ i).$$

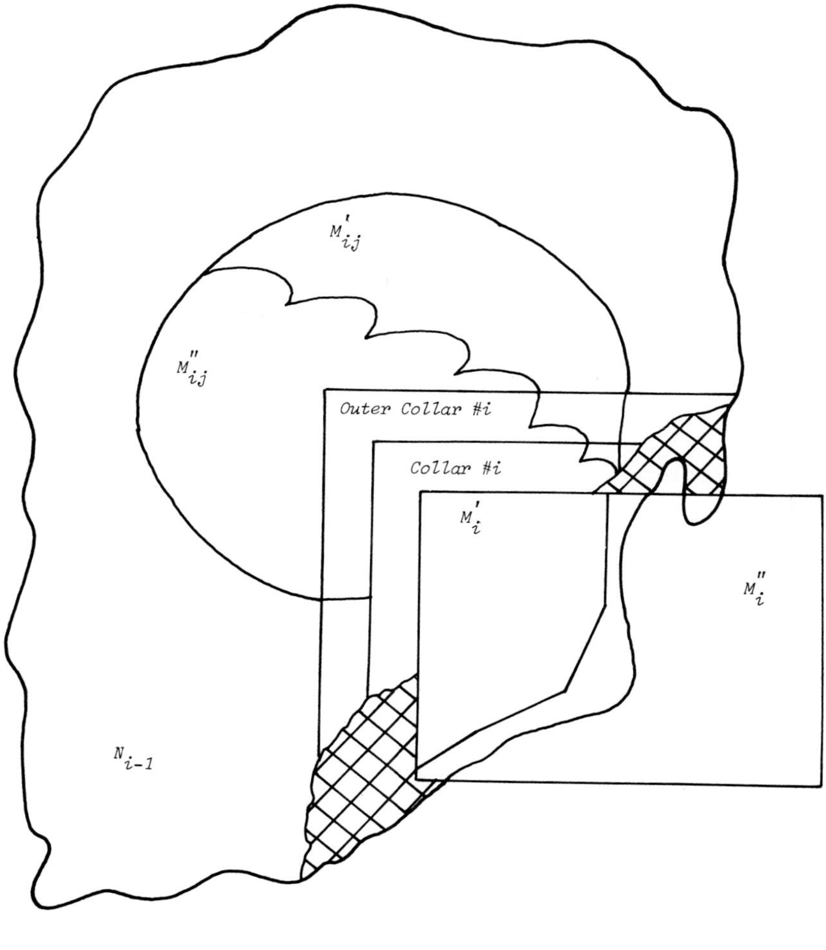

Fig. 2. Note: The cross hatched area is Buffer #i = $N_{i-1} - N_i$.

Let S_{i-1} denote the set of singularities of $g_{i-1}|N_{i-1}$ - Buffer # i. We note that by induction, S_{i-1} and $g_{i-1}(S_{i-1})$, are codimension ≥ 3, locally-tame polyhedra in M and Q, respectively.

Outline of Part (i). (a) By induction, the set of singularities S_{i-1} of $g_{i-1}|N_{i-1}$ is a codimension ≥ 3 locally-tame polyhedron in M. Hence a *Nbhd* $(S_{i-1} \cap M_i')$ is a tame polyhedron in E_i^m, a *Nbhd* of I_i. Therefore Taming Lemma 2 implies that the original coordinate chart $h : E_i^m \longrightarrow M$ may be replaced by a new chart $h': E_i^m \rightarrow M$, such that $S_{i-1} \cap M_i'$ is a subcomplex of both S_{i-1} and M_i'.

(b) By induction, the set $g_{i-1}(M_i')$ is a codimension ≥ 3 locally-tame polyhedron in E_i^q. Therefore, Taming Lemma 2 implies that the coordinate chart, $h : E_i^q \rightarrow Q$, may be replaced by a new chart, so that $g_{i-1} | : M_i' \longrightarrow E_i^q$ is now piecewise linear. The induction implies that $g_{i-1} | M_i' \longrightarrow E_i^q$ is a (piecewise linear) general position map.

(c) Extend this map $g_{i-1} | M_i'$ to a piecewise linear general position map $g_i | I_i \longrightarrow E_i^q$. In addition, put $g_i | (S_{i-1} \cap M_i') \cup M_i''$ into general position.

Background for Part (ii). The proof for Part (ii) is difficult. The proofs for section (a) and (b) both use "little inductions" on standard representations. Each step of these two little inductions uses the Lemma of Four Subcomplexes below.

Definition. A map $g : K \rightarrow Q$, with domain a simplicial complex, is *in general position* if there is a coordinate chart, $E^q \subset Q$ such that $g : K \longrightarrow\!\!\!\!\rightarrow g(K) \subset E^q$ is a (PL) general position map. Two maps $g_1 : K \longrightarrow Q$ and $g_2 : L \longrightarrow Q$, with domains simplicial complexes, are *in general position* if there is a coordinate chart $E^q \subset Q$, which contains $g_1(K)$ and $g_2(L)$, such that g_1 and g_2 are in piecewise linear general position with respect to the chart E^q.

LEMMA OF FOUR SUBCOMPLEXES. Let simplicial complex I be the union of subcomplexes M' and M''. Let simplicial complex K be the union of subcomplexes K' and K''. Suppose that $g_1 : K \longrightarrow E^q$ and $g_2 : I \longrightarrow E^q$ are general position maps. Suppose that the maps

(1) $g_1 \mid K'$ and $g_2 \mid I$ are in general position,

and

(2) $g_1 \mid K$ and $g_2 \mid M'$ are in general position.

also suppose that either

(a) $K \cap I = \phi$

or

(b) $K \cap M'' = \phi$ and $K \cap I$ is a subcomplex of both K and I.

Let Dim K *and* Dim $I \leq q - 3$. *Then there is a homeomorphism h of E^q onto itself such that*

(3) $h \circ g_1 \mid K$ and $g_2 \mid I$ are in general position

and such that

(4) $h \mid g_2(M') \cup g_1(K') = 1$

and h is the identity off a compact set.

Proof. Taming Lemma 2 and (1) implies that there is a piecewise linear triangulation of E^q for which $g_1 \mid K'$ and $g_2 \mid I$ are piecewise linear maps. Then Taming Lemma 1 and (2) will make $g_1 \mid K''$ into a piecewise linear map without changing $g_1 \mid K'$ and $g_2 \mid M'$. Now that $g_1 \mid K''$ is a piecewise linear map, it may be immediately changed to one that is in (piecewise linear) general position with respect to $g_2 \mid M''_2$. This, together with (1) and (2) implies (3) and (4). This establishes the Lemma of Four Subcomplexes.

Outline of Part (ii) (a). Condition (iii) for a topological general position map, together with compactness, implies that some neighborhood of the singular set S_{i-1} may be represented as:

$$NbhD\ S_{i-1} = \bigcup_{k=1}^{r} \text{Int}\ R_{ik},$$

where:

(i) each R_{ik} is a disjoint union of open m-cells;

(ii) each

$$g_{i-1} \mid : R_{ik} \twoheadrightarrow g_{i-1}(R_{ik})$$

is a piecewise linear map;

(iii) each $S_{i-1} \cap R_{ik}$ is a subcomplex of both S_{i-1} and R_{ik};

(iv) each set $g_{i-1}(R_{ik}) \cup g_i(I_i)$, $k=1, 2, \ldots, r$, is a locally-tame simplicial complex in E_i^q.

Consider R_{i1} and M_i''. The Taming Lemma implies that one may change $g_{i-1} \mid : R_{i1} \longrightarrow E_i^q$ into a piecewise-linear map which is in generaly position with respect to $g_i(M_i'')$ (and to $g_i(I_i)$), by an ambient isotopy which is fixed on $g_i(M_i')$.

If we now try to adjust the intersection of the images of R_{i2} and M_i'' in the same way, we will probably mess up the previous work on the images of R_{i1}.

Therefore we are forced to perform an induction on a "standard representation"

$$\{N_k^*, R_k, R_k', R_k'', k=1,2,\ldots,r\}$$

of $\cup R_k$ which is analogous to the standard representation described earlier. These R's are not cells, but this will be all right.

Little Induction A. For each integer, $k = 2,3,4,\ldots,r$, there is a homeomorphism h_k of E^q onto itself such that the maps $h_k \circ g_{i-1} \mid N_k^*$ and $g_i \mid I_i$ will be in topological general position, and such that $h_k \mid g_i(M_i') = 1$.

Outline of Proof. At each step $k = 2, 3, \ldots, r$, one simply uses the Lemma of Four Subcomplexes on the subcomplexes M_i' and M_i'' of I_i and R_k' and R_k'' of R_k. Some checking will show that the conditions of the standard representation and of the Lemma for Four Subcomplexes will establish Little Induction A.

The last step of Little Induction A establishes Part (ii)(a).

Outline of Part II(b). We shall make the intersection

$$g_i(M_i'') \cap g_{i-1}(N_{i-1} - I_i - \text{collar } \#i)$$

into a polyhedron using an induction on the N_{i-1} - part of the standard representation on M, namely

Little Induction B. We shall construct maps $g_{ij} : M \longrightarrow Q$, $j=1, 2, \ldots, i-1$, such that

$$g_{ij} | : N_j - I_i - \text{collar } \# \, i \longrightarrow E_i^q$$

is in topological general position *with respect* to $g_i(I_i)$. Furthermore g_{ij} will differ from $g_{i,j-1}$ by an ambient isotopy (Rel $g_i(M_i')$).

Here we are concerned with the part of N_{i-1} which misses $I_i \cup S_i$. Therefore we shall be concerned with the sets (see Figure 2)

$$M_{ij}' \approx M_j' - I_i - \text{Collar } \# \, i - \text{Nbhd } (S_i);$$

$$M_{ij}'' \approx M_j'' - I_i - \text{Collar } \# \, i - \text{Nbhd } (S_i),$$

for $j=1, 2, \ldots, i-1$; where M_{ij}' and M_{ij}'' are piecewise linear submanifolds of M_j' and M_j'', respectively; and where

$$\partial(M_{ij}' \cup M_{ij}'') \quad \subset \quad \partial I_j \ \cup \text{Collar } \# \, i \ \cup \ \text{Nbhd } S_i.$$

Proof of Little Induction B. At each step $j=1,2,3,\ldots,i-1$, one simply uses the Lemma of Four Subcomplexes on the subcomplexes M_i' and M_i'' of I_i and M_{ij}' and M_{ij}'' of

$$M_{ij}' \cup M_{ij}'' \approx I_j - I_i - \text{collar } \# \, i - Nbhd\ (S_i).$$

Some checking will show that the conditions of the standard representation and of the Lemma of Four Subcomplexes will establish Little Induction B.

The last step of Little Induction B establishes Part (ii)(b).

Outline of Part (ii)(c). What remains is to adjust the intersection of the images of Collar # i and M_i'', without "destroying" any of the work on $N_{i-1} - I_i$. In order to "protect" the earlier work of parts (a) and (b), we introduce another collar and two combinatorial m-manifolds, B_i'' and B_i'. Let *Outer Collar* # i be a collar on the side of inner collar # i. (The union of these two collars is just one large collar on $(\partial I_i) \cap N_{i-1}$). Let

$$B_i' \approx \text{Outer Collar } \# \, i - N(S_i)$$

$$B_i'' \approx \text{Collar } \# \, i - N(S_i).$$

Note: $I_i \cup B_i' \cup B_i''$ is a piecewise linear m-manifold.

To establish Part (ii)(c) one simply uses the Lemma of Four Subcomplexes on the subcomplexes M_i' and M_i'' of I_i and B_i' and B_i'' of $B_i' \cup B_i''$.

Thus, we have established Part (ii) and as noted earlier, this will establish the Topological General Position Lemma.

APPLICATIONS

Our initial interest in topological general position was motivated by a desire to establish embedding theorems for topological manifolds. Here we announce some of our results on this and results on separating topological submanifolds that we obtained using topological general position lemmas about the metastable range. We have not, as of today, worked out the corollaries and consequences of this codim 3 lemma.

EMBEDDING THEOREM 3. Let $f : M \to Q$ *be a properly* $(2m-q+1)$*-connected map of a topological m-manifold* M $(\partial M = \emptyset)$ *into a topological q-manifold* Q, $m \leq (2/3)q - (2/3)$. *Then f is properly homotopic to a locally flat embedding of M into Q.*

Definition. Let $f : X \to Y$ be a proper map with mapping cylinder C_f. Let K_1 be an $(r-1)$ subcomplex of a locally finite r-complex K. Then f is *properly r-connected* if for *each* pair (K,K_1) each proper map $g_0 : (K,K_1) \to (C_f,X)$ is properly homotopic (Rel K_1) to a map $g_1 : K \to X$.

Plan of Proof. In the previous large summer topology conference in Georgia (1969) we announced the analogous PL embedding theorem ([1]). The proof of Embedding Theorem 3 is a topological version of our combinatorial work, together with topological general position. (Here, the details are omitted.)

EMBEDDING THEOREM 4. Let $f : M \to Q$ *be a properly* $(2m-q)$*-connected map of a topological m-manifold M into a topological q-manifold Q. Let* $f|\partial M$ *be a locally flat embedding. Suppose that* $\partial Q = \emptyset$, $\partial M \neq \emptyset$ *and when M is not compact then* ∂M *is also not compact. Then f is properly homotopic* (Rel ∂M) *to a locally flat embedding of M into Q.*

Remark. Embedding Theorem 4 is ususual in that ∂M is embedded in the *interior* of Q. One consequence of this is that Embedding Theorem 4 is *false* when $m = (2/3)q$ even when f is a homotopy equivalence and M is compact. Another consequence of this is Corollary 6.

Remark. In order to establish Embedding Theorem 4, we use Lemma 5.

LEMMA 5. Let $f : M \to Q$ be as in the topological General Position Lemma. Let f be a locally flat embedding of a component M_1 of ∂M. (Let $m \leq (2/3)q - (1/3)$). Then the approximation g may be constructed so that g agrees with f on M_1.

Remark. Lemma 5 does *not* follow immediately from the proof of the Topological General Position Lemma when $f(M_1)$ is contained in the *interior* of Q. Lemma 5 was established as Corollaries 6.2 and 6.3 of [6]. Hopefully we will be able to establish Lemma 5 in codim 3, soon.

COROLLARY 6. Let h_0 and $h_1 : M \hookrightarrow Q$ be two locally-flat proper embeddings of a topological m-manifold M into a topological q-manifold Q, $m \leq (2/3)q - (5/3)$. Let h_0 be properly homotopic to h_1, and let h_1 be a properly $(2m-q+2)$-connected map. Suppose that $h_0(M) \cap h_1(M) = \emptyset$ (and that $\partial M = \emptyset = \partial Q$). Then there is a proper locally-flat embedding $H : M \times [0,1] \hookrightarrow Q$ which extends h_0 and h_1.

SEPARATING MANIFOLD THEOREM 7. Let M and V be topological m and v-submanifolds of a topological q-manifold Q,

$$2m + v + 3 < 2q \quad \text{and} \quad (m \leq (2/3)q - (4/3)).$$

Let $\partial M = \emptyset = \partial Q$. If M can be properly homotoped off V, in Q, then M can be ambiently isotoped off V in Q.

Remark. The proof of this Separating Manifold Theorem is basically a topological analog of similar work of Hatcher and Quinn, Laudenbach and Stallings on compact PL manifolds.

Proof. Lemma 5 provides a general position approximation of the homotopy on M, namely $g : M \times [0,1] \to Q$ such that $g|M \times 0$ is the inclusion map and $V \cap g(M \times 1) = \emptyset$. Let S_g be the set of singularities of g. Then S is a complex with Dim $S \leq 2m-q+2$. There is a natural (locally-tame) shadow Sh of S in $M \times [0,1]$:

$$Sh = (S \times [0,1])/(S \cap (M \times 1)).$$

The set $g(Sh)$ is a locally tame $(2m-q+3)$-complex in Q. Topological general position (Lemmas 5 and 19 of [5], or Corollary 6.4 of [6]) will move $g(Sh)$ off V (in Q). The Tietze Extension Theorem will provide a map $f : M \to [0,1]$ such that for each point $x \in M$,

$$(x,f(x)) \in [M \times 1 \cup Nbhd(Sh)] - Sh,$$

and

$$g(x,f(x)) \in Q - V.$$

Check that $\{g(x,f(x)),\ x \in M\}$ is an embedding of M into $Q - V$, which is isotopic to the original embedding. When M is compact, the Topological Isotopy Covering Theorem completes the proof. When M is not compact more work is needed (which we will omit here).

SEPARATING MANIFOLD THEOREM 8. Let M and V be proper locally-flat topological m- and v-submanifolds in the interior of a topological q-manifold Q,

$$\partial M = \emptyset \quad \text{and} \quad m \leq (2/3q - (4/3) \quad \text{and} \quad q \geq 3.$$

Suppose

$(V, \partial V)$ is $(2m+v+3-2q)$-connected when M is compact

and

$(V, \partial V)$ is properly $(2m+v+3-2q)$-connected when M is not compact.

If M can be properly homotoped off V, then M can be ambiently isotoped off V.

Remark. The only difference between the proofs of Separating Manifold Theorems 7 and 8 appears in the separation of g(Sh) and V. Here topological general position for locally tame complexes and locally flat submanifolds (Corollary 6.4 of [6]) will make $V \cap g(Sh)$ into a locally tame $(2m+v+3-2q)$-complex of V. Then the connectivity conditions on $(V, \partial V)$ enable us to engulf $V \cap g(Sh)$ from ∂V. Now, since ∂V is contained in the interior of Q, $V \cap g(Sh)$ may be "swept" off V, thereby making $V \cap g(Sh) = \emptyset$, as desired. (We note that topological engulfing is just as easy as PL engulfing when one uses General Position Lemma 2 of [6].) Actually, here the "engulfing" is accomplished by a proper, locally-tame embedding of the pair

$$(V \cap g(Sh)) \times ([0,1),1) \quad \text{into} \quad (V, \partial V).$$

This embedding will induce the desired ambient isotopy even in the non-compact case; (details omitted).

REFERENCES

1. Harry W. Berkowitz and Jerome Dancis, PL approximations and embeddings of manifolds in the 3/4-range, Topology of Manifolds, Ed. James C. Cantrell and C. H. Edwards, Jr., Markham (1970) 335-340.
2. J. L. Bryant, Approximating embeddings of polyhedra in codimension three, Trans. Amer. Math. Soc. 170 (1972), 85-95.
3. J. L. Bryant and C. L. Seebeck III, Locally nice embeddings in codimension three, Quart. J. Math. Oxford Ser. (2) 21 (1970), 265-272. MR 44 #7560.

4. A. V. Cernavskii, Piece-wise linear approximation of embeddings of cells and spheres in codimensions higher than two, Mat. Sb. 80 (122) (1969), 339-364, Math. USSR Sb. 9 (1969), 321-344. MR 40 #4957
5. J. Dancis, Approximations and isotopies in the trivial range, Topology Seminar (Wisconsin, 1965), Ann. of Math. Studies, no. 60, Princeton Univ. Press, Princeton, N.J., (1966), 171-187. MR 36 #7144.
6. ________, General Position Maps For Topological Manifolds in the 2/3 RDS Range, Transactions, A.M.S. 216 (1976) 249-266.
7. Tatsua Homma, On the imbedding of polyhedra in manifolds, Yokohama Math. J. 10 (1962), 5-10. MR 27 #4236.
8. Richard T. Miller, Approximating codimension 3 embeddings, Ann. of Math. (2) 95 (1972), 406-416. MR 46 #6366.

THE EQUIVARIANT SURGERY PROBLEM FOR INVOLUTIONS

Karl Heinz Dovermann

Department of Mathematics
Rutgers University
New Brunswick, N.J.

Department of Mathematics
University of Bonn
Bonn, Germany

This paper describes a complete solution of the Z_2*-surgery problem where the dimension restrictions are less stringent then in* [11]. *We have to investigate intersections of the fixed point set with spheres on which we have to do surgery. This means that the algebra involved no longer need to be quadratic. Because of this we have to define a new Wittgroup and compute it.*

Notation. Let G be a finite group and X and Y closed compact oriented and smooth G-manifolds. Let $f : X \to Y$ be a G-normal map of degree 1.

Definition. $f : X \to Y$ is a G-normal map if f is a G-map and for a given bundle $\eta \in KO_G(Y)$

$$f^*(TY - \eta) = TX \qquad \text{(in } KO_G(X))$$

The isomorphism should be incorporated in the notation, but it will be suppressed. A weaker condition could be used as well, but it would not change the result of this paper (compare [10]).

ISBN 0-12-158860-2

SURGERY PROBLEM. Can we perform equivariant surgery on (X,f) to get (X',f') such that f' is a homotopy equivalence? If so we say that the surgery problem is solvable.

f' need not be an equivariant homotopy equivalence as the homotopy inverse of f' need not be equivariant.

Definition [12]. f' with the above properties will be called a pseudoequivalence.

A general definition of G-surgery is given in [10]. If $G = Z_2$ equivariant surgery is described as follows:

1) Surgery on the fixed point set:

The attaching map of a handle is an imbedding

$$\phi : S^i \times D^{k-i} \times D(V) \to X$$

where Z_2 operates trivially on $S^i \times D^{k-i}$ and $D(V)$ is the unit disc in V where Z_2 operates by $-Id$ on V. V is the slice representation of the component of the fixed point set of dimension k on which we are doing surgery.

2) Surgery on the free part:

The attaching map of a handle is an imbedding

$$\phi : Z_2 \times S^i \times D^{k-i} \to X$$

with the obvious Z_2-action and $k = \dim X$. Let s be the generator of Z_2, X^s and Y^s the fixed point sets, and $f^s : X^s \to Y^s$ the induced map.

STATEMENT OF RESULTS. We will give a complete answer for the surgery problem in a specific case:

Let $\dim Y = 2m$ *and* $\dim Y^s = m$; *Y, X^s, Y^s 1-connected and $d = $ degree f^s; ($d \equiv 1 (\mathrm{mod}\ 2)$ [3]), $m = 2n \geq 6$.*

THEOREM 1. With the above assumptions the surgery problem is solvable iff

3) $\sigma_{Z_{(2)}}(f^s) = 0$

4) $\operatorname{sign}(Z_2, X) - \operatorname{sign}(Z_2, Y) = 0$

5) $(d^2 - 1)\operatorname{sign}(s, Y) = 0$

Remarks. $\sigma_{Z_{(2)}}(f^s)$ is defined in [10] and is the obstruction for coverting f^s into a Z_2-homology-equivalence. This is a necessary condition for a pseudo-equivalence by Smith theory [2]. More explicitly

$$\sigma_{Z_{(2)}}(f^s) = \begin{cases} \operatorname{sign} X^s - \operatorname{sign} Y^s & \text{if } n \equiv 0(2) \\ \textit{Arf} \text{ invariant} & \text{if } n \equiv 1(2) \end{cases}$$

$\operatorname{sign}(Z_2, \ )$ is defined as in [1].

This theorem also holds if we have more than one component in Y^s provided $f^s_\# : \Pi_0(X^s) \to \Pi_0(Y^s)$ is a homeomorphism, all components are 1-connected, exactly one of them has dimension $2n$, and the others have dimension less than $2n$ and greater or equal to 6. In this case we get an obstruction $\sigma_\alpha(f) \in L_{\dim \alpha}(Z_{(2)})$ for each component α. We neglect this here as it is already treated in [10].

LEMMA 2. If $f^*TY = TX$ (f is framed) (4) $\Longrightarrow$ (5).

Definition. We say that $f : X \to Y$ satisfies *condition P* if

6) $K_*(f^s, Z_2) = \operatorname{Ker}(f^s_* : H_*(X^s, Z_2) \to H_*(Y^s, Z_2)) = 0$

7) $K_i(f, Z) = \operatorname{Ker}(f_* : H_i(X, Z) \to H_i(Y, Z)) = 0$
for $i < m$ and $2m = \dim X$.

By [13] it follows that $K = K_m(f, Z)$ is a projective $Z[Z_2]$-module which implies that it is free. Let $m = 2n$. Then the kernel in the middle dimension gives rise to an object (K, λ, μ, s) :

8) K is $Z[Z_2]$free, $s : K \to K$ the induced involution.

9) $\lambda : K \times K \to Z$ is a symmetric nondegenerate bilinear form.

10) $\mu : K \to Z$ the induced quadratic form compare [14]; $\mu(x) = \frac{1}{2}\lambda(x,x)$ and the remark $\chi_N \equiv 0$

11) $\lambda(sx, sy) = \lambda(x,y)$

Definition. (K,λ,μ,s) is a quasiquadratic (q.q.)-form.

Note that $\lambda(x,sx)$ *need not be even, which means that* (K,λ,μ,s) *does not define an element in* $L_{2m}(Z[Z_2], 1)$. Even the *type* of $\lambda(.,s.)$ [7] is not an invariant.

Example:

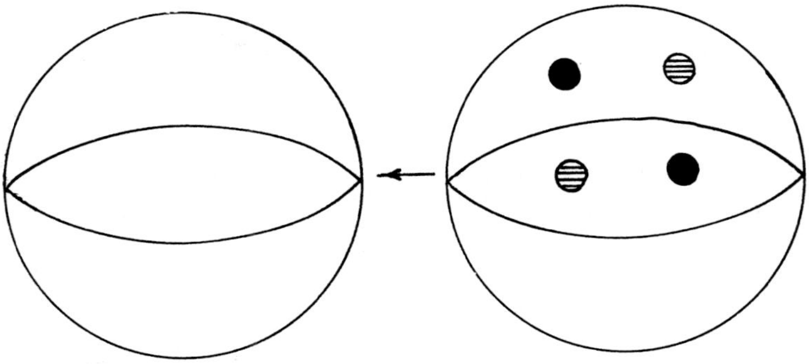

$$X = Y = S^2 \qquad (x,y,z) \xrightarrow{s} (x,y,-z) .$$

Doing surgery in the above picture on two copies $S^0 \times D^2$ which are interchanged by the involution (one is marked  the other one ) changes the type of $\lambda : K \times K \to Z$. This picture naturally generalizes to higher dimensions, and we do surgery on the boundary of a fiber of the normal bundle of the fixed point set.

Considering surgery in the middle dimension we give the following:

Definition. $(K,\lambda,\mu,s) \sim 0$ if there exists a free $Z[Z_2]$-module $N \subset K$ such that:

12) $\lambda|_{N \times N} = 0$

13) $N^{\perp} = N$

Then we call N a subkernel.

Definition. We can add q.q. forms by direct sum, and with the above equivalence relation "$\sim$" we get a group

$$\tilde{W}(Z[Z_2]).$$

Defining $\Lambda : K \times K \to Z[Z_2]$ by $\Lambda(x,y) = \lambda(x,y)1 + \lambda(x,sy)\, s$ (1 and s the elements in Z_2) we get the following:

THEOREM 3. $\tilde{W}(Z[Z_2]) \cong Z \oplus Z$. *The classification is given by the multisignature of* Λ. *The only relations for the invariants are:* sign $\lambda(.,.) \equiv 0$ mod 8 *and* sign $\lambda(,) \equiv$ sign $\lambda(,s)$ mod 2.

We want to give one more necessary obstruction in the case that $\dim X = 2m$, $\dim X^s \leq m$ and $\sigma_{Z_{(2)}}(f^s) = 0$. If Y^s is not connected we again have to express these obstructions in terms of components. Again assume 1-connectivity for components. Define $r(f)$ as follows: By our assumptions we can perform surgery on (X,f) to get (X',f') satisfying condition P. Let

14) $r(f) = rk_{Z[Z_2]} K_m(f', Z)$ modulo 2.

LEMMA 4. $r(f)$ *is well defined.* $r(f)$ *is an invariant for* Z_2-*surgery.*

One simple application is given as follows:

$$X = S^m \times S^m \qquad s(x,y) = (y,x)$$

$$Y = S^{2m} \qquad s(x_1,\dots,x_{2m+1}) = (-x_1,\dots,-x_m,x_{m+1},\dots,x_{2m+1})$$

and f is given as follows: Let $p \in X^s$ *and* D^{2m} *a small* Z_2*-invariant disc around p in X. Collapse* $X - D^{2m}$ *to one point. This gives a map* $f : X \to Y$ *with all our assumed properties including condition P. But* $H_m(X) = K_m(f) \cong Z[Z_2]$. $r(f) = 1$. *This implies that we cannot solve the surgery problem.*

The solution for the surgery problem $X \xrightarrow{f} Y$ where Z_2 operates freely on X and Y is given already by Wall. By looking at the orbit space we get a single obstruction in $L_{2m}(Z[Z_2],1)$ (compare theorem 1. 4)) If dim $X^s < \frac{1}{2}$ dim X, (this is the gaphypothesis in [10] and we get a complete answer for the surgery problem by the obstructions for the components of the fixed point set 3) and 4).) 5) vanishes automatically. (Here we assume that s preserves orientation). The gaphypothesis implies that Λ gives rise to a quadratic theory which allows us to describe the obstructions as elements in $L(Z[Z_2])$ and by general position we can always assume that $x \in K_m(f,Z)$ does not intersect X^s and sx.

Let dim $X^s = \frac{1}{2}$ dim X. A new problem arises if we want to do equivariant surgery in the middle dimension.

a.) (geometry). A sphere representing an element x in $K_m(f, Z)$ which we want to kill might intersect the fixed point set even if $\lambda(x,x) = \lambda(x,sx) = 0$. An obstruction for treating this problem is given the theorem 1. 5). It is remarkable that concrete geometric data are used.

b.) (algebra) New algebra is required as our theory is no longer quadratic (only quasiquadratic). This leads to the definition and computation of $\tilde{W}(Z[Z_2])$. Furthermore there arises a new invariant $r(f)$ which vanishes automatically for a quadratic theory. In the case that dim $X = 4n+2$ we can still solve the algebraic problem; $\tilde{W}_2(Z[Z_2]) \cong Z_2 \oplus Z_2$ if we use the appropriate definitions, but the geometric part of our argument can't be carried out as we don't have the G-signature theorem anymore.

If we increase the dimension of the fixed point set we encounter new problems. The kernels below the middle dimension are no longer classified as above, and we also get higher dimensional intersections. These problems seem to be unsolvable at the moment.

We have worked only with the most simple group Z_2, and we get into many more problems if we consider other groups. Some problems could be avoided by making more assumptions about $f : X \to Y$ than we did, like transversality, but we want a good G-surgery theory which is as broadly applicable as possible. This is the reason why it seems that we cannot avoid the gap-hypothesis in G-surgery.

This theory for differentiable manifolds has stronger results than we could expect in general by Smith theory.

Example. Consider $f : \mathbb{C}P^{2m} \to \mathbb{C}P^{2m}$, the fixed point set is $\mathbb{C}P^m \cup \mathbb{C}P^{m-1}$. Assume f is a pseudoequivalence and $d = \deg f_\alpha :$ $CP^m \to CP^m$. Then $d = \pm 1$ in our differientiable situation. Degree theory would only tell you $d \equiv 1 \bmod 2$ [3].

I want to thank T. Petrie for suggesting this problem and for many friendly and helpful discussions.

Proof of theorem 1.

"⟹" 3) is an immediate consequence of [10]

4) folklore

5) look at the commutative diagram

$$\begin{array}{ccc} H_m(X,\ Z) & \xrightarrow{\ f_*\ } & H_m(Y,\ Z) \\ {\scriptstyle i_*}\uparrow & & \uparrow{\scriptstyle i_*} \\ H_m(X^s,\ Z) & \xrightarrow{\ f^s_*\ } & H_m(Y^s,\ Z) \\ {\scriptstyle \cong}\uparrow & & \uparrow{\scriptstyle \cong} \\ Z & \xrightarrow{\ .d\ } & Z \end{array}$$

i_* is induced by the inclusion.

Compute sign$(s,.)$ in terms of the self intersection number λ of the fixed point set [8] then we get

$$\begin{aligned} \operatorname{sign}(s,X) - \operatorname{sign}(s,Y) &= \lambda([X^s],[X^s]) - \lambda([Y^s],[Y^s]) \\ &= \lambda(d[Y^s],\ d[Y^s]) - \lambda([Y^s],[Y^s]) \\ &= (d^2-1)\ \operatorname{sign}(s,Y) \end{aligned}$$

"$\Leftarrow$" $\sigma_{Z_{(2)}}(f^s) = 0$ implies that we can do surgery on the fixed point set such that f^s becomes a map inducing a Z_2-homology-equivalence. By [11] this is possible in X.

Now we do surgery in the free part and by transversality-arguments there is no problem below the middle dimension. Thus we can assume condition P. Furthermore theorem 2 and 5) tells us that $(K_m(f,\ Z),\lambda,\mu,s) = 0$ in $\tilde{W}(Z[Z_2])$. Let $\{e_1,\ldots,e_r\}$ be a $Z[Z_2]$-basis of N. By adding elements in this basis we can assume that $e_2,\ldots,e_r$ do not intersect the fixed point set. Thus we can represent them by imbedded spheres not intersecting the fixed point set or their image under s. [9]. This means we can kill them by surgery.

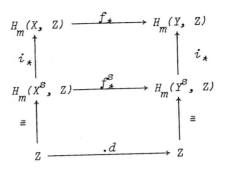

Assume now that N is generated by e. Choose $\tilde{f}$ such that $\lambda(e,\ \tilde{f}) = 1$ $\lambda(se,\ \tilde{f}) = 0$ ($\tilde{f}$ is dual to e) as in [14] and define $f := \tilde{f} - \frac{1}{2}\lambda(\tilde{f}, \tilde{f})e - \gamma se$ where $\lambda(\tilde{f}, \tilde{f}) = 2\gamma + a$. $a = 0$ or 1 depending on $\lambda(f, f') \equiv 0(2)$ or $\lambda(f, f') \not\equiv 0(2)$. Here $se = e'$ and $sf = f'$. In terms of the basis e, e', f, f' λ has the following matrix:

$$\lambda \hat{=} \begin{pmatrix} 0 & 0 & 1 & 0 \\ 0 & 0 & 0 & 1 \\ 1 & 0 & 0 & \ddot{a} \\ 0 & 1 & \ddot{a} & 0 \end{pmatrix}$$

Let (Π_1, Π_2) be the natural isomorphism:

$$(\Pi_1, \Pi_2) : H_m(X) \to H_m(Y) \otimes K$$

Then $i_* [X^s] = d\, i_* [Y^s] \oplus \alpha(e + e') + \beta(f + f')$ $\alpha, \beta \in Z$, and by an easy computation we get

$$0 = \text{sign}(s,X) - \text{sign}(s,Y) = (d^2-1)\ \text{sign}(s,Y) + 2\beta(2\alpha + a\beta) \Rightarrow$$
$$\beta(2\alpha + a\beta) = 0.$$

We want to show that we can do surgery to solve the surgery problem:

If $a = 0 \Rightarrow \alpha = 0$ or $\beta = 0 \Rightarrow$ we can do surgery on f and f' or on e and e'.

If $a \neq 0 \Rightarrow \beta = 0$ or $2\alpha + \beta = 0$.

If $\beta = 0$ we can do surgery on e and e'.

If $2\alpha + \beta = 0$ we can do surgery on $\tilde{e} := e - f - f'$ and $s\tilde{e}$.

This proves theorem 1.

Proof of Lemma 2. Computing the signature in terms of Hirzebruch L-classes we get:

$$\begin{aligned} \text{sign}(s,X) &= \langle L(TX^s, \nu(X^s, X)), [X^s]\rangle \\ &= \langle L(TY^s, \nu(Y^s, Y)), (f^s)^*[Y^s]\rangle \\ &= d \cdot \text{sign}(s,Y) \end{aligned}$$

$$(d^2-1)\ \text{sign}(s,Y) = (d+1)(d-1)\ \text{sign}(s,Y)$$

$$= (d+1)(\text{sign}(s,X) - \text{sign}(s,Y))$$
$$= 0$$

We need two more algebraic lemmas. Define the group $WG(.,.)$ as in [6]. Compared with the definition of $\tilde{W}$ we assume only that K and N are torsion free and $\lambda(x,x)$ need not be even. We have a split exact sequence [4]:

$$0 \to WG(Z_2, Z) \to WG(Z_2, Z[\tfrac{1}{2}]) \to W(Z_2) \to 0$$

With $\Lambda : K \times K \to Z[Z_2]$ as above we can define as in [15] $\Lambda\pm : K\pm \times K\pm \to Z$. This induces an isomorphism

$$\varphi : WG(Z_2, Z[\tfrac{1}{2}]) \to WG(1, Z[\tfrac{1}{2}]) \oplus WG(1, Z[\tfrac{1}{2}])$$
$$[(K,\lambda,s)] \longrightarrow [(K_+, \Lambda_+, Id)], \quad [(K_-, \Lambda_-, -Id)]$$

μ is not defined for elements in $WG(,)$.

$$0 \to WG(1,Z) \to WG(1, Z[\tfrac{1}{2}]) \to W(Z_2) \to 0$$

is split exact [4].

$$WG(1,Z) = Z \quad \text{and} \quad W(Z_2) = Z_2 \qquad [7]$$

This implies together

LEMMA 5. $WG(Z_2, Z) \cong Z \oplus Z \oplus Z_2$

Definition. Let $WG^{fr}(Z_2, Z)$ be those elements in $WG(Z_2, Z)$ which can be represented by (K,λ,s) where K is a free $Z[Z_2]$ module. Then we get a commutataive diagram:

$$\begin{array}{ccccc}
0 & \longrightarrow & WG^{fr}(Z_2, Z) & \longrightarrow & WG(Z_2, Z[\tfrac{1}{2}]) \\
 & & \downarrow \psi & & \downarrow \varphi \; \cong \\
 & & WG(1,Z) \oplus WG(1,Z) & \longrightarrow & WG(1,Z[\tfrac{1}{2}]) \oplus WG(1, Z[\tfrac{1}{2}])
\end{array}$$

ψ is defined as φ but it is defined only for free objects in $WG(Z_2, Z)$, being free is necessary to get nondegenerate forms. As ψ is injective and all signatures can be realized (the only restriction is $\mathrm{sign}\Lambda_+ \cong \mathrm{sign}\Lambda_-$ *(mod 2))* this proves

LEMMA 6. $WG^{fr}(Z_2, Z) \equiv Z \oplus Z$ *and the forms are classified by the multisignature.*

We have a projection map $\tilde{W}(Z[Z_2]) \to WG^{fr}(Z_2, Z)$. The image are the forms with $\lambda(x,x) \equiv 0 \bmod 2$, and we want to show injectivity.

Proof of theorem 3. If $\lambda(x, sx) \equiv 0(2)$ the induced quadratic form is defined and (K,λ,s) defines an element in $L_{4n}(Z[Z_2],1)$. Then the classification by the multisignature is given in [14], [15], which guarantees a $Z[Z_2]$-free subkernel.

Assume there exists $x \in K$ such that $\lambda(x, sx) \not\equiv 0(2)$. We have the exact sequence:

$$0 \longrightarrow N \longrightarrow K \longrightarrow \mathrm{Hom}(N, Z) \longrightarrow 0$$

and we can assume that $N = N_+ \oplus N_-$, where Z_2 operates by $\pm$ Id on $N_\pm$. If N contains a free part it can be split off. Write K as $\overline{K} + s\overline{K}$. Let $\{e_1^\pm, \ldots, e_r^\pm\}$ be a basis of $N_\pm$. Then $e_i^\pm = \bar{e}_i^\pm \pm s\bar{e}_i^\pm \in \overline{K} + s\overline{K}$ and $\{\bar{e}_i^\pm\}$ are a basis of $\overline{K}$. This is true as $\{\bar{e}_i^+, \bar{e}_i^-\}$ have to be linear independent, otherwise N would contain a free summand. They have to be a basis as otherwise K/N would contain torsion. By assumption

$$\lambda(\bar{e}_i^+ + s\,\bar{e}_i^+, \bar{e}_i^+ + s\,\bar{e}_i^+) = 0$$

$$\Rightarrow 2(\lambda(\bar{e}_i^+, \bar{e}_i^+) + \lambda(\bar{e}_i^+, s\,\bar{e}_i^+)) = 0$$

and as $\lambda(\bar{e}_i^+, \bar{e}_i^+) \equiv 0 \ (\mathrm{mod}\ 2)$ we get $\lambda(\bar{e}_i^+, s\,\bar{e}_i^+) \quad 0(\mathrm{mod}\ 2)$

Further more

$$\lambda(\overline{e_i^-} - s\,\overline{e_i^-},\ \overline{e_i^-} - s\,\overline{e_i^-}) = 0$$

$$\Rightarrow \lambda(\overline{e_i^-} m\ s\,\overline{e_i^-}) \equiv 0 (\text{mod } 2)$$

As $\lambda(\overline{e_i^{\pm}},\ s\,\overline{e_i^{\pm}}) \equiv 0(\text{mod } 2)$ it is true for all $x \in K$. ↯ This implies that N is a free $Z[Z_2]$-module.

Proof of Lemma 4. Assume (X,f) and (X',f') are as above satisfying condition P. Let (N,F) be a cobordism connecting them which is given by equivariant surgery. $\partial N = X \cup X'$. $F : (N,X,X') \to (Y \times I,\ Y \times 0,\ Y \times 1)$. Do surgery in the interior of N below the middle dimension. As in [14] you get the exact sequence

$$0 \longrightarrow K_m(N,\partial N) \longrightarrow K_m(\partial N) \longrightarrow K_m(N) \longrightarrow 0$$

and $\quad K_m(\partial N) \cong K_m(f) \oplus K_m(f')$

Furthermore $K_{m+1}(N,\ \partial N) \cong K_m(N)$ as $Z[Z_2]$ module.

$$\Rightarrow K_m(\partial N) \otimes \mathbb{Q} \cong (K_{m+1}(N,\partial N) \oplus K_m(N)) \otimes \mathbb{Q}$$

Applying Krull-Schmidt theorem [5] you see that

$$4 \mid rk_Z\, K_m(\partial N) \Rightarrow rk_{Z[Z_2]} K_m(f) \equiv rk_{Z[Z_2]} K_m(f') \ (\text{mod } 2)$$

REFERENCES

1. Atiyah, M. F. and Singer, I. M., The index of elliptic operators: III, Ann. of Math. 87 (1968), 546-604.
2. Bredon, G., Introduction to compact transformation groups, Academic Press (1972).
3. ____________, Fixed point sets of actions on Poincaré duality spaces, Topology 12 (1973), 159-175.
4. Conner, P. E., Writting invariants for periodic maps. (III) preprint.
5. Curtis, C. W. and Reiner, I., Representation Theory of finite groups and associative algebras, Interscience (1962).

6. Dress, A. W. M., Induction and structure theorems for orthogonal representations of finite groups, Ann. of Math. 102 (1975), 291-325.
7. Husemoller, D. and Milnor, J., Symmetric bilinear forms, Springer Verlag (1973).
8. Jänich, K. and Ossa, E., On the signature of an involution, Topology 8 (1969), 27-30.
9. Milnor, J., Lectures on the h-cobordism theory, Notes by L. Siebenmann and J. Sandow, Princeton (1965).
10. Petrie, T., G Surgery I, to appear, Proceedings Topology Conference, Santa Barbara, (1977).
11. ______________, G. Surgery II, to appear, (1977)
12. ______________, Pseudo equivalences of G manifolds, Proc. Sym. Pure Math. 32 (1977), 119-163.
13. ______________, G Maps and the projective class group, Comm. Math. Helv. 39 (1977), 611-626.
14. Wall, C.T.C., Surgery on Compact manifolds, Academic Press (1970).
15. ______________, Surgery of non-simply-connected manifolds, Ann. of Math. 84 (1966), 217-276.

THE DEGREE OF A BRANCHED COVERING OF A SPHERE[1]

Allan L. Edmonds[2]

Department of Mathematics
Cornell University
Ithaca, N. Y.

A new restriction on the degree of a branched covering of an n-manifold M^n over the n-sphere is discussed. In general the degree of such a branched covering is bounded below by the length of the longest non-trivial cup product in the reduced rational cohomology of M^n. Special cases of this result are proved here.

A well-known theorem of J. W. Alexander [1] states that any closed orientable piecewise linear manifold M^n is a piecewise linear branched covering of the n-sphere, i.e. there is a finite-to-one, open PL map $M^n \to S^n$. In this paper we discuss a new restriction on the degree of such a branched covering.

The restriction on degree is formulated in terms of the (complex) *cup length* of the connected manifold M^n. Define cup $M^n \geq r$ if there are homogeneous elements $u_1, u_2, \ldots, u_r \in \tilde{H}^*(M^n;\mathbb{C})$ such that the cup product $u_1 u_2 \ldots u_r \neq 0$.

THEOREM. If $\varphi : M^n \to N^n$ is a branched covering, then deg $\geq$ cup M^n/cup N^n.

[1] *This is a report of joint work with I. Berstein*

[2] *Supported in part by a National Science Foundation grant.*

ISBN 0-12-158860-2

This result is due to I. Berstein and the author and a complete proof will appear in [2]. Here we prove some special cases of this inequality which show why such a result should be true, without getting bogged down in the technical details of the general proof.

We prove this result when $N^n = S^n$ for cyclic branched coverings (Theorem 2) and for branched coverings φ with deg $\varphi \leq 3$ (Theorem 7). For degree 2 this was essentially observed by U. Hirsch and W. Newmann [8; Lemma].

Define the *height* of M^n by $ht\ M^n \geq r$ if there is a homogeneous class $x \in \tilde{H}^*(M^n;\mathbb{C})$ such that x^r is a nonzero element of $H^n(M^n;\mathbb{C})$. Of course cup $M^n \geq ht\ M^n$. We show (Theorem 8) that if $\varphi : M^n \to S^n$ is a branched covering, then deg $\varphi \geq ht\ M^n$. The proof nicely illustrates the idea behind the proof of the general result, but is considerably simpler.

In particular, the minimum all purpose degree in dimension n must increase with n. This puts in proper perspective the classical result that an orientable surface is a 2-fold branched covering of S^2 and the more recent result of H. Hilden [6], Hirsch [7], and J. Montesinos [9] that any orientable closed 3-manifold is a 3-fold branched covering of S^3.

Throughout M^n denotes a closed connected, orientable n-manifold, cohomology is taken with complex coefficients, and a branched covering is a finite-to-one open map.

The results here all depend upon the following formula [2]; If G is a finite group acting on a locally compact Hausdorff space X, then the orbit map $X \to X/G$ induces an isomorphism

$$H^*(X/G) \to H^*(X)^G. \qquad (*)$$

LEMMA 1. Let p be a positive integer and let $\{\lambda_1, \lambda_2, \ldots, \lambda_k\}$ *be a collection of pth roots of unity with* $k > p$. *Then some product* $\lambda_{i_1}\lambda_{i_2}\ldots\lambda_{i_s} = 1$, $s \leq p$.

Proof. Suppose not. For $i \leq p$, define $\mu_i = \lambda_1\lambda_2\cdots\lambda_i$. Then one easily checks that $\mu_i \neq \mu_j$ for $i \neq j$. But among the p distinct roots of unity $\mu_1,\mu_2,\ldots,\mu_p$, one must equal 1, a contradiction.

THEOREM 2. Let $\varphi : M^n \to S^n$ be a cyclic branched covering. Then $\deg \varphi \geq \mathrm{cup}\, M^n$.

Proof. Let $p = \deg \varphi$ and $k = \mathrm{cup}\, M^n$. Suppose $p < k$. There exist homogeneous classes $x_1,x_2,\ldots,x_k \in \tilde{H}^*(M^n)$ such that $\langle x_1x_2\cdots x_k, [M^n]\rangle \neq 0$.

Now φ can be identified with the orbit map for the cyclic group G of order p. By elementary representation theory $H^*(M^n)$ may be decomposed into the direct sum of the eigenspaces for the action of a fixed generator $g \in G$. Since each x_i is a linear combination of eigenvectors, we may assume that each x_i is actually a homogeneous eigenvector for g. Thus $g(x_i) = \lambda_i x_i$ for some p^{th} root of unity λ_i.

By Lemma 1, some product $\lambda_{i_1}\cdots\lambda_{i_s} \neq 0$, $s \leq p < k$. Therefore the non-zero element $x_{i_1}x_{i_2}\cdots x_{i_s} \in H^*(M^n)$, lies in $H^*(M^n)^G$, in some degree $r < n$. But then by (*), $H^r(S^n) \neq 0$, a contradiction.

REMARK 3. If $\varphi : M^n \to N^n$ is a cyclic branched covering, then essentially the same proof shows that $\deg \varphi \geq \mathrm{cup}\, M^n/\mathrm{cup}\, N^n$.

REMARK 4. If $\varphi : M^n \to N^n$ is a solvable branched covering -- that is φ can be factored as a composition of cyclic branched coverings -- then a trivial induction argument shows that $\deg \varphi \geq \mathrm{cup}\, M^n/\mathrm{cup}\, N^n$.

The following construction is the basis for our results on irregular branched coverings:

PROPOSITION 5. Let $\varphi : M^n \to N^n$ *be any branched covering of degree* p. *Then there is a space* X *with an action of a finite group* G *which fits into a commutative diagram*

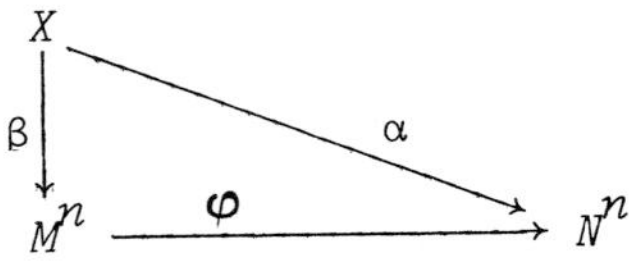

where α is identified with the orbit map for G, β is identified with the orbit map for a subgroup $H < G$ of index p and there is an embedding $G \subset S_p$ (the symmetric group on p letters) such that $H = G \cap S_{p-1}$.

Proof. According to Cernavskii [4] or Väisälä [11], the branch set B_φ has dimension $n-2$ and so neither separates nor locally separates N^n. Therefore φ is a branched covering in the sense of Fox [5], and so, according to Fox, is determined as an appropriate compactification of the associated unbranched covering $\varphi_0 : M^n - \varphi^{-1}B_\varphi \to N^n - B_\varphi$. This p-fold covering is determined by a homomorphism $\rho : \pi_1(N^n - B_\varphi, *) \to S_p$, with image G, say. By covering space theory we obtain a commutative diagram

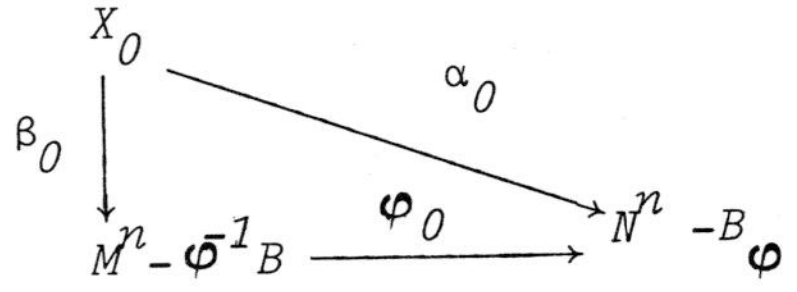

of covering spaces, where α_0 is the orbit map for an action of G on X_0 and α_0 is the orbit map for the action of $H = G \cap S_{p-1}$, where H necessarily has index p in G.

Then by [5] we may take the triple (X, α, β) to be a suitable compactification of (X_0, α_0, β_0) which makes the desired diagram commute. One easily sees that the action of G on X_0 extends uniquely over X.

REMARK 6. Of course, in general, the space X constructed above is not a manifold, but it is a "normal circuit". In particular $H^n(X) \approx \not{Q}$.

The subsequent work of this section is devoted to studying the G action on $H^*(X)$.

THEOREM 7. Let $\varphi : M^n \to S^n$ be a branched covering of degree ≤ 3. Then $\deg \varphi \geq cup\ M^n$.

Proof. By Theorem 2 it suffices to assume $\deg \varphi = 3$ and φ is irregular, since if $\deg \varphi = 2$ or if $\deg \varphi = 3$ and φ is regular, then φ is a cyclic branched covering. Let X, α, β, $G \subset S_3$. and $H = G \cap S_2$ be a guaranteed by Proposition 2.5. Since φ is irregular, one easily sees that $G = S_3$ and $H = S_2$.

Let G have the presentation $\{\sigma, \tau : \sigma^2 = 1, \tau^3 = 1, \sigma\tau\sigma = \tau^2\}$ where H is the subgroup generated by σ.

For any $x \in \tilde{H}^*(M^n)$, let x_1 denote its image in $\tilde{H}^*(X)$. Then by $(*)$, $\sigma(x_1) = x_1$. Let $x_2 = \tau x_1$. If $\dim x < n$, then $\tau x_2 = -x_1 - x_2$, since $x_1 + \tau x_1 + \tau^2 x_1$ is fixed by G, and so is 0. Also $\sigma x_2 = \sigma\tau x_1 = \tau^2 \sigma x_1 = \tau^2 x_1 = -x_1 - x_2$.

Now suppose that $cup\ M^n \geq 4$. Then, by Poincare duality, there are homogeneous elements $a, b, c, d \in \tilde{H}^*(M^n)$ such that $\langle abcd, [M^n]\rangle \neq 0$. Therefore $a_1 b_1 c_1 d_1 \in \tilde{H}^*(X)^G$.

We establish the following formulas

$$a_i b_j c_k d_\ell = \begin{cases} a_1 b_1 c_1 d_1 & \text{if } i = j = k = \ell, \\ -\frac{1}{2} a_1 b_1 c_1 d_1 & \text{otherwise.} \end{cases}$$

For the first part $a_2b_2c_2d_2 = \tau(a_1b_1c_1d_1) = a_1b_1c_1d_1$, since G acts trivially on $H^n(X)$. For the second part we prove one case and leave the remainder to the reader. For example $a_2b_1c_2d_1 = \varepsilon a_2c_2b_1d_1 = \varepsilon\sigma(a_2c_2b_1d_1) = \varepsilon(-a_1c_1-a_2c_2)b_1d_1 = -\varepsilon a_1c_1b_1d_1 - \varepsilon a_2c_2b_1d_1 = -a_1b_1c_1d_1 - a_2b_1c_2d_1$ (where $\varepsilon = (-1)^{\deg b \deg c}$), which implies the desired result.

Finally,

$$a_1b_1c_1d_1 = \tau^2(a_1b_1c_1d_1) = \tau a_2\tau b_2\tau c_2\tau d_2$$

$$= (-a_1-a_2)(-b_1-b_2)(-c_1-c_2)(-d_1-d_2)$$

$$= (a_1+a_2)(b_1+b_2)(c_1+c_2)(d_1+d_2)$$

$$= 2a_1b_1c_1d_1 + 14(\tfrac{1}{2})a_1b_1c_1d_1 \quad ,$$

since the indicated product contains 2 pure and 14 mixed terms. This implies $a_1b_1c_1d_1 = 0$, a contradiction.

THEOREM 8. If $\varphi : M^n \to S^n$ is a branched covering, then $\deg \varphi \geq \operatorname{ht} M^n$.

Proof. Let $\deg \varphi = p$ and again let X, α, β, $G \subset S_p$, and $H = G \cap S_{p-1}$ be as given by Proposition 5. Suppose $\operatorname{ht} M^n \geq r > p$ and let $x \in \tilde{H}^*(M^n)$ be a homogeneous class such that $\langle x^r, [M^n]\rangle \neq 0$. Let x_1 denote the image of x in $H^*(X)$. Then $x_1 \in H^n(X)^H$, while $x_1^r \in H^*(X)G$.

Let $\tau = (12\ldots p)$ in S_p. Since $p = (G:H) = (S_p:S_{p-1})$, for each τ^i there is a $g_i \in G$ such that $g_iS_{p-1} = \tau^iS_{p-1}$.

Consider the polynomial algebra $A = \mathbb{C}[a_1,a_2,\ldots,a_p]$, with its natural S_p action by permutations of subscripts, where S_{p-1} fixes a_1. Let $f : A \to H^*(X)$ be the algebra homomorphism given by $f(a_i) = g_ix_1$, where $g_iS_{p-1} = \tau^iS_{p-1}$ and $g_i \in G$. One easily verifies that f is a well-defined G-homomorphism.

Now observe that if $b = \frac{1}{p}(a_1^r + a_2^r + \ldots + a_p^r)$, then $b \in A^{S_p}$ and $f(b) = x_1^r$. By the symmetric function theorem (e.g., [10, 132ff]), b is a polynomial in the elementary symmetric functions in $a_1, a_2, \ldots, a_p$. In particular, b, and hence x_1^r, is a polynomial in elements fixed by G and of nonzero degree at most p. Therefore $H^i(X)^G \neq 0$ for some $i \leq p$ degree $x < n$. This implies $H^i(S^n) = 0$, a contradiction.

REFERENCES

1. J. W. Alexander, Note on Riemann spaces, Bull. Amer. Math. Soc. 26 (1920), 370-372.
2. I. Berstein and A. L. Edmonds, The degree and branch set of a branched covering, Invent. Math., to appear
3. A. Borel, Seminar on transformation groups, Annals of Math. Studies No. 46, Princeton University Press, Princeton, N.J., 1960.
4. A. V. Cernavskii, Finitely multiple open mappings of manifolds, Soviet Math. Doklady 4(1963), 946-949.
5. R. H. Fox, Coverings spaces with singularities, Algebraic Geometry and Topology (R.H. Fox, et. al., editors), pp. 243-257, Princeton University Press, Princeton, N.J. 1957.
6. H. M. Hilden, Three-fold branched coverings of S^3, Amer. J. Math. 98 (1976), 989-997.
7. U. Hirsch, Über offene Abbildungen auf die 3-sphäre, Math. Z. 140 (1974), 203-230.
8. U. Hirsch and W. D. Neumann, On cyclic branched coverings of spheres, Math. Ann. 215 (1975), 289-291.
9. J. M. Montesinos, Three-manifolds as 3-fold branched covers of S^3, Quart. J. Math. Oxford (2), 27 (1976), 85-94.
10. S. Lang, Algebra, Addison-Wesley, New York, 1965.
11. J. Väisälä, Discrete open mappings on manifolds, Ann. Acad. Sci. Fenn. Ser. AI, No. 392 (1966), 10 pp.

A UNIVERSAL 5-MANIFOLD WITH RESPECT TO SIMPLICIAL TRIANGULATIONS

D. Galewski

Department of Mathematics
University of Georgia
Athens, Georgia

R. Stern

Department of Mathematics
University of Utah
Salt Lake City, Utah

I. INTRODUCTION

One of the most important questions in geometric topology is whether or not every topological manifold has a locally finite simplicial triangulation. We first recall some results in this direction.

Let θ_3^H be the abelian group obtained from the set of oriented *PL* homology 3-spheres, under the operation of connected sum, modulo those which bound *PL* acyclic 4-manifolds. Also let $\mu : \theta_3^H \to \mathbb{Z}_2$ be the Kervaire-Milnor-Rohlin map given by $\mu[H^3] = I(W)/8 \bmod 2$ where $I(W)$ is the index of any parallelizable *PL* 4-manifold W which H^3 bounds. We note that μ is well defined and surjective.

ISBN 0-12-158860-2

THEOREM 1.1. (Galewsik-Stern [3], [4]). Let M^n be a compact topological n-manifold with $n \geq 6$ ($n \geq 5$ if ∂M simplicially triangulated). Then there exists an element $t_M \in H^5(M, \partial M; \ker \mu)$ such that $t_M = 0$ y and only if there exists a simplicial triangulation $K|\partial M$ compatible with the given triangulation on ∂M. Moreover, the number of concordance classes of simplicial triangulations rel ∂M *is in 1-1 correspondence with $H^4(M, \partial M; \ker \mu)$, where two triangulations K_1 and K_2 of M are concordant* rel ∂M *if there is a triangulation K of $M \times I$ such that $K|_{M\times 0}$, $K|_{M\times 1}$, and $K_{M\times I}$ are compatible with K_0, K_1 and $K|_{\partial M} \times I$, respectively.*

THEOREM 1.2. (Galewski-Stern [4]). If $\sigma_M \in H^4(M, \partial M; \mathbb{Z}_2)$ is the Kirby-Siebenmann obstruction [5] *to a PL triangulation of M* rel ∂M *then $\beta(\sigma_M) = t_M$ where β is the Bockstein associated with the exact sequence*

$$0 \to \ker \mu \to \theta_3^H \xrightarrow{\mu} \mathbb{Z}_2 \to 0.$$

THEOREM 1.3. (Galewski-Stern [3],[4]; T. Matumoto [7]). If there exists an element $x \in \theta_3^H$ with $\mu(x) = 1$ and $2x = 0$, then all compact topological n-manifolds with $n \geq 6$ ($n \geq 5$ if ∂M simplicially triangulated) have a simplicial triangulation.

Our original classification theorem in [3] had two possibly non-zero obstructions. However, the solution of the double suspension conjecture ([1],[2]) implies that one of these obstructions vanish.

In this paper we give a geometric construction of a closed non-orientable topological 5-manifold N^5 with the property that N^5 has a simplicial triangulation if and only if *every* compact topological n-manifold M^n $n \geq 6$ ($n \geq 5$ if ∂M simplicially triangulated) has a simplicial triangulation. Note that Siebenmann's Theorem B of [9] and the double suspension theorem ([1],[2]) show that all open or oriented closed 5-manifolds can be simplicially triangulated.

We would like to thank R. D. Edwards for suggesting the problem and for some helpful conversations.

II. PRELIMINARY RESULTS

Recall S_q^1 is the Bockstein associated with the short exact sequence

$$0 \longrightarrow \mathbb{Z}_2 \xrightarrow{\times 2} \mathbb{Z}_4 \xrightarrow{r} \mathbb{Z}_2 \longrightarrow 0.$$

THEOREM 2.1. If there exists a closed simplicially triangulated topological n-manifold N^n for any $n \geq 5$ with $S_1^1 \sigma_N \neq 0$ where σ_N is the Kirby-Siebenmann obstruction then all compact topological m-manifolds M^m with $m \geq 6$ ($m \geq 5$ if M simplicially triangulated) have a simplicial triangulation.

Proof. Let N^n be a closed simplicially triangulated topological n-manifold with $n \geq 5$ and assume that there exists a compact m-manifold M^m with $m \geq 6$ ($m \geq 5$ if ∂M simplicially triangulated) that does not have a simplicial triangulation. They by 1.3 there does not exist an element $x \in \theta_3^H$ with $\mu(x) = 1$ and $2x = 0$. Let θ be the finitely generated subgroup of θ_3^H generated by the 3-dimensional links of a triangulation of N^n. We now construct a homeomorphism $\gamma : \theta \to \mathbb{Z}_4$ so that the following diagram commutes

$$\begin{array}{ccccccccc} & & & & \theta & \xrightarrow{i} & \theta_3^H & & \\ & & & & \downarrow \gamma & & \downarrow \mu & & \\ 0 & \to & \mathbb{Z}_2 & \xrightarrow{\times 2} & \mathbb{Z}_4 & \xrightarrow{r} & \mathbb{Z}_2 & \to & 0 \end{array}$$

where r is reduction mod 2. By the fundamental theorem for finitely generated abelian groups there exists elements $x_1, x_2, \ldots, x_k$

of θ such that $\theta \approx \langle x_1\rangle \oplus \langle x_2\rangle \oplus \dots \oplus \langle x_k\rangle$ where $\langle x_i\rangle$ is the cyclic group generated by $\langle x_i\rangle$. If $\mu(x_i) = 0$, define $\gamma(x_i) = 0$; if $\mu(x_i) = 1$ and $\langle x_i\rangle \approx \mathbb{Z}$, define $\gamma(x_i) = 1$; and if $\mu(x_i) = 1$ and $\langle x_i\rangle \approx \mathbb{Z}_{2^j p}$ then $j \geq 2$ by our assumption, so define $\gamma(x_i) = 1$. It is easy to check that γ is well defined and that $\mu i = r\gamma$.

Note that N with its triangulation is a homology manifold so there exists an obstruction [6] $\tilde{\sigma}_N \in H^4(N;\ \theta_3^H)$ to resolving N to a *PL* manifold. Now $\tilde{\sigma}_m$ assigns to every 4-dimensional dual cell of N, its boundary, which is *PL* homeomorphic to a 3-dimensional link in N and hence in θ. So there exists a $\tilde{\sigma}_N \in H^4(N;\ \theta)$ so that $i_*\ \tilde{\sigma}_N = \tilde{\sigma}_N$. Also by [4] $\mu_*\ \tilde{\sigma}_N = \sigma_N$. Hence $S_q^1\ \sigma_N = S_q^1\ \mu_*\tilde{\sigma}_N = S_q^1\ \mu_* i_*\tilde{\sigma}_N = S_q^1 r_*\gamma_*\sigma_N = 0$ since $S_q^1 r_* = 0$. A contradiction to the assumption that there exists a manifold M^m, $m \geq 6$ ($m \geq 5$ if ∂M has a simplicial triangulation) which is not trianguable. Thus the theorem follows.

We note that Siebenmann [8] has shown the existence of a topological 5-manifold N with $S_q^1\ \sigma_N \neq 0$. In the next section we will explicitly construct such a manifold.

III. THE CONSTRUCTION

We first recall theorem 1.4 of [4].

THEOREM 3.1. A homology manifold H^n with $H = \emptyset$ and $n \geq 5$ is a topological n-manifold if and only if the links of vertices are 1-connected.

We note this theorem was a consequence of the double suspension conjecture recently proved by J. Cannon [1] and R. D. Edwards [2].

Now we use this result to geometrically construct a closed topological 5-manifold N with $S_q^1\ \sigma_N \neq 0$. Thus if there existed a simplicial triangulation of N, all compact topological m-manifolds with $m \geq 6$ ($m \geq 5$ if ∂M simplicially triangulated) would be simplicially triangulable!

Let H^3 be any oriented PL homology 3-sphere that bounds an oriented parallelizable PL 4-manifold W with index 8. Let $X = W \cup_H c(H)$ where $c(H)$ is the cone on H and x the cone point of $c(H)$. Attach a PL 1-handle $D^3 \times [0,1]$ to $c(H) \times 0 \cup c(H) \times 1 \subset X \times [0,1]$ so that $\partial S = H \# H$ (not $H \# - H$) where $S = c(H) \times 0 \cup_{D^3 \subset H^3} D^3 \times I \cup_{D^3 \subset H^3} c(H) \times 1$. Let $Y = X \times I \cup S \cup_{H\#H} c(H\#H)$ where z is the cone point of $c(H\#H)$. Note that the polyhedron Y contains the sub polyhedron $T = S \cup_{H\#H} c(H\#H)$ and is a homology 4-manifold with the same homotopy type as S^4. Let $P = Y \cup_T c(T)$ where y is the cone point of $c(T)$. Now P is a homology 5-manifold with ∂P PL homeomorphic to $W \quad W \cup c(H\#H)$, where y denotes connected sum along the boundary. Note that all of the Steiefel-Whitney numbers of ∂P are zero. Next add an exterior collar $C = \partial P \times [0,1)$ to P along ∂P and call the resulting homology 5-manifold Q.

We first observe that the only 4-dimensional links i.e. $z, y, x \times 0$ and $x \times 1$, which are not PL homeomorphic to S^3 are simply connected, hence by 3.1 Q is a simplicially triangulated 5-manifold.

We next observe that the only 3-dimensional links of Q which are not PL homeomorphic to S^3 occur as links of the subpolyhedron $L = x \times [0,1] \cup y*(x \times [0,1]) \overset{PL}{\approx} S^1$ and $M = y*z \cup z \times [0,1) \overset{PL}{\approx} [0,1)$ of Q. The links of 1-simplexes of L are PL homeomorphic to H and the links of 1-simplexes of M are PL homeomorphic to $H\#H$. Thus by Theorem C of [9] there exists a PL structure Σ of $Q - L$ since $\mu[H\#H] = 0$. Note that Σ does not agree with the polyhedral structure of Q.

We can now use PL transversality with respect to $\Sigma|_{\partial P} \times (0,1)$ to get a compact connected orientable 4-dimensional submanifold V in $\partial P \times (0,1)$, with trivial normal bundle, which separates $\partial P \times [0,1)$ into two components A and B, with the closure of one of them, say A, containing ∂P. Now $P \cup c\ell[A]$ is a topological manifold with $\partial(Q \cup c\ell[A]) = V$. Since all of the Stiefel-Whitney numbers of V are zero, V bounds a PL 5-manifold $\overline{W}$. Finally define $N^5 = P \cup_{\partial P} c\ell[A] \cup_V \overline{W}$.

Now since $N - L$ is a PL manifold it is clear that the Poincare dual of σ_N is represented by L. Also the Poincare dual of the first Stiefel-Whitney class of N, $w_1(N)$, restricted to P is represented by $X \times \frac{1}{2}$. Therefore by the $W\mu$ formula, $S_q^1\, \sigma_N = W_1(N) \cup \sigma_N$ = intersection number of L and $X \times \frac{1}{2}$ which is non-zero. Therefore N is the desired 5-manifold.

REFERENCES

1. Cannon, J., Shrinking cell-like decompositions of manifolds: Codimension three, preprint.
2. Edwards, R. D., Shrinking cell-like decompositions of manifolds, to appear.
3. Galewiski, D., and Stern, R., Classification of simplicial triangulations of topological manifolds, Bull. Amer. Math. Soc. 82 (1976), pp. 916-918.
4. ______________________________, Classification of simplicial triangulations of topological manifolds, (to appear).
5. Kirby, R., and Siebenmann, L., Foundational essays on topological manifolds, smoothings, and triangulations, Annals of Math Studies, No. 88, Princeton Univ. Press, 1977.
6. Martin, N., On the differences between homology and piecewise linear block bundles, J. London Math. Soc. 6 (1973), pp. 197-204.
7. Matumoto, T., Variétés simplicales d'homologie et variétés topologiques métrisables, Thesis, University de, Paris-Sud, 91405 Orsay, 1976.
8. Siebenmann, L., Topological manifolds, Proc. I.C.M. Nice (1970), Vol. 2, pp. 143-163, Gunthier-Villars, 1971.
9. Siebenmann, L., Are non-trianguable manifolds trianguable?, pp. 77-84 in Topology of Manifolds, ed. by J. C. Cantrell and C. E. Ewards, Markham Chicago, 1970.

ON π_i DIFF M^n

W.-C. Hsiang[1]

Department of Mathematics
Princeton University
Princeton, New Jersery

In this paper, we shall indicate some recent results on rational homotopy groups of the diffeomorphism groups of discs, spheres, aspherical manifolds and differentiable Spherical space forms.

I. INTRODUCTION AND STATEMENT OF RESULTS

Let M^n be a C^∞ manifold and let Diff M^n be the diffeomorphism group[2] of M^n endowed with C^∞ topology. Recently, topologists have become very interested in computing π_i Diff M^n [2] [3] [6] [8] [9] [10] [12] [13] [26]. The problem of calculating π_i Diff M^n is equivalent to classifying C^∞ fiber bundles over S^{i+1} with M^n as the fiber. In this lecture, I shall indicate some results on π_i Diff M^n obtained by B. Jahren and myself [12], F.T. Farrell and myself [6].

Before I state the theorems, let me recall some definitions and constructions in algebraic K-theory and surgery theory. Let

[1] *Partially supported by National Science Foundation grant GP 34324X1.*

[2] *If $\partial M^n \neq \emptyset$, the diffeomorphisms leave ∂M^n fixed*

ISBN 0-12-158860-2

$\Lambda = ZG$ be the integral group ring of a finitely presented group G and let $K_j(\Lambda) = \pi_j BGL^+(\Lambda)$ be Quillen's K-group of Λ [21]. Loday [16] constructs a homomorphism

$$(1) \qquad \lambda_* : h_*(BG; \underline{\underline{K}}_Z) \to K_*(ZG)$$

where $\underline{\underline{K}}_Z$ denotes the spectrum of the algebraic K-theory of Z. Let $\underline{\underline{S}}$ denote the sphere spectrum. We have an obvious map of spectra

$$(2) \qquad \varphi : \underline{\underline{S}} \to \underline{\underline{K}}_Z$$

and it induces a homomorphism of (generalized) homology theories

$$(3) \qquad \varphi_* : h_*(\ ; \underline{\underline{S}}) \to h_*(\ ; \underline{\underline{K}}_Z)$$

Since $h_*(\ ;\underline{\underline{S}}) \otimes Q$ is the ordinary homology theory, we have the composite

$$(4) \qquad \psi_n : H_n(BG;Q) \to h_n(BG;\underline{\underline{K}}_Z) \otimes Q \xrightarrow{\lambda_n \otimes id} K_n(ZG) \otimes Q.$$

We use $\overline{K}_n(G)$ to denote the cokernel of ψ_n. (Note that $\overline{K}_n(G)$ is a vector space over Q.)

Now, we recall the canonical involution '$\overline{}$' on $\Lambda = ZG$ given by sending $\Sigma\, n(g)g$ to $\Sigma\, n(g)g^{-1}$. We can define an involution on $GL_n(\Lambda)$ sending $A \in GL_n(\Lambda)$ to $(\overline{A}^*)^{-1}$ where $\overline{A}^*$ is the conjugate transpose of A; i.e., $(\overline{A}^*)_{ij} = \overline{A}_{ji}$. It induces an involution

$$(5) \qquad \overline{} : \underline{\underline{K}}_\Lambda \to \underline{\underline{K}}_\Lambda$$

at the spectrum level and hence the desired conjugations on $K_i(\Lambda)$ and $\overline{K}_i(G)$.

Let $K_i^{\pm}(G) = \{x \mid x = \pm\overline{x} \text{ for } x \in K_i(G)\}$; then $\overline{K}_i(G) = \overline{K}_i^+(G) \oplus \overline{K}_i^-(G)$. (Here, we use the fact that $K_i(G)$ is a vector space over Q.)

We also have the following map from surgery theory [27]

(6) $$\theta : [M^n \times D^i, \partial; G/\mathrm{Top}, *] \otimes Q \to L^h_{n+i}(\pi_1 M^n, w_1(M^n)) \otimes Q.$$

Novikov's conjecture on higher signature states that θ is a monomorphism [20] if M^n is a $K(\pi,1)$. The most interesting special case of Novikov's conjecture is when M^n is an orientable closed aspherical manifold; i.e., M^n is an orientable closed, $K(\pi,1)$ manifold. In this case, the conjecture is changed to the following:

CONJECTURE 1. *The homomorphism* θ *is an isomorphism, if* $i \geq 1$. *(Note that* θ *is a group homomorphism even when* $i = 0$ *[15].)*

In fact, there is no counterexample *even* when we *do not* tensor with Q; this strongest conjecture is equivalent to $S^{\mathrm{Top}}(M^n \times D^i, \partial) = 0$ $(n > 4)$ [15] [27]. Since $G/\mathrm{Top} \times Z$ may be identified with Quinn's spectrum $\underline{\underline{L}}_Z$, $[M^n \times D^i, \partial; G/\mathrm{Top}, *]$ may be viewed as the generalized cohomology groups with coefficents in $\underline{\underline{L}}_Z$. Hence, an algebraic K-theoretic analogue of Conjecture 1 can be formulated using (1).

CONJECTURE 2. *If* M^n *is a closed manifold which is a* $K(\pi,1)$, *then*

$$\lambda_* \otimes id : h_*(M^n; \underline{\underline{K}}_Z) \otimes Q \to K_*(Z\pi) \otimes Q$$

is an isomorphism.

These conjectures have been verified for various π: Novikov [20]; Rohlin [23]; Farrell-Hsiang [5] [6] [11]; Kasparov [12]; Cappell [4]; Wall [28]; Lusztig [17]; Miščenko [19]; Waldhausen [25] etc.

Conjecture 1 has a geometric meaning in terms of stable homeomoprhosms. So far, no additional geometric significance has been attached to Conjectures 1 and 2, although topologists consider them as very important problems. We shall indicate interesting applications of Conjectures 1 and 2 to diffeomorphism groups.

Let us now state our first main theorem.

THEOREM A [6]. *Assume that* $0 < i < \frac{n}{6} - 7$.

(i) $$\pi_i \mathrm{Diff}(D^n, \partial) \otimes Q = \begin{cases} Q & n \text{ odd and } i = 4k-1, \\ 0 & \text{otherwise.} \end{cases}$$

(ii) *Let* Σ^n *be an arbitrary n-dim homotopy sphere; then*

$$\pi_i \mathrm{Diff}\, \Sigma^n \otimes Q = \begin{cases} 0 & i \neq 4k-1, \\ Q & n \text{ even and } i = 4k-1, \\ Q \oplus Q & n \text{ odd and } i = 4k-1. \end{cases}$$

(iii) *Let* M^n *be an orientable closed manifold which is a* $K(\pi,1)$. *If Conjectures 1 and 2 are verified for* M^n, *then*

$$\pi_i \mathrm{Diff}\, M^n \otimes Q = \begin{cases} \text{Center}(\pi) \otimes Q & \text{for } i = 1, \\ \bigoplus_{j=1}^{\infty} H_{(i+1)-4j}(M^n; Q) & \text{for } i > 1 \text{ and } n \text{ odd}, \\ 0 & \text{otherwise.} \end{cases}$$

Let M^n be the quotient space of a free differentiable action of a periodic group G on a homotopy sphere Σ^n; i.e., $M^n = \Sigma^n/G$. We shall call such a manifold a *differentiable spherical space form*. If M^n is orientable, $n = 2m + 1$. Let $L_i^s(G)$ be the surgery group for $G = \pi_1 M^n$ and let $\tilde{L}_i^s(G)$ be the kernel of the augumentation map $L_i^s(G) \to L_i^s(1)$ [27]. (We assume that M^n is orientable.) We now state our second main theorem.

THEOREM B [12]. *Assume that* $0 < i < \frac{m}{3} - 4$. *Let* M^{2m+1} *be an orientable differential spherical space form; then*

$$\pi_i \text{ Diff } M^{2m+1} \otimes Q = \begin{cases} \overline{K}_{i+2}^{-}(G) \oplus \tilde{L}^{s}_{2m+i+3}(G) \otimes Q \\ \oplus Q \quad \text{if} \quad i = 4k-1, \\ \overline{K}_{i+2}^{-}(G) \oplus \tilde{L}^{s}_{2m+i+3}(G) \otimes Q \\ \qquad\qquad \text{otherwise} \end{cases}$$

The ingredients of the proofs come from Hatcher [8] [9], Waldhausen [26], Hsiang-Sharpe [13], Borel [1] and Garland-Hsiang [7]. We shall sketch the steps of the proofs in the next section.

II. INDICATION OF THE PROOFS

1. Let me first recall Hatcher's pseudo-isotropy theory [8] [9] and the stability theorem [8] [3]. Let M be a compact manifold and let $P(M)$ be the differential pseudo-isotopy space[3] of M; i.e., diffeomorphisms of $M \times I$ fixing $M \times 0$. (similarly, we may define PL and topological pseudo-isotopy spaces $P^{PL}(M)$ and $P^{\text{Top}}(M)$. In fact, if M is a PL manifold of dimension ≥ 5, the natural map $P^{PL}(M) \to P^{\text{Top}}(M)$ is a homotopy equivalence [15].)

There is a natural map from $P(M)$ to $P(M \times I)$ given by sending f to $f \times id_I$:

THEOREM 2.1 [3] [9]. *The natural map* $P(M) \to P(M \times I)$ *is* k*-connected for* $k \leq \dim M/6 - 7$. *In particular,*

$$\pi_i P(M) \to \pi_i P(M \times I) \to \pi_i P(M \times I^2) \to \cdots$$

is eventually an isomorphism.

[3] *If* $\partial M^n \neq \phi$, *the pseudo-isotopy space relative to the boundary* ∂M *is homotopy equivalent to the full pseudo-isotopy space* $P(M)$. *So, we shall not distinguish them.*

In fact, the space $\lim_{n\to\infty} P(M\times I^n)$ becomes a homotopy functor from the underlying space M to an infinite loop space. An appropriate double delooping of this space is denoted as $Wh^{\text{Diff}}(M)$ in [26]. (Similarly, we have $Wh^{PL}(M)$.)

We next replace $P(M)$ by the space $P'(M)$ consisting of diffeomorphisms of $(M\times I;\ M\times 0,\ M\times 1)$ which preserve projection to I over ∂M, modulo the subgroup of diffeomorphisms preserving projection to I over all of M. The natural map from $P(M)$ to $P'(M)$ is a homotopy equivalence. In $P'(M)$, we have the duality involution '$\overline{}$' defined as conjugation by $id_M \times r$, where $r : I \to I$ is the reflection through the midpoint. We also have the stabilization map $\Sigma : P'(M) \to P'(M\times I)$ compatible with the stablization map of $P(M)$. It was observed in [9] that Σ anti-commutes with $\overline{}$, up to homotopy.

Set $\pi_i^{\pm} P(M) = \{x = \pm\bar{x} \mid x \in \pi_i P(M)\}$. The following lemma is obvious.

LEMMA 2.2. $\pi_i P(M) \otimes Z[\frac{1}{2}] = \pi_i^{+} P(M) \otimes Z[\frac{1}{2}] \oplus \pi_i^{-} P(M) \otimes Z[\frac{1}{2}]$.

2. We next need the work of Waldhausen [26] which relates $\pi_i P(M)$ with Quillen's algebraic k-theory. In particular, he reduces the computation of the rational homotopy groups of the differentiable pseudo-isotopy space $P(D^n)$ of D^n to a question on the group cohomology of $GL_n(Z)$ with coefficients in $M_n(Q)$ under the adjoint action. Let us recall his result. Let $M_n(Q)$ be the vector space over Q of $(n\times n)$-Matrices and let $GL_n(Z)$ act on $M_n(Q)$ be the adjoint action; i.e.

$$(7) \qquad Ad(g)(x) = g\,x\,g^{-1}$$

for $g \in GL_n(Z)$ and $x \in M_n(Q)$. Let $GL_n(Z)$ act trivially on Q. There is an equivariant map

$$(8) \qquad tr : M_n(Q) \to Q$$

sending x to its trace. So, we have a homomorphism

$$(9)\quad tr_* : H_*(GL_n(Z);M_n(Q)) \to H_*(GL_n(Z);Q) .$$

Let us stablize $M_n(Q)$ by sending $x \in M_n(Q)$ to $\begin{pmatrix} x & 0 \\ 0 & 0 \end{pmatrix} \in M_{n+1}(Q)$. So, we have a limit homomorphism

$$(10)\quad tr_* : \lim_{n\to\infty} H_*(GL_n(Z);M_n(Q)) \to \lim_{n\to\infty} H_*(GL_n(Z);Q)$$

THEOREM 2.3 [26] . *Provided* $0 \leq i < \frac{n}{6} - 7$,

$$\pi_i P(D^n) \otimes Q \cong K_{i+2}(Z) \otimes Q$$

if and only if the homomorphism tr_* *of (10) is an isomorphism.*

By adapting the arguments in Borel [1] and Garland-Hsiang [7], the following lemma was proved in [6].

LEMMA 2.4. The homomorphism tr_* *of (10) is an isomorphism. Hence,*

$$\pi_i P(D^n) \otimes Q \cong K_{i+2}(Z) \otimes Q$$

for $0 \leq i < \frac{n}{6} - 7$.

In fact, the above lemma is generalized to the following more usable form.

LEMMA 2.5 [12]. *Let* M^n *be a compact orientable manifold such that* $\pi_i(M^n) = 0$ *for* $1 < i < \frac{n}{2}$. *Let* $0 < i < \frac{n}{6} - 7$; *then*

$$\pi_i P(M^n) \otimes Q \cong \overline{K}_{i+2}(\pi_1 M).$$

3. We also need the parametrized surgery theory of Sharpe and myself [13]. Let us recall the part needed here.

Let M^n be a n-dim $(n \geq 6)$ smooth manifold without boundary[4] Set Aut $M = \{f \in M^M \mid f$ is a simple homotopy equivalence of $M\}$. We have the following fibration

(11) $G(M) \to \text{Diff } M \to \text{Aut } M.$

Then, a point in $G(M)$ is represented by a pair (ϕ, ϕ_t) where $\phi \in \text{Diff } M$ and ϕ_t is a path in Aut M connecting ϕ to $id \in \text{Aut } M$. Set $M_\phi = M\text{x}I / \{(m,1) \sim (\phi(m),0)\}$, the mapping torus of ϕ, ϕ_t induces a simple homotopy equivalence.

(12) $F : (M_\phi, M\text{x}1) \to (M\text{x}S^1, M\text{x}1).$

One can construct a space $S(M\text{x}(S^1,1))$ of simple homotopy smoothings of $M\text{x}1$ [27]. We have a map

(13) $\tau : G(M) \to S(M\text{x}(S^1,1))$

defined by $\tau((\phi,\phi_t)) = F$. On the other hand, we have the map

(14) $\eta : S(M\text{x}(S^1,1)) \to G/0^{\Sigma M^+}$

where $\Sigma M^+ = M\text{x}S^1/M\text{x}1$. Let us consider the following diagram of fibrations

(15)
$$\begin{array}{ccccc} B(M) & \xrightarrow{=} & B(M) & & \\ \downarrow & & \downarrow & & \\ C(M) & \longrightarrow & G(M) & \xrightarrow{\eta \cdot \tau} & G/0^{\Sigma M^+} \\ \downarrow & & \downarrow{\scriptstyle\tau} & & \downarrow{\scriptstyle =} \\ L_2(M) & \longrightarrow & S(M\text{x}(S^1,1)) & \xrightarrow{\eta} & G/0^{\Sigma M^+} \end{array}$$

where $B(M)$, $C(M)$, $L_2(M)$ are the homotopy fibers of the obvious

[4] *Everything works for manifold with boundary, if the boundary is left fixed.*

maps. It is easy to see that $\pi_i(L_2(M)) = L^s_{n+i+2}(\pi_1 M, w_1(M))$. It follows from (15) that there is a braid

(16)

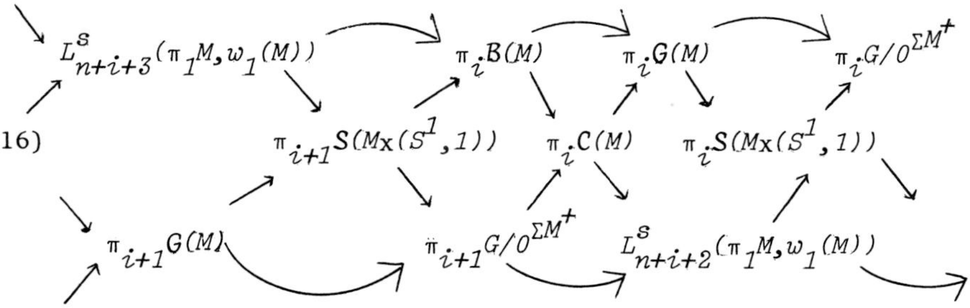

So, computing π_iDiff M is reduced to calculating the homotopy exact sequence of (11) together with the braid (16).

4. We have made some calculation of the braids in [12]. Let us recall the result. We first identify $B(M)$ as $\Omega(\widetilde{\text{Diff}}\, M^n/\text{Diff}\, M^n)$ where Diff M^n denotes the simi-simplicial group of block diffeomorphisms. Using this model for $B(M)$, it is not difficult to prove the following lemma.

LEMMA 2.6 [12]. *We have a homotopy fibration*

$$B(M\times I, \partial) \to P(M) \to B(M).$$

Localizing the above fibration away from 2, we actually decompose $P(M)$ as the product of $B(M)$ and $B(M\times I, \partial)$. In particular, we have

LEMMA 2.7. *Assume that* $0 \le i <$ dim $M/6-7$.

$$\pi_i P(M) \otimes Z[\tfrac{1}{2}] \cong \pi_i B(M) \otimes Z[\tfrac{1}{2}] \oplus \pi_i B(M\times I, \partial) \otimes Z[\tfrac{1}{2}] .$$

Comparing the decomposition $\overline{K}_i(G) = \overline{K}_i^+(G) \oplus \overline{K}_i^-(G)$ in the Introduction, Lemma 2.2, Lemma 2.5 and Lemma 2.7, one might suspect that there is a connection between $\pi_j B(M) \otimes Q$ and $\overline{K}_{j+2}^{\pm}(\pi, M)$. Here is the answer.

LEMMA 2.8. Let M^n be a compact orientable manifold such that $\pi_i(M^n) = 0$ for $1 < i < \frac{n}{2}$. Let $0 < j < \frac{n}{6} - 7$.

$$\pi_j B(M^n) \otimes Q = \begin{cases} \overline{K}_{j+2}^-(\pi_1 M) & \text{for } n \text{ odd,} \\ \overline{K}_{j+2}^+(\pi_1 M) & \text{for } n \text{ even.} \end{cases}$$

It follows from the Product Formula of $P(M)$ [9] and Lemma 2.7 that we have a Product Formula for $B(M)$.

LEMMA 2.9 [12]. *Suppose that $\ell >> k$ and $0 < k < \frac{n}{6} - 7$. Then, product with $S^{2\ell}$ induces an isomorphism*

$$\rho : \pi_k B(M) \otimes Z[\tfrac{1}{2}] \to \pi_k B(M \times S^{2\ell}) \otimes Z[\tfrac{1}{2}]$$

where ρ is induced by the map sending $f \in B(M)$ to $f \times id_{S^{2\ell}} \in B(M \times S^{2\ell})$.

In fact, we may consider the product of the whole braid diagram (16) with $S^{2\ell}$. Since the Produce Formula on $L^s_{n+i+2}(\pi_1 M, w_1(M))$ is given by multiplication of the index of $S^{2\ell}$ which is zero, we have the following result.

THEOREM 2.10 [12]. *For $0 < i < \frac{n}{6} - 7$,*

$$\pi_i C(M^n) \otimes Z[\tfrac{1}{2}] = \pi_i B(M^n) \otimes Z[\tfrac{1}{2}] \oplus L^s_{n+i+2}(\pi_1 M, w_1(M^n)) \otimes Z[\tfrac{1}{2}] .$$

5. Let us now prove Theorem A. We still need the following Lemma.

LEMMA 2.11 [6]. *For $x \in K_i(Z) \otimes Q$, we have*

$$\bar{x} = -x \quad \text{when } i > 0,$$

$$\bar{x} = x \quad \text{when } i = 0.$$

Let M^n be either (D^n, ∂) or Σ^n, a homotopy sphere. Since $\pi = \{1\}$, it follows from Lemma 2.8 and Lemma 2.11 that if $0 < i < \frac{n}{6} - 7$, then

$$\pi_i B(M^n, \partial) \otimes Q = \begin{cases} Q & \text{for } n \text{ odd and } i = 4k-1, \\ 0 & \text{otherwise.} \end{cases}$$

Since π_i Aut $M \otimes Q = 0$ for $i < n$, we have

(17) π_i Diff $M^n \otimes Q = \pi_i G(M) \otimes Q \ (i < n)$.

(i) *Let us first consider the case $M^n = (D^n, \partial)$.* In this case, $\pi_i S(M\times(S^1, 1)) \otimes Q = 0$ for $i \geq 0$. It follows from the exact sequence

$$(18) \quad \begin{aligned} &\to \pi_{i+1} S(M\times(S^1,1)) \to \pi_i B(M) \to \pi_i G(M) \\ &\to \pi_i S(M\times(S^1,1)) \end{aligned}$$

in the braid diagram (16) that

$$\pi_i B(M) \otimes Q = \pi_i G(M) \otimes Q.$$

So, we have case (i) of Theorem A.

(ii) *We next consider the case* $M^n = \Sigma^n$, *a homotopy sphere.*

Let us consider another exact sequence

$$(19)\quad \begin{aligned} &\to [\Sigma^{i+1}M^+, G/O] \to \pi_i C(M) \\ &\to \pi_i G(M) \end{aligned}$$

in the braid diagram (16). Since $\Sigma^{i+1}M^+$ is homotopy equivalent to $S^{i+1} \vee S^{n+i+1}$, it follows from Theorem 2.10 that

$$(20)\quad \pi_i T(M) \otimes Q = \pi_i B(M) \otimes Q \oplus \pi_{i+1} G/O \otimes Q$$

So, Case (ii) of Theorem A follows from (19) and (20).

(iii) *Let us now consider case (iii) of Theorem A.* Since we assume that Conjecture 2 for M^n is true, it follows from Lemma 2.11 that

$$(21)\quad \begin{aligned} \overline{K}_i^+(\pi_1 M^n) &= 0 \\ K_i^-(\pi_1 M) &= \bigoplus_{j=1}^{\infty} H_{(i-1)-4j}(M^n, Q) . \end{aligned}$$

Since M^n is a compact $K(\pi, 1)$, we have

$$(22)\qquad \pi_i \operatorname{Aut} M = \begin{cases} Center(\pi) & \text{for } i = 1, \\ 0 & \text{for } i > 1. \end{cases}$$

Hence, it follows from the fibration (11) that

$$\pi_i G(M) \otimes Q = \pi_i \operatorname{Diff} M^n \otimes Q \quad (i > 1),$$

$$(23)\quad \begin{aligned} &0 \to \pi_1 G(M) \to \pi_1 \operatorname{Diff} M^n \to Center\ (\pi) \\ &\to \pi_0 G(M). \end{aligned}$$

One of the exact sequences in the braid diagram (16) is

$$\begin{aligned} &\to \pi_i S(M^n \times (S^1,1)) \to [\Sigma^{i+1}M^+, G/O] \\ (24) \quad &\overset{\theta}{\to} L^s_{n+i+1}(\pi, w_1(M^n)) \to \pi_{i-1}S(M^n \times (S^1,1)) \\ &\to \end{aligned}$$

Since we assume that Conjecture 1 is verified for M^n, we conclude that $\pi_i S(M^n \times (S^1,1)) \otimes Q = 0$ for $i \geq 0$. Consequently, we have

$$(25) \quad \pi_i B(M) \otimes Q \cong \pi_i G(M) \otimes Q \text{ for } i > 0.$$

Since on the other hand, Conjecture 2 for M^n implies that $Wh_2(\pi) \otimes Q = 0$, it follows from Theorem 2.1 of [13] and (25) that $\pi_0 G(M) \otimes Q = 0$. So, (23) becomes

$$\pi_i B(M) \otimes Q \cong \pi_i \text{ Diff } M \otimes Q \quad (i > 1)$$

$$\pi_1 \text{ Diff}(M) \otimes Q \cong \pi_1 G(M) \otimes Q \oplus Center(\pi) \otimes Q.$$

Part (iii) of Theorem A follows from (21), (26) and Lemma 2.8.

6. Let us now prove Theorem B. We again have π_i Aut $M \otimes Q = 0$ for $i < n$, and hence (17) is valid. Let $\tilde{M}$ be the universal covering space of M. The braid diagram (16) for M is mapped into that for $\tilde{M}$. In particular, we have the following homomorphism of exact sequences

$$(27) \quad \begin{array}{ccccc} \to & [\Sigma^{i+1}M^+, G/O] & \to & \pi_i C(M) \\ & \downarrow & & \downarrow \\ \to & [\Sigma^{i+1}\tilde{M}^+, G/O] & \to & \pi_i C(\tilde{M}) \\ \\ \to & \pi_i G(M) & \to & \\ & \downarrow & & \\ \to & \pi_i G(\tilde{M}) & \to & \end{array} .$$

Since $\tilde{M}$ is a homotopy sphere, the lower exact sequence of (27) is just (19). It follows from Theorem A (ii), Theorem 2.10, Lemma 2.8 and diagram chasing that for $0 < i < \frac{n}{6} - 7$

$$\pi_i G(M) \otimes Q = \begin{cases} \overline{K}_{i+2}^{-}(G) \oplus \tilde{L}^s_{n+i+2}(G) \otimes Q \oplus Q & \text{if } i = 4k-1, \\ \overline{K}_{i+2}^{-}(G) \oplus \tilde{L}^s_{n+i+2}(G) \otimes Q & \text{otherwise.} \end{cases} \tag{28}$$

So, Theorem B follows from (17) and (28).

REFERENCES

1. A. Borel, Stable real cohomology of arithmetic groups, Ann. Sci. Ec. Norm. Supp. 4e série, t.7 (1974), 235-272.
2. J. Cerf, La stratification naturelle des espaces de fonctions différentiables réelles et le théorème de la pseudo-isoptopie, Publ. Math. I.H.E.S., 39 (1970).
3. D. Burghelea and R. Lashof, Stability of concordances and the suspension homomorphism, Ann. of Math. 105 (1977), 449-472.
4. S. E. Cappell, On homotopy invariance of higher signatures, Invent. Math., 33 (1976), 171-179.
5. F. T. Farrell and W. C. Hsiang, Manifolds with $\pi_1 = G \times_\alpha Z$, Ameri. J. Math., 95 (1973), 813-848.
6. F. T. Farrell and W. C. Hsiang, On the rational homotopy groups of the diffeomorphism groups of disc, spheres and aspherical manifolds (to appear).
7. H. Garland and W. C. Hsiang, A square integrability criterion for the cohomology of arithmetic groups, Proc. Nat. Acad. Sci. U.S.A., 59 (1968), 354-360.
8. A. Hatcher, Higher simple homotopy theory, Ann. of Math. 102 (1975), 101-137.
9. A. Hatcher, Concordance Spaces, Higher Simple-Homotopy Theory, and Applications (to appear).
10. A. Hatcher and J. Wagoner, Pseudo-isotopies of compact manifolds, Asterisque 6, Soc. Math. de France, 1973.
11. W. C. Hsiang, A splitting theorem and the Künneth formula in algebraic K-theory, Algebraic K-theory and its geometric applications, Lecture Notes in Math., 108, Springer, Berlin, 1969, 72-77.
12. W. C. Hsiang and B. Jahren, On the homotopy groups of the diffeomorphism groups of spherical space forms (to appear).

13. W. C. Hsiang and R. W. Sharpe, Parametrized surgery and isotopy, Pacific Jour. of Math., 67 (1976), 401-459.
14. G. G. Kasparov, On the homotopy invariance of the national Pontrjagin numbers, Dokl. Akad. Nauk SSSR, 190 (1970), 1022-1025.
15. R. C. Kirby and L. C. Siebenmann, Foundational Essays on Topological Manifolds, Smoothings, and Triangulations, Annals of Math. Studies, Princeton Univ. Press, 1977.
16. J-L. Loday, K-théorie algébrique et représentations de groupes, Ann. Sci. Ec. Norm. Sup. 4e série, t.9 (1976), 309-377.
17. G. Lusztig, Novikov's higher signature and families of elliptic operators, J. Diff. Geom., 7 (1972), 229-256.
18. Y. Matsushima and S. Murakami, Vector bundle valued harmonic forms and automorphic forms on symmetric Riemannian manifolds, Ann. of Math., 78 (1963), 365-416.
19. A. S. Miščenko, Infinite-dimensional representations of discrete groups and higher signature, Izv. Akad, Nauk, SSSR, Ser. Mat., 38 (1974), 81-106.
20. S. P. Novikov, Homotopic and topological invariance of certain rational classes of Pontrjagin, Dokl. Akad. Nauk SSSR, 162 (1965), 1248-1251.
21. D. Quillen, Higher algebraic K-theory I, Algebraic K-Theory I, Lecture Notes in Math., 341, Springer, Berlin, 1973 85-147.
22. M.S. Raghunathan, Cohomology of arithmetic subgroups of algebraic groups II, Ann. of Math., 87 (1968), 279-304.
23. V. A Rohlin, The Pontrjagin-Hirzebruch class of codimensionality 2, Izv. Akad, Nauk SSSR, Ser. Mat., 30 (1966), 705-718.
24. R. G. Swan, K-Theory of Finite Groups and Orders, Lecture Notes in Math., 149, Springer, Berlin, 1970.
25. F. Waldhausen, Algebraic K-theory of generalized free products, Ann. of Math. (to appear).
26. F. Waldhausen, Algebraic K-theory of topological spaces, I (to appear).
27. C. T. C. Wall, Surgery on Compact Manifolds, Academic Press, 1970.
28. C. T. C. Wall, The topological space-form problems, Topology of Manifold (Georiga University, edited by J.C. Cantrell and C. H. Edwards, Jr.) 319-331, Markham 1970.

CONSTRUCTION OF SURGERY PROBLEMS

Lowell Jones

Department of Mathematics
State University of New York
Stony Brook, New York

K denotes a triangulated set in the n-ball B^n. If $(K, K\cap\partial B^n)$ is a Z_2-homology manifold pair, $\overline{H}_*(K,Z_2) = 0$, and codimension of K is even and greater than 5, then $K \subset B^n$ is the fixed point of an involution on B^n. Thus the converse to P.A. Smith's fixed point theorem is true.

I. INTRODUCTION

Notation.

$\mathbb{Z}_n$: cyclic multiplicative group of order n.

Z_n : cyclic addative group of order n.

$\mathbb{Z}_n \times D^m \to D^m$: an orientation preserving semi-free PL action of the group $\mathbb{Z}_n$ on the m-dimensional disc D^m.

$K \subset D^m$: the fixed point set of this action.

P. A. Smith has proven three things about K:

0.1. $\overline{H}_*(K,Z_n) = 0$

0.2. $(K, K\cap \partial D^m)$ is a Z_n-homology manifold pair.

0.3. $\dim(D^m) - \dim(K) = 0 \bmod 2$.

In this paper I try to prove the converse to Smith's result. The theorem is

ISBN 0-12-158860-2

THEOREM 1.4. Let $K \subset D^m$ *be a compact PL subset of* D^m *which satisfies* 0.1, 0.2, 0.3. *If in addition,* $\dim(D^m) - \dim(K) \geq 6$, *and* $\overline{H}_*(K, Z_2) = 0$, *then* $K \subset D^m$ *is the fixed point set of a PL action* $\mathbb{Z}_n \times D^m \to D^m$.

To construct $\mathbb{Z}_n \times D^m \to D^m$ having $K \subset D^m$ for fixed point set we first prove that, in the Poincare duality category, there exists a semi-free symmetry $\mathbb{Z}_n \times D^m \to D^m$ having K for fixed point set, and that the Spivak fibration for this symmetry has a canonical PL reduction covered by the inverse to the tangent bundle of D^m (see Chapt. 4): the key argument for this step is the generalization of Chapt. 2 in [5] to deal with maps (see Chapt. 2). Now we have a surgery problem, the domain of which will be the desired *PL* symmetry $\mathbb{Z}_n \times D^m \to D^m$, if surgery can be completed.

This paper is one section of a long paper by the author [9], which has been announced under different titles in Bull. Amer. Math. Soc. 78 (1972) pp. 234-235, and 79 (1973) pp. 167-169.

II. KEY HOMOTOPY PROPOSITIONS

In 2.1 and 2.2 below, we assume $f : X \to Y$ is a map between finite *CW* complexes such that $\pi_i(X) = \pi_i(Y) = 0$ for $i = 1,2$ and $f_* : H_*(X, Z_n) \to H_*(Y, Z_n)$ is an isomorphism.

PROPOSITION 2.1. Let X' *be a subcomplex of* X; *and let* $r_y : \mathbb{Z}_n \times Y \to Y$, $r_{x'} : \mathbb{Z}_n \times X' \to X'$ *be free, cellular actions commuting with* $f_{|X'}$. *Then after replacing* X (mod X') *by a homotopy equivalent finite CW complex, if need be,* $r_{x'}$ *will extend to a free cellular action* $r_x = \mathbb{Z}_n \times X \to X$ *which acts trivially on* $H_*(X, Z)$ *and satisfies* $f \circ r_x = r_y \circ f$.

PROPOSITION 2.2. Let $r_x : \mathbb{Z}_n \times X \to X$ *be a free cellular action which acts trivially on* $H_*(X, Z)$. *Then, after replacing* Y *by a* *homotopy equivalent finite CW complex, if need be, there will be a free cellular action* $r_y : \mathbb{Z}_n \times Y \to Y$ *which acts trivially on* $H_*(Y,Z)$ *and satisfies* $r_y \circ f = f \circ r_x$.

We think of $f : X \to Y$ as a fibration with fiber F. By 2.1, F is seen to be a simply connected Z_n-homology disc. Now the proof divides into three steps.

Step 1. In this step we suppose that F has the homotopy type of a finite *CW* complex. Let *CS* denote the classifying space for homotopy orientable fibrations having F for fiber: orientable means the bundle is fiber homotopy trivial over any one-skeleton. Choosing an orientation for $F \to X \xrightarrow{f} Y$ yields a classifying map $g : Y \to CS$.

An Eilenberg-Whitney obstruction argument will show that $\pi_i(CS)$ is all n*-torsion for each $i \geq 1$ (n^* = set of primes, prime to n). The case $\pi_1(CS)$ is discussed in the appendix to [5]; see also 5.2 in [1].

Let Y_0 denote the orbit space for r_y. Since $\mathbb{Z}_n$ acts trivially on $H_*(Y,Z)$, the covering projection $p : Y \to Y_0$ induces an isomorphism $H_*(Y,Z_{n^*}) \to H_*(Y_0,Z_{n^*})$. So the homotopy commutative diagram

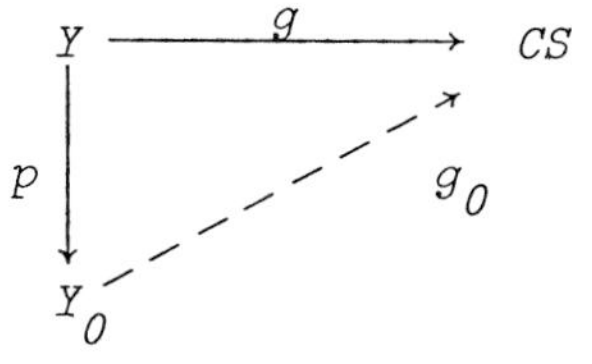

can be completed uniquely up to homotopy. Let $f_0 : X_0 \to Y_0$ denote the CS-fibration classified by g_0. Then X_0 is the orbit space of a free action $r_x : \mathbb{Z}_n \times X \to X$ satisfying $f \circ r_x = r_y \circ f$.

We consider X' as a subcomplex of Y. Then the embedding $X' \subset X$ becomes a bundle cross section

$$\begin{array}{ccc} X & \xrightarrow{\quad f \quad} & Y \\ & \nwarrow^{C} & \cup \\ & & X' \end{array}$$

over X'. This extends uniquely to a bundle cross section

$$\begin{array}{ccc} X_0 & \xrightarrow{\quad f_0 \quad} & Y_0 \\ & \nwarrow^{C_0} & \cup \\ & & X_0' \end{array}$$

over X_0', because $p : X' \to X_0'$ induces isomorphisms $H_*(X', Z) \cong H_*(X_0', Z)$, $\pi_1(X') \cong \pi_1(X_0')$ *mod n-torsion*, and the homotopy groups of F are all n^*-torsion groups. Thus $r_x : \mathbb{Z}_n \times X \to X$ can be chosen to extend the given action $r_{x'} : \mathbb{Z}_n \times X' \to X'$.

Step 2. Note that there is a spectral sequence[1] $E^r_{*,*}$, of $Z(\mathbb{Z}_n)$-modules satisfying:

a) $\mathbb{Z}_n$ acts trivially on each $E_{*,*}$ having $r \geq 2$,

b) $E^r_{*,*}$ converges to the quotients for a fibration of $H_*(X, Z)$.

Now (modulo n-torsion), that $r_x : \mathbb{Z} \times X \to X$ constructed in Step 1 acts trivially on $H_*(X,Z)$, follows from a), b) and

LEMMA 2.3. Let $0 \to A \to B \to C \to 0$ be an exact sequence of $Z(\mathbb{Z}_n)$-modules. If $\mathbb{Z}_n$ acts trivially on A, C, then modulo n-torsion, $\mathbb{Z}_n$ acts trivially on B.

[1] *Consider the universal cover of the fibration $F \to X_0 \to Y_0$ constructed in Step 1.*

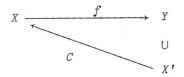

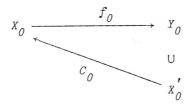

Proof of 2.3. For $x \in B$, and $t \in \mathbb{Z}_n$, we have $t \cdot x = x+y$, where $y \in A$. But then $x = t^n \cdot x = x+ny \Rightarrow ny = 0$. Q.E.D.

Finally, modulo n^*-torsion, $r_x : \mathbb{Z}_n \times X \to X$ acts trivially on $H_*(X,Z)$, because $f : X \to Y$ is a Z_n-homology equivalence. Thus $r_x : \mathbb{Z}_n \times X \to X$ acts trivially on $H_*(X,Z)$.

Step 3. In general the fiber F of Step 1 is not a finite *CW* complex, but it always has the homotopy type of a *CW* complex with finitely many cells at each dimension. Thus, for any positive integer k, the previous argument (Step 1) can be applied to get a *k-truncated* projection of fibrations

$$\begin{array}{ccccc} F & \longrightarrow & X & \xrightarrow{\ f\ } & Y \\ \downarrow & & \downarrow{\scriptstyle p_1} & & \downarrow \\ H & \longrightarrow & J_0 & \longrightarrow & Y_0 \end{array} \quad :$$

that is, p_1 is defined only on some finite subcomplex of X, it preserves the fiber blocks over each cell in Y and Y_0, and when restricted to any one of these blocks it is a k-equivalence. Let J denote the $\mathbb{Z}_n$-covering space for J_0.

Choose $k > \max(\dim(X) + 3, \dim(Y) + 3)$. Then calculations show

1) $H_i(J^{k-1}, Z) \cong H_i(X,Z)$ for $i \neq k-1$
2) $H_{k-1}(J^{k-1}, Z) \overset{\partial}{\cong} H_k((Y,J^{k-1}), Z)$, where ∂ preserves $Z(\mathbb{Z}_n)$-module structures.

For 1), 2) we see there is a finitely generated free $Z(\mathbb{Z}_n)$-chain complex

$$0 \to C_k \overset{\partial_k}{\to} C_{k-1} \to C_{k-2} \to \cdots \to C_0 \to 0$$

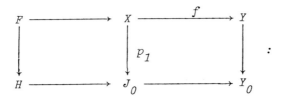

with $\ker(\partial_k) = H_{k-1}(J^{k-1}, Z)$ and with $H_j(C_*)$ a finite n_*-torsion module for all $j \leq k-1$: namely take C_* equal the cellular chain complex for (Y, J^{k-1}). Thus by the appendix $H_{k-1}(J^{k-1}, Z)$ is a stably free $Z(\mathbb{Z}_n)$-module. If need be, add a wedge of $(k-1)$-spheres to J_0^{k-1} so that $H_{k-1}(J^{k-1}, Z)$ becomes free. Represent a $Z(\mathbb{Z}_n)$-basis for $H_{k-1}(J^{k-1}, Z)$ by maps $g_i : S^{k-1} \to J^{k-1}$: the g_i exist because the inclusion $J^{k-1} \to J$ kills $H_{k-1}(J^{k-1}, Z)$. Add k-cells equivariantly along these g_i to obtain the desired action $\mathbb{Z}_n \times Y \to Y$.

This completes the proof of 2.1. Q.E.D.

Proof of 2.2. Choose an equivariant *localization at n*

$$\begin{array}{ccc} \mathbb{Z}_n \times X & \xrightarrow{r_x} & X \\ & \downarrow & \\ \mathbb{Z}_n \times X_{(n)} & \xrightarrow{r_{x(n)}} & X_{(n)} \end{array}$$

for the action r_x: that is, g commutes with $r_x, r_{x(n)}$, and $g_* : \pi_*(X) \to \pi_*(X_{(n)})$, $g_* : H_*(X,Z) \to H_*(X_{(n)}, Z)$ are just the algebraic localizations at n of $\pi_*(X)$, $H_*(X,Z)$. The localization construction given in [2] is particularly suited to our purpose. This construction adds a cone to X along every map $f : M_\ell^j \to X$ with ℓ prime to n, where M_ℓ^j denotes the co-Moore space formed by adding D^j to S^{j-1} along a degree $-\ell$ map $\partial D^j \to S^{j-1}$. Let X_1 be the resulting space. Now repeat the construction for X_1 obtaining X_2, and so forth. $X_{(n)}$ is the direct limit ($\varinjlim_k X_k$). Since X is a finite CW complex only a countable number of cone-glueings will suffice to get $X_{(n)}$ from X. Moreover these countable set of cone-glueings can be chosen equivariantly with respect to $\mathbb{Z}_n \times X \to X$.

So [2] assures that $X_{(n)}$ is the direct limit of an increasing sequence

$$A_1 \subset A_2 \subset \ldots \subset A_i \subset A_{i+1} \subset \ldots \subset X_{(n)}$$

satisfying

1) Each A_i is a finite, simply connected, subcomplex of $X_{(n)}$ left invariant by $r_{x_{(n)}}$

2) $\mathbb{Z}_n$ acts trivially on each image $(H_*(A_i, Z) \to H_*(A_{i+1}, Z))$.

3)

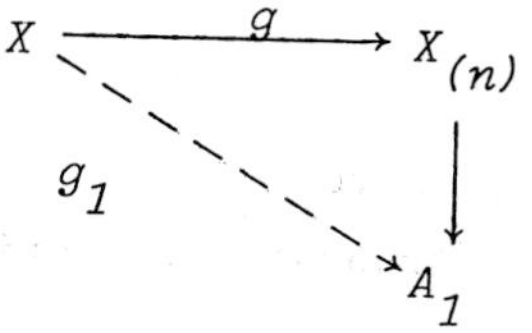

can be equivariantly completed and each of $X \xrightarrow{g_1} A_1$, $A_i \longrightarrow A_{i+1}$ is a Z_n-homology equivalence.

4) The infinite, commulative diagram

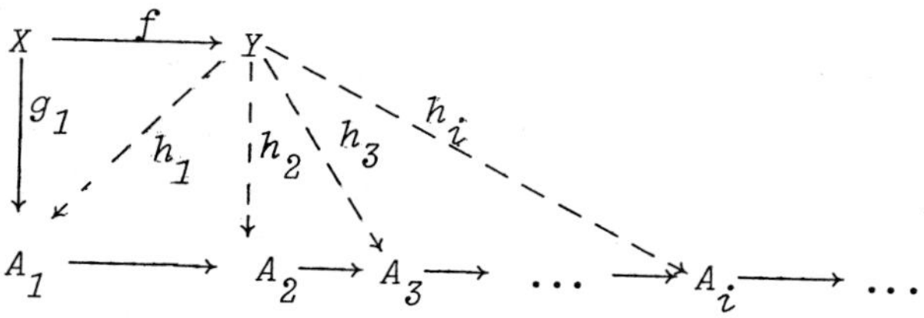

can be completed.

If $\mathbb{Z}_n$ acted trivially on the homology of some A_i, 2.1 would apply to pull $r_{x_{(n)}} : \mathbb{Z}_n \times A_i \to A_i$ back to the desired $\mathbb{Z}_n$-action on Y. However [2] doesn't provide us with this desired homological triviality of the action $r_{x_{(n)}}\big|_{A_i}$. So instead, use (2) to deduce that each projection $p_i : A_i \to A_{i,0}$ (to the orbit space $A_{i,0}$) induces an isomorphism mod n-torsion.

$$\text{image}(H^*(A_{i+1}, Z) \to H^*(A_i, Z)) \cong \text{image}(H^*(A_{i+1,0}, Z) \to H^*(A_{i,0}, Z)).$$

Choose $m > \dim(A_1)$, and let $F_m \to E_{i,m} \to A_i$ denote the restriction of the fibration

$$F_m \to Y \xrightarrow{h_i} A_m \text{ to } A_i \ (i \leq m) .$$

We claim that any finite truncation of $E_{1,m} \to A_1$[1] extends to a truncated fibration $E_{1,m,0} \to A_{1,0}$ over the orbit space $A_{1,0}$. To see this, suppose by induction a truncation of $E_{i+1,m} \to A_{i+1}$ extends to the $(m-(i+1)$-skeleton of $A_{i+1,0}$. Then the obstruction to extending the restriction to $A_{i,0}^{m-(i+2)} \cup A_i$ of this latter truncated fibration, to a truncated fibration over $A_{i,0}^{m-i} \cup A_i$, lies in

$$\text{image}(H^{m-i}((A_{i+1,0} \cup A_{i+1}, A_{i+1}), L)$$
$$\to H^{m-i}((A_{i,0} \cup A_i, A_i), L)),$$

where L is a finite n^*-torsion group; but this image group is zero. Let $B_{1,0}$ denote the k-skeleton of $E_{1,m,0}$, where k and the truncated fibration $E_{1,m,0} \to A_{1,0}$ are chosen such that

$$\begin{array}{ccc} Y \overset{\sim}{\to} E_{1,1} & \to & E_{1,m} \\ & \overset{t}{\searrow} & \cup \\ & & B_1 \end{array}$$

can be completed.

In the same way a sequence $B_{1,0} \subset B_{2,0} \subset B_{3,0} \subset \ldots$ of finite CW orbit spaces can be constructed, converging to the orbit space Y_0' (not necessarily finite) of a free action $r_{y'}: \mathbb{Z}_n \times Y' \to Y'$, in such a way that the composite $Y \xrightarrow{t} B_1 \to Y'$ is a homotopy equivalence. If in addition to the

[1] *A finite truncation of $E_{1,m} \to A_1$ is a subspace of $E_{1,m}$ which is a finite CW complex "over" each cell of A_1. We have in mind an approximation argument as in step 3 of the proof of 2.1, and desire that all finite truncations of fibrations be "good" approximations to those fibrations.*

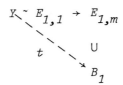

conditions placed on B_1 in the preceding paragraph, we require that $(k \equiv)\dim(B_1) > \max(\dim(A_1) + 2, \dim(Y) + 2)$, then the homological kernel of $B_1 \to Y'$, denoted $K_j(B_1 \to Y')$, will be a finite n^*-torsion module for all $j < \dim(B_1)$, and a stably free $Z(\mathbb{Z}_n)$-module for $j = \dim(B_1)$: in fact module n-torsion we have $K_*(B_1 \to Y') \cong K_*(B_1 \to A_1)$; and $K_j(B_1 \to Y') \cong K_j(B_1 \to A_1) \cong H_{j+1}((A_1, B_1), Z)$ on the nose whenever $j > \max(\dim(A_1), \dim(Y))$; but $H_{k+1}((A_1, B_1), Z)$ is a stably free $Z(\mathbb{Z}_n)$-module when $k = \dim(B_1)$ by the appendix.

Thus $K_*(B_1 \to Y')$ can be killed by adding a finite number of cells equivariantly to the domain of $B_1 \to Y_1$[1], proving that Y_0' has the homotopy type of a finite CW complex.

Finally, as in the proof for 2.1, one sees that the constructed action $r_{y'} : \mathbb{Z}_n \times Y' \to Y'$ acts trivially on $H_*(Y, Z)$, and extends $r_x : \mathbb{Z}_n \times X \to X$.

This completes the proof of 2.2. Q.E.D.

III. BLOCKED NORMAL MAPS

In this section we discuss the type of surgery problem which shall concern us in later chapters. F. Quinn's thesis [12] has been of great help in shaping the point of view described below.

A *blocked space* ξ consists of an underlying space $E(\xi)$ together with a finite collection of subspaces $\{b_j(\xi) : J \in J_\xi\}$ called the *blocks* of ξ, satisfying

3.1 a) $E(\xi)$ is a finite polyhedron. The $\{b_j(\xi) : j \in J_\xi\}$ are non-empty compact $p\ell$ manifolds contained in $E(\xi)$ as subpolyhedra.

b) $E(\xi) = \bigcup_{j \in J} b_j(\xi)$.

[1] *The existence of the composite equivalence $Y \overset{t}{\to} B_1 \to Y'$ assures that kernel homology elements can be represented by Kernel homotopy elements.*

c) For $j, i \in J_\xi$ if $j \neq 1$ then

$(b_j(\xi)-\partial b_j(\xi)) \cap (b_i(\xi)-\partial b_i(\xi)) = \phi.$

d) For $j, i \in J_\xi$, $b_j(\xi) \cap b_i(\xi)$ and $\partial b_j(\xi)$ both equal the union of blocks in ξ.

ξ' is a *blocked subcomplex* of ξ if $E(\xi') \subset E(\xi)$, $J_{\xi'} \subset J_\xi$, $b_j(\xi) = b_j(\xi')$ for all $j \in J_{\xi'}$, and $b_j(\xi) \subset b_{j'}(\xi)$ for $j' \in J_{\xi'}$ implies that $j \in J_{\xi'}$. This relationship shall be denoted by $\xi' \leq \xi$.

Given two blocked spaces ξ, η the cartesian product $\xi x \eta$ is defined: $J_{\xi x \eta} = J_\xi x J_\eta$, $E(\xi x \eta) = E(\xi) x E(\eta)$, $b_{(j,j')}(\xi x \eta) = b_j(\xi) x b_{j'}(\eta)$ for all $(j,j') \in J_\xi x J_\eta$.

ξ and ξ' are isomorphic if there exists a $p\ell$ homeomorphism $f : E(\xi) \to E(\xi')$ mapping each block of ξ onto a block of ξ'. ξ is blocked cobordant to ξ' if there exists a blocked space $\overline{\xi}$ satisfying

a) There are $\overline{\xi}_- \subset \overline{\xi}$, $\overline{\xi}_+ \subset \overline{\xi}$ and isomorphisms $f_- : \xi \to \overline{\xi}_-$, $f_+ : \xi' \to \overline{\xi}_+$.

b) $J_{\overline{\xi}_-} = J_{\overline{\xi}} = J_{\overline{\xi}_+}$

c) $b_j(\overline{\xi})$ is a $p\ell$ cobordism from $b_j(\overline{\xi}_-)$ to $b_j(\overline{\xi}_+)$ for all $j \in J_{\overline{\xi}}$.

In this case $\overline{\xi}$ is called a blocked cobordism from ξ to ξ', or symbolically we write $\overline{\xi}_- = \xi$, $\overline{\xi}_+ = \xi'$.

A *correspondence* $c : \xi \to \eta$ between two blocked spaces ξ, η consists of an embedding $c : J_\xi \to J_\eta$ satisfying

3.2 a) For $j,i \in J_\xi$, $j \leq i$ if and only if $c(j) \leq c(i)$

b) If $i \geq c(j)$ for $(i,j) \in J_\eta x J_\xi$, then there exists $j' \in J_\xi$ with $c(j') = i$.

The *parity* of $c : \xi \to \eta$ is the integer $\dim(b_j(\xi)) - \dim(b_{c(j)}(\eta))$ $(j \in J_\xi)$. It is straightforward to check that the parity is well defined whenever $E(\xi)$ is connected and both

ξ,η are closed. In the future this shall be included in the definition of correspondence as

3.2' The parity of a correspondence is well defined.

We think the essence of a blocked space is a closed PL manifold, because that is the nature of a blocked space with only one block. Likewise, the essence of a *blocked normal map* is a PL normal cobordism class represented by a degree one normal map

$$\begin{array}{ccc} f^*(\tau) & \xrightarrow{f^*} & \tau \\ \downarrow & & \downarrow \\ (N,\partial N) & \longrightarrow & (X,\partial X) \end{array}$$

with $(X,\partial X)$ a Poincare pair, $(N,\partial N)$ a PL manifold pair, f^* a PL mapping of PL bundles, and $f^*|_{\partial N} : \partial N \to \partial X$ a homotopy equivalence.

3.3 SURGERY GROUPS

For a blocked space ξ let $\xi^0 \subset \xi^1 \subset \xi^2 \subset \xi^3 \subset \ldots \subset \xi^k \ldots$ denote the filtration by blocked subcomplexes $\xi^k \equiv \{\bigcup_j b_j(\xi) \mid \dim(b_j(\xi)) \leq k\}$. There is the *blocked chain complex* $BC_*(\xi)$:

$$\xrightarrow{\partial_{\ell+1}} H_\ell\left(\frac{E(\xi^\ell)}{E(\xi^{\ell-1})}, Z\right) \xrightarrow{\partial_\ell} H_{\ell-1}\left(\frac{E(\xi^{\ell-1})}{E(\xi^{\ell-2})}, Z\right) \xrightarrow{\partial_{\ell-1}}$$

$$\begin{array}{cc} \parallel\!\mid & \parallel\!\mid \\ BC_\ell(\xi) & BC_{\ell-1}(\xi) \end{array}$$

where ∂_ℓ is the composition $H_\ell\left(\frac{E(\xi^\ell)}{E(\xi^{\ell-1})}, Z\right) \xrightarrow{\partial} H_{\ell-1}(E(\xi^{\ell-1}), Z) \to H_{\ell-1}\left(\frac{E(\xi^{\ell-1})}{E(\xi^{\ell-2})}, Z\right)$. The cohomology of $BC_*(\xi)$ with coefficients in A is denoted $BH^*(\xi,A)$. For any $\xi' \leq \xi$, $BH^*((\xi,\xi'),A)$ is defined, and the usual long exact sequence connects all $BH^i(\xi,A)$, $BH^k(\xi',A)$, $BH^q((\xi,\xi'),A)$.

A correspondence $c : (\xi,\xi') \to (\eta,\eta')$ is *orientable* if it satisfies

a) every block of ξ or η is connected and orientable.

b) there are orientations $[b_j(\xi)]$, $[b_i(\eta)]$ so that $[b_j(\xi)] \to [b_{c(j)}(\eta)]$ defines a chain map $BC_*(\xi,\xi') \to BC(\eta,\eta')$.

An *orientation* for $c : (\xi,\xi') \to (\eta,\eta')$ is a particular selection of orientations for the blocks of ξ and η as in b). Two orientations are equivalent if they induce the same chain map $BC_*(\xi,\xi') \to BC(\eta,\eta')$. If $E(\xi)$ is connected then c has precisely two orientations, obtained from each other by multiplying by -1 the orientations of ξ-blocks while retaining the same orientation for η-blocks.

Let ξ' denote a blocked subcomplex of ξ, π denotes a group. There is a surgery group $L_k^h((\xi,\xi'),\pi)$ defined as follows. Every element $\alpha \in L_k^h((\xi,\xi'),\pi)$ is represented by an oriented correspondence $c_\alpha : (N_\alpha,N_\alpha') \to (\xi,\xi')$ of parity k, where (N_α,N_α') denotes a blocked normal map pair having a specified global fundamental group $r_\alpha : R(N_\alpha) \to K(\pi,1)$ and a specified completion of surgery on N_α'. Here $R(N_\alpha)$ denotes the range blocked space for N_α. $-\alpha$ is represented by c_α with its negative orientation. Addition comes from disjoint union. c_α and c_α' both represent the same $\alpha \in L_k^h((\xi,\xi'),\pi)$ if there is a blocked normal cobordism connecting c_α to c_α' which preserves orientations and global fundamental groups.

Now let $c : (\beta,\beta') \to (\xi,\xi')$ denote a correspondence between blocked space pairs, having parity equal p, such that $c_* : BC_*(\beta,\beta') \to BC_{*-p}(\xi,\xi')$ is an isomorphism. Then it is geometrically clear that each normal map over (ξ,ξ') having parity equal k is also a normal map over (β,β') having parity equal k-p, and that each normal map with parity $k - p$ over (β,β') can be regarded as one over (ξ, ξ') with parity k.

Hence

LEMMA 3.4. There is a canonical isomorphism

$$L_k^h((\xi,\xi'),\pi) \cong L_{k-p}^h((\beta,\beta'),\pi).$$

LEMMA 3.5. Suppose $BH^*((\xi,\xi'),Z)$ is a torsion group prime to q. Then $L_k^h((\xi,\xi'),\pi)$ is a torsion group prime to q for all π, provided $k \geq 5$.

Proof. Suppose surgery has been completed on N_α^k mod N_α'. Then there is an Eilenberg obstruction $\alpha_{k+1}(\alpha) \in BH^{k+1}((\xi,\xi' \cup \xi^{k-1}), L_{k+1}^h(\pi))$ to completing surgery on N_α^{k+1} mod $N_\alpha' \cup N_\alpha^{k-1}$ (here $k \geq 5$ is required). Let p = order $(\sigma_{k+1}(\alpha))$, and choose a representative $(N_{p\cdot\alpha}, c_{p\cdot\alpha}, r_{p\cdot\alpha})$ for $p\cdot\alpha$ as described above. After alterations as in Chapter 9 of [14], it may be assumed that $r_\alpha : R(N) \to K(\pi,1)$ is two connected on each block of $R(N)$, and that surgery is still completed on $N_{p\cdot\alpha}^k \cup N_{p\cdot\alpha}'$. A calculation shows $\sigma_{k+1}(p\cdot\alpha) = p\cdot\alpha_{k+1}(\alpha) = 0$. So surgery can be completed on $N_{p\cdot\alpha}^{K+1} \cup N_{p\cdot\alpha}'$ mod $N_{p\cdot\alpha}^{k-1} \cup N_{p\cdot\alpha}'$. When $k+1$ reaches $\dim(N)$ we are through.[1]

IV. REDUCTION OF THEOREM A TO A SURGERY PROBLEM

Let ∂R denote the (topological) boundary for a regular neighborhood R for K in D^m. Use the Spanier-Whitehead dual of the Serre-Hurewicz Theorem to find $g : \partial R \to S^k$ inducing an isomorphism $H_*(\partial R, Z_n) \cong H_*(S^k, Z_n)$ (recall the hypothesis concerning $K \subset D^m$

[1] *For a complete account of this Eilenberg-Whitney (surgery) obstruction argument see either the thesis of D. Sullivan or J. Wagoner (Princeton University 1966) where it was first used.*

in Theorem A). Let $c : \partial R \to K$ denote the collapsing map. Then $f \equiv c \times g : \partial R \to K \times S^k$ respects the *blocked structures* given to ∂R and K by their intersections with cells dual in D^m to simplices of K. Let $r : \mathbb{Z}_n \times (K \times S^k) \to K \times S^k$ be a free, cellular action, acting trivially on the factor K but freely on S^k. Apply 2.1 inductively to the blocks of f to get a free, cellular action $r' : \mathbb{Z}_n \times \partial R \to \partial R$ preserving the dual cell blocks of ∂R.

Let ξ denote the blocked space obtained by adding $D^m - R$, $\partial D^m - R$ to the blocks of ∂R. Use 2.1, 2.2 to extend r' to a free, cellular action $r'' : \mathbb{Z}_n \times \xi \to \xi$ preserving blocks.

The orbit blocked space ξ_0 has Poincaré duality spaces (with respect to any $Z(\mathbb{Z}_n)$-module) for blocks.[1] Let ψ denote the Spivak fibration for ξ_0. Since any minimal block, $b_j(\xi_0)$, is homotopy equivalent to a lens space L^k, $\psi|_{b_j(\xi_0)}$ has a PL reduction. The covering map $\xi \to \xi_0$ induces an isomorphism $H_*((\xi, b_j(\xi)), Z) \cong H_*((\xi_0, b_j(\xi_0)), Z)$; so this PL reduction extends uniquely to one for all ψ, which is covered by the inverse to the PL tangent bundle of ξ.

We conclude that there is a PL, blocked, normal map

$$\begin{array}{ccc} h^*(\gamma) & \xrightarrow{h^*} & \gamma \\ \downarrow & & \downarrow \\ M_0 & \xrightarrow{h} & \xi_0 \end{array}$$

having a fixed completion of surgery on its $\mathbb{Z}_n$-universal covering, and on the minimal block $b_j(M_0) \xrightarrow{h} b_j(\xi_0)$ as well.

Suppose there is a completion of surgery for h compatible with that given for $\hat{h}$. Then M_0 would be the orbit space for a free, PL, $\mathbb{Z}_n$-action on $D^m - R$, which preserved the dual cell

[1] *In general, a finite complex P has spherical bundle for normal bundle in S^L if and only if some finite cover of P does.*

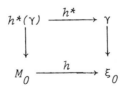

blocks of R. Now fill in an action for R by inductively *coning*[1] the action on each dual cell block of R. This gives the desired $\mathbb{Z}_n \times D^m \to D^m$ having $K \subset D^m$ for fixed point set.

V. COMPLETING SURGERY

We will show that there exists a completion of surgery for h of Chapt. 4 above, compatible with the given completion of $\hat{h}$. For the moment we only assume that $(K, \partial K)$ is a Z_n-homology manifold pair satisfying $\overline{H}_*(K, Z_n) = 0$.

Consider the diagram

$$M_0' \xrightarrow{h'} \xi_0' \xrightarrow{t} \xi_0,$$

where $M_0' = \underbrace{(M_0 \cup \dots \cup M_0)}_{(n+1)\text{-copies}} \cup (-M)$, $\xi_0' = \underbrace{(\xi_0 \cup \dots \cup \xi_0)}_{(n+1)\text{-copies}} \cup (-\xi)$,

$h' = \underbrace{(h \cup \dots \cup h\,)}_{(n+1)\text{-copies}} \cup (-h)$, t restricts on each ξ_0 to 1_{ξ_0} and on ξ to the covering map : toh' and h' are PL blocked normal maps, t is a blocked normal map in the Poincaré category.

There are these formulae in $L^h_*(\xi, \mathbb{Z}_n)$:

(a) $h' = (n+1)\cdot h$.

(b) $toh' = h$ if n = odd,

$toh' = h$ mod 2-torsion, if n = even.

(c) $toh' = t+h'$.

(d) $\ell\cdot h = 0$ for some positive integer ℓ prime to n.

Addition in $L^h_*(\ ,\mathbb{Z}_n)$ is just disjoint union, so (a) is a matter of definition. In [3] it is proven that any surgery problem having a closed manifold for range, $\mathbb{Z}_n (n$ = odd$)$ for fundamental

[1] *Assume all blocks of dimension* $\leq k$ *have been coned. Let* b_j *denote a* $k+1$*-dimensional block, and* $(\partial b_j)^C$ *the coned structure associated to* ∂b_j. *Then cone* $(b_j \cup (\partial b_j)^C)$ *is the coned structure associated to* b_j.

group, and a given completion of surgery on the universal cover, has a completion of surgery. Now note that the difference of surgery problems having the same closed Poincare space for range, can be represented by a normal map with closed manifold for range. This argument works for blocked normal maps also, proving (b) for n = odd. Modulo 2-torsion the same argument works for n = even. (d) depends on $H_*(K, Z_n) = 0$ and 3.5 above.[1] Finally, (c) follows from the surgery composition formula (see [6]), provided the framing information for the two maps t and h' are preserved by h'. To show that the framings are so preserved requires a little discussion. Recall, that in the Poincaré category a surgery problem consists of a degree one fiber map

$$\begin{array}{ccc} \tau & \xrightarrow{h^*} & \Upsilon \\ \downarrow & & \downarrow \\ X & \xrightarrow{h} & Y \end{array}$$

between stable spherical fibrations over Poincaré spaces X, Y together with framing information which can be given either as a spherical class representing the positive generator of the top dimensional homology of the Thom space $T(\tau)$, or as a homotopy commutative diagram

$$\begin{array}{ccc} \tau & \xrightarrow{g} & \dot{R} \\ \downarrow & & \downarrow C \\ X & \xrightarrow{1_X} & X \end{array}$$

[1] $\overline{H}_*(K, Z_n) = 0$ *implies that the "*Z_n*-blocked homology" of* ξ *comes from a minimal block. But surgery can be completed on any minimal block of h.*

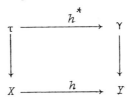

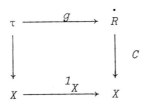

Where $\dot{R}$ denotes the boundary of a regular neighborhood, R, of X in high dimensional Euclidean space, C is the collapsing map, and g is a homotopy equivalence. The first type of framing information I'll call *homotopy framing*, the second type *geometric framing*. These two types of framings are seen to be equivalent under the Thom construction. Now let's return to the construction at the beginning of Chapt. 5 above. Let

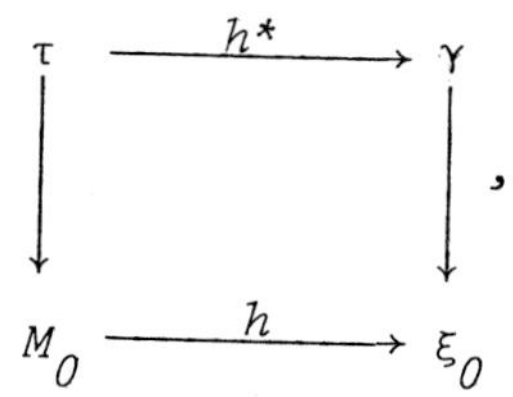

with framing information $f \in \pi_N(T(\tau), T(\tau|_{\partial M_0}))$, denote the blocked surgery problem constructed in Chapt. 4. By applying the construction of page 15 to (h, h^*), f, we get a diagram

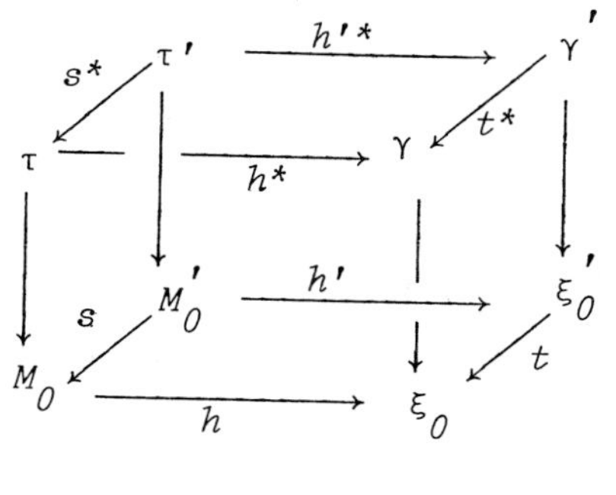

and a framing $f' \in \pi_N(T(\tau'), T(\tau'|_{\partial M_0'}))$.[1] Each of (t, t^*), (s, s^*) can also be equipped with framing information $f_t \in \pi_N(T(\gamma'), T(\gamma'|_{\partial \xi_0'}))$, $f_s \in \pi_N(T(\tau'), T(\tau'|_{\partial M_0'}))$, by first

[1]*Since the construction on page 15 is geometric, first express* f *as a geometric framing* $G(f)$, *then construct* $G(f)$ *and let* f' *be the homotopic form of* $G(f)'$.

choosing geometric framings for the bundle maps $(1_{\xi_0},1_\gamma)$, $(1_{M_0},1_\tau)$, applying the construction of page 15 to get geometric framings for $(1_{\xi_0'}, 1_{\gamma'})$, $(1_{M_0'}, 1_{\tau'})$, then converting to homotopic framings f_t, f_s. The framings f_t, f_s, of course, depend on the initial geometric framings of $(1_{\xi_0},1_\gamma)$, $(1_{M_0},1_\tau)$, but the surgery problems $(t,t^*;f_t)$, $(s,s^*;f_s)$ are well defined up to a fiber homotopy equivalence which preserves the framing information. To complete this discussion relating to the equivality (c) above, it suffices to show that f_t can be chosen so that $T(h'^*)$ maps f' to f_t. Construct f_t from the geometric framing equivalent to the image under $T(h^*)$ of f. Construct f_s from the geometric framing equivalent to f. Then $f_s = f'$, and by the commutativity in the diagram 5.0 above, f_t is the image of f_s under $T(h'^*)$.

Computing from (a)-(d), one sees that h = 2-torsion in $L^h_*(\xi,\mathbb{Z}_n)$ follows from this lemma.

LEMMA 5.1. t is a 2-torsion element in $L^h_*(\xi,\mathbb{Z}_n)$.

Proof of 5.1. It suffices to prove that the restriction of t to $L^H_*(\partial R,\mathbb{Z}_n)$ is a 2-torsion element (the restriction map $L^H_*(\xi,\mathbb{Z}_n) \to L^h_*(\partial R,\mathbb{Z}_n)$ is an isomorphism).

There is a commutative diagram

$$\begin{array}{ccc} \partial R_0' & \xrightarrow{\quad t \quad} & \partial R_0 \\ {\scriptstyle f'}\big\downarrow & & \big\downarrow{\scriptstyle f} \\ K\times(L^k)' & \xrightarrow{\quad 1\times\gamma \quad} & K\times L^k \end{array},$$

where $(L^k)' \equiv \underbrace{(L^k \cup\cdots\cup L^k)}_{(n+1)\text{-copies}} \cup (-\hat{L}^k)$, and γ restricts on each L^k[1] to 1_{L^k}, but on $\hat{L}^k$ to the covering projection. Note that

[1] L^k *is the k-dimensional lens space of Chapt. 4 above.*

f, f' are blocked preserving, and are $Q(\mathbb{Z}_n)$-homology equivalence on each block. So $t = 1 \times \gamma$ in $L^h_*(\partial R, Q(\mathbb{Z}_n))$.[1] But $\gamma = 0$ in $L^h_*(\mathbb{Z}_n)$ for n = odd γ = 2-torsion if n =. even [3] so $t = 0$ in $L^h_*(\partial R, Q(\mathbb{Z}_n))$.

We complete the proof of 5.1 by showing that the canonical map $L^h_*(\partial R, Z(\mathbb{Z}_n)) \to L^h_*(\partial R, Q(\mathbb{Z}_n))$ is an isomorphism modulo the class, $\mathcal{S}$, of 2-torsion groups.

First consider the canonical map $L^h_*(Z(\mathbb{Z}_n)) \to L^h_*(Q(\mathbb{Z}_n))$. Calculations in [4, 11] show that this is a mod $\mathcal{S}$ isomorphism in all odd dimensions.[2] At even dimensions we can say this. $Q(\mathbb{Z}_n)$ splits as the sum of two fields $Q \oplus Q(\theta)$, where θ denotes a primitive n^{th} root of unity (n = prime). So $L^h_{2i}(Q(\mathbb{Z}_n)) = L^h_{2i}(Q) \oplus L^h_{2i}(Q(\theta))$. Mod $\mathcal{S}$, the first summand equals the one element group for i = odd, and equals Z for i = even: this is immediate from the classification of non-singular quadratic forms over the rationals by the index, discrimanent, and Hasse-Symbol invariants. The second summand, $L^h_{2i}(Q(\theta))$, is just the Witt group of non-singular forms $\emptyset$ over $Q(\theta)$ which are hermetian (i = even) or skew Hermetian (i = odd) with respect to the conjugation field automorphism of $Q(\theta)$ which sends θ to θ^{-1}. Skew Hermetian forms $\emptyset$ correspond bijectively to hermetian forms by the rule $\emptyset \to \alpha\cdot\emptyset$, where "$\alpha\cdot$" denotes multiplication by any non-zero $\alpha \in Q(\theta)$ satisfying $\bar{\alpha} = -\alpha$. So it suffices to consider Hermetian forms. There are $\frac{n-1}{2}$ topologically distinct field imbeddings of $Q(\theta)$ in the complex field $\mathbb{C}$. So there are indices $\lambda_i(\emptyset)$, $i = 1, 2, \ldots, \frac{n-1}{2}$, associated to any non-singular Hermetian forms $\emptyset_i$, $i = 1, 2, \ldots, \frac{n-1}{2}$ and setting $\lambda_i(\emptyset) \equiv \text{index}(\emptyset_i)$. The $\lambda_i(\emptyset)$, the rank of $\emptyset$, plus a discriminant invaa-riant give a complete set of invariants for $\emptyset$ over $Q(\theta)$ (see [10]).

[1] *$L^h_*(\xi, Q(\mathbb{Z}_n))$ is defined just as $L^h_*(\xi, \mathbb{Z}_n)$, but that each block is normal map only for $Q(\mathbb{Z}_n)$ coefficients (see 7.11 in [7]). In this notation we have $L^h_*(\xi, \mathbb{Z}_n) \equiv L^h_*(\xi, Z(\mathbb{Z}_n))$.*

[2] *See in particular Theorems A, B, C of [4].*

Thus, mod $\mathfrak{L}$, $L^h_{2i}(Q(\theta))$ is a free abelian group having rank $\leq \frac{n-1}{2}$. So, mod $\mathfrak{L}$, $L^h_{2i}(Q(\mathbb{Z}_n))$ is a free abelian group having rank $\leq \frac{n-1}{2}$ for i = odd, and rank $\leq \frac{n-1}{2} + 1$ for i = even.

Using Theorem 13A.4 on page 168 of [14], and the fact that there exists even, unimodular, symmetric matrices over the integer having index 8, one gets a commutative diagram

$$\begin{array}{ccc} L^h_{2i}(Z(\mathbb{Z}_n)) & \longrightarrow & L^h_{2i}(Q(\mathbb{Z}_n)) \\ \downarrow I_{\mathbb{Z}_n} & \swarrow I^Q_{\mathbb{Z}_n} & \\ Z^{\frac{n-1}{2}} \oplus Z & & \end{array}$$

where $I_{\mathbb{Z}_n}$, $I^Q_{\mathbb{Z}_n}$ are equivariant index invariants, and $I_{\mathbb{Z}_n}$ is a mod $\mathfrak{L}$ isomorphism for i = even, but only a mod $\mathfrak{L}$ isomorphism onto the factor $Z^{\frac{n-1}{2}}$ i = odd. But now the restrictions determined for $L^h_{2i}(Q(\mathbb{Z}_n))$ in this last paragraph show that $L^h_*(Z(\mathbb{Z}_n)) \to L^h_*(Q(\mathbb{Z}_n))$ must be a mod $\mathfrak{C}$ isomorphism in all even dimensions.

Now by induction over the dual cell blocked structure for ∂R, using the mod $\mathfrak{L}$-isomorphism $L^h_*(Z(\mathbb{Z}_n)) \cong L^h_*(Q(\mathbb{Z}_n))$ at every induction step, argue that the canonical map $L^h_*(\partial R, Z(\mathbb{Z}_n)) \to L^h_*(\partial R, Q(\mathbb{Z}_n))$ is a mod $\mathfrak{L}$ isomorphism.

This completes the proof of 5.1. Q.E.D.

If $\overline{H}_*(K, Z_2) = 0$ holds, as hypothesized in Theorem A above, then h has odd order in $L^h_*(\xi, \mathbb{Z}_n)$ by (d) above. But then $h = 0$ in $L^h_*(\xi, \mathbb{Z}_n)$ (5.1).

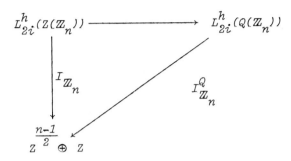

To complete the proof of Theorem A it remains to find a completion of surgery for h compatible with that given for $\hat{h}$. There is a surgery obstruction group $L^{\Delta h}_*(\xi, \mathbb{Z}_n)$ defined as before, but where "Δ" means each representative surgery problem has a given completion of surgery on its $\mathbb{Z}_n$-universal covering. h, and the given completion of surgery for h, determine $\alpha(h,\hat{h}) \in L_*^{\Delta h}(\xi,\mathbb{Z}_n)$. We wish to show $\alpha(h,\hat{h}) = 0$. If ξ_j denotes a minimal block of ξ, there is a "relative surgery group" $L_*^{\Delta h}((\xi,\xi_j),\mathbb{Z}_n)$. Since the Z_n-blocked homology of (ξ,ξ_j) is zero, an Eilenberg-Whitney (surgery) obstruction argument[1] shows $L_*^{\Delta h}(\xi,\xi_j),\mathbb{Z}_n)$ is all torsion prime to n. By construction, it is clear that there is a completion of surgery for h over ξ_j consistent with the completion of surgery on $\hat{h}$: this gives $\alpha'(h,\hat{h}) \in L^{\Delta h}((\xi,\xi_j),\mathbb{Z}_n)$. Since $\alpha'(h,\hat{h})$ has order prime to n, $\alpha(h,\hat{h})$ must also. Now $L_*^h(\xi,\mathbb{Z}_n) \underset{i}{\overset{\tau}{\rightleftarrows}} L^h(\xi,\{1\})$ represents $L^h(\xi,\{1\})$ as a retract (modulo the class of n-torsion groups) of $L_*^h(\xi,\mathbb{Z}_n)$, where τ is the transfer map and i is the inclusion: in fact the composite $\tau\circ i$ equals multiplication by n in the abelian group structure of $L_*^h(\xi,\{1\})$. Thus if $C(h)$ denotes any completion of surgery for h, and $\widehat{C(h)}$ its $\mathbb{Z}_n$-universal cover, by varying $\widehat{C(h)}$ through a blocked surgery cobordism representing any element in $i(L_*^h(\xi,\{1\}))$ of dimension one larger than $\hat{h}$, $C(h)$ will be varied through any element (mod n-torsion) of $L_*^h(\xi,\ 1\ )$ of dimension one larger than $\hat{h}$. But then $\alpha(h,\hat{h})$ has order equal both a power of n, and a prime to $n \Rightarrow \alpha(h,h') = 0$.

This completes the proof of Theorem 1.1. We have as a corollary of this discussion:

THEOREM 5.2. Let $K \subset D^m$ be a PL subset having even codimension ≥ 6, and n odd integer. Suppose $(K, K \cap \partial D^m)$ a Z_n-homology manifold pair and that $\bar{H}_*(K,Z_n) = 0$. Then K is the fixed point of a semi-free combinatorial symmetry $\mathbb{Z}_n \times D^m \to D^m$, having some fake lens space in the same normal cobordism class as L^k as the

[1] *See the proof of 3.5 above.*

orbit of a PL slice, if a certain surgery obstruction $\sigma(k, L^k) \in L_*^h(\xi, \mathbb{Z}_n)_2$ *vanishes.*

Remark 5.4. Let $K \subset M$ be a PL embedding of the Z_n-homology manifold pair $(K, K \cap \partial M)$ in the Z_n-homology manifold pair $(M, \partial M)$. Suppose there is a regular neighborhood R for K in M, so that each of $\partial R \cap D$, $\partial R \cap \partial M \cap D$ is a simply connected PL manifold for any "cell" D dual in M to a simplex of K. Then theorem A and 5.2 generalize to the embedding $K \subset R$. If $M - R$, $\partial M - R$ are also simply connected, PL, manifolds, with M a Z_n-homology disc, then Theorems 1.1 and 5.2 generalize to the embedding $K \subset M$. In fact the proof goes through exactly as before. A point the reader should convince himself of is that the existence of a regular neighborhood R, with the properties listed above, depends only on the PL embedding $K \subset M$.

APPENDIX

Let $0 \to C_m \xrightarrow{\partial m} C_{m-1} \xrightarrow{\partial m-1} C_{m-2} \xrightarrow{\partial m-2} \cdots C_0$ be a finitely generated free $Z(\mathbb{Z}_n)$- chain complex. Suppose the homology groups $H_i(C)$ are torsion prime to n for i ≠ j, and the torsion part of $H_j(C)$ -- denoted $\mathrm{Tor}_Z(H_j, Z)$ -- is also prime to n. It is a well-known fact that under these restrictions the quotient $H_j(C)/\mathrm{Tor}_Z(H_j, Z)$ must be a projective $Z(\mathbb{Z}_n)$- module.

THEOREM 1.0. $H_j(C)/\mathrm{Tor}_Z(H_j, Z)$ *is a stably free* $Z(\mathbb{Z}_n)$*- module if one of the following holds:*

(a) $\mathbb{Z}_n$ *acts trivially on each of the modules* $H_i(C)$ $(i \neq j)$, $\mathrm{Tor}_Z(H_j, Z)$.

(b) n *is a prime;* p *is a prime which generates the group of units in the field* Z_n*; each of the modules* $H_i(C)$ $(i \neq j)$, $\mathrm{Tor}_Z(H_j, Z)$, *has order equal to a power of* p.

Proof.

Step 1. For M equal any of the modules $H_i(C)(i \neq j)$, $\mathrm{Tor}_Z(H_j, Z)$, there is an exact sequence of $Z(Z_n)$-modules

$$0 \to F \to F' \to M \to 0$$

where F, F' are finitely generated stably free $Z(Z_n)$-modules.

If the chain complex satisfies (a), then this follows Lemma 1.1 [5].

Suppose (b) is satisfied. First consider the case when $H_i(C)(i \neq j)$, and $\mathrm{Tor}_Z(H_i, Z)$, are Z_p-vector spaces. Set $\Gamma \equiv Z_p(\mathbb{Z}_n)$. Γ is the direct sum of two fields $Z_p \oplus Z_p(\theta)$ where θ is a primitive n-th root of unity : $Z_p(\theta)$ is a field because by the hypothesis of (b) $1 + x + x^2 + \ldots + x^{n-1}$ is an irreducible polynomial over the field Z_p. So as Γ-modules $H_i(C)(i \neq j)$ and $\mathrm{Tor}_Z(H_j, Z)$ are the direct sum of a finite number of copies of Z_p and $Z_p(\theta)$. The kernels of the first two of the following natural projections

$$Z(Z_n) \to Z_p$$
$$Z(Z_n) \to \Gamma$$
$$Z(Z_n) \to Z_p(\theta)$$

are free $Z(Z_n)$-modules (see Section 1 in [5]); hence the kernel of the third projection must be stably free (Schanuel's Lemma). Now it is seen that the kernel of $F' \xrightarrow{\alpha} H_i(C)$ (or of $F' \xrightarrow{\alpha} \mathrm{Tor}_Z(H_j, Z))$ must be stably free, where α is any surjection from a finitely generated, stably free, $Z(\mathbb{Z}_n)$ -module F' (again, Schanuel's Lemma).

To establish Step 1 in general, split $H_i(C)$ (or $\mathrm{Tor}_Z(H_j, Z))$ into its primary components, filter each of these by submodules $A_p^1 \subset A_p^2 \subset A_p^3 \subset \ldots \subset H_i(C)_p$ where A_p^{i+1}/A_p^i is a Z_p-vector space. Using the induction, and the special case just considered, argue that each A_p^i has a length two finitely generated free resolution. Clearly then the same holds for $H_i(C) = \oplus_p H_i(C)_p$.

Step 2. There is no loss in supposing $j \neq m$. Consider the exact sequence

$$0 \to K_i \to C_i \to K_{i-1} \to H_{i-1}(C) \to 0,$$

where $K_i \equiv \text{kernel}(\partial_i)$. Using this sequence, Schanuel's Lemma, and Step 1, argue that K_i is stably free for all $i \leq j$. In particular, $H_j(C)$ has a finitely stably free resolution

$$0 \to C_m \to \cdots \to C_{j+1} \to K_j \to H_j(C) \to 0.$$

$\text{Tor}_Z(H_j, Z)$ has a finite stably free resolution by Step 1. It follows that the quotient of these two modules, $H_j(C)/\text{Tor}_Z(H_j, Z)$, also has a finite stably free resolution. But since this last module is torsion free, and with finite projective dimension, it must be projective.[1] Finally a projective module with a finite stably free resolution must be stably free, so $H_j(C)/\text{Tor}_Z(H_j, Z)$ is stably free as claimed.

REFERENCES

1. Arbowitz, M., and Curjel, C., Groups of Homotopy Classes, Springer-Verlag (1964) Berlin.
2. Anderson, D. W., Localizing CW Complexes.
3. Browder, W., Free $\mathbb{Z}_n$ - actions on Homotopy Spheres, The Proceedings of the Georgia Conference on Topology of Manifolds (1969).
4. Connoly, Francis X., Linking Numbers and Surgery (preprint).
5. Jones., The Converse to the fixed point theorem of P. A. Smith: I, Annals of Math. 94 (1971) 52-68.
6. _____., Corrections to Patch Spaces, Ann. Math. 102 (1975) 102 (1975) 103-185.

[1]*According to the Theorem on page 145* [17], *every finitely generated torsion free* $Z(\mathbb{Z}_p)$*-module is the direct sum of 3 types of modules: 1) projective in* $Z(\mathbb{Z}_p)$*-modules; 2) projective* $Z(\theta)$*-modules (which are* $Z(\mathbb{Z}_p)$*-modules via the augmentation* $Z(Z_p) \to Z(\theta)$*); 3)* Z *with trivial* $\mathbb{Z}_p$*-action.*

7. Jones., Patch Spaces: A geometric representation for Poincaré spaces, Ann. Math. 97 (1973) 306-343.
8. _____., Construction of $\mathbb{Z}_n$-action on Manifolds, preprint.
9. _____., Combinatorial symmetries of the m-disc, preprint (Berkeley) 1971-72.
10. Land Herr, W., Aquivalence Hermitscher Formen uber einem belilbigen algebraischen zahlkörper, Abh. Math. Sem. Hamburg 11 (1936) 245-248.
11. D. S. Passman and T. Petrie, Surgery with coefficients in a field, Ann. Math. 95 (1972) 385-405.
12. Quinn, F., Thesis (Princeton, 1969).
13. Smith, P. A., Fixed Points of Periodic Transformation, Amer. Math. Soc. Coll. Pub. XXVII (1942) 350-373.
14. Wall, C. T. C., Surgery on Compact Manifolds.
15. ____, Finiteness conditions for CW-complexes, Annals of Math. 81 (1965) 56-69.
16. ____, On the classification of Hermitian forms I, Compositio Mathematica 22 (1970) 425-451.
17. Reiner, I., Integral representations of cyclic groups of prime order, Proc. Amer. Math. Soc. 8(1957) 142-146.

WILD EMBEDDINGS OF PIECEWISE LINEAR MANIFOLDS IN CODIMENSION TWO

Yukio Matsumoto[1]

Department of Mathematics
The Institute of Advanced Study
Princeton, N. J.

Department of Mathematics
College of General Education
University of Tokyo
Komaba, Meguroku, Tokyo

We shall construct infinitely many topological embeddings of non-simply connected piecewise linear (=PL) 2n-manifolds into the 2n+2-space, which cannot be approximated by PL embeddings. The existence of such embeddings is in sharp contrast with Homma and Miller's codimension ≥ 3 *approximation theorems. By Kirby-Siebenmann's theorems, it is shown that if the source manifolds* M^{2n} *satisfy* $H^3(M^{2n}; \mathbb{Z}_2) = 0$*, then the above embeddings are not approximated by locally flat topological embeddings. Thus they also contrast with Ancel and Cannon's recent approximation theorem in codimension 1.*

The main idea is a generalization of Giffen and Eaton, Pixley and Venema's, and to apply Giffen's shift spinning construction to PL spineless manifolds. The existence of PL spineless manifolds was proved by Cappell and Shaneson, and also by the author. Chappell-Shaneson gave a large class of them.

[1]*Supported by the National Science Foundation.*

ISBN 0-12-158860-2

I. INTRODUCTION

The problem we are concerned with is the PL approximation problem of topological embeddings: *Given PL manifolds* M^m, Q^q *and a topological embedding* $f : M^m \to \text{Int}\, Q^q$, *is it possible to approximate* f *by PL embeddings arbitrarily closely?*

For the history, refer to Rushing [25]. Notably, Homma [9] proved a PL approximation theorem in the metastable range $(m \leqq \frac{2}{3}q - 1)$, see also [25, §5.4], and Miller [23] established the PL approximation theorem in codimension $(= q - m) \geqq 3$.

However, in codimension 2, few have been known.

Recently, Giffen [8] and Eaton, Pixley and Venema [4] suggested a certain pessimistic foreboding by constructing a wild topological embedding of $T^2 = S^1 \times S^1$ into $\mathbb{R}^4$ which cannot be approximated by a PL embedding. The purpose of this paper is to justify it to a large extent.

We shall construct infinitely many topological embeddings of non-simply connected, closed, even dimensional PL manifolds in codimension 2, which cannot be approximated by any PL (even non-locally flat PL) embeddings. (Thms. 5.1, 5.2, 7.1, 7.3.) Our method, which is a generalization of [4], [8], only works for non-simply connected, even dimensional closed manifolds. Thus the PL approximation problem for topological embeddings of simply-connected or odd dimensional PL manifolds is still open.

Venema proved a PL approximation theorem for topological embeddings of disks [30], [31].

Although Cappell and Shaneson's homology surgery [2] would apply equally well (in fact, we will follow some of their arguments in §7), we shall use, in this paper, codimension 2 surgery developed in [10][17]. It is mainly because our method is more familiar to the author than theirs. We have given a summary of codimension 2 surgery in the appendix, which will be referred to as [A].

Some results of this paper were announced in [21].

II. A BASIC CONSTRUCTION

Let $2n \geq 4$.

Notation: $\mathbb{R}_{+}^{2n+2} = \{(x_1,\ldots,x_{2n+2}) \mid x_{2n+2} \geq 0\}$,

$\mathbb{R}_{-}^{2n+2} = \{(x_1,\ldots,x_{2n+2}) \mid x_{2n+2} \leq 0\}$,

$\mathbb{R}^{2n+1} = \{(x_1,\ldots,x_{2n+1},0)\} \subset \mathbb{R}^{2n+2}$

Embeded $M^{2n} = S^1 \times D^{2n-1} \# S^n \times S^n$ (an interior connected sum) into $\mathbb{R}_{-}^{2n+2}$ so that $M^{2n} \cap \mathbb{R}^{2n+1} = \partial M^{2n} = S^1 \times S^{2n-2}$ (identifying M^{2n} with its embedded image by abuse of notation). We assume that the embedding is sufficiently nice, in particular,

1) it is locally flat PL,
2) $\pi_1(\mathbb{R}_{-}^{2n+2} - M^{2n}) = \mathbb{Z} = \langle t \rangle$, and
3) there exist disjoint $n+1$-disks D_a^{n+1}, $D_b^{n+1} \subset \mathbb{R}_{-}^{2n+2}$ such that $D_a^{n+1} \cap M^{2n} = \partial D_a^{n+1} = S_a^n$, $D_b^{n+1} \cap M^{2n} = \partial D_b^{n+1} = S_b^n$, where $S_a^n = S^n \times \{pt\}$, $S_b^n = \{pt\} \times S^n$ in the $S^n \times S^n$-component of M^{2n}.

Draw two arcs A, B ($\subset$ Int M^{2n}) as indicated by Figure 1. Push S_a, S_b by regular homotopies along these arcs A, B. The resulting immersed spheres S_a', S_b' are shown in Figure 2. The immersed sphere S_a' (or S_b') has two self-intersection points p_+^a, p_-^a (or p_+^b, p_-^b), $\pm$ indicating the sign of the intersections with respect to the order of the branches (coming back branch, stationary branch). They have a mutual intersection point p.

Let N be a regular (or tubular) neighborhood of M^{2n} in $\mathbb{R}_{-}^{2n+2}$. N may be considered as a trivial D^2-bundle over M^{2n}: $N = M^{2n} \times D^2$. Let FN be the frontier: $M^{2n} \times \partial D^2$. (The product structures are specified by requiring that a horizontal lift $C \times \{\xi\}$ $(\xi \in \partial D^2)$ of any loop C of M be null-homotopic in $\mathbb{R}_{-}^{2n+2} - M^{2n}$.)

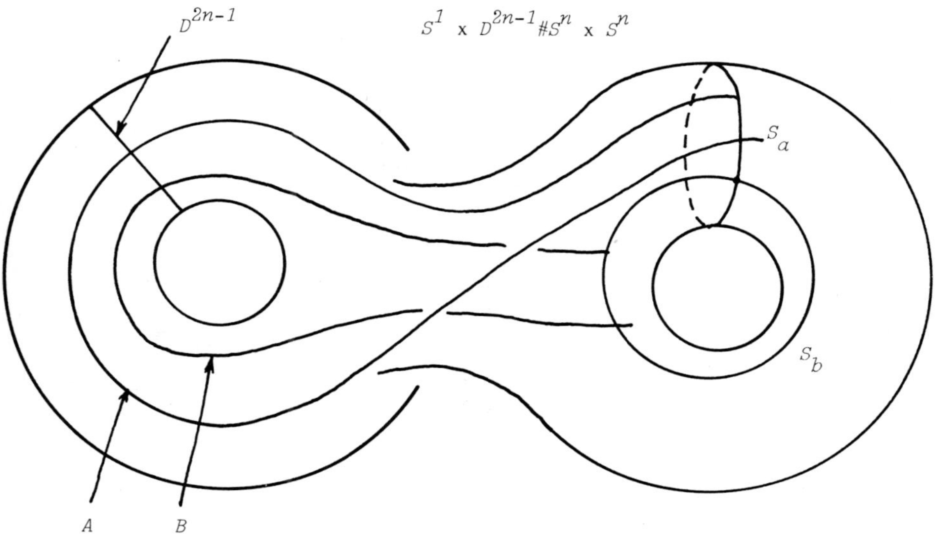
D^{2n-1}
$S^1 \times D^{2n-1} \# S^n \times S^n$
S_a
S_b
A
B

Fig. 1.

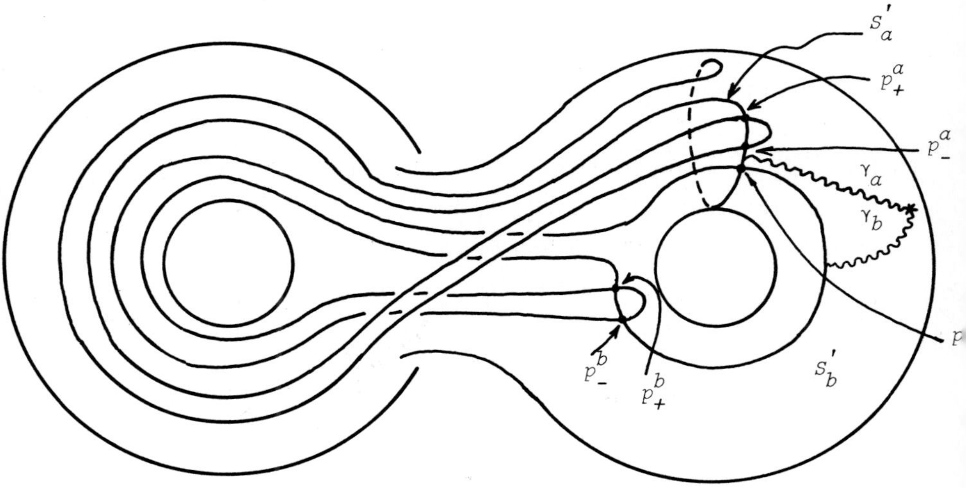
S'_a
p^a_+
p^a_-
γ_a
γ_b
p^b_-
p^b_+
S'_b

Fig. 2.

Consider horizontal lifts $S'_a \times \{\xi_a\}$, $S'_b \times \{\xi_b\}$ of S'_a, S'_b into different levels ξ_a, ξ_b. Modify these lifts slightly in neighbourhoods of the self-intersection points as indicated by Figure 3. (For the choice of the positive direction of $S^1 = \partial D^2$ in Fig. 3, see orientation conventions in the latter part of §2.) Then one obtains embedded lifts S''_a, S''_b.

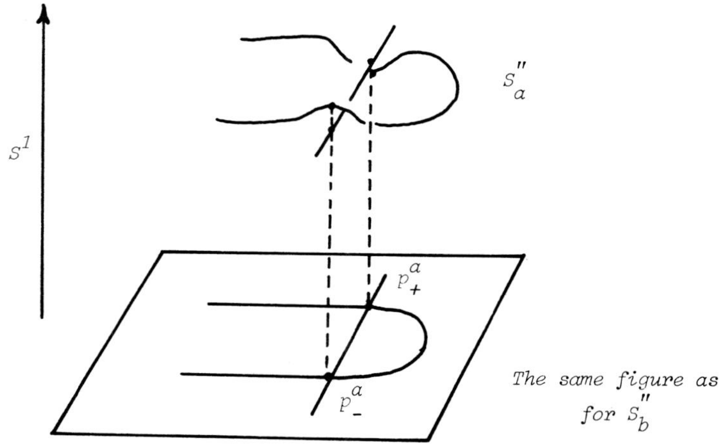

Fig. 3.

We assume that ξ_b is slightly 'higher' than ξ_a : $\xi_b = \xi_a e^{\sqrt{-1}\,\varepsilon}$ ($\varepsilon > 0$, very small).

These embedded spheres S''_a, S''_b are (disjointly) regularly homotopic in FN to horizontal lifts $S_a \times \{\xi_a\}$, $S_b \times \{\xi_b\}$ of the standard spheres S_a, S_b. See Figure 4.

Embed a collar $FN \times [0,1]$ in closure $(\mathbb{R}_-^{2n+2} - N)$ so that $FN \times \{0\}$ identified with FN. Then, considering $[0,1]$ as a set of time parameters, the above regular homotopies trace out disjointly immersed annuli $\approx S^n \times [0,1]$. Denote them by A_a, A_b. By Figure 4, A_a, A_b have self-intersection points q_a, q_b. By the niceness assumption on M^{2n}, (3), we can find disjoint $n{+}1$-disks in closure $(\mathbb{R}_-^{2n+2} - N \cup FN \times [0,1])$, whose boundaries are

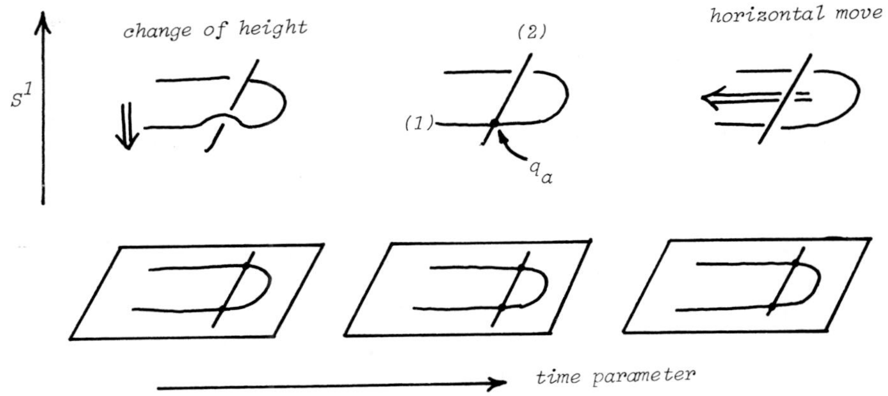

Fig. 4.

$A_a \cap FN \times \{1\} = S_a \times \{\xi_a\} \times \{1\}$ and $A_b \cap FN \times \{1\} = S_b \times \{\xi_b\} \times \{1\}$, respectively. Attaching these disks to A_a, A_b, we obtain disjointly immersed $n+1$-disks D'_a, D'_b such that $\partial D'_a = S''_a$, $\partial D'_b = S''_b$. D'_a (or D'_b) has self-intersection point q_a (or q_b). We will apply Whitney's method to eliminate it.

A loop on D'_a starting at a point on D'_a reaching q_a via the 'coming back branch' (1) (Fig. 4) then passing to the 'stationary branch' (2) (Fig. 4), coming back to the starting point, is null-homotopic in closure $(\mathbb{R}_-^{2n+2} - N)$, by the construction of D'_a. Thus we can apply Whitney's trick. Suppose the sign $(\pm)$ of q_a is ε with respect to a fixed order of branches, say, branch (1) and branch (2) in this order. Introduce a double point q'_a with sign $(-\varepsilon)$, locally to D'_a. Then Whitney's process pairwise cancels q_a, q'_a to give an embedded $n+1$-disk D''_a. The same argument can be applied to D'_b. Thus we have obtained disjointly embedded $n+1$-disks D''_a, D''_b such that $\partial D''_a = S''_a$, $\partial D''_b = S''_b$.

Let W^{2n+2} be a regular neighbourhood of $N \cup D''_a \cup D''_b$ $\mathbb{R}^{2n+2}$ meeting $\mathbb{R}^{2n+1}$ regularly.

LEMMA 2.1. W^{2n+2} *is PL homeomorphic to* $S^1 \times D^{2n+1}$.

Proof. It is easy to see that $W \simeq S^1$ and $\pi_1(\partial W) = \mathbb{Z}$. Let $C \hookrightarrow \mathrm{Int}(W^{2n+2})$ be an embedding of the circle which is a homotopy equivalence. Delete the interior of a regular neighbourhood T of C. Then we have a cobordism V^{2n+2} between ∂T and ∂W. Note that $\pi_1(\partial T) = \pi_1(\partial W) = \pi_1(V) \cong \pi_1(S^1)$. By excision, $H_*(V, \partial T; \Lambda) = H_*(W, T; \Lambda) = 0$, Λ being $\mathbb{Z}[\pi_1 S^1]$. This implies $H^*(V, \partial T; \Lambda) = 0$. By the Poincare duality [32], $H_*(V, \partial W; \Lambda) = 0$. Thus V is an s-cobordism between $\partial T(\cong S^1 \times S^{2n})$ and ∂W (for $Wh(\pi_1 S^1) = 0$). 2.1 follows from the s-cobordism theorem, for $2n+2 \geq 6$.

Recall $W^{2n+2} \cap \mathbb{R}^{2n+1} = \partial M^{2n} \times D^2$. Thus we have a framed embedding $S^1 \times S^{2n-2} \overset{\cong}{\to} \partial M^{2n} \times \{0\} \subset \partial W^{2n+2}$. Since the pair $(W^{2n+2}, \partial M^{2n})$ is proved to be homotopically equivalent to $(M^{2n} \cup D_a \cup D_b, \partial M^{2n}) \simeq (S^1 \times D^{2n-1}, S^1 \times S^{2n-2})$, it is a global knot in the sense of [A]. The homomorphism associated with it[1] is $\{\pi_1(M \times \partial D^2) \to \pi_1(M)\} \cong \{\mathbb{Z} \times \mathbb{Z} = \langle t \rangle \times \langle s \rangle \to \langle s \rangle = \mathbb{Z}\}$. To compute the Seifert form[2] of it, let E = closure $(W-N)$, the exterior of N in W. Clearly $\pi_i(E, \dot{F}N) = 0$, $i \leq n$, and $\pi_{n+1}(E, \dot{F}N)$ is a free Λ-module $(\Lambda = \mathbb{Z}[t, t^{-1}, s, s^{-1}])$ with basis Δ_a, Δ_b represented by D''_a, D''_b. To make D''_a, D''_b pathed disks, assign γ'_a, γ'_b to them, which are horizontal lifts $\gamma_a \times \{\xi_a\}$, $\gamma_b \times \{\xi_b\}$ of paths γ_a, γ_b in Figure 2. By [A], $\pi_{n+1}(E, \dot{F}N)$ carries a $(-1)^n$ - Seifert form (λ, μ).

COMPUTATION OF (λ, μ). From now on, the notation and the terminology are those of [A]-(7). To compute $\mu(\Delta_a)$, $\mu(\Delta_b)$, arbitrarily fix an order of branches at self-intersection points p^a_+, p^a_-, p^b_+, p^b_-, say, the 'coming back' branch and the 'stationary' branch in this order. Then

[1] *See Appendix (5).*
[2] *See Appendix (7).*

$$(*)\quad \begin{cases} g(S''_a)(p^a_+) = s, & \varepsilon(S''_a)(p^a_+) = 1 \\ g(S''_a)(p^a_-) = st, & \varepsilon(S''_a)(p^a_-) = -1. \end{cases}$$

Here s is a generator of $\pi_1 M^{2n} = \mathbb{Z}$, in the direction of which A,B in Figure 1 traverse, and t is represented by the fiber in the positive direction.

By $(*)$ we have

$$\alpha(S''_a) = s - st,$$

and similarly

$$\alpha(S''_b) = s - st.$$

By the construction of D'_a, it is clear that $\mathcal{O}(D'_a) = 0$. D''_a is regularly homotopic to D'_a plus a Whitney's double point q'_a with sign $(-\varepsilon)$. Thus by Assertion 2.8.1 of [17], we have

$$\mathcal{O}(D''_a) = \begin{cases} 2\varepsilon & (n+1 : \text{even}), \\ 0 & (n+1 : \text{odd}). \end{cases}$$

To compute ε, we recall our orientation conventions from [17].

ORIENTATION CONVENTIONS. For an oriented manifold V, $[V]$ denotes the orientation.

The orientation $[E^{2n+2}]$ is induced from the natural orientation of $\mathbb{R}^{2n+2}$. The orientation $[FN]$ is given by

$$[E] = [FN] \times \nu,$$

ν being the inward normal of E. An orientation $[M^{2n}]$ of M^{2n} is assumed to be given, then the orientation of ∂D^2 (recall $FN = M \times \partial D)$ is given by

$$[FN] = [M^{2n}] \times [\partial D^2].$$

Also, recall the 'curious incidence relation'

$$[D^{n+1}] = [S^n] \times [\text{inward normal}].$$

Now near q_a, the 'coming back' branch (1) of D''_a is oriented as $[S''_a]_{(1)} \times [-\partial D^2]$. Another branch (2) is oriented as $[S''_a]_{(2)} \times \nu$, ν being identified with the increasing direction of the time parameter $[0,1]$, see Fig. 4. Thus

$$\varepsilon = \frac{[D''_a]_{(1)} \times [D''_a]_{(2)}}{[E]} = \frac{[S''_a]_{(1)} \times [-\partial D^2] \times [S''_a]_{(2)} \times \nu}{[M^{2n}] \times [\partial D^2] \times \nu}$$

$$= (-1)^{n+1} \frac{[S''_a]_{(1)} \times [S''_a]_{(2)}}{[M^{2n}]} = (-1)^n ,$$

because q_a is a lift of p^a_-. Therefore, $\mathcal{O}(D''_a) = -2$ if $n+1$: *even*, $\mathcal{O}(D''_a) = 0$ if $n+1$: *odd*. The computation of $\mathcal{O}(D''_b)$ is similar.

By the definition of μ, (see [A]), we have

$$\mu(\Delta_a) = \mu(\Delta_b) = \begin{cases} s(1-t)-2 & n : \text{odd}, \\ s(1-t) & n : \text{even}. \end{cases}$$

The computation of $\lambda(\Delta_a, \Delta_b)$ is as follows: We assume $\varepsilon(S'_a, S'_b)(p) = 1$. Then it is easy to see $\alpha(S''_a, S''_b) = 1$, $\beta(D''_a, D''_b) = 0$. Therefore, $\lambda(\Delta_a, \Delta_b) = 1$.

Thus we have proved

LEMMA 2.2. *The* $(-1)^n$*-Seifert form of the global knot* $(W^{2n+2}, \partial M)$ *is given by* $(\Lambda\Delta_a \oplus \Lambda\Delta_b, \lambda, \mu)$: $\lambda(\Delta_a, \Delta_b) = 1$, $\lambda(\Delta_b, \Delta_a) = (-1)^n t$, $\mu(\Delta_a) = \mu(\Delta_b) = \begin{cases} s(1-t)-2, & n : \text{odd} \\ s(1-t), & n : \text{even} \end{cases}$.

III. A PL-SPINELESS MANIFOLD

By definition, a PL-*spine* of a PL-manifold V^{2n+2} is a closed PL submanifold $L^{2n} \subset V^{2n+2}$ such that the inclusion is a simple homotopy equivalence. L^{2n} is not necessarily locally flat.

Let W^{2n+2} ($\subset \mathbb{R}^{2n+2}$) be the manifold constructed in §2. Embed $S^1 \times D^{2n-1}$ nicely into $\mathbb{R}^{2n+2}_+$ so that $S^1 \times D^{2n-1} \cap \mathbb{R}^{2n+1} = S^1 \times S^{2n-2} = \partial M^{2n}$, M^{2n} being the same manifold as in §2. Take a regular neighbourhood R of the embedded $S^1 \times D^{2n-1}$ in $\mathbb{R}^{2n+2}_+$ which meets $\mathbb{R}^{2n+1}$ regularly. Clearly $R \cong S^1 \times D^{2n-1} \times D^2$. Let $\overline{W}^{2n+2} = W^{2n+2} \cup R$. $\overline{W}^{2n+2}$ has the same homotopy type as $S^1 \times S^{2n-1}$.

LEMMA 3.1. $\overline{W}^{2n+2}$ *is PL-spineless, i.e.,* $\overline{W}^{2n+2}$ *does not admit any PL-spine.*

Proof. The homomorphism associated with $\overline{W}^{2n+2}$ (c.f. [A]) is $\mathbb{Z} \times \mathbb{Z} \to \mathbb{Z}$ (the projection onto the second factor). Denote it simply by $(\mathbb{Z} \to 1) \times \mathbb{Z}$. Let t, s be the fixed generators of the first and the second $\mathbb{Z}$ (considered as multiplicative groups). The $(-1)^n$-Seifert form (G, λ, μ) of $\overline{W}^{2n+2}$ was calculated in §2 as follows (c.f. (9) of [A]) :

$$G = \Lambda e_1 \oplus \Lambda e_2,\ \Lambda = \mathbb{Z}[t, t^{-1}, s, s^{-1}],$$

$$\lambda(e_1, e_2) = 1,\ \lambda(e_2, e_1) = (-1)^n t,$$

$$\mu(e_1) = \mu(e_2) = \begin{cases} s(1-t)-2, & n : odd \\ s(1-t), & n : even. \end{cases}$$

ASSERTION. The class $\eta(\overline{W}^{2n+2}) \in P_{2n}((\mathbb{Z} \to 1) \times \mathbb{Z} ; t)$ *represented by the above* (G, λ, μ) *cannot come from* $P_{2n}(\mathbb{Z} \to 1; t)$ *through* $i_* : P_{2n}(\mathbb{Z} \to 1; t) \to P_{2n}((\mathbb{Z} \to 1) \times \mathbb{Z} ; t)$, *where* i_* *is induced from the inclusion* $1 \hookrightarrow \mathbb{Z} = \langle s \rangle$.

Proof. Let $(\Lambda e_1 \oplus \cdots \oplus \Lambda e_r, \lambda, \mu)$ be a $(-1)^n$-Seifert form representing a class $\eta \in P_{2n}((\mathbb{Z} \to 1) \times \mathbb{Z}\ ; t)$. Let $(\lambda(e_i, e_j))$ be the associated matrix. The entries are elements of Λ. Thus substitutions $t = -1$, $s = \zeta$ (a root of unity) give a (skew -) hermitian matrix, whose signature (or the signature of $\sqrt{-1}$ x the skew-hermitian matrix, in the skew-hermitian case) is denoted by $I_{(-1,\zeta)}(\eta)$. It defines a homomorphism $I_{(-1,\zeta)}$: $P_{2n}((\mathbb{Z} \to 1) \times \mathbb{Z}) \to \mathbb{Z}$. (Proofs of the well-definedness of $I_{(-1,\zeta)}$ are found in [3], [20].) If $\eta(\overline{W}^{2n+2})$ were in the image $\mathrm{Im}(i_*)$, the associated matrix $(\lambda(e_i, e_j))$, $i, j = 1, 2$, would be conjugate to a matrix all of whose entries are in $\mathbb{Z}[t, t^{-1}]$. Thus $I_{(-1,\zeta)}(\eta(\overline{W}^{2n+2}))$ must be independent of ζ. As we see below, this is not the case. Thus $\eta(\overline{W}^{2n+2}) \notin \mathrm{Im}(i_*)$ as asserted.

Case (i), n : odd.

$$(\lambda(e_i, e_j)) = \begin{bmatrix} (1-t)(s+s^{-1}-2), & 1 \\ -t, & (1-t)(s+s^{-1}-2) \end{bmatrix},$$

$$I_{(-1,1)}(\eta(\overline{W})) = \mathrm{sign} \begin{bmatrix} 0 & 1 \\ 1 & 0 \end{bmatrix} = 0,$$

$$I_{(-1,-1)}(\eta(\overline{W})) = \mathrm{sign} \begin{bmatrix} -8 & 1 \\ 1 & -8 \end{bmatrix} = -2.$$

Case (ii), n : even.

$$(\lambda(e_i, e_j)) = \begin{bmatrix} (1-t)(s-s^{-1}), & 1 \\ t\ \ , & (1-t)(s-s^{-1}) \end{bmatrix},$$

$$I_{(-1,1)}(\eta(\overline{W})) = \mathrm{sign}\ \sqrt{-1} \begin{bmatrix} 0 & 1 \\ -1 & 0 \end{bmatrix} = 0,$$

$$I_{(-1,\sqrt{-1})}(\eta(\overline{W})) = \mathrm{sign}\ \sqrt{-1} \begin{bmatrix} 4\sqrt{-1}, & 1 \\ -1, & 4\sqrt{-1} \end{bmatrix} = -2.$$

This completes the proof of Assertion. (Similar arguments were done in [3], [19], [20].)

Now Suppose $\overline{W}^{2n+2}$ ever had a PL spine L^{2n}. Apply Noguchi's obstruction theory [24] to making the embedding $L^{2n} \hookrightarrow \overline{W}^{2n+2}$ locally flat. Since $L^{2n} \simeq S^1 \times S^{2n-1}$ and C_j (the knot cobordism group of j-knots) $= 0$ for j : *even*, the homology obstruction groups $H_i(L^{2n};\ C_{2n-i-1}) = 0$ for $i > 0$. Thus L^{2n} can be made 1-flat in the sense of [24], i.e., one can assume that L^{2n} is locally flat except at one point p. The star pair $(St(p,\overline{W}), St(p,L^{2n}))$ is a cone pair over a knot $(Lk(p,L^{2n}) \hookrightarrow Lk(p,\overline{W}))$. Let $\xi_0 \in C_{2n-1}$ be the knot cobordism class of the knot. It is proved that $P_{2n}(\mathbb{Z} \to 1) \cong C_{2n-1}$ (See, [A], [17, 6.2]). By naturality of the η-obstruction, see [A]-(9), we have $\eta(\overline{W}) = i_*(\xi_0)$ (See [17, 6.3] for an analogous argument). This contradicts the above assertion and completes the proof of 3.1.

IV. EXTENSION THEOREMS

In this section we generalize Giffen's construction [7] to obtain

THEOREM 4.1. Let $f : S^1 \times S^m \to S^1 \times S^{m+c}$ *be a topological embedding* ($c \geq 1$) *which induces an isomorphism of* π_1.

(i) *If* $m \geq 3$, f *extends to a topological embedding* $\tilde{f} : S^1 \times D^{m+1} \to S^1 \times D^{m+c+1}$.

(ii) *If* $m = 2$, *suppose* f *is a CAT embedding, then* f *extends to a topological embedding* $\tilde{f} : V^4 \to S^1 \times D^{3+c}$, *where* V^4 *is a CAT manifold with* $\partial V = S^1 \times S^2$, *whose homotopy type is that of* S^1. *CAT = TOP, PL, DIFF.*

In case $c = 0$, Max proved an extension theorem [22] :

THEOREM (Max [22, Th. 2]) Every homeomorphism $f : S^1 \times S^m \to S^1 \times S^m$ *(*$m \geqq 2$*) extends to a homeomorphism* $\tilde{f} : S^1 \times D^{m+1} \to S^1 \times D^{m+1}$.

Notation:

X = a space,

$C(X)$ = the open cone over X

$= [0,\infty) \times X/\{0\} \times X$,

p_X = the vertex of $C(X) \sim \{0\} \times X$,

$C_r(X)$ = the closed cone corresponding to

$[0,r] \times X \subset C(X)$.

X is identified with $\{1\} \times X \subset C(X)$.

The following is the fundamental technical lemma (Generalized Giffen construction):

LEMMA 4.2. Let X, Y be arcwise-connected, simply connected spaces. Let $f : S^1 \times X \to S^1 \times Y$ *be a topological embedding inducing an isomorphism of* π_1. *Then, for a fixed* $t \neq 0 \in \mathbb{R}$, *there exists a topological embedding* $F_t : S^1 \times C(X) \to S^1 \times C(Y)$ *such that*

(i) $F_t(S^1 \times \{p_X\}) = S^1 \times \{p_Y\}$,

(ii) $F_t(S^1 \times C(X)) \cap S^1 \times Y = f(S^1 \times X)$,

(iii) there exists a topological flow $\{\Phi_\theta^{(t)}\}_{\theta \in \mathbb{R}}$ *(with* $\Phi_0^{(t)} = id.$*) on* S^1 $C(Y)$ *which preserves the image* $F_t(S^1 \times C(X))$ *and maps* $S^1 \times C_r(Y)$ *homeomorphicaly onto* $S^1 \times C_{r'}(Y)$, $r' = e^{\frac{t}{\theta}} r$, *for* $\forall\ r > 0$.

COROLLARY 4.3. $F_t(S^1 \times C(X)) \cap S^1 \times C_r(Y)$ *has the same homotopy type as* S^1, $\forall\ r > 0$.

Proof. By *(iii)* we may assume $r = 1$. Suppose $t > 0$. Let $g : S^1 \times X \times [0,\infty) \to S^1 \times C(Y)$ be defined by

$$g(\zeta,x,s) = \Phi_s^{(t)} f(\zeta,x), (\zeta,x,s) \in S^1 \times X \times [0,\infty).$$

Then by *(ii)* and *(iii)*, g is an embedding such that $g|S^1 \times X \times \{0\} = f$. One has $F_t(S^1 \times C(X)) = \{F_t(S^1 \times C(X)) \cap S^1 \times C_1(Y)\} \cup g(S^1 \times X \times [0,\infty))$, i.e., $g(S^1 \times X \times [0,\infty))$ is an 'open collar'. Thus $F_t(S^1 \times C(X)) \cap S^1 \times C_1(Y) \simeq F_t(S^1 \times C(X)) \simeq S^1$ as asserted. The proof for $t < 0$ is the same.

Proof of 4.2. Following Giffen [7], let $\{\varphi_t\}_{t\in\mathbb{R}}$ be the topological flow on $S^1 \times (C(Y) - \{p_Y\}) = S^1 \times (0,\infty) \times Y$ defined by

$$\varphi_t(\zeta,r,y) = (\zeta e^{\sqrt{-1}\, t \log r}, r,y).$$

φ_t is the identity on $S^1 \times Y$. Let $F_0 = S^1 \times \{p_Y\} \cup \{(\zeta,r,y) | 0 < r, (\zeta,y) \in f(S^1 \times X)\}$ be the open 'fringe' in $S^1 \times C(Y)$. We will show that for any $t \neq 0 \in \mathbb{R}$, $F_t = \varphi_t(F_0 - S^1 \times \{p_Y\}) \cup S^1 \times \{p_Y\}$ is homeomorphic with $S^1 \times C(X)$. The main idea is to cut $F_t - S^1 \times \{p_Y\}$ into slices by each $\{\xi\} \times C(Y)$ $(\xi \in S^1)$ and show that each section $S_\xi = (F_t - S^1 \times \{p_Y\}) \cap \{\xi\} \times C(Y)$ homeomorphic with $(0,\infty) \times X$. In fact, each section is a universal covering of $f(S^1 \times X)$. Figure 5 illustrates the main idea.

To be precise, let $\pi_{\xi,t} : \{\xi\} \times (C(Y) - p_Y) \to S^1 \times Y$ be defined by

$$\pi_{\xi,t}(\xi,r,y) = (\xi e^{-\sqrt{-1}\, t \log r}, y)$$

$(t, \xi,$ fixed). Then $\pi_{\xi,t}$ is a universal covering map. We have

$$S_\xi = (F_t - S^1 \times \{p_Y\}) \cap \{\xi\} \times C(Y)$$
$$= \varphi_t((F_0 - S^1 \times \{p_Y\}) \cap \varphi_t^{-1}(\{\xi\} \times (0,\infty) \times Y))$$

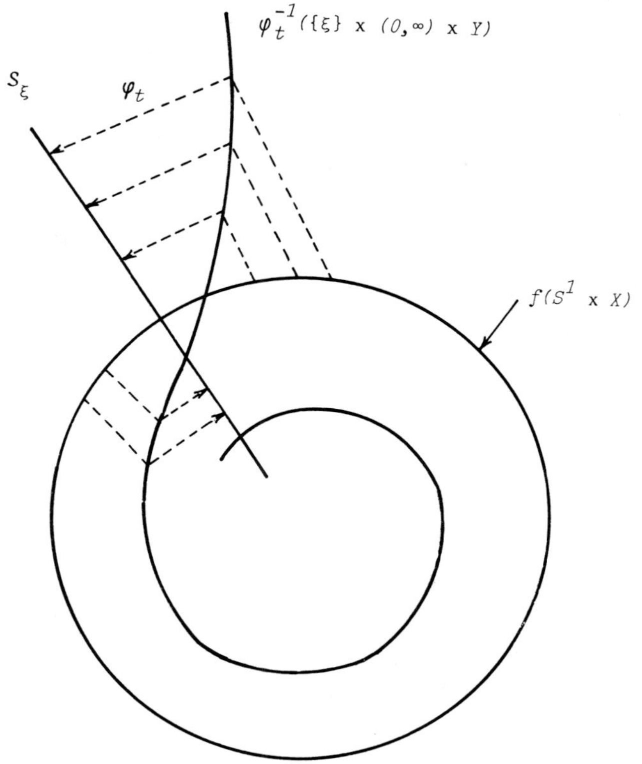

Fig. 5.

$$= \varphi_t\{(\zeta,r,y) \mid r > 0,\ \zeta = \xi e^{-\sqrt{-1}\, t \log r} \text{ and } (\zeta,y) \in f(S^1 \times X)\}$$

$$= \{(\xi,r,y) \mid r > 0,\ (\xi e^{-\sqrt{-1}\, t \log r}, y) \in f(S^1 \times X)\}$$

$$= \pi_{\xi,t}^{-1}(f(S^1 \times X)).$$

Since f induces an isomorphism of π_1, S_ξ is a universal covering space of $f(S^1 \times X) \approx (0,\infty) \times X$ as asserted. Let $\{\Phi_\theta^{(t)}\}_{\theta\in\mathbb{R}}$ be the topological flow on $S^1 \times C(Y)$ defined by

$$\Phi_\theta^{(t)}(\zeta,r,y) = (\zeta e^{\sqrt{-1}\,\theta},\ e^{\frac{\theta}{t}} r,\ y),$$

$$(\zeta,\ r,\ y) \in S^1 \times [0,\infty) \times Y$$

with t fixed.

Then $\Phi_\theta^{(t)}\ \varphi_t(\zeta,\ r,\ y) = \varphi_t(\zeta,\ e^{\frac{\theta}{t}} r,\ y)$. This implies $\Phi_\theta^{(t)}(F_t) = F_t$, because $(\zeta,r,y) \in F_0$ if $(\zeta,e^{\theta/2}r,\ y) \in F_0$. Since $\Phi_\theta^{(t)}$ maps $\{\xi\} \times C(Y)$ homeomorphically onto $\{\xi'\} \times C(Y)$ $(\xi' = \xi e^{\sqrt{-1}\,\theta})$, $\Phi_\theta^{(t)}$ gives a homeomorphism $\Phi_\theta^{(t)} : S_\xi \to S_{\xi'}$. Thus the flow describes $F_t - S^1 \times \{p_Y\}$ as a mapping cylinder of $\Phi_{2\pi}^{(t)}|S_\xi : S_\xi \to S_\xi$. But

$$\Phi_{2\pi}^{(t)}(\xi,\ r,\ y) = (\xi,\ e^{\frac{2\pi}{t}} r,\ y),$$

and this is exactly the restriction of the covering transformation for $\pi_{\xi,t} : \{\xi\} \times (C(Y) - \{p_Y\}) \to S^1 \times Y$ to $S_\xi = \pi_{\xi,t}^{-1}(f(S^1 \times X))$. This restriction induces a covering transformation for the universal covering $S_\xi \to f(S^1 \times X)$. By the uniqueness of the universal covering, it is conjugate to the translation : $\mathbb{R} \times X \to \mathbb{R} \times X$ defined by $(m,x) \mapsto (m+1,x)$, thus isotopic to the identity. Therefore $F_t - S^1 \times \{p_Y\}$ is homeomorphic with $S^1 \times (0,\infty) \times X$. One can easily extend the homeomorphism to a homeomorphism $S^1 \times C(X) \to F_t$ to obtain a required topological embedding $F_t : S^1 \times C(X) \to S^1 \times C(Y)$ with the image F_t. We have already seen property (iii) Properties (i), (ii) are easy to verify.

Proof of 4.1. Fix a $t \neq 0$. By 4.2, there exists a topological embedding $F_t : S^1 \times C(S^m) (= S^1 \times \mathbb{R}^{m+1}) \to S^1 \times C(S^{m+c})$. The existence of the flow $\{\Phi_\theta^{(t)}\}$ which preserves the image F_t assures that $F_t^{-1}(S^1 \times S^{m+c})$ is a flat $(m+1)$-submanifold. The closure of the bounded component of $S^1 \times \mathbb{R}^{m+1} - F_t^{-1}(S^1 \times S^{m+c})$ is a compact submanifold V^{m+2} with the boundary homeomorphic to $S^1 \times S^m$. By 4.3, $V^{m+2} \approx F_t(S^1 \times C(S^m)) \cap S^1 \times D^{m+c+1}$ has the homotopy type of S^1. (Here, $D^{m+c+1} = C_1(S^{m+c})$.)

(i) Suppose $m \geq 4$. Then by the TOP s-cobordism theorem [27, §7.4], V^{m+2} is homeomorphic to $S^1 \times D^{m+1}$, see the proof of 2.1. Let $H : S^1 \times D^{m+2} \to V^{m+2}$ be the homeomorphism. Consider the diagram

$$\begin{array}{ccccccc}
S^1 \times D^{m+1} & \overset{G}{\dashrightarrow} & S^1 \times D^{m+1} & \xrightarrow{F_t \circ H} & F_t(V^{m+2}) & \subset & S^1 \times D^{m+c+1} \\
\cup & & \cup & & \cup & & \cup \\
S^1 \times S^m & \xrightarrow[H^{-1}\circ(F_t|\partial V)^{-1}\circ f]{} & S^1 \times S^m & \xrightarrow[(F_t|\partial V)\circ H]{} & F_t(\partial V) & = & f(S^1 \times S^m) .
\end{array}$$

By Max's theorem, $H^{-1}\circ(F_t|\partial V)^{-1}\circ f$ extends to a homeomorphism G. The desired embedding $\tilde{f}$ is defined to be $F_t \circ H \circ G$.

In case $m = 3$, the above argument is valid except for $V^5 \approx S^1 \times D^4$. However, Fukuhara's TOP version [6] of Shaneson's s-cobordism theorem between $S^1 \times S^3$ and its self [26] can apply to show the existence of the homeomorphism.

(ii) Suppose $m = 2$. We have only to verify that V^4 is a CAT manifold, assuming that f is CAT. By the construction, $F_t|S^1 \times (0,\infty) \times S^m : S^1 \times (0,\infty) \times S^m \to S^1 \times (0,\infty) \times S^{m+c}$ is considered as a CAT embedding. Thus $F_t^{-1}(S^1 \times S^m) = \partial V^{m+2}$ is a flat CAT submanifold of $S^1 \times C_1(S^m) = S^1 \times \mathbb{R}^{m+1}$. V^{m+2} inherits a CAT structure from $S^1 \times \mathbb{R}^{m+1}$.

V. WILD EMBEDDINGS OF $S^1 \times S^{2n-1}$ INTO $\mathbb{R}^{2n+2}$

Suppose $2n \geq 4$. Let W^{2n+2} be the manifold constructed in §2. Recall that the product $S^1 \times S^{2n-2}$ is PL embedded in ∂W^{2n+2} as the boundary of M^{2n}. The PL embedding $S^1 \times S^{2n-2} \to W^{2n+2}$ induces a π_1-isomorphism, and by 2.1, $W^{2n+2} \approx S^1 \times D^{2n+1}$. Thus we have the setting of §4. By 4.1, the embedding $S^1 \times S^{2n-2} \to \partial W^{2n+2}$ is extended to a (wild) topological embedding $h : S^1 \times D^{2n-1} \to W^{2n+2}$ (if $2n \geq 6$) or $h : V^4 \to W^6$ (if $2n = 4$), V^4 being a PL manifold homotopically equivalent to S^1. Gluing h with a 'standard' embedding $s : S^1 \times D^{2n-1} \to \mathbb{R}^{2n+2}_+$ considered at the beginning of §1, one obtains a topological embedding $g = h \cup s : S^1 \times S^{2n-1} \to \overline{W}^{2n+2}$ $(2n \geq 6)$ or $L^4 \to \overline{W}^6$ $(2n = 4)$, where L^4 is a PL 4-manifold whose homotopy type is that of $S^1 \times S^3$. Compose it with $i : \overline{W}^{2n+2} \subset \mathbb{R}^{2n+2}$, then we have a topological embedding $f = i \circ g : S^1 \times S^{2n-1} \to \mathbb{R}^{2n+2}$ or $L^4 \to \mathbb{R}^6$.

THEOREM 5.1. f cannot be approximated by any PL embedding.

Proof. If f could be approximated by a PL embedding f' so closely that the image $f'(S^1 \times S^{2n-1})$ (or $f'(L^4)$) and the trace of a homotopy between f and f' are contained in $\overline{W}^{2n+2}$, then $\overline{W}^{2n+2}$ would have a PL spine, because $g : S^1 \times S^{2n-1} \to \overline{W}^{2n+2}$ (or $L^4 \to \overline{W}^6$) is a homotopy equivalence. However, this contradicts 3.1.

Starting with a 'standard' embedding of $S^1 \times D^{2n-1} \# S^n \times S^n \# \ldots \# S^n \times S^n$ (N copies of $S^n \times S^n$) into $\mathbb{R}^{2n+2}_-$, one can similarly construct a PL spineless manifold $\overline{W}^{2n+2}_{(N)}$ whose $(-1)^n$-Seifert form is an orthogonal sum of N copies of (G, λ, μ) of §3, $N = 1, 2, \cdots$. The signature homomorphism of §3 distinguishes these $(-1)^n$-Seifert forms for different values of N, as elements of $P_{2n}((\mathbb{Z} \to 1) \times \mathbb{Z})$. By §4 each $\overline{W}^{2n+2}_{(N)}$ has a topological spine $S^1 \times S^{2n} \to \overline{W}^{2n+2}_{(N)}$, thus composing with $i : \overline{W}^{2n+2}_{(N)} \subset \mathbb{R}^{2n+2}$ we have a wild embedding $f_{(N)} : S^1 \times S^{2n} \to \mathbb{R}^{2n+2}$ (or $L^4 \to \mathbb{R}^6$). None of these $f_{(N)}$'s is approximated by a PL embedding.

THEOREM 5.2. Suppose $2n \geq 6$. The equivalence classes[1] *of wild embeddings $f_{(N)}$ $(N = 1, 2, \ldots)$ are mutually distinct.*

Proof. Looking at the construction of §4, and §5, one can find an infinite sequence of PL submanifolds

[1] *In the sense of* [25 ,*p.*2].

$\overline{W}_{(N)} \supset \overline{W}^{(1)}_{(N)} \supset \ldots \supset \overline{W}^{(i)}_{(N)} \supset \ldots$ such that (i) every inclusion $\overline{W}^{(i)}_{(N)} \supset \overline{W}^{(i+1)}_{(N)}$ is a homotopy equivalence and (ii) $\bigcap_i \overline{W}^{(i)}_{(N)} = f_{(N)}(S^1 \times S^{2n-1})$. If there were a topological homeomorphism $H : \mathbb{R}^{2n+2} \to \mathbb{R}^{2n+2}$ such that $H \circ f_{(N)}(S^1 \times S^{2n-1}) = f_{(M)}(S^1 \times S^{2n-1})$ for some different N, M, then we would have $H(\overline{W})^{(i)}_{(N)} \subset \text{Int } \overline{W}_{(M)}$ for some $i > 0$, the inclusion being a homotopy equivalence. Since $H^3(\text{Int } \overline{W}^{(i)}_N : \mathbb{Z}_2) = H^3(S^1 \times S^{2n-1} : \mathbb{Z}_2) = 0$, the restriction $H|\text{Int } \overline{W}^{(i)}_{(N)} : \text{Int } \overline{W}^{(i)}_{(N)} : \text{Int } \overline{W}^{(i)}_{(N)} \to \text{Int } \overline{W}_{(M)}$ is ε-isotopic to a PL embedding ($\varepsilon > 0$, small) by Kirby-Siebenmann's theorem [14]. Thus a PL-embedding $\overline{W}^{(i+1)}_{(N)} \to \text{Int } W_{(M)}$ exists, and it is a homotopy equivalence. By naturality of the η-obstruction ([A]-(9)) one has $\eta(\overline{W}_{(N)}) = \eta(\overline{W}^{(i+1)}_{(N)}) = \eta(\overline{W}_{(M)})$. But this is a contradiction if $N \neq M$.

THEOREM 5.3. Any wild embedding $f_{(N)}$ cannot be approximated by a locally flat topological embedding. ($2n \geq 6$.)

Proof. Suppose $f_{(N)}$ were approximated by a locally flat topological embedding $f' : S^1 \times S^{2n-1} \to \mathbb{R}^{2n+2}$ sufficiently closely. By Kirby-Siebenmann [15], the image $f'(S^1 \times S^{2n-1})$ has a normal bundle which is trivial in the present case. Thus f' extends to a topological embedding $f'' : S^1 \times S^{2n-1} \times \mathbb{R}^2 \to \mathbb{R}^{2n+2}$. Since $H^3(S^1 \times S^{2n-1}; \mathbb{Z}_2) = 0$, we can apply Kirby-Siebenmann's Classification theorem [14] to deform f'' by an ε-isotopy to a PL embedding $S^1 \times S^{2n-1} \times \mathbb{R}^2 \to \mathbb{R}^{2n+2}$. The $\varepsilon > 0$ can be taken arbitrarily small. Thus $f_{(N)}$ can be approximated by a PL embedding, a contradiction.

In contrast with 5.3, Ancel and Cannon [1] recently proved the locally flat approximation theorem for topological embeddings in codimension 1.

VI. WILD EMBEDDINGS OF CLOSED SURFACES INTO $\mathbb{R}^4$.

A wild embedding $T^2 \to \mathbb{R}^4$ which cannot be approximated by a PL embedding was constructed in [4], [8]. For any closed orientable surface F_g^2 of genus $g \geq 1$, the same method can apply to give such a wild embedding $F_g^2 \to \mathbb{R}^4$:

Considering F_g^2 as the boundary of a handle-body, take a meridian circle m. Remove an open tubular neighbourhood $m \times (0,1)$ from F_g^2, then we have a 2-manifold A_g^2 with boundary : $\partial A_g^2 = m \times \{0\} \cup m \times \{1\}$. Let $\bar{h} : m \times \{1\} \times D^2 \to \mathrm{Int}(m \times \{0\} \times D^2)$ be the 0-framed Mazur embedding ([19]). Identify each point $p \in m \times \{1\} \times D^2 (\subset \partial(A_g^2 \times D^2))$ with $\bar{h}(p) \in m \times \{0\} \times D^2 \ (\subset \partial(A_g^2 \times D^2))$. Then we have a 4-manifold $W_g^4 (= A_g^2 \times D^2/p \sim h(p))$ homotopically equivalent to F_g^2.

W_g^4 is easily seen to be piecewise linearly embedded in $\mathbb{R}^4$. On the other hand, Giffen's construction gives a wild embedding $F_g^2 \to \mathrm{Int}(W_g^4)$. Composing the two embeddings, we have a wild embedding $F_g^2 \to \mathbb{R}^4$.

To show that this embedding is not approximated by any PL embedding, we have only to prove that W_g^4 is PL spineless. But this is done in the same way as in [19].

Remark. Kawauchi [11] gave an elementary proof of the spinelessness of W_1^4. His method seems applicable to general W_g^4 also.

VII. EXTENSION OF THE RESULTS

Cappell and Shaneson constructed a large class of PL-spineless manifolds [3] (totally spineless manifolds in their terminology). If one can find a topological spine of such a manifold, then the inclusion map cannot be approximated by any PL embedding. We are tempted to say that almost all totally spineless manifolds admit topological spines so that they potentially provide abundant counterexamples to PL approximation conjecture in codimension 2. It is partially justified by the following.

THEOREM 7.1. Let C be the class of finitely presented (briefly f.p. -) groups defined below. If L^{2n} is a closed orientable PL manifold with $\pi_1(L^{2n}) \in C$, $2n \geq 4$, then there exists a PL-spineless manifold V^{2n+2} which is simple homotopy equivalent to L^{2n} and admits a (wild) topological spine.

Definition of C. C is the least class of f.p. groups satisfying *C.1 - C.3.*

C.1. C contains every non-trivial cyclic group.

C.2. Suppose π' is a normal subgroup of π of finite index. Suppose there exists a set of representatives $\{x_1, \ldots, x_r\}$ of π/π' such that each x_i commutes with any element of π', i.e., $x_i \in C_\pi(\pi')$, the centralizer. Then $\pi' \in C$ implies $\pi \in C$.

C.3. Let $\pi \to \pi'$ be an onto homomorphism whose kernel is the normal closure of a finite subset. Then $\pi' \in C$ implies $\pi \in C$.

The class C slightly generalizes Cappell-Shaneson's class of fundamental groups [3].

Now consider the following $(-1)^n$-Seifert form:

$$(\Lambda e_1 \oplus \Lambda e_2, \lambda, \mu), \ \Lambda = \mathbb{Z}[t, t^{-1}, s, s^{-1}],$$

$$\lambda(e_1, e_2) = 1 \quad , \ \lambda(e_2, e_1) = (-1)^n t,$$

$$\mu(e_1) = \mu(e_2) = (1-t)(1-s)t \ .$$

The associated matrix $(\lambda(e_i, e_j))$ is

$$\begin{bmatrix} (1-t)\{(1-s)t - (-1)^n(1-s^{-1})t^{-1}\}, & 1 \\ (-1)^n t, & (1-t)\{(1-s)t - (-1)^n(1-s^{-1})t^{-1}\} \end{bmatrix} \ .$$

The element of $P_{2n}((\mathbb{Z} \to 1) \times \mathbb{Z}\, ;\, t)$ represented by the above form is denoted ξ_0.

An element $\eta_0 \in P_{2n}((\mathbb{Z} \to 1) \times \pi; t)$ is said to be *spineless* if "$\eta(W^{2n+2}) = \eta_0$" implies that W^{2n+2} is PL-spineless for any $2n$-Poincaré thickening (in the sense of [A]) with the associated homomorphism $(\mathbb{Z} \to 1) \times \pi$, $2n \geq 4$.

LEMMA 7.2. The element $\xi_0 \in P_{2n}((\mathbb{Z} \to 1) \times \mathbb{Z}\, ;\, t)$ is spineless. Moreover, if $\pi \in$ C, then there exists a homomorphism $h : \mathbb{Z} \to \pi$ such that the induced homomorphism $h_* : P_{2n}((\mathbb{Z} \to 1) \times \mathbb{Z}\, ;\, t) \to P_{2n}((\mathbb{Z} \to 1) \times \pi;\, t)$ carries ξ_0 to a spineless element.

Sketch of proof. The proof is almost identical with the argument in Cappell and Shaneson [3] in which their finiteness criterion plays a central role. We state it in terms of our P-groups.

FINITENESS CRITERION (Cappell and Shaneson). Let W^{2n+2} be a $2n$-Poincare thickening with the associated homomorphism $(\mathbb{Z} \to 1) \times \pi$ $(2n \geq 4)$. Let $\alpha : \pi \to \pi'$ be any homomorphism of π to a finite group π'. Then if W^{2n+2} admits a PL spine, the induced homomorphism $\alpha_* : P_{2n}((\mathbb{Z} \to 1) \times \pi) \to P_{2n}((\mathbb{Z} \to 1) \times \pi')$ followed by the projection $P_{2n}((\mathbb{Z} \to 1) \times \pi') \to P_{2n}((\mathbb{Z} \to 1) \times \pi')/ i_*(P_{2n}(\mathbb{Z} \to 1))$ carries $\eta(W^{2n+2})$ to a torsion element.

With the criterion, one can show that $q_*(\xi_0)$ is spineless, where q_* is the induced homomorphism by the quotient map $\mathbb{Z} \to \mathbb{Z}_p$ (for any $p \neq 1$), by calculating appropriate signature invariants for $q_*(\xi_0)$. Thus the lemma is true for any non-trivial cyclic group π.

Now suppose $i : \pi' \subset \pi$ is an extension satisfying C.2. Suppose we have a homomorphism $h' : \mathbb{Z} \to \pi'$ with the property required by 7.2. Then the composite map $i \circ h : \mathbb{Z} \to \pi$ has the same property. The proof is done by passing to the covering space corresponding to $\pi' \subset \pi$ and applying the transfer homomorphism in

P-groups, see [A]. Refer to [3]. Finally suppose an onto homomorphism $\omega : \pi \to \pi'$ satisfies Condition C.3 and suppose we have a desired homomorphism $h' : \mathbb{Z} \to \pi'$. Then any lift $h : \mathbb{Z} \to \pi$ of h' satisfies the requirement of 7.2. This is verified as follows: If there were a $2n$-Poincaré thickening W^{2n+2} whose associated homomorphism is $(\mathbb{Z} \to 1) \times \pi$ and whose obstruction $\eta(W^{2n+2})$ is equal to $h_*(\xi_0)$ but yet admitting a PL spine L^{2n}, then passing to a regular neighbourhood, we may assume that W^{2n+2} is itself a regular neighbourhood of L^{2n} (cf. (9) of [A]). By doing surgery on a finite set of circles embedded in the locally flat part of L^{2n}, one can obtain a regular neighbourhood $W^{2n+2\,\prime}$ of the resulting $L^{2n\,\prime}$ whose associated homomorphism is $(\mathbb{Z} \to 1) \times \pi'$ and whose obstruction satisfies $\eta(W^{2n+2\,\prime}) = \omega_* h_*(\xi_0) = h'_*(\xi_0)$. However, this contradicts the assumption on h'.

Now 7.2 follows inductively.

Proof of 7.1. Starting with $M^{2n} = S^1 \times D^{2n-1} \# S^n \times S^n$ and proceeding analogously to §2 (but managing absolutely rather than working in $\mathbb{R}_-^{2n+2}$), one can construct a PL manifold U^{2n+2} satisfying the following:

(1) U^{2n+2} PL homeomorphic to $S^1 \times D^{2n+1}$,

(2) U^{2n+2} contains M^{2n} as a proper submanifold, and

(3) the $(-1)^n$- Seifert form of U^{2n+2} is the same form as given in this section before 7.2.

The embedding $S^1 \times S^{2n-1} = \partial M^{2n} \to \partial U$ induces a π_1-isomorphism. The pair $(U, \partial M)$ has the same homotopy type as $(S^1 \times D^{2n-1}, S^1 \times S^{2n-2})$, thus it is a global knot in the sense of [A]. By 4.1, the embedding $S^1 \times S^{2n-1} \to \partial U$ extends to a topological embedding $K^{2n} \to U$, K^{2n} being a PL manifold $\simeq S^1$.

Take a closed orientable PL manifold L^{2n} with $\pi_1(L^{2n}) \in C$. By 7.2, there is a homomorphism $h : \mathbb{Z} \to \pi_1(L^{2n})$ such that $h_*(\xi_0) \in P_{2n}((\mathbb{Z} \to 1) \times \pi_1 L^{2n})$ is spineless. Embed a circle in L^{2n} so that it represents $h(s)$, where $\langle s \rangle = \mathbb{Z}$. Take a regular neighbourhood T of the circle. Then the desired manifold V^{2n+2} is defined by replacing $T \times D^2$ in $L^{2n} \times D^2$ with U^{2n+2} : $V^{2n+2} = (L^{2n} - \mathrm{Int}T) \times D^2 \underset{\varphi}{\cup} U$, where φ identifies $\partial T \times D^2$ with a tubular neighbourhood of $\partial M \subset \partial U$, and $\partial T \times \{0\}$ with ∂M. V^{2n+2} has the same simple homotopy type as L^{2n} and contains a topological spine $\approx (L^{2n} - \mathrm{Int}T) \cup K^{2n}$. Since $\eta(V^{2n+2}) = h_*\eta(U^{2n+2}, \partial M) = h_*(\xi_0)$, V^{2n+2} is PL-spineless by 7.2.

Remark. The topological spine is homeomorphic to L^{2n} if $2n \geqq 6$.

WILD EMBEDDINGS INTO EUCLIEAN SPACE. If V^{2n+2} of 7.1 can be embedded piecewise linearly into $\mathbb{R}^{2n+2}$, the composite maps : (the topological spine) $\hookrightarrow V^{2n+2} \hookrightarrow \mathbb{R}^{2n+2}$ would give many examples of wild embeddings into euclidean space. The author does not know whether V^{2n+2} is embeddable into $\mathbb{R}^{2n+2}$ or not. So, to prove the following theorem, we shall construct PL spineless manifolds *within* $\mathbb{R}^{2n+2}$ using W^{2n+2} of §2 (just as we did in §3).

Let $\mathcal{D}$ be the subclass of $\mathcal{C}$ satisfying C.2, C.3, and the following D.1 in place of C.1.

D.1. *Every non-trivial cyclic group whose order is not a power of 2 belongs to* $\mathcal{D}$.

THEOREM 7.3. *Let L^{2n} be a closed orientable PL manifold, $2n \geqq 6$. Suppose*

(i) L^{2n} can be piecewise linearly embedded into $\mathbb{R}^{2n+2}$ so that $\pi_1(\mathbb{R}^{2n+2} - L^{2n}) = \mathbb{Z}$, and

(ii) $\pi_1(L^{2n}) \in \mathcal{C}$ if n = odd, or $\pi_1(L^{2n}) \in \mathcal{D}$ if n = even.

Then there exist infinitely many topological embeddings $L^{2n} \to \mathbb{R}^{2n+2}$ which cannot be approximated by PL embeddings. Moreover, if $H^3(L^{2n}; \mathbb{Z}_2) = 0$, they are not approximated by locally flat topological embeddings.

Proof. Let $\eta_0 \in P_{2n}((\mathbb{Z} \to 1) \times \mathbb{Z}; t)$ be the element represented by (G, λ, μ) of §3.1, i.e., $\eta_0 = \eta(\overline{W}^{2n+2})$. The same argument as in 7.2 can apply to show that for any $\pi \in \mathcal{C}$ (if $n = odd$) or $\pi \in \mathcal{D}$ (if $n = even$), there exists a homomorphism $h : \mathbb{Z} = \langle s \rangle \to \pi$ such that $h_*(\eta_0) \in P_{2n}((\mathbb{Z} \to 1) \times \pi; t)$ is spineless. (The only reason for the restriction from $\mathcal{C}$ to $\mathcal{D}$ in case $n = even$ is that we failed to detect the spinelessness of $q_*(\eta_0)$, $q : \mathbb{Z} \to \mathbb{Z}_2$ the quotient map.)

Let $\Sigma(L^{2n})$ be the subcomplex of the non-locally flat points of $L^{2n} \subset \mathbb{R}^{2n+2}$. Embed a circle in $L^{2n} - \Sigma(L^{2n})$ so that it represents $h(s) \in \pi_1(L^{2n})$. Let C be the image. Condition (i) on L^{2n} assures the existence of an embedded 2-disk $\Delta^2 \subset \mathbb{R}^{2n+2}$ such that $\Delta^2 \cap L^{2n} = \partial\Delta^2 = C$.

Let N^{2n+2} be a regular neighbourhood of $L^{2n} \subset \mathbb{R}^{2n+2}$. If it is 'restricted' to $L^{2n} - \Sigma(L^{2n})$, it is considered as a product bundle. In particular, let T be a tubular neighbourhood of $C \subset L^{2n}$, then $N|T$ is identified with $T \times D^2$. We may (and will) consider that L^{2n} is far apart from $\overline{W}^{2n+2}$ of §3. Then using the existence of Δ^2, one can construct a PL spineless manifold V^{2n+2} within $\mathbb{R}^{2n+2}$ similarly to 7.1, but by using W^{2n+2} of §1 instead of U^{2n+2} of this section, and N^{2n+2} instead of $L^{2n} \times D^2$. (The details will be left to the reader.) The resulting V^{2n+2} is PL homeomorphic to $(N^{2n+2} - (\text{Int } T) \times D^2) \cup_\varphi {}^{2n+2}$, where φ identifies $\partial T \times D^2$ with $W^{2n+2} \cap \mathbb{R}^{2n+1} = \partial M \times D^2$ (in the same notation as in §2) and $\partial T \times \{0\}$ with $\partial M \times \{0\}$. Evidently, V^{2n+2} admits a wild topological spine homeomorphic to L^{2n}. The PL spinelessness of V^{2n+2} follows from the fact that $\eta(V^{2n+2}) = h_*(\eta_0) + \eta(N^{2n+2})$. See [17, p. 305] , or Appendix (9)). (Note that the elements of $P_{2n}((\mathbb{Z} \to 1) \times \pi)$ which are *not* spineless such as $\eta(N^{2n+2})$ form a subgroup, for the subset coincides with the image $\Omega_{2n}(G_2/RN_2 \times K(\pi,1)) \to P_{2n}((\mathbb{Z} \to 1) \times \pi)$. Cf. [3]) Therefore, the topological embedding $L^{2n} \approx$ (the topological spine) $\hookrightarrow V^{2n+2} \hookrightarrow \mathbb{R}^{2n+2}$ cannot be approximated by any PL embedding. If we use $W^{2n+2}_{(N)}$ $(N = 1,2,\ldots)$ of §5, we will obtain infinitely many such wild embeddings. The proof of the last assertion of 7.3 is the same as 5.3.

WILD EMBEDDINGS OF HOMOLOGY SPHERES. Let M^{2n} be a $\mathbb{Z}$-homology sphere with $\pi_1(M^{2n}) = \tilde{I}$ (the binary icosahedral group). For the existence of such a homology sphere, see [13], [28]. Suppose $2n \geq 6$, $n = odd$. Then as a special case of 7.3, there are many topological embeddings $M^{2n} \to \mathbb{R}^{2n+2}$ which cannot be approximated by PL embeddings. (Note that $\tilde{I}$ contains a center of order 2.)

APPENDIX

Codimension 2 surgery. We shall summarize some of the main results of [10], [17]. Throughout this appendix, we shall work in the PL category, for the sake of simplicity.

1) *Definitions.* Let W^{m+2} be a compact connected oriented $m+2$-manifold, K^{m-1} a locally flat closed $(m-1)$-submanifold of ∂W^{m+2}. The pair (W^{m+2}, K^{m-1}) is called an *m-Poincare thickening pair* if it has the same simple homotopy type as an oriented finite Poincaré pair $(X^{(m)}, Y^{(m-1)})$ of formal dimension m.

The case $K^{m-1} = Y^{(m-1)} = \emptyset$ will not be excluded. In that case, $W^{m+2} (\simeq X^{(m)})$ is called an *m-Poincaré thickening*.

Let $\Sigma^{m-1} \subset \partial D^{m+2}$ be a locally flat $(m-1)$-knot. Then the pair (D^{m+2}, Σ^{m-1}) has the same homotopy type as $(D^m, \partial D^m)$. Therefore, the notion of an m-Poincaré thickening is a generalization (or a 'globalization') of that of an $(m-1)$-knot. For this reason, an m-Poincare thickening will be sometimes called a *global knot*.

2) *Spines.* A proper m-submanifold L^m of W^{m+2} is a *homotopy spine* (or simply, a *spine*) of (W^{m+2}, K^{m-1}) if $\partial L^m = K^{m-1}$ and the inclusion $(L^m, \partial L^m) \to (W^{m+2}, K^{m-1})$ is a simple homotopy equivalence. (In case $K^{m-1} = \emptyset$, L^m is a closed submanifold of Int W^{m+2}, of course.) The main objective of our codimension 2 surgery is to describe the obstruction to finding a *locally flat* spine of a given pair (W^{m+2}, K^{m-1}) : a generalization of the knot cobordism theory of Kervair [12] and Levine [16].

3) *Surgery below the middle dimension.* By transversality, we can find a locally flat PL m-submanifold M^m with $\partial M^m = K^{m-1}$,

which represents the fundamental class $[X^{(m)}, Y^{(m-1)}] \in H_m(W^{m+2}, K^{m-1}; \mathbb{Z})$. We modify M^m by a sequence of ambient codimension 2 surgeries (in Int W^{m+2}) so that the inclusion $(M^m, \partial M^m) \to (W^{m+2}, K^{m-1})$ becomes as close as possible to a simple homotopy equivalence. Below the middle dimension, one meets no obstruction:

THEOREM A.1. ([10, Lemma 3.4]) M^m can be made exterior $[\frac{m}{2}]$-connected.

Exterior - k - connectivity is defined as follows: Let N be a regular neighbourhood of $M^m \subset W^{m+2}$ which meets ∂W regularly, E the exterior of N, E = closure$(W^{m+2} - N)$. Let FN be the frontier of N, $FN = N \cap E$. Then M^m is *exterior k-connected* if $\pi_i(E, FN) = 0$ for $\forall i \leq k$. (Thomas and Wood [29] called an exterior $[\frac{m}{2}]$- connected submanifold a *taut* submanifold.)

The middle dimensional surgery splits into two cases, i.e., the odd or even dimensional cases.

4) *The odd dimensional case.* If m = odd ≥ 5, an ambient middle dimensional surgery can be performed if and only if it is performed as an abstract Wall surgery.

THEOREM A.2. ([10], see also [2]). There exists a unique obstruction $\eta(W^{m+2}, K^{m-1}) \in L_m(\pi_1 W)$ which vanishes if (W^{m+2}, K^{m-1}) has a locally flat spine.

5) *The even dimensional case.* This case occupies the rest of the appendix. Suppose $m = 2n \geq 4$. Let M^{2n} be an exterior n-connected, locally flat submanifold with $\partial M^{2n} = K^{2n-1}$. Then the onto homomorphism $\pi_1(W - M) \to \pi_1(W)$ is identified with the onto homomorphism $\pi_1(FN) \to \pi_1(M)$ induced by the projection of the circle bundle $FN \to M$. This homomorphism is proved to be independent of the choice of M^{2n} and will be called the *homomorphism associated with* (W^{2n+2}, K^{2n-1}). We will denote it by

$h : \pi \to \pi'$ ($\pi = \pi_1(FN)$, $\pi' = \pi_1(M)$). The kernel of h is generated by a central element t, which is the homotopy class of a circle fiber in its positive direction (defined as in §2). The element t is independent of the choice of a fiber. We will define, in the next paragraph, abelian groups $P_{2n}(h; t)$ depending functorially on h, t, and $2n$ modulo 4.

THEOREM A.3. ([17, Thms 4.12, 5.10]) *There exists a unique obstruction* $\eta(W^{2n+2}, K^{2n-1}) \in P_{2n}(h ; t)$ *which vanishes if* (W^{2n+2}, K^{2n-1}) *admits a locally flat spine. If* $2n \geq 6$, *the converse is true.*

6) *Seifert forms: Algebraic preliminaries.* Let $h : \pi \to \pi'$ be an onto homomorphism, t an element of $\mathrm{Ker}(h) \cap \mathrm{Center}(\pi)$. ($t$ need not generate $\mathrm{Ker}(h)$.) Let $\Lambda = \mathbb{Z}\pi$ and $\Lambda' = \mathbb{Z}\pi'$. A $(-1)^n$*-Seifert form over* $(h ; t)$ is a triple $X = (G, \lambda, \mu)$, where G is a finitely generated left Λ-module and λ, μ are functions $\lambda : G \quad G \to \Lambda$, $\mu : G \to Q_n^t(\mu) \equiv \Lambda/\{a-(-1)^n\bar{a}\, t \mid a \in \Lambda\}$. The anti-involution $- : \Lambda \to \Lambda$ is defined by $\overline{\Sigma m(g)g} = \Sigma m(g)g^{-1}$. We require the following six conditions:

(i) $\lambda(x,y) = (-1)^n \overline{\lambda(y,x)}t$.

(ii) $\lambda(x,y)$ is Λ-linear in x.

(iii) $\lambda(x,x) \equiv \mu(x) + (-1)^n\overline{\mu(x)}\ t$.

(iv) $\mu(x+y) - \mu(x) - \mu(y) \equiv \lambda(x,y)$.

(v) $\mu(ax) = a\mu(x)\bar{a}$, $a \in \Lambda$.

(vi) Let $G' = \Lambda' \otimes_\Lambda G$, $\lambda' = h_* \circ \lambda$, $\mu' = h_* \circ \mu$.

Then G' is a stably free, s-based Λ'-module, and $\Lambda' \otimes X = (G', \lambda', \mu')$ is non-singular in Wall's sense [33].

As one sees, the definition of Seifert forms is almost the same as that of Cappell-Shaneson's special η-forms [2, p. 286] except for the appearance of t.

A $(-1)^n$-Seifert form (G, λ, μ) is *null-cobordant* if there exists a Λ-submodule $H \subset G$, not necessarily a direct summand, such that $\lambda(H \times H) = 0$, $\mu(H) = 0$ and which is mapped through the canonical map $G \to G'$ onto a subkernel of (G', λ', μ') in Wall's sense [32, p. 47]. We will call such H a *pre-subkernel* after Cappell-Shaneson's terminology [2].

The *standard plane* $S = (\Lambda e_1 \oplus \Lambda e_2, \lambda, \mu)$ is defined by $\lambda(e_1, e_2) = 1$, $\lambda(e_2, e_1) = (-1)^n t$, $\mu(e_1) = \mu(e_2) = 0$. A $(-1)^n$-Seifert form X is *stably null-cobordant* if an orthogonal sum $X \perp S_1 \perp \cdots \perp S_r$ is null-cobordant, where each S_i is a copy of S. Two $(-1)^n$-Seifert forms (G_1, λ_1, μ_1) and (G_2, λ_2, μ_2) are *stably cobordant* if $(G_1, \lambda_1, \mu_1) \perp (G_2, -\lambda_2, -\mu_2)$ is stably null-cobordant. This relation is shown (non-trivially!) to be an equivalence relation ([18, §4], [2]).

Let $P_{2n}(h\,;\,t)$ be the totality of the stable cobordism classes of $(-1)^n$-Seifert forms over $(h\,;\,t)$. $P_{2n}(h\,;\,t)$ is an abelian group under the sum $\perp$. If h and t are evident in the context, it is sometimes denoted by $P_{2n}(\pi \to \pi')$.

Example. ([17, Prop. 6.2]) If $2n \geq 6$, $P_{2n}(\mathbb{Z} \to 1\,;\,t)$ is isomorphic to C_{2n-1}, the knot cobordism group of $(2n-1)$-knots. Here $\mathbb{Z}$ is considered as a multiplicative group $\langle t \rangle$. If A is a Seifert matrix of a knot, then $\lambda = A + (-1)^n tA'$ gives the corresponding $(-1)^n$-Seifert form.

Functorial property. Let $(h_i\,;\,t_i)$ be pairs satisfying the requirement of this paragraph, $i = 1, 2$. By a *morphism* $(h_1; t_1) \Rightarrow (h_2; t_2)$ we mean a pair of homomorphisms $f : \pi_1 \to \pi_2$, $f' : \pi_1' \to \pi_2'$ such that $f(t_1) = t_2$, which makes the diagram commute:

$$\begin{array}{ccc} \pi_1 & \xrightarrow{\quad f \quad} & \pi_2 \\ h_1 \downarrow & & \downarrow h_2 \\ \pi_1' & \xrightarrow{\quad f' \quad} & \pi_2' \end{array}$$

A morphism $(f, f') : (h_1 ; t_1) \Rightarrow (h_2 ; t_2)$ induces a homomorphism $P_{2n}(h_1 ; t_1) \to P_{2n}(h_2 ; t_2)$.

Remark. Define $FP_{2n}(h ; t)$ in a similar way to $P_{2n}(h ; t)$ but using $(-1)^n$-Seifert forms (G, λ, μ) in which G is Λ-*free*. Then the natural map $FP_{2n}(h ; t) \to P_{2n}(h ; t)$ is an isomorphism. This assertion (Prop. 2 of [18]) is proved by Lemma 4 of [18] and by using the notion of a free core [18, p. 586]. The 'free care'-device of [17, §5] was considered because we did not know this isomorphism there.

7) *Seifert forms: Geometric construction.* Let M^{2n} be an exterior n-connected, locally flat submanifold of a global knot (W^{2n+2}, K^{2n-1}) with $\partial M^{2n} = K^{2n-1}$ and representing the fundamental class $[X^{(m)}, Y^{(m-1)}]$. Then the homotopy group $\pi_{n+1}(E, FN)$ is a left Λ-module (through the left action of $\pi_1(E)$ on the universal covering $\tilde{E}$). We can define a $(-1)^n$-Seifert form (λ, μ) on $\pi_{n+1}(E, FN)$.

Let $f, g : (D^{n+1}, S^n) \to (E^{2n+2}, FN)$ be pathed immersions with images $D_f^{n+1}, S_f^n; D_g^{n+1}, S_g^n$. Let $w : FN \to M$ be the projection of the circle bundle. We assume D_f^{n+1} and D_g^{n+1} (or $w(S_f^n)$ and $w(S_g^n)$) meet in general position.

Let $\{p_1, \ldots, p_r\} = w(S_f^n) \cap w(S_g^n)$. With each p_i, we assign the sign $\varepsilon_i = \varepsilon(w(S_f^n), w(S_g^n))(p_i) (= \pm 1)$ and the group element $g_i = g(S_f^n, S_g^n)(p_i) \in \pi_1(FN) = \pi$ as follows:

The sign ε_i is the sign of the intersection of $w(S_f^n)$ and $w(S_g^n)$ (in this order) at p_i. Note that $w^{-1}(p_i) \cap S_f^n$ (or $w^{-1}(p_i) \cap S_g^n$) consists of a single point p_i^f (or p_i^g). Then g_i is the class of a loop starting at the base point $*$ along the assigned path $\gamma(f)$ ($\subset FN$) to a point on S_f^n, rounding S_f^n to p_i^f, then traversing the circle fiber $w^{-1}(p_i)$ in the positive direction, reaching p_i^g (for the first time), rounding S_g^n and coming back to $*$ along $\gamma(g)^{-1}$.

Define $\alpha(S_f^n, S_g^n) \in \Lambda$ as $\alpha(S_f^n, S_g^n) = \sum_{i=1}^{r} \varepsilon_i g_i$.

Let $\{q_1,\ldots,q_s\} = D_f^{n+1} \cap D_q^{n+1}$. Let $\varepsilon_i' = \varepsilon'(D_f^{n+1}, D_g^{n+1})(q_i)$ be the sign (± 1) of the intersection of D_f^{n+1} and D_g^{n+1} (in this order) at q_i. Define $g_i' = g'(D_f^{n+1}, D_g^{n+1})(q_i)$ to be the class of a loop starting at $*$ along $\gamma(f)$ then rounding D_f^{n+1} to q_i, rounding D_f^{n+1} and coming back to $*$ along $\gamma(q)^{-1}$.

Define $\beta(D_f^{n+1}, D_g^{n+1}) \in \Lambda$ as $\beta(D_f^{n+1}, D_g^{n+1}) = \sum_{i=1}^{s} \varepsilon_i' g_i'$.

Now λ is a combination of α and β:

$$\lambda(f,g) = \alpha(S_f^n, S_g^n) + (-1)^{n+1}(1-t)\beta(D_f^{n+1}, D_g^{n+1}).$$

It is shown that $\lambda(f,g)$ depends only on the homotopy classes of f,g. [17, §2].

To define μ, let $f : (D^{m+1}, S^n) \to (E^{2n+2}, FN)$ be a pathed immersion. Let $\{p_1,\ldots,p_r\}$ and $\{q_1,\ldots,q_s\}$ be the sets of self-intersection points of $w(S_f^n)$ and D_f^{n+1}, respectively. At each p_i choose an order of the branches, say $(w(S_f^n)_{(1)}, w(S_f^n)_{(2)})$, and define $\varepsilon_i = \varepsilon(S_f^n)(p_i)$ to be $\varepsilon(w(S_f^n)_{(1)}, w(S_f^n)_{(2)})(p_i)$ in the previous notation. Also define $g_i = g(S_f^n)(p_i)$ to be $g((S_f^n)_{(1)}, (S_f^n)_{(2)})(p_i)$. Let $\alpha(S_f^n) = \sum_{i=1}^{r} \varepsilon_i g_i$. Similarly, by choosing an order of branches at each q_i, we can define $\beta(D_f^{n+1}) = \sum_{i=1}^{s} \varepsilon_i' g_i'$. The ambiguities of the choices of orders of branches amount to elements of $\{a - (-1)^n \bar{a}\, t \mid a \in \Lambda\}$. Let $\mathcal{O}(D_f^{n+1}) \in \mathbb{Z}$ be the obstruction to extending a normal vector field v on $S_f^n \subset FN$, to a non-zero normal vector field over $D_f^{n+1} \subset E$. Here, v is the restriction of the vector field on FN whose vectors are tangent to circle fibers in the positive direction. The function μ is a combination of α, β, $\mathcal{O}$:

$$\mu(f) = \alpha(S_f^n) + (-1)^{n+1}(1-t)\beta(D_f^{n+1}) + (-1)^{n+1}\ (D_f^{n+1}).$$

μ depends only on the homotopy class of f. [17, §2]. The triple $(\pi_{n+1}(E,FN), \lambda, \mu)$ is proved to be a $(-1)^n$-Seifert form in the sense of (6).

Freedman [5] independently found Seifert forms.

Remark. In [17], and also here, we adopted the *curious incidence relation,* according to which an orientation $[V]$ of a manifold is compatible with an orientation $[\partial V]$ of the boundary if and only if $[V] = [\partial V] \times$ (inward normal). If we adopt the *natural incidence relation,* $[V] =$ (outward normal) $\times [\partial V]$, then the sign $(-1)^{n+1}$ in the definition of λ, μ might be changed.

8) *Transfer homomorphism.* Let $(h : \pi \to \pi' ; t)$ be as in (6). Let $\sigma' \subset \pi'$ be a normal subgroup of finite index. Let $\sigma = h^{-1}(\sigma')$. Then the *transfer homomorphism* $\tau : P_{2n}(h ; t) \to P_{2n}(h|\sigma ; t)$ is defined as follows:

Choose a set of representatives $\{x_1, \ldots, x_r\}$ of π/σ. Suppose $(\Lambda e_1 \oplus \cdots \oplus \Lambda e_s, \lambda, \mu)$ represents an $\eta \in P_{2n}(h ; t)$. Then $\tau(\eta)$ is represented by $(\Gamma x_1 e_1 \oplus \Gamma x_2 e_1 \oplus \cdots \oplus \Gamma x_r e_s, \bar{\lambda}, \bar{\mu})$, where

$$\Gamma = \text{the group ring } \mathbb{Z}[\sigma],$$
$$\bar{\lambda}(x_i e_j, x_k e_l) = [\lambda(x_i e_j, x_k e_l)]$$
$$\bar{\mu}(x_i e_j) = [\mu(x_i e_j)].$$

The symbol $[\ \ ]$ denotes the Γ-homomorphism $\Lambda \to \Gamma$ which is the unique extension of $[\ \ ] : \pi \to \Gamma$ defined by $[g] = \begin{cases} g & , g \in \sigma, \\ 0 & , g \notin \sigma. \end{cases}$

The homomorphism τ is independent of the choice of $\{x_1, \ldots, x_r\}$.

LEMMA A.4. *Let $i_* : P_{2n}(h|\sigma ; t) \to P_{2n}(h ; t)$ be the homomorphism induced by the 'inclusion' $h|\sigma \Rightarrow h$. If $\sigma \subset \pi$ be an extension satisfying C.2 of §7, then $\tau \cdot i_*(\eta) = |\pi'/\sigma'| \cdot \eta$ for any $\eta \in P_{2n}(h|\sigma; t)$.*

The homomorphism τ corresponds to taking a covering space. Let (W^{2n+2}, K^{2n-1}) be a global knot whose associated homomorphism is $h : \pi \to \pi'$. Let $\sigma' \subset \pi' = \pi_1(W)$ be a normal subgroup of finite index, $w : \tilde{W} \to W$ the corresponding covering. Let $\tilde{K} = w^{-1}(K)$.

The pair $(\tilde{W}, \tilde{K})$ is a global knot, and its associated homomorphism is identified with $h|\sigma : \sigma \to \sigma'$, $(\sigma = h^{-1}(\sigma'))$.

LEMMA A.5. $\eta(\tilde{W}, \tilde{K}) = \tau(\eta(W, K))$.

One can prove A.5 by a direct geometric observation.

Remark. The materials of this paragraph (8) appears here for the first time.

9) *Additivity and naturality.* ([17, p. 305]) Let (W_i^{2n+2}, K_i^{2n-1}), $i = 1,2$, be global knots, $2n > 4$. Let Q^{2n-1} be a compact connected PL manifold, U^{2n+1} the total space of a -bundle over Q^{2n-1} which contains Q^{2n-1} as the zero-section. Suppose that there are PL embeddings $g_i : U^{2n+1} \to \partial W_i^{2n+2}$ such that $g_i(U^{2n+1}) \cap K_i^{2n-1} = g_i(Q^{2n-1})$. By identifying $g_1(x)$ with $g_2(x)$ for every $x \in U^{2n+1}$, we glue W_1 and W_2 to obtain W^{2n+2}. Let $K^{2n-1} = (K_1^{2n-1} - g_1(\text{Int } Q^{2n-1})) \cup (K_2^{2n-1} - g_2(\text{Int } Q^{2n-1}))$. Then (W^{2n+2}, K^{2n-1}) is again a global knot. (We assume that W_1, W_2, W have compatible orientations.) Let $(h_1; t_1)$, $(h_2; t_2)$, $(h ; t)$ be respective associated pairs. We have the inclusion morphisms $(h_i; t_i) \Rightarrow (h ; t)$, $i = 1,2$, and they induce the homomorphisms $j_i : P_{2n}(h_i; t_i) \to P_{2n}(h ; t)$.

THEOREM A.6. $\eta(W, K) = j_1\eta(W_1,K_1) + j_2\, \eta(W_2,K_2)$.

COROLLARY A.7. If (W_2^{2n+2}, K_2^{2n-1}) admits a locally flat spine, then $\eta(W, K) = j_1\eta(W_1, K_1)$.

There is another type of naturality. Let V^{2n+2} be a codimension 0 submanifold of W^{2n+2} such that $V^{2n+2} \cap W^{2n+2}$ is a regular neighbourhood of K^{2n-1}. Suppose the inclusion $(V^{2n+2}, K^{2n-1}) \to (W^{2n+2}, K^{2n-1})$ is a simple homotopy equivalence.

Then the pair $(h\ ;\ t)$ associated with (V, K) is identified with that of (W, K), and we have

THEOREM A.8. $\eta(V, K) = \eta(W, K)$.

This follows from the cobordism invariance of η-obstruction [17, Thm. 3.5, Prop. 4.11].

10) *Geometric periodicity.* ([17, Thm. 5.12]) By the definition of P-groups, there is the natural identification $\rho : P_{2n}(h\ ;\ t) \underset{\rightarrow}{\cong} P_{2n+4}(h\ ;\ t)$. This periodicity has a geometric interpretation as the Wall surgery. Let (W^{2n+2}, K^{2n-1}) be a global knot with the associated pair $(h\ ;\ t)$. Then the product with the complex projective plane $(W^{2n+2} \times \mathbb{C}P_2, K^{2n-1} \times \mathbb{C}P_2)$ is again a global knot with the same associated pair $(h\ ;\ t)$.

THEOREM A.9. $(W^{2n+2} \times \mathbb{C}P_2, K^{2n-1} \times \mathbb{C}P_2)$ *is equal to* $\rho(\eta(W^{2n+2}, K^{2n-1}))$.

REFERENCES

1. Ancel, F. D. and Cannon, J. W., The locally flat approximation of cell-like embedding relations, preprint.
2. Cappell, S. E., and Shaneson, J. L., The codimension two placement problem and homology equivalent manifolds, Ann. of Math., 99 (1974), 277-348.
3. ______, Totally spineless manifolds, Illinois J. Math., 21 (1977) pp. 231-239.
4. Eaton, W. T., Pixley, C. P., and Venema, G. A., A topological embedding which cannot be approximated by a piecewise linear embedding, Notice Amer. Math. Soc., 24 (1977), A-302.
5. Freeman, M. H., Surgery on codimension-2 submanifolds, Memoirs Amer. Math. Soc., 12, #191 (1977). See also "On the classification of taut submanifolds", Bull. Amer. Math. Soc., 81 (1975) pp. 1067-1068.
6. Fukuhara, S., On the hauptvermutung of 5-dimensional manifolds and s-cobordisms, J. London Math. Soc., 5 (1972), pp. 549-555.
7. Giffen, C. H., Disciplining dunce hats in 4-manifolds, to appear.
8. ______, F-isotopy and I-equivalence of links, to appear.

9. Homma, T., Piecewise linear approximations of embeddings of manifolds, Mimeographed Notes. Florida State Univ., Florida, 1965.
10. Kato, M., and Matsumoto, Y., Simply connected surgery of submanifolds in codimension two I, J. Math. Soc. Japan, 24 (1972), pp. 586-608.
11. Kawauchi, A., Several topics on quadratic forms of 3-manifolds, R.I.M.S. (Kyoto Univ.) 1976.
12. Kervaire, M. A., Les noeds de dimensions supérieures, Bull. Soc. Math. de France, 93 (1965), pp. 225-271.
13. ______________, Smooth homology spheres and their fundamental groups, Trans. Amer. Math. Soc., 144 (1969), 67-72.
14. Kirby, R. C., and Siebenmann, L. C., On the triangulation of manifolds and the Hauptvermutung, Bull. Amer. Math. Soc., 75 (1969), pp. 742-749.
15. ______________, Normal bundles for codimension 2 locally flat imbeddings, Proc. Topology Conf., Park City, Utah, 1974, Springer Lecture Notes in Math. 438 (1975).
16. Levine, J., Knot cobordism in codimension two, Comment. Math. Helv., 44 (1969), pp. 229-244.
17. Matsumoto, Y., Knot cobordism groups and surgery in codimension two, J. Fac. Sci. Univ. Tokyo, Sec. IA. 20, (1973), pp. 253-317.
18. ______________, Some relative notions in the theory of Hermitian forms, Proc. Japan Acad., 49, (1973), pp. 583-587.
19. ______________, A 4-manifold which admits no spine, Bull. Amer. Math. Soc., 81 (1975), 467-470.
20. ______________, Some counterexamples in the theory of embedding manifolds in codimension two, Sci. Papers College Gen. Ed. Univ. Tokyo, 25 (1975), pp. 49-57.
21. ______________, Some wild embeddings of $S^1 \times S^{m+1}$ into $\mathbb{R}^{m+4}$, preprint, 1977.
22. Max, N. L., Homeomorphisms of $S^n \times S^1$, Bull. Amer. Math. Soc., 73 (1967), pp. 939-942.
23. Miller, R. T., Approximating codimension 3 embeddings, Ann. of Math., 95 (1972), pp. 406-416.
24. Noguchi, H., Obstructions to locally flat embeddings of combinatorial manifolds, Topology, 5 (1966), pp. 203-213.
25. Rushing, T. B., Topological Embeddings, Pure and Apply. Math., 52 (1973), Academic Press.
26. Shaneson, J. L., Embeddings of spheres in codimension two and h-cobordism of $S^1 \times S^3$, Bull. Amer. Math. Soc., 74 (1968), pp. 972-974.
27. Siebenmann, L. C., Topological manifolds, Proc. I.C.M. Nice (1970), Vol. 2, pp. 143-163.
28. Tamura, I., The diversity of manifolds, (in Japanese) Sûgaku 21 (1969), 275-285.
29. Thomas, E., and Wood, J., On manifolds representing homology classes in codimension 2, Invent. Math., 25 (1974) pp. 63-89.

30. Venema, G. A., A topological disk in a 4-manifold can be approximated by piecewise linear disks, Bull. Amer. Math. Soc., 83 (1977), pp. 386-387.
31. ____________, Approximating disks in 4-space, Michigan Math. J. (to appear).
32. Wall, C. T. C., Surgery on compact manifolds, 1970, Academic Press.
33. ____________, On the axiomatic foundation of the theory of Hermitian forms, Proc. Camb. Phil. Soc., 67 (1970), pp. 243-250.

FOLIATIONS AND MONOIDS OF EMBEDDINGS

Dusa McDuff

Mathematics Institute
University of Warwick
Coventry, England

This paper deals with the relationship between the classifying space for foliations and monoids of embeddings. If U is an open subset of the n-dimensional manifold V, there is a "restriction map" from $\overline{B}\mathrm{Emb}_0 V$ *to* $\overline{B}\mathrm{Emb}_0 U$ *which behaves like a fibration as far as homology is concerned. (Here* $\overline{B}\mathrm{Emb}_0 V$ *denotes the homotopy fibre of the map* $B\mathrm{Emb}_0 V \to B\mathrm{Emb}_0 V$*, where* $\mathrm{Emb}_0 V$ *is the discrete monoid which has the same elements as the topological monoid,* $\mathrm{Emb}_0 V$*, of all* C^r*-self-embeddings of V which are isotopic to the identity, with the* C^r*-topology.) This yields a proof of Thurston's Theorem, which says that* $\overline{B}\mathrm{Diff}_c \mathbb{R}^n$ *has the same homology as the n-fold loop space of Haefliger's classifying space* $\overline{B}\Gamma^r_n$*, (where* $\mathrm{Diff}_c \mathbb{R}^n$ *is the group of all* C^r*-diffeomorphisms of* $\mathbb{R}^n$ *with compact support, with the usual direct limit topology). It also clarifies the relation between this theorem and Segal's result that* $\overline{B}\mathrm{Emb}_0 \mathbb{R}^n \simeq \overline{B}\Gamma^r_n$.

I. INTRODUCTION

This paper is a report on joint work with Graeme Segal. In it, I describe an immersion-theoretic method (see [3] V) for dealing with the homology of monoids of embeddings and use it to prove

ISBN 0-12-158860-2

the theorem of Mather and Thurston [15] which relates the homology of the group of diffeomorphisms of a manifold to the homology of Haefliger's classifying space for foliations. The codimension 1 case of this theorem is due to Mather [5]. He also wrote up the details of Thurston's arguments for the general case in [7]. The techniques used in this paper are essentially due to Segal, who in [14] used them to give a very simple proof in the codimension 1 case.

The general idea is as follows. Let X be an n-dimensional C^r-manifold without boundary, where $r = 1,\ldots,\infty$, and denote by $\mathcal{E}mb X$ the monoid of all C^r-self-embeddings of X with the compact-open C^r-topology. Thus, when X is compact, $\mathcal{E}mb X$ is simply the group of all C^r-diffeomorphisms of X.

We write $\mathcal{E}mb_0 X$ for the identity component of $\mathcal{E}mb X$, and $\mathrm{Emb}_0 X$ for $\mathcal{E}mb_0 X$ considered with the discrete topology. We will be concerned mostly with $\overline{B}\mathcal{E}mb_0 X$, the homotopy fibre of the map $B\,\mathrm{Emb}_0 X \to B\mathcal{E}mb_0 X$. Now, the manifold X has a unique, codimension n foliation, namely the foliation by points, and, by Haefliger's theorem [2], this gives rise to a commutative diagram

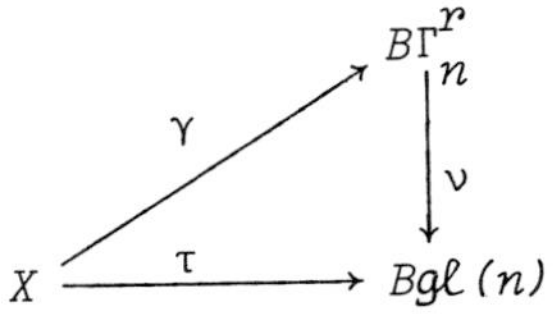

where τ classifies the tangent bundle, $\tau(X)$, to X. Let $L \to X$ be the pull back by τ of the Hurewicz fibration associated to ν, and let $S(X)$ be the space of continuous sections of $L \to X$ with the compact-open topology. $S(X)$ may also be thought of as the space of all pairs (g,h), where $g : X \to B\Gamma_n^r$ and h is a homotopy from τ to $\nu \circ g$. It thus has a base point $s_0 = (\gamma, h_0)$ where h_0 is the trivial homotopy. We will show that the product $(\overline{B}\,\mathcal{E}mb_0 X) \times X$ also has a natural codimension n "foliation", which is obtained by pulling back the point foliation on X by the evaluation map

$\in : (m,x) \mapsto m(x)$ from $(\mathrm{Emb}_0 X) \times X$ to X. This foliation is transverse to the spaces $\{b\} \times X$, where $b \in \overline{B}\ \mathrm{Emb}_0 X$, and so its normal bundle is the pull back of $\tau(X)$ by the projection $\pi : (\overline{B}\ \mathrm{Emb}_0 X) \times X \to X$. Therefore, the foliation is classified by a homotopy commutative diagram

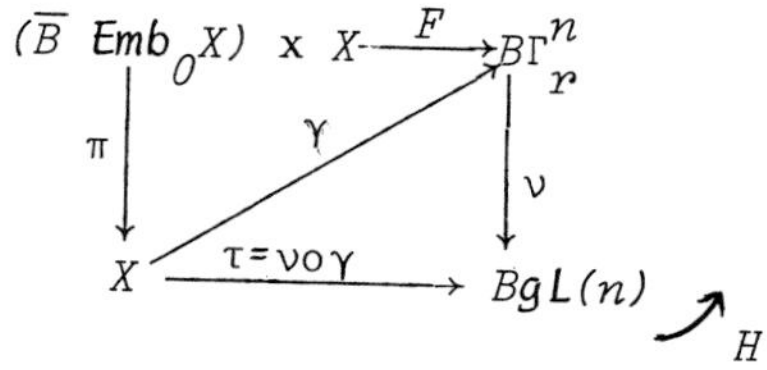

(*)

We will construct this diagram in such a way that there is a canonical choice of homotopy from $\tau \circ \pi$ to $\nu \circ F$. This homotopy, H say, determines a map $f_X : \overline{B}\ \mathrm{Emb}_0 X \to S(X)$. In fact f_X assigns to each $b \in \overline{B}\ \mathrm{Emb}_0 X$ the pair (F_b, H_b), where $F_b(x) = F(b,x)$ and H_b is the homotopy induced by H from τ to $\nu \circ F_b$.

We will prove the following theorems.

THEOREM A. *If X is the interior of a compact manifold $\overline{X}$ then f_X induces an isomorphism $H_*(\overline{B}\ \mathrm{Emb}_0 X) \cong H_*(S(X))$, where H_* denotes integral homology.*

COROLLARY. *If $\overline{X}$ is connected and $\partial\overline{X} \neq \phi$, then f_X is a weak homotopy equivalence.*

Proof of Corollary. Since the fibre, $\overline{B\Gamma}^r_n$, of $L \to X$ is n-connected ([2], p. 154), the hypotheses on $\overline{X}$ imply that $\pi_1(S(X)) = 0$. They imply also that $\pi_1(B\ \mathrm{Emb}_0 X) = 0$, (see [8]). (The analogous statement for a closed manifold would, of course, be false.) Thus, $\pi_1 \overline{B}\ \mathrm{Emb}_0 X$ is a quotient of $\pi_2 B\ \mathrm{Emb}_0 X$, and so is abelian. But, in that case, f_x must induce an isomorphism on π_1. Therefore, both spaces are simply connected, and the result follows from Whitehead's theorem.

(Notice that, although $\overline{B}\,\mathrm{Emb}_0 X$ has the homotopy type of a *CW* complex see [9], $B\Gamma_n^r$, and hence also $S(X)$, probably does not. Therefore f_X need not be a homotopy equivalence.)

There is a similar result for the group $\mathcal{D}i\!f\!f_c X$ of all C^r-diffeomorphisms of X which equal the identity outside a compact subset of X, with the usual direct limit topology. Let $\mathcal{D}i\!f\!f_{co} X$ be the identity component of $\mathcal{D}i\!f\!f_c X$. It is well known that, because $\mathcal{D}i\!f\!f_c X$ is a group, $\overline{B}\,\mathcal{D}i\!f\!f_{co} X \simeq \overline{B}\,\mathcal{D}i\!f\!f_c X$. (A proof is given in [8].) The map f_X is defined so that it takes $\overline{B}\,\mathcal{D}i\!f\!f_{co} X$ into $S_c(X)$, the space of sections over X which equal the base section, s_o, outside a compact subset of X, with the direct limit topology.

THEOREM B. *The map* $f_X : \overline{B}\,\mathcal{D}i\!f\!f_{co} X \to S_c(X)$ *induces an isomorphism on* H_*.

This is Thurston's theorem. (Of course, when X is compact it coincides with Theorem A.) When $X = \mathbb{R}^n$, it may be reformulated as:

COROLLARY (*Thurston*). *There is a isomorphism*

$$H_*(\overline{B}\,\mathcal{D}i\!f\!f_{co}\mathbb{R}^n) \xrightarrow{\cong} H_*(\Omega^n \overline{B\Gamma}_n^r) ,$$

where Ω^n *denotes the n-fold loop space functor.*

Notice that $\overline{B}\,\mathcal{D}i\!f\!f_{co}\mathbb{R}^n$ is not homotopy equivalent to $\Omega^n \overline{B\Gamma}_n^r$. For, $\pi_1(\Omega^n \overline{B\Gamma}^r)$ is abelian. (In fact, except possibly when $r = n+1$, it is 0, see [15], [6]). On the other hand, $\pi_1 \overline{B}\,\mathcal{D}i\!f\!f_{co}\mathbb{R}^n$ maps onto the enormous discrete group $\mathrm{Diff}_{co}\mathbb{R}^n$.

These theorems are proved using a modification of the familiar induction-over-handles procedure used for instance by Gromov. The starting point is the proof by Segal of Theorem A in the case when $X = \mathbb{R}^n$, see [14].

We begin by constructing the map f_X. Because we need the diagram (†) of Proposition 1 to commute, this has to be done rather carefully. The inductive procedure is described in Chapt. 3.

The present arguments go through in the case $r = 0$ (that is, for topological embeddings) almost without change. However the topological category $g\ell(W)$ which occurs in the construction of f_x must be replaced by a certain quasi-topological category (compare Haefliger's description of $BTop_n$ in [2]), and some argument is needed to show that its realization is equivalent to $BTop_n$. It is perhaps easier to prove the topological case of Theorems A and B by a slightly different approach. For, because the group of homeo- of $\mathbb{R}_n$ with compact support has trivial integral homology, see Mather [4], the theorems degenerate in this case, and both $\overline{B}Emb_0X$ and $S(X)$ are acyclic. (In fact, $\overline{B}\Gamma_0^n$, and hence also $S(X)$, are contractible.) These results can be proved by the methods of Chapt. 3, without it being necessary to construct the map f_X. Details of this argument are given in [8]. That paper also contains the proofs of the main technical results used here.

II. THE CONSTRUCTION OF f_X.

We will consider a (fixed) n-dimensional C^r-manifold, X, without boundary, and (variable) n-dimensional C^r-submanifolds, Y and Z, of X which will be either compact and possibly with corners, or the interior of such a compact submanifold. The subset A of X will always be closed. We will denote by $EmbY$ the monoid of all C^r-self-embeddings of Y with the compact-open topology. If $Z \subseteq Y$, we write $Emb_0(Y,Z,\text{rel } A)$ for the identity component of the monoid $\{m \in EmbY : m(Z) \subseteq Z \text{ and } m = id \text{ near } Y \cap A\}$. This is given the direct limit topology, where the limit is taken over all open neighborhoods of $Y \cap A$ in Y. Similarly, $Emb_0(X,Y,Z,\text{rel } A)$ is the identity component of the monoid $\{m \in EmbX : m(Y) \subseteq Y, m(Z) \subseteq Z, m = id \text{ near } A\}$. The corresponding discrete monoids are called $\text{Emb}_0(Y,Z,\text{rel } A)$ and $\text{Emb}_0(X,Y,Z,\text{rel } A)$. Of more interest to us is the discrete monoid $\mathrm{Emb}_0^g(Y,Z,\mathrm{rel}\ A)$ of germs of self-embeddings of Y, which is defined to be the quotient $\mathrm{Emb}_0(X,Y,Z,\mathrm{rel}\ A)/\sim$, where $m \sim m'$ if $m = m'$ near Y. Thus, if Y is an open subset of X, $\mathrm{Emb}_0^g Y = \mathrm{Emb}_0 Y$. Notice also, that because elements in $\mathrm{Emb}_0^g(Y,Z,\mathrm{rel}\ Y \cap A)$ are the identity near $Y \cap A$, it follows from the isotopy extension theorem that $\mathrm{Emb}_0^g(Y,Z,\mathrm{Rel}\ A) = \mathrm{Emb}_0^g(Y,Z,\mathrm{rel}\ Y \cap A)$. We will write $\overline{B}\,\mathcal{E}mb_0$, resp. $\overline{B}\,\mathcal{E}mb_0^g$, for a particular model of the homotopy fibre of the map $B\ \mathrm{Emb}_0 \to B\ \mathcal{E}mb_0$, resp. $B\ \mathrm{Emb} \to B\ \mathcal{E}mb_0$, which will be described later.

Let $\rho : \mathcal{E}mb_0(Y,Z,\mathrm{rel}\ A) \to \mathcal{E}mb_0(Z,\mathrm{rel}\ A)$ be the restriction homomorphism. Clearly, $\mathcal{E}mb_0(Y,\mathrm{rel}\ Z \cup A)$ is in the kernel of ρ. Since the same is true for Emb_0^g, it follows that the induced mapping $\overline{B}\ \mathcal{E}mb_0^g(Y,Z,\mathrm{rel}\ A) \to \overline{B}\ \mathcal{E}mb_0^g(Z,\mathrm{rel}\ A)$, which we denote by $\overline{\rho}$, maps $\overline{B}\ \mathcal{E}mb_0^g(Y,\mathrm{rel}\ Z \cup A)$ to $\overline{B}\{e\}$, where e is the identity element of $\mathcal{E}mb_0(Z,\mathrm{rel}\ A)$. We will define $\overline{B}$ in such a way that $\overline{B}\{e\}$ is a single point, $*$ say. Then $\overline{B}\ \mathcal{E}mb_0^g(Y,\mathrm{rel}\ Z \cup A)$ will be contained in the fibre of $\overline{\rho}$ at $*$.

Correspondingly, we define $S(Y,\mathrm{rel}\ A)$ to be the space of continuous sections of $L \to X$ over Y which equal the base section, s_0, near $Y \cap A$, with the usual direct limit topology.

The rest of Chapt. 2 is taken up with proving

PROPOSITION 1. For each Y, there is a map $f_Y : \overline{B}\ \mathcal{E}mb_0^g Y \to S(Y)$ which takes the subspace $\overline{B}\ \mathcal{E}mb_0^g(Y,\mathrm{rel}\ A)$ to $S({}^{v},\mathrm{rel}\ A)$ for all A, and is such that for all $Z \subseteq Y$, the diagram

$$\begin{array}{ccccc} \overline{B}\ \mathcal{E}mb_0^g(Y,\mathrm{rel}\ Z \cup A) & \to & \overline{B}\,\mathcal{E}mb_0^g(Y,Z,\mathrm{rel}\ A) & \overset{\overline{\rho}}{\to} & \overline{B}\,\mathcal{E}mb_0^g(Z,\mathrm{rel}\ A) \\ \downarrow f'_Y & & \downarrow f_Y & & \downarrow f_Z \\ S(Y,\mathrm{rel}\ Z \cup A) & \longrightarrow & S(Y,\mathrm{rel}\ A) & \longrightarrow & S(Z,\mathrm{rel}\ A) \end{array} \qquad (\dagger)$$

commutes strictly.

Proof. We will construct a map $f_X : \overline{B}\ \mathcal{E}mb_0 X \to S(X)$ which induces all the maps f_Y.

First, let us describe the particular model chosen for $\overline{B}\ \mathcal{E}mb_0 X$. We will always use the "thin" realisation for our categories, where degeneracies are collapsed. Since all the monoids we consider are "good" (see [12] Appendix A), this realisation is equivalent to the "thick" one. If M is a topological monoid (i.e. a strictly associative H-space with two-sided identity) which acts on the left on a space S, we will write $C(M\backslash\backslash S)$ for the topological category whose space of objects is S and whose space of morphisms is $M \times S$, where (m,s) corresponds to the morphisms $M : s \to ms$. Its realisation will be called $M \backslash\backslash S$. Thus $M\backslash\backslash *$ is simply BM, where $*$ denotes the one-point space. We will use $E^g \backslash\backslash \mathcal{E}$ as our model for $\overline{B}\ \mathcal{E}mb_0^g$. (Here we write E^g, for instance, instead of Emb_0^g.) To justify this, recall that, if the map $s \mapsto ms$ from S to S is a homotopy equivalence for each $m \in M$, then S is weakly equivalent to the homotopy fibre of $M\backslash\backslash S \to M\backslash\backslash *$. (See [12] (1-6) or [8].) In particular, since $\mathcal{E} = \mathcal{E}mb_0$ is connected, $\mathcal{E}$ is equivalent to the fibre of $\mathcal{E}\backslash\backslash\ \mathcal{E} \to \mathcal{E}\backslash\backslash *$. Because $\mathcal{E}\ \backslash\backslash\ \mathcal{E}$ is contractible [11], it follows that $\mathcal{E} \simeq \Omega(\mathcal{E}\backslash\backslash *)$. Also, since $E^g = \mathrm{Emb}_0^g$ consists of elements which are isotopic to the identity, $\mathcal{E}$ is equivalent to the fibre of $E^g \backslash\backslash\ \mathcal{E} \to E^g\backslash\backslash *$. Therefore, by comparing the sequence $\mathcal{E} \to E^g\backslash\backslash\ \mathcal{E} \to E^g\backslash\backslash *$ with the fibration sequence $\Omega(\mathcal{E}\backslash\backslash *) \to F \to E^g\backslash\backslash * \to \mathcal{E}\backslash\backslash *$, where F is the homotopy fibre of $E^g\backslash\backslash * \to \mathcal{E}\backslash\backslash *$, one sees that $E^g\backslash\backslash \mathcal{E} \simeq F$ as required. Note that $\{e\}\backslash\backslash\{e\}$ is a single point, $*$ say, which serves as a base point for $E^g\backslash\backslash \mathcal{E}$.

The "foliation" on $(\overline{B}\ \mathcal{E}mb_0 X) \times X$ which gives rise to the map f_X may be described intuitively as follows. (It will be defined more formally later by means of a certain functor.) Let us write E for $\mathrm{Emb}_0 X$ and $\mathcal{E}$ for $\mathcal{E}mb_0 X$. Then E acts on $\mathcal{E} \times X$ by $p : (m,x) \mapsto (pm,x)$, and it is not difficult to see that the spaces $E\backslash\backslash(\mathcal{E}\times X)$ and $(E \backslash\backslash\ \mathcal{E}) \times X$ are homeomorphic. Therefore, it suffices to foliate $E\backslash\backslash(\mathcal{E}\times X)$. Let $\mathcal{F}$ be the pull-back by the evaluation map $\in\ : (m,x) \mapsto m(x)$ from $\mathcal{E} \times X$ to X of the foliation of X by points. Then, because the action of E on $\mathcal{E} \times X$ preserves $\mathcal{F}$, there is an induced foliation on $E \backslash\backslash (E \times E)$. This is clear if one thinks of a foliated space as a space with two topologies, for, since the action of E on $E \times X$ is continuous with respect to both topologies on $E \times X$, the space $E \backslash\backslash E \times X$ has two topologies also. If X is compact, this is the same as the foliation on $(\overline{B}\,\mathcal{D}iff\,X) \times X$ described in [15].

In order to define this foliation precisely, it is convenient to use special models for $B\Gamma_n^r$ and $Bg\ell(n)$, which we now describe. If W is any n-dimensional C^r-manifold without boundary, let $\Gamma(W)$ be the category whose space of objects is W with its usual topology, and where the space of morphisms from x to y is the discrete set of germs at x of local C^r-diffeomorphisms of W which take x to y. The space, Mor $\Gamma(W)$, of morphisms is given the sheaf topology. Thus, $|\Gamma(\mathbb{R}^n)|$ is the model for $B\Gamma_n^r$ which is described in [1], Chapt. 8. It will be more convenient for us to use the category $\Gamma(X)$, whose realisation is also a model for $B\Gamma_n^r$, by the lemma below.

LEMMA 1. Any embedding $i : \mathbb{R}^n \hookrightarrow W$ induces an equivalence $|\Gamma(\mathbb{R}^n)| \to |\Gamma(W)|$.

Proof. Let $\mathcal{U}$ be an open covering of W by sets which are diffeomorphic to subsets of $\mathbb{R}^n$, and suppose that the image $i(\mathbb{R}^n)$ is one of the sets, U_0 say, in $\mathcal{U}$. Clearly, there is a projection functor $\tilde{Q} : \Gamma(\coprod_{U\in\mathcal{U}} U) \to \Gamma(W)$. Now $\Gamma(\coprod_{U\in\mathcal{U}} U)$ is simply the "disintegration" of $\Gamma(W)$ by $\mathcal{U}$, see [13], and it follows from [13] Proposition (2.8) that $Q : |\Gamma(\coprod_{U\in\mathcal{U}} U)| \to |\Gamma(X)|$ is shrinkable, and hence a homotopy equivalence. (We will always write the realisation of a functor $\tilde{Q}$ as Q.) On the other hand, the realisation of the inclusion $\tilde{J} : \Gamma(U_0) \to \Gamma(\coprod_{U\in\mathcal{U}} U)$ is also an equivalence. To see this, choose, for each $U \in \mathcal{U}$, an embedding, f_U, of U into U_0, with $f_{U_0} = id$. (Such embeddings exist by the choice of $\mathcal{U}$.) Then, let $\tilde{F}$ be the functor from $\Gamma(\coprod_{U\in\mathcal{U}} U)$ to $\Gamma(U_0)$ which takes the object (x,U) to $f_U(x)$, and the morphism $g : (x,U) \to (y,V)$ to the morphism

$f_V g f_U^{-1} : f_U(x) \to f_V(y)$. Clearly, $\tilde{F} \circ \tilde{J} = \tilde{Id}$. Moreover, there is a natural transformation $\tilde{T} : \tilde{Id} \to \tilde{J} \circ \tilde{F}$ which takes the object (x,U) of $\Gamma(\coprod_{U \in \mathcal{U}} U)$ to the morphism (germ of f_U) $: (x,U) \to (f_U(x), U_0)$. This implies, by [11], Chapt. 2, that F is homotopy inverse to J. Since the map from $|\Gamma(\mathbb{R}^n)|$ to $|\Gamma(W)|$ which is induced by i may be identified with $Q \circ J$, this map is an equivalence, as required.

Since there is no differential $\Gamma(W) \to C(g\ell(n) \backslash\backslash *)$, we replace $C(g\ell(n) \backslash\backslash *)$ by a category $g\ell(W)$ which also has W as its space of objects. Its space of morphisms is the bundle over $W \times W$, whose fibre at (x,y) is the space of all linear isomorphisms from T_x to T_y, with its usual topology, where T_x and T_y are the tangent spaces to W at x and y, respectively. Thus, for each $x_0 \in W$, the space of morphisms in $g\ell(W)$ from x_0 to x_0 is homeomorphic to $g\ell(n)$, and so, by identifying $*$ with x_0 we get an inclusion $C(g\ell(n) \backslash\backslash *) \to g\ell(W)$. Also, just as before, any embedding $i : \mathbb{R}^n \hookrightarrow X$ induces an inclusion $i : g\ell(\mathbb{R}^n) \hookrightarrow g\ell(W)$.

LEMMA 2. *These inclusions induce homotopy equivalences*

$$B\, g\ell(n) \hookrightarrow |g\ell(\mathbb{R}^n)| \to |g\ell(W)| .$$

Proof. This is very similar to the proof of Lemma 1, and will be omitted.

Notice that there is a differential $\tilde{\nu} : \Gamma(W) \to g\ell(W)$. It is the identity on objects and it takes the morphism $g : x \to y$ to its derivative $dg_x : x \to y$. Since the diagram

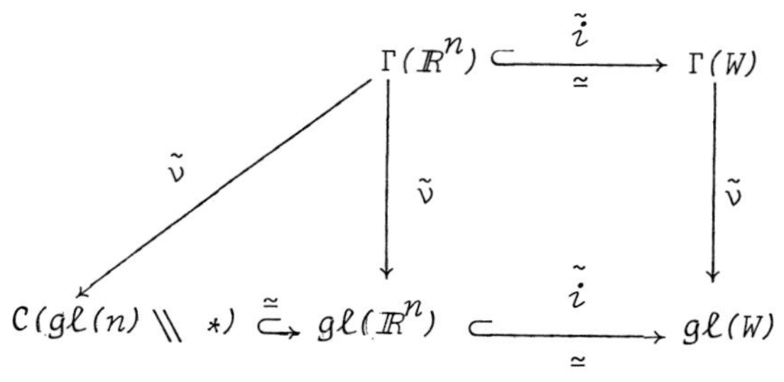

commutes, where $\tilde{\nu} : \Gamma(\mathbb{R}^n) \to C(g\ell(n)\backslash\backslash *)$ is the differential defined in [1], Chapt. 8, we may use $|\Gamma(X)| \overset{\nu}{\to} |g\ell(X)|$ as our model for $B\Gamma_n^r \overset{\nu}{\to} Bg\ell(n)$.

We will now define the "foliation" on $E\backslash\backslash E \times X$, as a certain functor $\tilde{F} : C(E\backslash\backslash E \times X) \to \Gamma(X)$. Recall that E acts on $E \times X$ by $p : (m,x) \mapsto (pm,x)$. Therefore there is a functor $\tilde{F} : C(E\backslash\backslash E{\times}X) \to \Gamma(X)$ which takes the object (m,x) to $m(x)$, and the morphism $p : (m,x) \to (pm,x)$ to the germ of p at $m(x)$. (It is easy to see that this does correspond to the foliation described earlier.) Consider the diagram

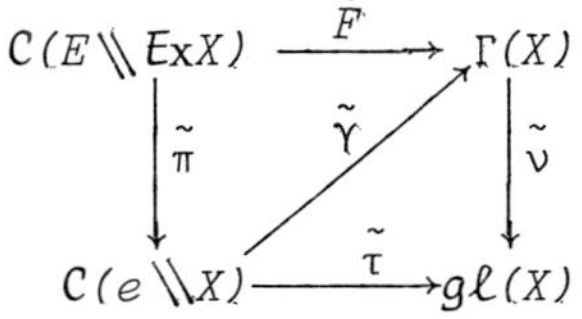

(**)

where $C(e\backslash\backslash X)$ is the category formed from the action of the trivial group $\{e\}$ on X, $\tilde{\pi}$ is the obvious projection, and $\tilde{\gamma}$ and $\tilde{\tau}$ are induced by the identity map on X. Clearly, $\tilde{\tau} = \tilde{\nu} \circ \tilde{\gamma}$. On the other hand, $\tilde{\tau} \circ \tilde{\pi} \neq \tilde{\nu} \circ \tilde{F}$. However, there is a natural transformation $\tilde{H} : \tilde{\tau} \circ \tilde{\pi} \to \tilde{\nu} \circ \tilde{F}$ which takes the object (m,x) of $C(E\backslash\backslash E \times X)$ to the morphism

$$dm_x \; : \; x = \tilde{\tau} \circ \tilde{\pi}(m,x) \;\to\; m(x) = \tilde{\nu} \circ \tilde{F}(m,x)$$

of $g\ell(X)$. (Notice that, because a natural transformation from $\tilde{F}_1 : C \to C'$ to $\tilde{F}_2 : C \to C'$ is a *continuous* map from ObjC to MorC', $\tilde{H}$ does not lift to give a natural transformation from $\tilde{\gamma} \circ \tilde{\pi}$ to $\tilde{F}$.)

Now, $|C(e\backslash\backslash X)|$ is homeomorphic to X. Therefore, the realisation of diagram (**) is the diagram (*) of Chapt. 1:

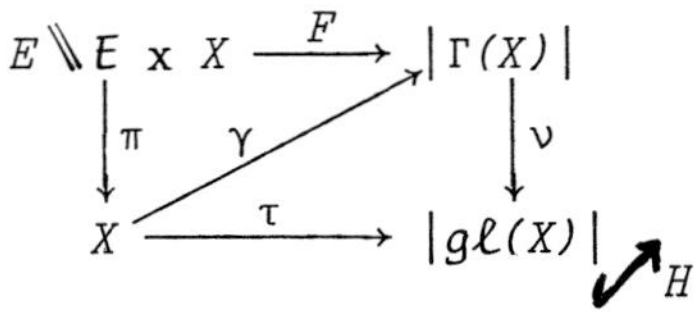

Let $L \to X$ be the pullback by τ of the Hurewicz fibration associated to $\nu : |\Gamma(X)| \to |g\ell(X)|$. By [11], Chapt. 2, the realisation, H, of $\tilde{H}$ determines a homotopy from $\tau \circ \pi$ to $\nu \circ F$. Therefore, F induces a map $f_X : E\backslash\backslash E \to S(X)$, as described in Chapt. 1. The base section, s_0, is determined by $\gamma : X \to |\Gamma(X)|$.

It is immediate from the definitions that, for any Y, the restriction of $\tilde{F}$ to $E(X,Y)\backslash\backslash E(X,Y) \times Y$ factors through the quotient category $E^g(Y)\backslash\backslash E(Y) \times Y$, where $E(X,Y)$, for instance, denotes $\mathrm{Emb}_0(X,Y)$. Therefore, F defines a map $f_Y : E^g(Y)\backslash\backslash E(Y) \to S(Y)$. Because the restrictions of $\tilde{F}$ and $\tilde{\gamma} \circ \tilde{\pi}$ to $E(X, \mathrm{rel}\ A) \backslash\backslash E(X, \mathrm{rel}\ A) \times A$ are equal, one sees easily that f_Y takes $E(Y, \mathrm{rel}\ A)\backslash\backslash E(Y, \mathrm{rel}\ A)$ to $S(Y, \mathrm{rel}\ A)$ for any A. Therefore diagram $(\dagger)$ exists, and is clearly strictly commutative.

III. THE IMMERSION-THEORETIC METHOD

A pair (Y,A) will be called admissible if Y is a compact, n-dimensional C^r-submanifold of X, with boundary but no corners, and if A is any closed subset of X. We will prove the following version of the theorems in Chapt. 1.

THEOREM C. For all admissible pairs (Y,A), f_Y induces an isomorphism

$$H_*\overline{B}\ \mathrm{Emb}_0^g(Y, \mathrm{rel}\ A) \xrightarrow{\cong} H_*S(Y, \mathrm{rel}\ A).$$

This includes Theorem B, but does not cover the case of non-compact X in Theorem A. However, it is proved in [8] that if Y is the interior of a compact manifold, $\overline{Y}$ say, then $\overline{B}\ \mathrm{Emb}_0^g(\overline{Y})$ and $\overline{B}\ \mathrm{Emb}_0^g(Y) = \overline{B}\ \mathrm{Emb}_0 Y$ have the same homotopy type. Since $L \to X$ is a locally trivial bundle, $S(\overline{Y})$ and $S(Y)$ also have the same homotopy type. Therefore, it follows from Theorem C that f_Y induces an isomorphism $H_*\overline{B}\ \mathrm{Emb}_0 Y \to H_*S(Y)$ when $Y = \mathrm{Int}\ \overline{Y}$, and so all cases of Theorems A and B can be deduced from Theorem C.

In order to carry through the immersion-theoretic method of proof, (which is explained, for instance, in [3] V), one has to know how $\overline{B}\ Emb_0^g$ and S behave under restriction. Now, the spaces $S(Y)$ behave as nicely as one could wish. Let us say that a pair (Y, Z) of compact n-dimensional submanifolds of X is a nice manifold pair if Y has no corners and if Z is cleanly embedded in Y, see [3], p. 13. Then the following result is well known.

LEMMA 3. If (Y,Z) is a nice manifold pair, the restriction map $r : S(Y,\text{rel } A) \to S(Z,\text{rel } A)$ is a Serre fibration. Moreover, the inclusion of $S(Y,\text{rel } Z \cup A)$ into the fibre $r^{-1}(s_0)$ is a homotopy equivalence, and $\pi_1(S(Z,\text{rel } A))$ acts trivially on $H_*(S(Y,\text{rel } Z \cup A))$.

Proof. Let K be any finite complex. Then a map $f : K \times I \to S(Z,\text{rel } A)$ may be considered as a section of the locally trivial bundle $L_Y \times K \times I \to Y \times K \times I$ over $(Z \cup V) \times K \times I$, where L_Y is the part of L over Y and V is a suitable neighborhood of $Y \cap A$ in Y. Because $Y \times K \times \{0\} \cup (Z \cup V) \times K \times I$ is a deformation retract of $Y \times K \times I$, any extension of f to $Y \times K \times \{0\} \cup (Z \cup V) \times K \times I$ may be extended to the whole of $Y \times K \times I$. Therefore, r is a Serre fibration. Its fibre $r^{-1}(s_0)$ is $\{s \in S(Y) : s = s_0 \text{ on } Z \text{ and near } A\}$. Since Z is cleanly embedded in Y, this may be deformed into $S(Y,\text{rel } Z \cup A\} = \{s \in S(Y) : s = s_0 \text{ near } Z \cup A\}$.

To see that $\pi_1(S(Z,\text{rel } A))$ acts trivially on $H_*(S(Y,\text{rel } Z \cup A))$, let A' be the union with A of all components of $Z - A$ which do not intersect ∂Z, and let $A'' = (Z-A') \cup A$. Then $S(Z,\text{rel } A)$ is the product $S(Z,\text{rel } A') \times S(Z,\text{rel } A'')$, and so $\pi_1(S(Z,\text{rel } A)) \cong \pi_1(S(Z,\text{rel } A')) \times \pi_1(S(Z,\text{rel } A''))$. Because $\partial Z \subseteq A''$, the sections in $S(Z,\text{rel } A'')$ may be extended to Y by putting them equal to s_0 outside Z. This clearly implies that $\pi_1(S(Z,\text{rel } A''))$ acts trivially on $H_*S(Y,\text{rel } Z \cup A)$. On the other hand, because $S(Z,\text{rel } A') = \varinjlim S(Z,\text{rel } U)$, where U runs through all neighborhoods of A' in Z we may suppose that (Z,A') is a nice manifold pair such that every component of $Z - A'$ meets ∂Z. But the fibre, $\overline{B\Gamma}^r_n$, is n-connected [2]. Therefore, since Z may be got from A' by adding handles of index $<n$, $\pi_1(S(Z, \mathrm{rel}\ A')) = 0$.

It is more difficult to deal with $\overline{B}\ \mathrm{Emb}_0^g$, for there is no restriction map $\overline{B}\ \mathrm{Emb}_0^g Y \to \overline{B}\ \mathrm{Emb}_0^g Z$. However, the following proposition shows that the restriction map $\overline{\rho} : \overline{B}\ \mathrm{Emb}_0^g(Y,Z) \to \overline{B}\ \mathrm{Emb}_0^g Z$ does just as well.

PROPOSITION 2. If (Y,Z) is a nice manifold pair, the inclusion $\overline{B}\ \mathrm{Emb}_0^g(Y,Z,\mathrm{rel}\ A) \to \overline{B}\ \mathrm{Emb}_0^g(Y,\mathrm{rel}\ A)$ is a homotopy equivalence.

The proof of this, which is rather difficult, is given in [8]. Now consider the restriction map $\overline{\rho} : \overline{B}\ \mathrm{Emb}_0^g(Y,Z,\mathrm{rel}\ A) \to \overline{B}\ \mathrm{Emb}_0^g(Z,\mathrm{rel}\ A)$. Since $\overline{\rho}$ takes $\overline{B}\ \mathrm{Emb}_0^g(Y,\mathrm{rel}\ Z \cup A)$ to $\overline{B}\{e\}$, which is a single point $*$, $\overline{B}\ \mathrm{Emb}_0^g(Y,\mathrm{rel}\ Z \cup A)$ is contained in the actual fibre, $\overline{\rho}^{-1}(*)$, of $\overline{\rho}$ at $*$. It therefore may be considered as a subset of the homotopy fibre, $F(\overline{\rho},*)$, of $\overline{\rho}$ at $*$.

PROPOSITION 3. If (Y,Z) is a nice manifold pair, this inclusion of $\overline{B}\ \mathrm{Emb}_0^g(Y,\mathrm{rel}\ Z \cup A)$ into $F(\overline{\rho},*)$ induces an isomorphism on H_*. Moreover, $\pi_1\overline{B}\ \mathrm{Emb}_0^g(Z,\mathrm{rel}\ A)$ acts trivially on $H_*(F(\overline{\rho},*))$.

This is also proved in [8]. As an immediate corollary, we have:

PROPOSITION 4. Let (Y,Z) be a nice manifold pair. Then, if any two of the maps f'_Y, f_Y and f_Z in diagram (†) of Proposition 1

$$\begin{array}{ccccc}
\overline{B}\ \mathrm{Emb}_0^g(Y,\mathrm{rel}\ Z \cup A) & \longrightarrow & \overline{B}\ \mathrm{Emb}_0^g(Y,Z,\mathrm{rel}\ A) & \xrightarrow{\overline{\rho}} & \overline{B}\ \mathrm{Emb}_0^g(Z,\mathrm{rel}\ A) \\
\downarrow{\scriptstyle f'_Y} & & \downarrow{\scriptstyle f_Y} & & \downarrow{\scriptstyle f_Z} \\
S(Y,\mathrm{rel}\ Z \cup A) & \longrightarrow & S(Y,\mathrm{rel}\ A) & \xrightarrow{r} & S(Z,\mathrm{rel}\ A)
\end{array}$$

induce isomorphisms on H_*, the third does too.

Proof. This follows by comparing the homology spectral sequences for the Hurewicz fibrations associated to $\bar{\rho}$ and r, and using Lemma 3 and Proposition 3.

COROLLARY. Suppose that (Y,A_1) is a nice manifold pair. Then, if Theorem C holds for the pairs (Y,A_1), (Y,A_2) and $(Y,A_1 \cap A_2)$, it holds also for $(Y,A_1 \cup A_2)$. Similarly, if Theorem C holds for (Y,A_1), (Y,A_2) and $(Y,A_1 \cup A_2)$, it holds for $(Y,A_1 \cap A_2)$.

Proof. Let us compare the diagram:

$$\begin{array}{ccccc}
\bar{B}\,\mathrm{Emb}_0^g(Y,\mathrm{rel}\ A_1 \cup A_2) & \to & \bar{B}\,\mathrm{Emb}_0^g(Y,A_1,\mathrm{rel}\ A_2) & \longrightarrow & \bar{B}\,\mathrm{Emb}_0^g(A_1,\mathrm{rel}\ A_2) \\
\downarrow & & \downarrow & & \downarrow \\
\bar{B}\,\mathrm{Emb}_0^g(Y,\mathrm{rel}\ A_1) & \longrightarrow & \bar{B}\,\mathrm{Emb}_0^g(Y,A_1,\mathrm{rel}\ A_1 \cap A_2) & \to & \bar{B}\,\mathrm{Emb}_0^g(A_1,\mathrm{rel}\ A_1 \cap A_2)
\end{array}$$

with the corresponding diagram for S. To prove the first statement, notice that our hypotheses together with Proposition 2 tell us that in the bottom row f'_Y and f_Y induce isomorphisms on H_*. Therefore $f_Z : \bar{B}\,\mathrm{Emb}_0^g(A_1,\mathrm{rel}\ A_1 \cap A_2) \to S(A_1,\mathrm{rel}\ A_1 \cap A_2)$ does too. But, as was remarked at the beginning of Chapt. 2, $\bar{B}\,\mathrm{Emb}_0^g(A_1,\mathrm{rel}\ A_1 \cap A_2) = \bar{B}\,\mathrm{Emb}_0^g(A_1,\mathrm{rel}\ A_2)$. Therefore, in the top row we know that f_Y and f_Z induce isomorphisms on H_*. Hence $f'_Y : \bar{B}\,\mathrm{Emb}_0^g(Y,\mathrm{rel}\ A_1 \cup A_2) \to S(Y,\mathrm{rel}\ A_1 \cup A_2)$ does too. The proof of the second statement is similar.

The next lemma says intuitively that Theorem C holds whenever A is a deformation retract of Y.

LEMMA 4. Let (Y,A) be an admissible pair such that, for all neighborhoods U of $Y \cap A$ in Y there is an element m in $\mathrm{Emb}_0^g(Y,\mathrm{rel}\ A)$ which takes Y into U. Then Theorem C holds for the pair (Y,A).

Proof. Since $\mathrm{Emb}_0^g(Y, \mathrm{rel}\ A)$ consists of embeddings isotopic to the identity it is not difficult to see that both $S(Y, \mathrm{rel}\ A)$ and $\mathrm{Emb}_0(Y, \mathrm{rel}\ A)$ are contractible. Therefore, it suffices to prove that $B\mathrm{Emb}_0^g(Y, \mathrm{rel}\ A)$ is contractible. But the category $C(\mathrm{Emb}_0^g(Y, \mathrm{rel}\ A) \backslash\backslash *)$ is filtering, that is, for all $m_1, m_2 \in \mathrm{Emb}_0^g(Y, \mathrm{rel}\ A)$, there is $m \in \mathrm{Emb}_0^g(Y, \mathrm{rel}\ A)$ such that $m_1 m = m_2 m = m$. (Choose m so that the image $m(Y)$ lies so close to A that both m_1 and m_2 equal the identity on $m(Y)$.) Therefore, $B\ \mathrm{Emb}_0^g(Y, \mathrm{rel}\ A)$ is contractible by [10], p. 84.

LEMMA 5. If Theorem C holds for all nice manifold pairs (Y,A) where $\partial Y \subseteq A$, then it holds for all admissible pairs (Y,A).

Proof. Suppose that (Y,A) is a nice manifold pair in which A is a product near ∂Y. This means that there is a collar neighborhood $\partial Y \times [0,1]$ of $\partial Y \cong Y \times \{0\}$ in Y such that $A \cap (\partial Y \times [0,1]) = (A \cap \partial Y) \times [0,1]$. Put $Y_1 = Y - \partial Y \times [0,1)$ and $Y_2 = \partial Y \times [0,\frac{1}{2}]$ Since the pair $(Y, Y_1 \cup A)$ satisfies the condition in Lemma 4, Theorem C holds for $(Y, Y_1 \cup A)$. By hypothesis, it holds for the nice manifold pairs $(Y, Y_2 \cup A)$ and $(Y, Y_1 \cup Y_2 \cup A)$. Therefore, by the corollary to Proposition 4, it holds for (Y,A) as well. The result now follows because any admissible pair (Y,A) is the limit of a decreasing sequence, (Y, A_n), of nice manifold pairs in which A_n is a product near ∂Y. Therefore, as $H_*(\overline{B}\ \mathrm{Emb}_0^g(Y, \mathrm{rel}\ A_n)) \cong H_*(S(Y, \mathrm{rel}\ A_n))$ for each n, this is true also for $\overline{B}\ \mathrm{Emb}_0^g(Y, \mathrm{rel}\ A) = \lim_{\overrightarrow{n}} \overline{B}\ \mathrm{Emb}_0^g(Y, \mathrm{rel}\ A_n)$ and $S(Y, \mathrm{rel}\ A) = \lim_{\overrightarrow{n}} S(Y, \mathrm{rel}\ A_n)$.

Proof of Theorem C. Segal showed in [14] that, if D^n is the closed n-disc, $f_{D^n} : \overline{B}\ \mathrm{Emb}_0^g D^n \to S(D^n) \simeq \overline{B}\Gamma_n^r$ is a weak homotopy equivalence. It follows by induction on k that Theorem C holds for the pairs (D^n, S^{k-1}), for $k = 0,1,\ldots,n$, where S^{k-1} is embedded in D^n by the standard embedding of $\mathbb{R}^k$ into $\mathbb{R}^n$. (To prove this, thicken S^{k-1} slightly to make (D^n, S^{k-1}) a nice manifold pair, and then use Lemma 4 and the first part of the corollary to Proposition 4.) Similiarly, by using the second part of this corollary and induction over k, one can show that Theorem C holds for the pairs $(S^k \times D^{n-k}, S^k \times \partial D^{n-k})$, for $k = 0,1,\ldots,n$. Finally, given a pair (Y,A), which by Lemma 5 we can assume to be a nice manifold pair with $\partial Y \subseteq A$, one proves the theorem for (Y,A) by downwards induction over a handle decomposition $A + Y_0 \subset Y_1 \subset \ldots \subset Y_m = Y$ for the pair (Y,A). For more details, see [8].

REFERENCES

1. Bott, R., Lectures on characteristic classes and foliations, in Springer Lecture Notes #279 (1972), 1-80.
2. Haefliger, A., Homotopy and integrability, in Springer Lecture Notes #197, (1971), 133-163.
3. Kirby, R and Siebenmann, L., Foundational essays on topological manifolds, Ann. of Math Studies #88, 1977, Princeton U. P.
4. Mather, J. : The vanishing of the homology of certain groups of homeomorphisms, Topology 10 (1971), 297-298.
5. Mather, J., Integrability in codimension 1, Comm. Math. Helv. 48 (1973), 195-233.
6. Mather, J., Commutators of diffeomorphisms, I, II, Comm. Math. Helv. 49 (1974), 512-528 and 50 (1975), 33-40.
7. Mather, J., On the homology of Haefliger's classifying space, Summer School at Varenna, 1976.
8. McDuff, D., On the homology of some groups of diffeomorphisms, to appear.
9. Palais, R., Homotopy theory of infinite dimensional manifolds, Topology 5 (1966), 1-16.
10. Quillen, D., Higher algebraic K-theory, in Springer Lecture Notes #341 (1972), 75-148.
11. Segal, G. B., Classifying spaces and spectral sequences, Publ. IHES 34 (1968), 105-112.
12. Segal, G. B., Cateogries and cohomology theories, Topology 13 (1974), 293-312.
13. Segal, G. B., Configuration spaces and iterated loop spaces, Invent. Math. (1973), 213-221.
14. Segal, G. B., The classifying space for foliations, preprint c. 1974.
15. Thurston, W., Foliations and groups of diffeomorphisms, B.A.M.S. 80 (1974), 304-307.

ON THE TOPOLOGY OF NON-ISOLATED SINGULARITIES

Richard Randell[1]

School of Mathematics
Institute for Advanced Study
Princeton, N.J.

It is well-known that isolated singularities of complex analytic (or algebraic) varieties possess interesting topological properties. Here, after a brief survey of the isolated case, we present some basic facts about non-isolated singularities. Our main result is a method for computing the homology and connectivity of certain Stein manifolds which for weighted homogeneous singularities are diffeomorphic to the fiber of the Milnor fibration. These computations are closely related to similar results for singular projective varieties.

I. INTRODUCTION

Suppose we are given two sets $\{f_i\}$, $i \in I$, and $\{g_j\}$, $j \in J$ of holomorphic functions (or complex polynomials) defined on some neighborhood U of $\underline{0}$ in $\mathbb{C}^{n+1}$. A basic problem is: Describe the topology of the pair $(V_I, V_{I \cup J})$, where $V_I = \{\underline{z} \in U | f_i(\underline{z}) = 0, i \in I\}$, $V_{I \cup J} = \{\underline{z} \in V_I | g_j(\underline{z}) = 0, j \in J\}$. That is, describe the local topology of pairs of varieties over $\mathbb{C}$. Since the relevant rings are Noetherian we may assume I and J are actually finite sets. In this paper we will make the further assumption that $I = \emptyset$ and J has one element, so that we consider points on

[1]*Partially supported by the National Science Foundation.*

ISBN 0-12-158860-2

hypersurfaces $V(f) = f^{-1}(0)$ in $\mathbb{C}^{n+1}$ This assumption is naturally somewhat restrictive; the most general situation for which similar techniques work seems to be $V_{I \cup J}$ a complete intersection with respect to V_I [6], [9]. That is, J consists of k elements, and $\dim_{\mathbb{C}} V_I = \dim_{\mathbb{C}} V_{I \cup J} + k$. In any case our situation, the study of $(U, U \cap V(f))$, illustrates much of the general behavior.

In the first section we state some basic results, and provide a quick survey of what is known when the singularity is isolated. This is useful for comparison with the subsequent sections. Next we prove a number of basic results for the non-isolated case. Finally, in the last sections we show how to compute the connectivity and homology of the Milnor fiber for a large class of examples, and we relate these computations to global properties (i.e., homology) of singular projective varieties.

II. A REVIEW OF BASIC RESULTS

Let $f : (\mathbb{C}^{n+1}, \underline{0}) \to (\mathbb{C}, 0)$ be a germ of an analytic function. That is, f is an equivalence class of functions defined on neighborhoods of 0 in $\mathbb{C}^{n+1}$, two such functions being equivalent if they agree on some neighborhood of $\underline{0}$, and $f(\underline{0}) = 0$. The singular set S of f (again, a germ in the appropriate sense) is the set $S = \{\underline{z} \in \mathbb{C}^{n+1} \mid (\text{grad } f)(\underline{0}) = \underline{0}\}$. Henceforth we will drop the language of germs; the reader should remember that we always work in some neighborhood of $\underline{0}$. Let $\dim_{\mathbb{C}} S = s$, and assume $\underline{0} \in S$. There is the following basic result, proved by Milnor [18, Theorem 2.10] with $s = 0$ and Burghelea and Verona [4], $s > 0$.

CONE THEOREM. Let S_ε^{2n+1} be the $(2n+1)$-sphere of radius ε centered at $\underline{0}$ in $\mathbb{C}^{n+1}$, and let $K_\varepsilon = V(f) \cap S_\varepsilon^{2n+1}$. Then for all $\varepsilon > 0$ sufficiently small, $(D_\varepsilon^{2n+2}, V(f))$ is homeomorphic to $C(S_\varepsilon^{2n+1}, K_\varepsilon)$, the cone over $(S_\varepsilon^{2n+1}, K_\varepsilon)$ with cone point $\underline{0}$.

Thus one studies K and its embedding in the sphere. Actually, one may include S in the cone theorem, using a proof similar to [4], to get a homeomorphism of $(D_\varepsilon^{2n+2}, V(f), S)$ with $C(S_\varepsilon^{2n+1}, K_\varepsilon, S_\varepsilon^{2n+1} \cap S)$.

A second fundamental result is the following.

MILNOR FIBRATION THEOREM [18, Theorem 4.8]. *For ε sufficiently small, $f/|f| : S_\varepsilon^{2n+1} - K_\varepsilon \to S^1$ is the projection of a smooth fiber bundle with fiber F. Furthermore,*

a) F is an open parallelizable manifold of real dimension $2n$.

b) F has the homotopy type of an n-dimensional CW complex.

c) K has real dimension $2n-1$, and $\pi_i(K) \cong 0$, $i \leq n-2$.

d) F $(= \{\underline{z} \in S_\varepsilon^{2n+1} | f(\underline{z})$ real and $f(\underline{z}) > 0\})$ is diffeomorphic to $F' = \{\underline{z} \in \mathbb{C}^{n+1} | f(z) = \delta, \ |\underline{z}| < \varepsilon$, for $0 < |\delta| << \varepsilon\}$, and has the homotopy type of the compact manifold - with-boundary F_c = Closure F' in $\mathbb{C}^{n+1}$

For the rest of this section let us assume that $\underline{0}$ is an *isolated* singularity of V, i.e. $s = 0$. Then one may add the following to the conclusions of the fibration theorem.

e) $\partial F_c = F_c - F' = \{\underline{z} \in \mathbb{C}^{n+1} | f(\underline{z}) = \delta, \ |\underline{z}| = \varepsilon\}$ is a closed manifold diffeomorphic to K. Thus,

f) K is a stably parallelizable manifold.

g) F (or F' or F_c) has the homotopy type of a bouquet of μ n-spheres.

The number μ, called the Milnor number, pops up however one considers the situation. For instance, Milnor showed [18, Theorem 7.2] $\mu = \deg(\text{grad } f/||\text{grad } f||) : S_\varepsilon^{2n+1} \to S_1^{2n+1}$ and Palamodov [23] (following a suggestion in [18]) showed $\mu = \dim_{\mathbb{C}} \mathbb{C}[[z_0, \ldots, z_n]]/(f_0, \ldots, f_n)$. (Here $(f_0, \ldots, f_n)$ denotes the Jacobian ideal J

generated by the partials in the formal power series ring. That the quotient is a finite-dimensional vector space over $\mathbb{C}$ is the content of the Nullstellensatz: Rad $\mathcal{J}$ = the maximal ideal.]

Sometimes one can determine μ directly, as in the following examples. *(i)* $f(x, y) = x^3 - xy^4$.

Here one may find a deformation retraction of F to the one-complex Ω_1:

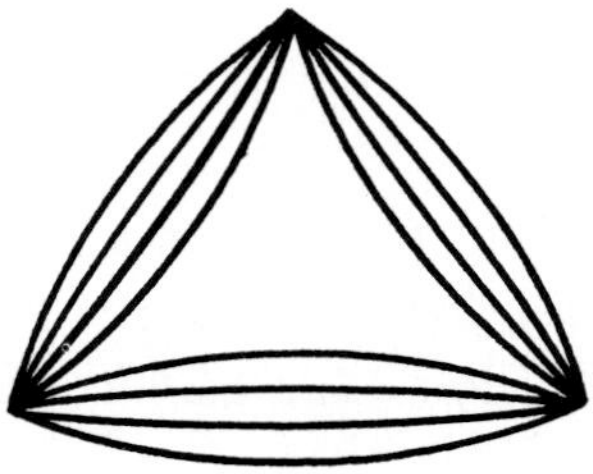

Also $f(x, y) = x(x^2 - y^4)$ implies that K is the 3-component link of $x = 0$, $x + y^2 = 0$, $x - y^2 = 0$ in $S^3 \subset \mathbb{C}^2$:

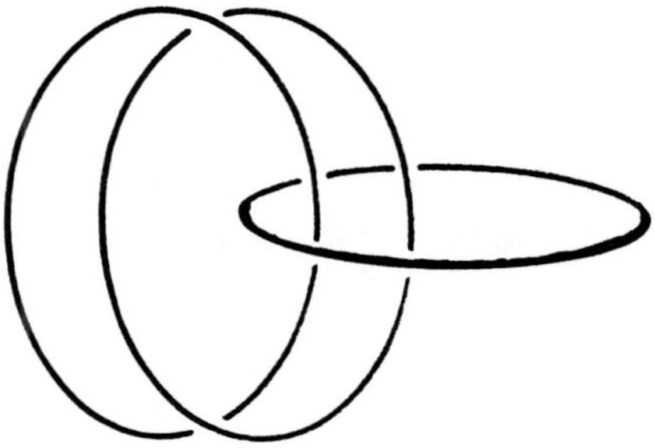

But in the example *(ii)*, $g(x, y, z) = x^3 - xy^4 + yz^2$ it is difficult to find an explicit retraction to a 2-complex Ω_2, even though one knows that Ω_2 should have the homotopy type of a complex formed from Ω_1 by attaching 24 $(= 3 \cdot 4 \cdot 2)$ 2-cells in a symmetric manner.

In general one cannot deal with F so explicitly, of course, and so other methods are sought.

Now the fibration theorem shows that in the case of an isolated singularity, K is a simple fibered knot. ("Simple" means K and F are highly-connected.) Then Durfee [5] and Kato [11], using results of Levine [16], show how such knots are classified:

THEOREM ([5], [11]). *Let $n \geq 3$. There is a one-to-one correspondence of smooth isotopy classes of simple fibered knots in S^{2n+1} and equivalence classes of unimodular bilinear forms. The correspondence associates to each knot its Seifert form.*

The Seifert form L is bilinear on the free abelian group $H_n(F; \mathbb{Z})$ of rank μ. One represents classes a and b by embedded spheres in the fiber over $1 \in S^1$, pushes a into a nearby fiber, and computes linking in S^{2n+1}. See [5] for details.

There are two other forms on $H_n(F; \mathbb{Z})$ associated to an isolated singularity. First, there is the $(-1)^n$-symmetric bilinear intersection form S. Since F is not closed, S may fail to be unimodular. Second, there is the linear monodromy H, defined as follows. The total space $S^{2n+1} - K$ of the bundle over S^1 may be formed by identifying the ends of $F \times [0, 2\pi]$ via a diffeomorphism h. Let H be the matrix of $h_n : H_n(F; \mathbb{Z}) \to H_n(F; \mathbb{Z})$. Then in [5], [11] it is shown that (fixing a basis of $H_n(F; \mathbb{Z})$) we have $S = L + (-1)^n L^t$ and $H = (-1)^{n+1} L^{-1} L^t$. Durfee then showed [5] that for certain bases ("distinguished") L is upper triangular with $(-1)^{n(n+1)/2}$ on the diagonal. In this case one may solve uniquely for L if one knows S or H.

Thus one may determine if two isolated singularities give the same topology by computing and comparing. Much work has been done on such computations. (See [22] for a recent bibliography.)

But suppose, as is usually the case, that explicit computations can't be performed. The following result tells how to compare singularities without computing, and again emphasizes the importance of μ.

Definition. Suppose we are given holomorphic germs $f, g : (\mathbb{C}^{n+1}, 0) \to (\mathbb{C}, 0)$. We say that f and g are μ-homotopic ($f \underset{\mu}{\sim} g$) if there is a holomorphic germ $F : (\mathbb{C}^{n+1} \times \mathbb{C}, (\underline{0}, 0)) \to (\mathbb{C}, 0)$ so that $F(\underline{0}, t) = 0$, $F_0 = f$, $F_1 = g$ (with $F_t(\underline{z}) = F(\underline{z}, t)$), and every F_t, $t \in [0, 1]$ has an isolated singularity with $\mu(F_t) = \mu(f)$.

Examples. (i) $x^2 + y^9$ and $x^3 + y^5$ have $\mu = 8$, but they are not μ-homotopic. One sees this by computing the respective intersection forms and applying the corollary below. The natural candidate for F, $(1 - t)(x^2 + y^9) + t(x^3 + y^5)$, has $\mu(F_0) = \mu(F_1) = 8$, but $\mu(f_t) = 4$, $t \in (0, 1)$.

(ii) The germs $x^3 + xy^2$ and $x^3 + y^3$ are μ-homotopic, but $x^2 + y^5$ is μ-homotopic to neither. Each germ has $\mu = 4$.

THEOREM (Lê and Ramanujam [15], Timourian [29]). *Suppose $n \neq 2$ and $f \underset{\mu}{\sim} g$. Then f and g have the same topological type.*

Remarks. (i) To say f and g have the same topological type means that there is a diffeomorphism $\varphi : (S_\varepsilon^{2n+1}, K(f)) \to (S_\varepsilon^{2n+1}, K(g))$ so that $f(\underline{z})/|f(\underline{z})| = g\varphi(\underline{z})/|g\varphi(\underline{z})|$.

(ii) H. King [13] has shown that if f and g have the same topological type, then there is a germ of a homeomorphism $\psi : (\mathbb{C}^{n+1}, \underline{0}) \to (\mathbb{C}^{n+1}, 0)$ so that $g = f\psi$.

(iii) The condition $n \neq 2$ arises because the h-cobordism theorem in real dimension $2n$ is used in the proof. One suspects the result holds for $n = 2$ as well. In any case, the singularities are homotopically alike when $n = 2$.

(iv) However, there is an example of Briançon and Speder [2] $F(x, y, z, t) = x^5 + txy^6 + z^{15} + zy^7$, of a 1-parameter family so that $(\mathbb{C}^4, V(F))$ admits no Whitney stratification in which projection onto t is a submersion on each stratum. The existence of such a stratification would give the conclusion of the above theorem, so that the use of h-cobordism is perhaps essential.

(v) If we are interested only in comparing L, S, or H there is the following corollary.

COROLLARY. If f and g are μ-homotopic, then L, S, and H are equivalent.

The hypothesis $n \neq 2$ is removed by noting that adding a nondegenerate quadratic Q in four new variables to f does not change these forms (and $f + Q \underset{\mu}{\sim} g + Q$).

In the next sections we will compare the above results with the non-isolated case. We will see in section V that isolated singularities occur naturally in the study of non-isolated ones.

III. BASIC RESULTS FOR NON-ISOLATED SINGULARITIES

In this section we state a basic result of Kato and Matsumoto and derive some straightforward consequences of standard exact sequences. We no longer assume s $(= \dim_{\mathbb{C}} S)$ is zero.

Recall that we do not have properties *(e)* - *(g)* of the fibration theorem. We do know that F has the homotopy type of the manifold with boundary F_c, and F and F_c have the homotopy of a CW complex of dimension n. However, F may not be $(n-1)$-connected.

THEOREM. (Kato and Matsumoto [12]). *If $\dim_{\mathbb{C}} S = s$, then F is $(n-s-1)$-connected.*

In general, this is the best one can say. For instance, $f(z_0, \ldots, z_{s+1}) = z_0 \ldots z_{s+1}$ yields $F' \sim \{z_0 \ldots z_{s+1} = 1\} \sim S^1 \times \ldots \times S^1$ ($(s+1)$ times). By adding a non-degenerate quadratic in $n-s-1$ variables, one obtains a germ so that $\dim_{\mathbb{C}} S = s$ and F is not $(n-s)$-connected.

There are now several proofs of this result. In particular, we will later see how to actually improve the result in some cases, thus giving a partial answer to Kato and Matsumoto's question: For what germs is F $(n-1)$-connected?

For the rest of this section we assume that F is $(n-r-1)$-connected, where by the above we may take $r \leq s$. The results below show that we should focus our attention on F; this is done in the following sections.

Definition. For all $\eta > 0$ sufficiently small we define $T_\eta(K)$, *the* η*-tube around* K, by $T_\eta(K) = \{\underline{z} \in S_\varepsilon^{2n+1} \mid |f(\underline{z})| < \eta\}$.

Since f has 0 as critical value and since s may be positive so that $S \cap S_\varepsilon^{2n+1} \neq \emptyset$, we cannot immediately conclude that $T_\eta(K)$ is a disc bundle over K or a regular neighborhood of K.

PROPOSITION (3.1). For η *sufficiently small,* $i : K \to T_\eta(K)$ *is a homotopy equivalence.*

Proof. By an argument analogous to [17, 2.8], $0 \in \mathbb{C}$ is an isolated crital value of $f|_{S^{2n+1}}$. Thus $f : T_{\eta_0}(K) - K \to D_{\eta_0} - 0$ is a bundle projection for η_0 small, and given $\eta_0 > \eta_1 > \eta_2$ there is a deformation retraction $r_{12} : T_{\eta_1}(K) \to T_{\eta_2}(K)$ such that $r_{12} : T_{\eta_2}(K) = \text{id}$. Fix η_0 and η_1. Then there is an $\omega > 0$ so that any two maps $\varphi_1, \varphi_2 : T_{\eta_1}(K) \to T_{\eta_1}(K)$ which are ω-close are homotopic. Now since K is triangulable [17], it is an ANR, so there is a retraction $r_1 : T_{\eta_1}(K) \to K$, which we may assume to be uniformly continuous. Thus there is a $\delta > 0$ so that $d(x, y) < \delta$ implies $d(r_1(x), r_1(y)) < \omega/2$.

Next, there is an $\eta_2 < \eta_1$ so that for any element x of $T_{\eta_2}(K)$, there is a $y_x \in K$ such that $d(x, y_x) < \rho = \min\{\omega/2, \delta\}$. For if not one may construct a sequence $x_n \in T_{\eta_1}(K)$ with a subsequence converging to x_0 such that $f(x_0) = 0$, but $d(x_0, K) \geq \rho > 0$, a contradiction. Fix this η_2, and let $r_2 = r_1|T_{\eta_2}(K)$. Then for $x \in T_{\eta_2}(K)$ we have $d(x, r_2(X)) \leq d(x, y_x) + d(y_x, r_2(x)) < \omega/2 + d(y_x, r_2(x)) = \omega/2 + d(r_2(y_x), r_2(x)) < \omega$, by the uniform continuity of r_2. Then $r_2 \circ r_{12}$ is the required deformation retraction, since it is within ω of r_{12}, which is homotopic to the identity of $T_{\eta_1}(K)$.

Now let $\partial T_\eta = \{z \in S_\varepsilon^{2n+1} \mid |f(z)| = \eta\}$. Then ∂T_η is independent of η for η small, and f is the projection of a bundle $F_\eta \longrightarrow \partial T_\eta \longrightarrow S^1_\eta$. It is easily seen that the fiber F_η is diffeomorphic to ∂F_c. We will henceforth suppress η.

To summarize, we have

$$K \underset{r}{\overset{i_1}{\rightleftarrows}} T(K) \xleftarrow{\;i_2\;} \partial T(K) \xleftarrow{\;i_3\;} \partial F_c$$

where r is the retraction of (3.1). Let $r' : \partial F_c \longrightarrow K$ be the obvious composition, $r' = r \circ i_2 \circ i_3$. Then the objects of interest are (i) K, (ii) $S^{2n+1} - K$ ($\cong S^{2n+1} - T(K)$), (iii) $(F_c, \partial F_c)$, where F_c is an $(n-s-1)$-connected parallelizable $2n$-manifold, (iv) the monodromy $h : F \longrightarrow F$, (v) the intersection form on $H_n(F)$, and (vi) $r' : \partial F_c \longrightarrow K$. (If $s = 0$, r' can be taken as a diffeomorphism but it certainly is not in general; see (3.4.1).)

The next several results are easy consequences of some standard exact sequences. Suppose F is $(n-r-1)$-connected, $r \leq s$.

PROPOSITION (3.2). $i_* : H_q(\partial F_c) \to H_q(F_c)$ *is an isomorphism,* $q < n-1$, *and an epimorphism if* $q = n-1$.

Proof. We have $H_q(F_c, \partial F_c) \cong H^{2n-q}(F_c)$, which is 0 if $q < n$, so the result follows from the exact homology sequence of $(F_c, \partial F_c)$.

PROPOSITION (3.3). *Let* $r' = \max\{1, r\}$ *and suppose* $s \leq n-3$. *Then* ∂F_c *is* $(n-r'-1)$*-connected.*

Proof. This follows from (3.2), since F_c is $(n-r-1)$-connected, once we observe that the proof ([12] for instance) that F_c is $(n-s-1)$-connected can be generalized to show ∂F_c is $(n-s-2)$-connected, hence simply-connected.

Since ∂F_c is a $(2n-1)$-manifold its homology will then be determined (in terms of $H_*(F_c)$) once we know how to compute $H_{n-1}(\partial F_c)$. Let S denote the intersection form on $H_n(F_c)$. The following is then a direct generalization of Milnor [18, Lemma 8.3].

PROPOSITION (3.4). *There is a short exact sequence*

$$0 \longrightarrow \text{coker } S \longrightarrow H_{n-1}(\partial F_c) \longrightarrow H_{n-1}(F_c) \longrightarrow 0 .$$

(3.4.1) Since ∂F_c may have homology below dimension $n-1$, while K does not, we see that the map $r' : \partial F_c \to K$ must have interesting point inverses.

PROPOSITION (3.5). $H^q(K) \cong 0$, $q > n + r$.

Proof. The exact Wang sequence of the Milnor fibration is

$$\to H_{q+1}(S^{2n+1} - K) \to H_q(F) \xrightarrow{h_q - I} H_q(F) \to H_q(S^{2n+1} - K) \to .$$

Since $H_q(F) \cong 0$, $q \leq n-r-1$, the above sequence and Alexander duality imply $H_q(S^{2n+1} - K) \cong H^{2n-q}(K) \cong 0$, $q \leq n-r-1$, i.e. $2n - q \geq 2n - (n-r-1) = n + r + 1$.

Since K is $(n-2)$-connected, we see that failure of Poincaré duality for K is measured by the homomorphisms $h_q - I$. In particular we have, as a generalization of [18, Theorem 8.5] ,

PROPOSITION (3.6). For $n \neq 2$ ($n = 2$) the simplicial complex K is homotopy (homology) equivalent to S^{2n-1} if and only if $h_q - I : H_q(F) \to H_q(F)$ is an isomorphism for $n - r \leq q \leq n$.

Finally, if f has an isoloated singularity it is easy to show that the map $h : F_c \to F_c$ giving the monodromy may be taken as the identity on ∂F_c. (Sketch: $f|T_\eta(K)$ is actually a projecion of a bundle over D_η with fiber K. Thus $\partial T_\eta(K) = K \times S^1$.) This does not hold in the non-isolated case. First, we have

PROPOSITION (3.7). If $q \leq n - 2$, then $h_q - I$ and $h_q|_{\partial F_c} - I$ have isomorphic kernel and cokernel.

Proof. We have a commuting diagram with exact rows:

$$\begin{array}{ccccccc}
\to H_{q+1}(F_c, \partial F_c) & \to & H_q(\partial F_c) & \to & H_q(F_c) & \to & H_q(F_c, \partial F_c) \to \\
 & & \downarrow{\scriptstyle h_q|\partial F_c - I} & & \downarrow{\scriptstyle h_q - I} & & \\
\to H_{q+1}(F_c, \partial F_c) & \to & H_q(\partial F_c) & \to & H_q(F_c) & \to & H_q(F_c, \partial F_c) \to
\end{array}$$

Since $H_{q+1}(F_c, \partial F_c) \cong H_q(F_c, \partial F_c) \cong 0$ (by duality) if $q \leq n - 2$, the proof is complete.

Now consider the simple example $f = z_0 z_1 z_2 z_3 + z_4^3$. By [19] it is seen that f_c has the homotopy type of the join of $\{z_0 z_1 z_2 z_3 = 1\}$

and $\{z_4^3 = 1\}$. Thus $F_c \sim (S^1 \times S^1 \times S^1) * (3 \text{ points})$. So $H_2(F) \simeq H_1(S^1 \times S^1 \times S^1) \otimes \tilde{H}_0(3 \text{ pts}) \simeq (\mathbb{Z} \oplus \mathbb{Z} \oplus \mathbb{Z}) \otimes (\mathbb{Z} \oplus \mathbb{Z})$. Also by [19], the monodromy $h_2 : H_2(F_c) \to H_2(F_c)$ has matrix

$$\begin{pmatrix} 1 & 0 & 0 \\ 0 & 1 & 0 \\ 0 & 0 & 1 \end{pmatrix} \otimes \begin{pmatrix} 0 & -1 \\ 1 & -1 \end{pmatrix}, \quad \text{so that } h_2 \neq \text{Id, and thus } h_2|\partial F_c \neq \text{Id.}$$

IV. $\chi(F)$ AND $\pi_1(F)$

The preceding gives ample evidence that F and the monodromy h govern the topology of the situation. In this and the following section we will study the algebraic invariants of F.

First, let us briefly discuss two ways of determining $\chi(F)$. First there is a beautiful formula of A. N. Varčenko [30]. We will omit the details of this, and instead briefly give the flavor of the result. Given f, one marks an integer lattice point in $\mathbb{R}^n$ for each monomial. From these points one constructs the Newton polyhedron. Then

$$(-1)^n \chi(F) = n!V_n - (n-1)!V_{n-1} + \dots + (-1)^{n-1} V_1 ,$$

where V_j is the sum of the j-dimensional volumes of the intersection of this Newton polyhedron with j-dimensional coordinate planes.

This formula works for almost all f, in the following sense: For a given set of monomials, the formula holds for a Zariski open set of the possible coefficients. This open set is explicitly defined in [30].

In fact, Varčenko proves more, in that he gives a formula of similar type for the zeta function

$$\zeta(t) = \prod_{q \geq 0} \text{Det}(I - th_q)^{(-1)^{q+1}} .$$

These formulas are easily computable for small n.

Here is a second method ("Polar curves"), suggested by Thom and developed by Lê [14] and Teissier [27]. Take a generic hyperplane H of $\mathbb{C}^{n+1}$ and consider the inclusion $H \cap F \to F$. In [14] Lě shows that up to homotopy F is formed by adding r_n n-cells to $H \cap F$. The number r_n is independent of the generic H chosen, and may be computed as the intersection multiplicity at $\underline{0}$ of $\{f = 0\}$ and a certain curve Γ (the polar curve). Thus we have a canonical sequence of integers $(r_0, \ldots, r_n) = r^*$ attached to F (or f), and $\chi(F) = \sum(-1)^i r_i$. This technique works for any f, but it may only occasionally be done explicitly. See [22] for one example.

Teissier [27] has shown that r^* is important in the deformation theory of isolated singularities. One might suspect that r^*-constant might be a substitute for the hypothesis μ-constant in a non-isolated version of the Lê-Ramanujam theorem mentioned earlier, since Teissier showed [27] that r^*-constant implied Whitney conditions for isolated singularities. In fact, the following simple example suggests that the situation is more complicated.

Let $f_0 = w^5 + x^5 + xy^4 + y^3z^3$ and $f_1 = w^5 + x^5 + xy^4 + y^2z^3 + y^3z^3$. Both germs have $\pi_1(F) \cong 0 \cong H_2(F)$ (the latter using (5.14)), and $H_3(F) \cong \mathbb{Z}^{176}$. Furthermore the singularity set $S = \{w = x = y = 0\}$ in both cases. Consider the one-parameter family $f_\rho = w^5 + x^5 + xy^4 + \rho y^2z^3 + y^3z^3$. For any ρ, $r^*_\rho = (5, 20, 80, 240)$. However, f_0 and f_1 have different characteristic polynomials:

$$\text{Det}\,(tI - h_3) = (t^{15} - 1)^{16}(t^5 - 1)^{-13}(t - 1) \quad \text{for } f_0 ,$$

$$\text{Det}\,(tI - h_3) = (t^5 - 1)^{35}(t - 1) \quad \text{for } f_1 .$$

These characteristic polynomials may be computed via Varčenko's formula for $\zeta(t)$, or by polar curves, using the fact that there is a coordinate change making f weighted homogeneous (for each ρ separately), and then applying [18, Chap. 9] to get h explicitly.

The difficulty here may be looked at in (at least) two different ways. First, there is no stratification of $V(f_\rho)$ which has all strata transverse to the generic hyperplanes chosen, and which satisfies an important technical property ("bonne" in the sense of [7]). Or, the behavior transverse to the singular set changes at $\rho = 0$. To see this, fix $z = 1$. Then at $(w, x, y, 1, \rho)$ we have a "transverse isolated singularity"

$$f_\rho^t : w^5 + x^5 + xy^4 + \rho y^2 + y^3 .$$

With $\rho = 0$, we get $w^5 + x^5 + xy^4 + y^3$, while with $\rho \neq 0$ we may make a coordinate change to see that f_ρ^t behaves like $w^5 + x^5 + xy^4 + y^2$. These have distinct topological qualities.

We will see transverse singularities again in (5.10).

Next we turn to $\pi_1(F)$. Of course, the Kato-Matsumoto result tells us that $\pi_1(F) = 0$ unless $s = n - 1$. In this case one computes $\pi_1(F)$ as follows.

1) Take a suitable generic three-dimensional subspace H in $\mathbb{C}^{n+1}$, and consider $f|H$. In fact by a simple consequence of [7, Théorème 0.2.1], $i_* : F \cap H \to F$ induces an isomorphism on fundamental groups. So we need only consider $f \cdot (\mathbb{C}^3, 0) \to (\mathbb{C}, 0)$ with one-dimensional singularity set.

2) Take a suitable projection of F to $\mathbb{C}$ and use Van Kampen's theorem. In fact, this is analogous to what was done many years ago (see [31]) to compute $\pi_1(\mathbb{P}^2 - C)$, where C is a (complex) curve. Van Kampen developed a version of the theorem now bearing his name to attack this problem. The point is that such a projection is almost a bundle map, but one must patch together the behavior around the exceptional fibers.

Often for a given germ f or curve C the above process can be carried out explicitly. Let us now, however, see some difficulties with this in general, and at the same time see a connection of the local situation we have been studying with global projective varieties.

Let $f(x, y, z)$ be a homogeneous polynomial of degree d $(f(tx, ty, tz) = t^d f(x, y, z))$. Then $V = f^{-1}(0)$ [1] $\subset \mathbb{C}^3$ is invariant under the $\mathbb{C}^* = \mathbb{C} - \{0\}$ action on $\mathbb{C}^3$ given by $t(x, y, z) = (tx, ty, tz)$, and $C = (V - \{0\})/\mathbb{C}^*$ is a curve in $\mathbb{P}^2 = (\mathbb{C}^3 - \{0\})/\mathbb{C}^*$. Furthermore, we may take F' for f as $F' = \{f(x, y, z) = 1\}$ in $\mathbb{C}^3$, and it is not hard to see that F' is a d-fold cover of $\mathbb{P}^2 - C$, via $(x, y, z) \mapsto (x : y : z) \in \mathbb{P}^2$. Using this, Oka [21] showed the following.

THEOREM. Let $f(x, y, z)$ be a homogeneous polynomial with Milnor fiber F and associated projective curve C. Suppose C is irreducible. Then $\pi_1(F) = 0$ if and only if $\pi_1(\mathbb{P}^2 - C)$ is abelian.

This relates to the

ZARISKI CONJECTURE. Suppose $C \subset \mathbb{P}^2$ has only double points as singularities, i.e. there are local coordinates (x, y) in $\mathbb{P}^2$ so that $C = \{xy = 0\}$ locally. Then $\pi_1(\mathbb{P}^2 - C)$ is abelian.

This was "proved" by Zariski (see [31, p.210], using a lemma of Severi which had an error in the proof. The problem is still open. The relevance of this to our discussion is that it shows that even for homogeneous F the problem of studying $\pi_1(F)$ is quite difficult.

Example. Let $f(x, y, z) = x^2y^2 + y^2z^2 + z^2x^2$. Then C has genus zero and has three ordinary double points as singularities. These points are $(0 : 0 : 1)$, $(0 : 1 : 0)$, and $(1 : 0 : 0)$ in $\mathbb{P}^2$ corresponding to the singular set S of $f^{-1}(0) \subset \mathbb{C}^3$, $S = \{x = y = 0\} \cup \{x = z = 0\} \cup \{y = z = 0\}$. Consider $(0 : 0 : 1)$. In $\mathbb{P}^2$, $z \neq 0$ gives the coordinate patch $(x : y : 1) = \mathbb{C}^2$. At the origin,

[1]*Since f is homogeneous, we take the entire set in $\mathbb{C}^3$, not just the germ.*

i.e. $(0 : 0 : 1)$, in this patch, C looks like $\{x^2y^2 + x^2 + y^2 = 0\}$. There are coordinate changes so that C looks like $\{x^2 + y^2 = 0\}$, and then $\{xy = 0\}$. Now one may compute directly that $\pi_1(F) \cong 0$ and $\pi_1(\mathbb{P}^2 - C) \cong \mathbb{Z}_4$. Also, $\chi(F) = 16$, so that $H_2(F) \cong \mathbb{Z}^{15}$.

V. THE HOMOLOGY OF F

In this section we give a technique which allows one to compute and prove theorems about the homology of F. We give a number of examples at the end. Similar techniques are used by Oka [21] in his work on the Zariski conjecture.

Actually, what we will show is how to determine the homology of sets of the form $A_\delta = \{\underline{z} \in \mathbb{C}^{n+1} | f(\underline{z}) = \delta\}$, where δ is a regular value of the polynomial f. When f is weighted homogeneous (below) this set is actually diffeomorphic to the Milnor fiber [18, Lemma 9.4]. In any case A_δ is a Stein manifold, and so its cohomology (over $\mathbb{C}$) is given by a holomorphic De Rham theorem. Thus our results say something about holomorphic forms on A_δ. This and the fact that our homology calculations relate to the homology of singular projective varieties add interest to the results.

(5.1) Before beginning this discussion, let us note several other ways to study F or $H_*(F)$ for non-isolated singularities.

(5.1.1) One may deformation retract F to some space with recognizable cohomology. This of course depends on the form of f and is generally difficult. For example, try $\{x^py^p + y^pz^p + z^px^p = 1\}$ or $\{x^2y + y^3z + z^4x = 1\}$ as exercises.

(5.1.2) Break f down into $g + h$, with g and h polynomials in disjoint sets of variables. Then $F(f)$ has the homotopy type of the join of $F(g)$ and $F(h)$, [26], [19], [25].

(5.1.3) One may present F as a set related to a resolution of $\{f = 0\} \subset \mathbb{C}^{n+1}$. This technique has been fruitful in computations with zeta functions or complex coefficients [1], [30] and may yet yield interesting results over $\mathbb{Z}$.

(5.1.4) Map F to another space (typically D^2 or D^{2n}) and analyze the spectral sequence of the map. For instance, one may always find a projection $F \to D^2$ so that in the spectral sequence one has $E_2^{p,q} \cong 0$ unless $(p, q) = (0, 0)$ or $(1, k)$, $k \leq n - 1$. This is one way of showing $H_2(F) \cong 0$ for $w^5 + x^5 + xy^4 + y^3z^3$, for example.

Now recall that F is diffeomorphic to $F' = \{\underline{z} \in \mathbb{C}^{n+1} | f(\underline{z}) = \delta, |\underline{z}| < \varepsilon\}$, for $0 < |\delta| << \varepsilon$. Let $A_\delta = f^{-1}(\delta) \subset \mathbb{C}^{n+1}$ as before. Then $F' \subset A_\delta$, and for the class defined below, $F' \cong A_\delta$.

Definition. A polynomial $f(z_0, \ldots, z_n)$ is said to be *weighted* (or quasi-) *homogeneous* of type $(d; q_0, \ldots, q_n)$ if $t^d f(z_0, \ldots, z_n) \equiv f(t^{q_0} z_0, \ldots, t^{q_n} z_n)$ with $d, q_0, \ldots, q_n \in \mathbb{Z}^+$ and g.c.d. $(q_0, \ldots, q_n) = 1$.

The integer d will be called the polynomial degree and the q_i will be called the coordinate degrees. The rational numbers d/q_i are called the weights.

PROPOSITION [18, Lemma 9.4]. *If f is a weighted homogeneous polynomial then $F' = \{\underline{z} \in \mathbb{C}^{n+1} | f(\underline{z}) = \delta, |\underline{z}| < \varepsilon\}$ is diffeomorphic to $A_\delta = \{\underline{z} \in \mathbb{C}^{n+1} | f(\underline{z}) = \delta\}$.*

For an arbitrary polynomial this may not hold. In comparing F' and A one must take into account the possibility that $f^{-1}(0)$ may not be contractible and the possibility that the singular set S may have components not passing through 0.

So we now study the set A_δ. We assume that δ is a regular value of f, so A_δ is a smooth real $2n$-manifold (or, an n-dimensional Stein manifold). By the proposition above, if there is some coordinate change making f weighted homogeneous, then $A_\delta \tilde{=} F' \tilde{=} F$, the Milnor fiber.

There is a standard technique for embedding A_δ in $\mathbb{P}^{n+1}$. Define f_h, the homogenization of $f - \delta$, by $f_h(z_0, \ldots, z_{n+1}) = z_{n+1}^d(f(z_0/z_{n+1}, \ldots, z_n/z_{n+1}) - \delta)$, where d is the degree of f. (Here f is not assumed homogeneous.) Then f_h is homogeneous of degree d, and hence defines a hypersurface B in $\mathbb{P}^{n+1}$. In $\mathbb{P}^{n+1}$ we have coordinates $(z_0 : z_1 : \ldots : z_{n+1})$, where $(z_0 : z_1 : \ldots : z_{n+1}) \sim (tz_0 : tz_1 : \ldots : tz_{n+1})$, $t \in \mathbb{C}^*$. In the set where $z_{n+1} \neq 0$, we may express each point as $(z_0 : \ldots : z_n : 1)$, and thus this set is a coordinate patch, say $\mathbb{C}_0^{n+1}$. Clearly, $\mathbb{C}_0^{n+1} \cap B = A_\delta$.

Furthermore, if we set $z_{n+1} = 0$ we obtain a projective hyperplane $\mathbb{P}_0^n \subset \mathbb{P}^{n+1}$, and $\mathbb{P}_0^n \cap B = C$, where C is the projective hypersurface of $\mathbb{P}_0^n$ defined by the sum of those monomials of f of degree d. Thus if f is already homogeneous, C is just the projective hypersurface it defines.

Finally, we observe that B is non-singular on $\mathbb{P}^{n+1} - \mathbb{P}_0^n = \mathbb{C}_0^{n+1}$ (since $B \cap \mathbb{C}_0^{n+1} \tilde{=} A_\delta$). Let σ_B and σ_C denote the complex dimension of the singular sets S_B in B and S_C in C, respectively. If f is homogeneous $S_B = S_C$, and $\sigma_B = \sigma_C = s - 1$. We refer the reader to the examples of Section VI.

We now give a series of lemmas which will show that the lower-dimensional homology groups of F depend on the homology of the boundary of regular neighborhoods of S_B and S_C

Definition. Let X be a subset of $\mathbb{P}^m$, for some m. Then the *projectively reduced homology and cohomology* groups of X, $rH_*(X)$ and $rH^*(X)$, are defined by

$$rH_*(X) = \ker i_* : H_*(X) \to H_*(\mathbb{P}^m)$$
$$rH^*(X) = \operatorname{coker} i^* : H^*(\mathbb{P}^m) \to H^*(X).$$

Here any coefficient ring may be used. (When we wish to specify coefficients, we mention them explicitly.)

LEMMA (5.2). There is an integral exact sequence $(j \geq 1)$

$$\longrightarrow rH^{n+j-1}(C) \xrightarrow{\ \delta\ } \tilde{H}_{n-j}(A_\delta) \longrightarrow rH^{n+j}(B) \longrightarrow \ .$$

With $\mathbb{Z}$ *coefficients,* $\ker \delta$ *is a finite* $\mathbb{Z}_d$ *module, so that over any ring R in which d is a unit, there are short exact sequences*

$$0 \longrightarrow rH^{n+j-1}(C) \xrightarrow{\ \delta\ } \tilde{H}_{n-j}(A_\delta) \longrightarrow rH^{n+j}(B) \longrightarrow 0.$$

Proof. Since the singular points of B are contained in C, duality gives $H_{n-j}(A_\delta) = H_{n-j}(B - C) \cong H^{n+j}(B, C)$. Thus we consider the long exact sequence

$$\longrightarrow H^{n+j-1}(C) \xrightarrow{\ \delta\ } H^{n+j}(B, C) \longrightarrow H^{n+j}(B) \longrightarrow \ .$$

This remains exact if we use projectively reduced cohomology, except at $j = n$, where replacing $H^{2n}(B, C)$ by $\tilde{H}_0(A_\delta)$ gives exactness.

To obtain the last statement we show that $\delta : rH^{n+j-1}(C) \to H^{n+j}(B, C)$ is almost injective. Consider the commuting diagram

$$\begin{array}{ccccc}
rH^{n+j-1}(C) & \xrightarrow{\ \delta\ } & H^{n+j}(B, C) & & \\
\downarrow \delta_C & & \cong \downarrow D & & \\
H^{n+j}(\mathbb{P}^n, C) & & H_{n-j}(A_\delta) & & \\
\cong \downarrow D_C & & \cup & & \\
H_{n-j}(\mathbb{P}^n - C) & \underset{u_*}{\overset{p_*}{\rightleftarrows}} & \ker(T_* - I) : H_{n-j}(A_\delta) & \longrightarrow & H_{n-j}(A_\delta)
\end{array} ,$$

where δ_C from the sequence of $(\mathbb{P}^n, C)$ is injective by the definition of projective reduction, and D_C and D are duality isomorphisms. The map u_* is the transfer map coming from the d-fold cover $p : A_\delta = B - C \to \mathbb{P}^n - C$, $p((z_0 : \ldots : z_n : 1)) = (z_0 : \ldots : z_n)$. The map $T : A_\delta \to A_\delta$ is just the covering translation. It is a standard fact that in this case the kernel of such a transfer map u_* consists of a direct sum of finite cyclic groups, each of order dividing d.

Example (6.2) shows ker $\delta \neq 0$ in general.

LEMMA (5.3). $rH_q(B^n) \cong 0,\ q \leq n - 1.$

$rH_q(C^{n-1}) \cong 0,\ q \leq n - 2.$

Proof. We prove the statement for B. The proof for C is entirely similar.

Let $K = S_1^{2n+3} \cap f_h^{-1}(0)$. Then there is the obvious S^1-action on K given by multiplication by $t \in S^1$ in each coordinate, and $K/S^1 = B$. The homology Gysin sequences for this action on S^{2n+3} and K yield

$$\begin{array}{ccccccccc}
\longrightarrow & H_q(S^{2n+3}) & \longrightarrow & H_q(\mathbb{P}^{n+1}) & \longrightarrow & H_{q-2}(\mathbb{P}^{n+1}) & \longrightarrow & H_{q-1}(S^{2n+3}) & \longrightarrow \\
 & \uparrow & & \uparrow & & \uparrow & & \uparrow & \\
\longrightarrow & H_q(K) & \longrightarrow & H_q(B) & \longrightarrow & H_{q-2}(B) & \longrightarrow & H_{q-1}(K) & \longrightarrow
\end{array}$$

Since K^{2n+1} is $(n-1)$-connected the maps in the inner square are isomorphisms for $1 < q < n$, and the result follows.

Recall that S_B denotes the singular set of B. Let E_B be a closed regular neighborhood of S_B in $\mathbb{P}^{n+1}$, and let $B_p = B - E_B$ (= "punctured B"). Make similar definitions for C, using a regular neighborhood in $\mathbb{P}^n$. Then we will see that ∂E_B and ∂E_C are crucial.

LEMMA (5.4). Suppose $2\sigma_B < n - j$ (Resp. $2\sigma_C < n - j - 1$). Then $i : H_{n-j}(\partial E_B) \to rH_{n-j}(B_p)$ (Resp. $i : H_{n-j-1}(\partial E_C) \to rH_{n-j-1}(C_p)$) is an isomorphism for $j \geq 2$, and an epimorphism for $j = 1$.

Remark (5.5). The hypothesis $2\sigma_B < n - j$ causes no difficulty in practice, since adding z_{n+2}^d to f_h merely yields a new A'_δ which is of the homotopy type of A_δ joined with d points [19]. Thus $H_{n-j}(A_\delta) \otimes \mathbb{Z}^{d-1} \cong H_{n-j+1}(A'_\delta)$, and we have now increased the right side of the inequality by one, while leaving σ_B and σ_C unchanged. This remark applies to the following results as well, so that by adding enough d-th powers of new variables, these results may be applied to any polynomial.

Proof of (5.4). A Mayer-Vietoris sequence yields

$$\to H_{n-j+1}(B) \to H_{n-j}(\partial E_B) \to H_{n-j}(B_p) \oplus H_{n-j}(E_B) \to H_{n-j}(B) \to \quad .$$

But for $n - j > 2\sigma_B$, we have $H_{n-j}(E_B) \cong 0$, since E_B has the homotopy type of a complex of real dimension $2\sigma_B$. Also, $rH_{n-j}(B) \cong 0$ for $n - j \leq n - 1$, by (5.3). Finally, $H_{n-j+1}(B) \cong \mathbb{Z}$, for $j \geq 2$, $n - j + 1$ even, and $i_* : H_{n-j+1}(B_p) \to H_{n-j+1}(B)$ is onto in this case.

LEMMA (5.6). If $2\sigma_B < n - j$, then $rH^{n+j}(B_p) \simeq 0$, $j \geq 1$. If $2\sigma_C < n-j-1$, then $rH^{n+j-1}(C_p) \cong 0$, $j \geq 1$.

Proof. $(B_p, \partial E_B)$ is a manifold with boundary. Thus $H^{n+j}(B_p) \cong H_{n-j}(B_p, \partial E_B)$, and the result follows from (5.4).

THEOREM (5.7). Suppose $2\sigma_B < n - j$ ($2\sigma_C < n-j-1$). Then there is a homomorphism $\Delta_j : H^{n+j-1}(\partial E_B) \to rH^{n+j}(B)$ ($\Delta_j : H^{n+j-2}(\partial E_C) \to rH^{n+j-1}(C)$) which is an isomorphism for $j \geq 2$ and an epimorphism for $j = 1$. (Unless $n = j$ ($n - 1 = j$ for C), when an isomorphism holds with non-reduced range.)

Proof. There is a Mayer-Vietoris sequence

$$\to H^{n+j-1}(B_p)\oplus H^{n+j-1}(E_B)\to H^{n+j-1}(\partial E_B)\xrightarrow{\Delta_j} H^{n+j}(B)\to H^{n+j}(B_p)\oplus H^{n+j}(E_B)\to$$

Now $H^q(E_B) \cong 0$ for $q > 2\sigma_B$, since E_B is just a regular neighborhood of S_B. Note that $n+j-1 \geq n-j > 2\sigma_B$. By Lemma (5.6) $rH^{n+j}(B_p) \cong 0$, $j \geq 1$, and the result follows.

It is often possible to make explicit computations using (5.7). One is able to compute $H_q(A_\delta)$, $q \leq n-2$, and find an upper bound on the size of $H_{n-1}(A_\delta)$. Since $\chi(A_\delta)$ may be computed as in section IV, this gives a lower bound for the rank of $H_n(A_\delta)$ (and we know A_δ has the homotopy type of an n-dimensional CW complex so that $H_n(A_\delta; \mathbb{Z})$ is free). We state a number of corollaries and then give several examples. These results are stated for the Stein manifolds $A_\delta = f^{-1}(\delta)$ but as we have mentioned, for f homogeneous or weighted homogeneous, A_δ is diffeomorphic to the Milnor fiber.

COROLLARY (5.8). Suppose $s < n-1$, $2\sigma_B < n-s$, and $2\sigma_C < n-s-1$. Then A_δ is $(n-1)$-connected if (i) $H_q(\partial E_B) \cong 0$, $n-s \leq q \leq n-1$ and $H_q(\partial E_C) \cong 0$, $n-s-1 \leq q \leq n-2$; or more generally, (ii) $H_q(\partial E_B) \cong 0$, $n-s \leq q < n-1$ and $H_q(\partial E_C) \cong 0$, $n-s-1 \leq q < n-2$, and $rH^n(C) \cong rH^{n+1}(B) \cong 0$.

By remark (5.5), the hypotheses on σ_B and σ_C aren't crucial. One may easily state similar conditions which are necessary and sufficient for $\tilde{H}_{n-j}(A_\delta; \mathbb{Q}) \cong 0$, $j \geq 1$. Also, one may think of (5.8) as a computable version (see examples (6.1) - (6.5)) of the following.

COROLLARY (5.9). Suppose B and C are R-homology manifolds. Then $\tilde{H}_{n-j}(A_\delta; R) \simeq 0$, $j \geq 1$. If d is a unit in R, then $\tilde{H}_{n-j}(A_\delta; R) \cong$

0, $j \geq 1$ implies B and C have the global homology of manifolds, i.e. homology and cohomology in complementary dimensions agree.

Proof. This follows from (5.2) with R coefficients.

See example (6.5).

For the next two results we need the concept of transverse singularity and transverse link. Suppose S_B and S_C are smooth. Let B_i be a component of B. Pick local coordinates $(\zeta_0, \ldots, \zeta_n)$ in $\mathbb{P}^{n+1}$ at $\zeta' \in B_i$ so that $S_B = \{\zeta_0 = \ldots = \zeta_{n-\sigma_B} = 0\}$ locally. Let L be the linear space $\{\zeta_{n-\sigma+1} = \ldots = \zeta_n = 0\}$. Then at $\zeta' \in B \cap L$ there is an isolated singularity. We will call this the "*transverse isolated singularity*" at ζ'. By [27, Lemma 1.4] it is well-defined. Let $K_{B,\zeta'} = \{\zeta \in B \cap L \mid |\zeta - \zeta'| = \varepsilon\}$ for ε sufficiently small. Then $K_{B,\zeta'}$ will be called the "*transverse link at ζ'.*" Make similar definitions for C.

COROLLARY (5.10). Suppose for each component D of S_B and S_C there is a map $\partial E_D \to D$ which is the projection of a bundle with fiber K_D, where K_D is a transverse link at some ζ' in D. If K_D is an R-homology sphere for all D, then $\tilde{H}_{n-j}(A_\delta; R) \simeq 0$, $j \geq 1$, and $H_n(A_\delta; R)$ is determined by $\chi(A_\delta)$.

Proof. This follows from (5.9).

If B and C have only isolated critical points, as will occur for example if f is homogeneous, and $s = \dim_{\mathbb{C}} S = 1$, then (5.10) automatically applies. For such f we recover the following special case of a theorem proved by Iomdin [10] for *all* f with $s = 1$.

THEOREM (5.11), [10]. *Let f be a homogeneous polynomial of degree d with 1-dimensional singularity set. Then*

$$\chi(F) = 1 + (-1)^n[(d-1)^{n+1} - d \sum_{i=1}^{r} \mu_i] \quad ,$$

where μ_i *is the Milnor number of the i-th transverse singularity.*

Proof. B has r isolated singular points. Let F_i^B be the Milnor fiber at the i-th singular point, and let $\tilde{B} = B_p \cup (\bigcup_{i=1}^{r} F_i^B)$. Then $\tilde{B}$ is diffeomorphic to the non-singular hypersurface of degree d.

It is easy to show that (with b_i = betti number),

$$(5.11.1) \qquad \chi(B) + (-1)^n \sum_{i=1}^{r} b_n(F_i^B) = \chi(\tilde{B}), \qquad \text{and}$$

$$(5.11.2) \qquad \chi(C) + (-1)^{n-1} \sum_{i=1}^{r} b_{n-1}(F_i^C) = \chi(\tilde{C}) \quad .$$

Also, it is well-known that

$$(5.11.3) \qquad \chi(\tilde{B}) - \chi(\tilde{C}) = 1 + (-1)^n(d-1)^{n+1} \quad .$$

Finally, $b_{n-1}(F_i^C) = \mu_i$, and since F_i^B is just the join of F_i^C with d points (up to homotopy; if $F_i^C = \{|z| < \varepsilon, g(z_0, \ldots, z_{n-1}) = \delta\}$, then $F_i^B = \{|z| < \varepsilon, g(z_0, \ldots, z_{n-1}) + z_n^d = \delta\}$) we have $b_n(F_i^B) = (d-1)\mu_i$. From this and (5.11.1) - (5.11.3) the result follows, since $\chi(F) = \chi(B) - \chi(C)$.

Remark (5.12). Iomdin's full result may be obtained by polar curve methods. There is no good generalization to higher-dimensional singular sets because one doesn't know $\tilde{B}$, and because the transverse singularities may vary along S_B and S_C.

VI. EXAMPLES

Throughout this section we use integral coefficients unless otherwise noted.

We have chosen weighted homogeneous examples so that A_δ and F are diffeomorphic.

Example (6.1). Let $f = z_0^5 + z_1^5 + z_2^5 + z_3^5 + z_3^2 z_4^3 + z_3^2 z_5^3$. Then $S = \{z_0 = z_1 = z_2 = z_3 = 0\}$, and $s = 2$. The singular set $S_C \subset C \subset \mathbb{P}^5$ is $\{(0 : 0 : 0 : 0 : z_4 : z_5)\}$, and $\sigma_C = 1$. Also, $B = \{f + z_6^5 = 0\} \subset \mathbb{P}^6$, and $S_B = \{(0 : 0 : 0 : 0 : z_4 : z_5 : 0)\}$, so that $\sigma_B = 1$. We know $H_{n-j}(F) \simeq 0$, for $n - j = 1, 2$ (i.e. $j = 4, 3$). For $j = 1$ or 2, $2\sigma_B < n - j$ and $2\sigma_C < n-j-1$, so the lemmas of section V apply, except that $2\sigma_C = 2 = n-j-1$, when $j = 2$. This causes no difficulty.

To determine the behavior in C transverse to S_C in $\mathbb{P}^5$, we fix $z_4 = a_4$ and $z_5 = a_5$ in f, obtaining $z_0^5 + z_1^5 + z_2^5 + z_3^5 + a_4^3 z_3^2 + a_5^3 z_3^2$. Since a_4 and a_5 are not both zero, we see that this germ is equivalent (via analytic coordinate change) to $z_0^5 + z_1^5 + z_2^5 + z_3^2$. Then by [24], $H_2(K_C^5) \cong \mathbb{Z}_2^{12}$. We suppress the point $(0 : 0 : 0 : 0 : a_4 : a_5 : 0)$ in S_C, since the transverse link is independent of the point chosen. Similarly one may compute $H_3(K_B^7) \cong \mathbb{Z}_2^{52}$.

Applying the Wang sequences of the bundles $K_B^7 \to \partial E_B \to \mathbb{P}^1 = S_B$ and $K_C^5 \to \partial E_C \to \mathbb{P}^1 = S_C$, we obtain $H^5(\partial E_B) \cong 0$, $H^6(\partial E_B) \cong Z_2^{52}$, $H^4(\partial E_C) \cong 0$, and $H^5(\partial E_C) \cong \mathbb{Z} \oplus \mathbb{Z}_2^{12}$. Thus $rH^6(B) \cong H^5(C) \cong 0$, and so $H_4(F) \cong 0$. Also, $H^7(B) \cong \mathbb{Z}_2^{52}$ and $rH^6(C) \cong \mathbb{Z}_2^{12}$, so that we have an exact sequence $0 \to \mathbb{Z}_2^{12} \to H_3(F) \to Z_2^{52} \to 0$. Finally, $H_5(F) \cong \mathbb{Z}^{\mu}$, where $\mu = 1 - \chi(F)$ may be computed as in section IV.

Notice that in this example the transverse links have two-torsion in homology, while $d = 5$. Thus it follows that (5.2) yields short exact sequences.

(6.2) Let $f = z_0^6 + z_1^6 + z_1^3 z_2^3 + z_1^3 z_3^3$. Here $S = \{z_0 = z_1 = 0\}$, $s = 2$, $C = \{f = 0\} \subset \mathbb{P}^3$, and $B = \{f + z_4^6 = 0\} \subset \mathbb{P}^4$. Also, $S_C = S_B$ is a rational curve, $\{z_0 = z_1 = z_4 = 0\} \subset \mathbb{P}^4$. At any point of S_C the transverse isolated singularity is equivalent to $z_0^6 + z_1^3$. Thus K_C^1 is three linked, unknotted circles. Similarly, $H_1(K_B^3) \cong \mathbb{Z}^8 \oplus \mathbb{Z}_3$.

Now $n = 3$, $s - 1 = \sigma_B = \sigma_C = 1$, so we may not apply our lemmas directly to compute $H_1(F)$ since $2\sigma \not< 2 = n - j$. So we let $g = f + z_5^6 + z_6^6$, and compute $H_3(F(g))$. We can then recover $H_1(F)$, since $H_3(F(g)) \cong H_1(F) \otimes \mathbb{Z}^{25}$. For g, $n = 5$, $s - 1 = \sigma = 1$, $j = 2$, and we proceed as in (6.1).

Using [24], $H_2(K_C^5(g)) \cong H^3(K_C^5(g)) \cong \mathbb{Z}^{42}$. Also, $H_3(K_B^7(g)) \cong H^4(K_B^7(g)) \cong \mathbb{Z}^{208}\quad \mathbb{Z}_3$. We have a bundle $K_C^5(g) \to E_C(g) \to \mathbb{P}^1$, and the Wang sequence yields

$$\to H^4(K_C^5(g)) \to H^3(K_C^5(g)) \to H^5(\partial E_C(g)) \to H^5(K_C^5(g)) \to H^4(K_C^5(g)) \to .$$

so that $H^5(\partial E_C(g)) \cong \mathbb{Z}^{43}$. Similarly, $H^6(\partial E_B(g)) \cong \mathbb{Z}^{208}\quad \mathbb{Z}_3$. Thus we have

$$\to H^6(B) \to \mathbb{Z}^{43} \xrightarrow{\delta} H_3(F(g)) \to \mathbb{Z}^{208}\quad \mathbb{Z}_3 \to H^7(C) \xrightarrow{\delta} 0.$$

Therefore, $H_3(F(g)) \cong \mathbb{Z}^{250}$. Thus $H_1(F) \cong \mathbb{Z}^{10}$. A similar example, $h = z_0^6 + z_1^6 + z_1^3 z_2^3 + z_1^3 z_3^3 + z_4^6$ has ker $\delta : H^5(C) \to H_2(F(h)) \cong \mathbb{Z}_3$. (Compare lemma (5.2).)

Example (6.3). $f = z_0^6 + z_1^6 + z_1^3 z_2^3 + z_1^2 z_3^4$. This example is much like the preceding one. However, at $(0 : 0 : 1 : 0) \in S_C$ the transverse singularity is $z_0^6 + z_1^3$, while at any other point $(0 : 0 : z_2 : 1)$ it is $z_0^6 + z_1^2$. Thus ∂E_C is not naturally a bundle over $S_C \cong \mathbb{P}^1$.

Example (6.4). $f = z_0^8 + z_0 z_1^4 + z_1^3 z_2^5$. This polynomial is weighted homogeneous of type $(160; 20, 35, 11)$. Clearly $S = \{z_0 = z_1 = 0\}$. We will show $H_1(F; \mathbb{Z}[1/15]) \cong 0$. Now, $f_h = z_0^8 + z_0 z_1^4 z_3^3 + z_1^3 z_2^5 + \delta z_3^8$ and $S_B = (0 : 0 : 1 : 0) \cup (0 : 1 : 0 : 0)$ in $\mathbb{P}^3$. The curve C in $\mathbb{P}^2$ is given by $z_0^8 + z_1^3 z_2^5$, and $S_C = (0 : 1 : 0) \cup (0 : 0 : 1) \subset \mathbb{P}^2$. Furthermore, $H^2(C) \cong \mathbb{Z}$, and $\delta : H^2(C) \to H^3(B, C) \cong H_1(F)$ is the zero map. Thus in this special case, as in [21] we need only consider $H^3(B)$. We apply (5.7) with $n = 2$, $\sigma = 0$, and $j = 1$, to obtain $H^2(\partial E_B) \to H^3(B) \to 0$. Now at

$(0:0:1:0) \in S_B$ one has a singularity of the form $z_0^8 + z_0 z_1^4 + z_1^3 + z_3^8$, which is equivalent by change of coordinate to $z_0^8 + z_1^3 + z_3^8$. Similarly, at $(0:1:0:0)$ we have $z_0^8 + z_0 z_3^3 + z_2^5$. Thus $H^2(\partial E_B) \cong H^2(K_3^B) \cong \mathbb{Z}_3^7 \oplus \mathbb{Z}_5$, by [24], and $H_1(F; \mathbb{Z}[1/15]) \cong 0$.

A very similar example is the homogeneous $f = z_0^5 + z_0^3 z_1^2 + z_1 z_2^4$. But here there is one singular point $(0:1:0:0) \in B \subset \mathbb{P}^3$, and at this point the transverse link K_B^3 is defined by $z_0^3 + z_2^4 + z_3^5$, so that K_B^3 is an integral homology sphere, and $H_1(F; \mathbb{Z}) \cong 0$.

In fact, by (5.1.4) one may see in both cases that $H_1(F; \mathbb{Z}) \cong 0$.

Example (6.5). $f = \left(\sum_{i=1}^{6} z_i^5\right) + z_7^3 \left(\sum_{i=1}^{5} z_i^2\right) + z_7^2 z_6^3$. Then $S = \{z_1 = \ldots = z_6 = 0\}$, $S_C = S_B = (0 : \ldots : 1 : 0) = P \in \mathbb{P}^7$. At P, setting $z_7 = 1$ shows that the singularity in C is equivalent after coordinate change to $\sum_{i=1}^{5} z_i^2 + z_6^3$, and thus the link of P in C is the Kervaire 9-sphere [3]. Similarly at P in B we have $\sum_{i=1}^{5} z_i^2 + z_6^3 + z_8^5$, the Milnor 11-sphere. Both C and B are singular projective varieties which are PL manifolds, and $F(f)$, a 12-dimensional manifold, is 5-connected. Finally, $H_6(F(f)) \cong \mathbb{Z}^{16,374}$, by section IV or (5.11).

REFERENCES

1. N. A'Campo, La fonction zeta d'une monodromie, Comment. Math. Helv. 50 (2) (1975), 233-248.
2. J. Briançon and J. P. Speder, La trivialité topologique n'implique pas les conditions de Whitney, C. R. Acad. Sci. Paris 280 (1975), 365-367.
3. E. Brieskorn, Beispiele zur Differentialtopologie von Singularitäten, Inventiones math. 2 (1966), 1-14.
4. D. Burghelea and A. Verona, Local homological properties of analytic sets, Manuscr. Math. 7 (1972), 55-66.

5. A. Durfee, Fibered knots and algebraic singularities, Topology 13 (1974), 47-60.
6. H. Hamm, Lokale Topologische Eigenschaften komplexer Räume, Math. Ann. 191 (1971), 235-252.
7. H. Hamm and Lê Dũng Tráng, Un Théorème de Zariski du type de Lefschetz, Ann. Sci. L'Ecole Nor. Sup. 6 (1973), 317-366.
8. I. N. Iomdin, The Euler characteristic of the intersection of a complex surface with a disk, Siberian Math. J. 14 (1973), 222-232.
9. I. N. Iomdin, Local topological properties of complex algebraic sets, Siberian Math. J. 15 (1974), 558-572.
10. I. N. Iomdin, Complex surfaces with a one-dimensional set of singularities, Siberian Math. J. 15 (1974), 748-762.
11. M. Kato, A classification of simple spinnable structures on a 1-connected Alexander manifold, J. Math. Soc. Japan 26 (1974), 454-463.
12. M. Kato and Y. Matsumoto, On the connectivity of the Milnor fiber of a holomorphic function at a critical point, Proc. of 1973 Tokyo Manifolds Conf., pp. 131-136.
13. H. King, Topological type of isolated singularities. preprint.
14. Lê Dũng Tráng, Calcul de nombre de cycles evanouissants d'une hypersurface complexe, Ann. Inst. Fourier Grenoble 23 1973), 261-270.
15. Lê Dũng Tráng and C. P. Ramanujam, The invariance of Milnor's number implies the invariance of the topological type, Amer. J. Math. 98 (1976), 67-78.
16. J. Levine, An algebraic classification of some knots in codimension two, Comment. Math. Helv. 45 (1970), 185-198.
17. S. Lojasiewicz, Triangulation of semi-analytic sets, Annali Scu. Norm. Sup. Pisa, Sc. Fis, Math. Ser. 3, v. 18, fasc. 4 (1964), 449-474.
18. J. Milnor, Singular points of complex hypersurfaces, Ann. Math. Studies 61, Princeton, 1968.
19. M. Oka, On the homotopy types of hypersurfaces defined by weighted homogeneous polynomials, Topology 12 (1973). 19-32.
20. M. Oka, on the cohomology structure of projective varieties, Proc. of 1973 Tokyo manifolds conf., pp. 137-143.
21. M. Oka, The monodromy of a curve with ordinary double points, Inventiones math. 27 (1974), 157-164.
22. P. Orlik and R. Randell, The monodromy of weighted homogeneous singularities, Inventiones math. 39 (1977), 199-211.
23. V. P. Palamodov, Multiplicity of a holomorphic transformation, Func. Analysis and Appl. 1 (1967), 218-226.
24. R. Randell, The homology of generalized Brieskorn manifolds, Topology 14 (1975), 347-355.
25. K. Sakamoto, The Seifert matrices of Milnor fiberings defined by holomorphic functions, J. Math. Soc. Japan 26 (1974), 714-721.
26. M. Sebastiani and R. Thom, Un resultat sur la monodromie, Inventiones math. 13 (1971), 90-96.

27. B. Teissier, Cycles évanescents, section planes, et conditions de Whitney, Astérisque 7-8 (1973), 285-362.
28. B. Teissier, Introduction to equisingularity problems, Proc. Sym. Pure Math. 29 (1975), Amer. Math. Soc., Providence, pp. 593-632.
29. J. G. Timourian, The invariance of Milnor's number implies topological triviality, Amer. J. Math. 99 (1977), 437-446.
30. A. N. Varčenko, Zeta-function of monodromy and Newton's diagram, Inventiones math. 37 (1976), 253-262.
31. O. Zariski, Algebraic Surfaces, 2nd Supplemented Edition, Springer, Berlin-Heidelberg-New York, 1971.

APPROXIMATING CAT CE MAPS BY CAT HOMEOMORPHISMS

Martin Scharlemann[1]

Department of Mathematics
University of California
Santa Barbara, California

Suppose $f : M \to N$ is a CE map between manifolds of dimension $n \geq 6$. Siebenmann shows that f can be approximated by a homeomorphism. Here we show that if f is CAT(=DIFF or PL) then f can be approximated through CAT maps by a CAT homeomorphism if and only if an obstruction in $H^3(N;\theta_3^h)$ vanishes. The theory in the PL category is analogous to (but differs slightly from) that of Cohen-Sullivan, which predates Siebenmann's theorem.

I. INTRODUCTION

Define a *CE* map to be a proper continuous map $f : X \to Y$ such that, for each y in Y, $f^{-1}(y)$ has the shape of a point. Siebenmann has shown [Si_1] that, given $f : M \to N$ a *CE* map of metric topological m-manifolds, $m \geq 6$ (or $m = 5$ and $f|\partial M$ a homeomorphism) there is a level-preserving *CE* map $F : M \times [0,1] \to N \times [0,1]$ such that $F|M \times \{1\} = f$ and $F|M \times \{t\}$ is a homeomorphism for $0 \leq t < 1$. Furthermore F can be chosen so that, for all t, $F|M \times \{t\}$ is arbitrarily close to f.

Earlier, Cohen and Sullivan showed the following [2] [3] : Let $f : M \to N$ be a *PL CE* map between *PL* manifolds. There is an

[1] *Research supported in part by an NSF grant.*

ISBN 0-12-158860-2

obstruction in $H^3(N;\theta_3^h) \cong H^4(N \times I, N \times \partial I;\theta_3^h)$ to extending the PL CE map $f \cup id: M \cup N \to N \times \partial I$ to a PL CE map $F: W \to N \times I$, W a PL manifold. Then W is an s-cobordism, hence, if $m \geqq 5$, the inclusion $M \hookrightarrow W$ extends to a PL homeomorphism $H: M \times I \to W$, and $FH: M \times I \to N \times I$ is a PL CE map such that $F|M \times \{1\} = f$ and $F|M \times \{0\}$ is a PL homeomorphism. In this sense, Cohen's theorem may be viewed as a PL version of Siebermann's theorem, but a version in which an obstruction may arise.

Here we give a proof of the analogous theorem in the smooth category. Since the proof also works in the PL category and shows in an elementary way that F may be level-preserving, we will in fact prove theorems in both the PL and $DIFF$ category.

In order to state theorems and proofs in a form applicable to both $DIFF$ and PL, it will be convenient to use the following notation:

$$D^k = \{(x_1,\ldots,x_k) \mid x_1^2 + x_2^2 + \cdots + x_k^2 \leqq 1)\}$$

$$S^{k-1} = \partial D^k$$

$$I^k = D^1 \times \cdots \times D^1,\ k \text{ factors}$$

$$B^k = D^k \text{ or } I^k \text{ dependening on whether } CAT \text{ is } DIFF \text{ or } PL$$

$$id_X : X \to X \text{ is the identity map.}$$

Two maps $f_0, f_1 : X \to Y$ are concordant if there is a map $F: (X \times I, X \times 0, X \times 1) \to (Y \times I, Y \times 0, Y \times 1)$ such that for some $\varepsilon > 0$, $F = f_0 \times id_I$ over $N \times [0,\varepsilon)$ and $F = f_1 \times id_I$ over $N \times (1-\varepsilon, 1]$. If $F(x,t) = (f_0(x), t)$ for all t, F is fixed on x. The support of F is the set of points in X on which F fails to be fixed. If, for all t, $F(X \times \{t\}) \subset Y \times \{t\}$ F is called level-preserving and we denote $F \mid X \times \{t\}$ by f_t.

Throughout the paper, CAT will denote either $DIFF$ or PL. In almost all cases the PL theory is easier, because PL homeomorphisms are PL isomorphisms, whereas a $DIFF$ homeomorphism may fail to have a smooth inverse. The *PL* specialist will, therefore, find many arguments irrelevant.

THEOREM A. The following statements are equivalent:

1) For any CAT homotopy k-sphere M^k, $k = 3,4$, there is a CAT homeomorphism $M \to \partial B^{k+1}$.

2) Let $f : M^n \to N^n$ be a CAT CE map between two metric CAT n-manifolds such that $f \mid \partial M \to \partial N$ is also a CE map of manifolds. Let $\varepsilon : M \to (0,\infty)$ be continuous. Then there is a CAT CE level-preserving concordance $F : M \times I \to N \times I$ from $f = f_0$ to a (CAT) homeomorphism. Furthermore $d(f(x), f_t(x)) < \varepsilon(x)$, $0 \leqq t \leqq 1$, and if $f \mid \partial M$ is a homeomorphism then F fixes ∂M.

Let $\theta_3^h(CAT)$ denote the group of all homotopy 3-spheres (with operation connected sum) modulo those bounding *CAT* contractible 4-manifolds. By classical smoothing theory, $\theta_3^h(DIFF) = \theta_3^h(PL)$ so we abbreviate by $\theta_3^h(CAT) = \theta_3^h$.

THEOREM B: Let $f : M^n \to N^n$ be a CAT CE map between two metric CAT n-manifolds such that $f : \partial M \to \partial N$ is a CE map. Let $\varepsilon : M \to (0,\infty)$ be a continuous map.

1) If $n \geqq 6$ there is an obstruction α in $H^3(N, \theta_3^h)$ which vanishes if and only if there is a CAT CE level-preserving concordance $F : M \times I \to N \times I$ from $f = f_0$ to a CAT homeomorphism f_1 such that for all t, $d(f_t(x), f_0(x)) < \varepsilon(x)$.

2) If $n \geqq 5$ and $f \mid \partial M$ is a homeomorphism, then there is an obstruction α in $H^3(N, \partial N; \theta_3^h)$ which vanishes if and only if a concordance F as above exists and fixes ∂M.

Let $\rho : \theta_3^h \to Z_2$ be the homomorphism which assigns to any homotopy 3-sphere its Rohlin invariant.

THEOREM C: *Let* $f : M^n \to N^n$, $\varepsilon : M \to (0,\infty)$ *and* α *in* $H^3(N, \theta_3^h)$ *(resp.* $H^3(N, \partial N; \theta_3^h)$*) be as in Theorem B. Then* $\rho_*(\alpha)$ *in* $H^3(N, Z_2)$ *(resp.* $H^3(N, \partial N, Z_2)$*) is trivial if and only if there is a (non-CAT) CE level-preserving concordance* $F : M \times I \to N \times I$ *from* f *to a CAT homeomorphism (which fixes* ∂M*) such that* $d(f_t(x), f_0(x)) < \varepsilon(x)$ *for all* t.

COROLLARY: *If every homotopy 3-sphere has trivial Rohlin invariant then any CAT CE map* $f : M^n \to N^n$, $n \geq 6$, *is approximable by a PL homeomorphism.*

Remarks: The source of the obstruction, which fails to arise in the topological case, is the glut of regular values with which *CAT* maps are encumbered. Indeed, topological maps between manifolds may have *no* regular values (cf. [1]).

The obstruction $\rho_*(\alpha)$ of Theorem C has another interpretation. By [9]. f can be approximated by a (non-*CAT*) homeomorphism f'. Then $\rho_*(\alpha)$ is the obstruction to isotoping f' to a *PL* homeomorphism.

For any α in θ_3^h, there is an example $f : M \to S^3 \times T^2$ such that the obstruction of Theorem B is α in $H^3(S^3 \times T^2; \theta_3^h) \simeq \theta_3^h$.

This paper was inspired by a remark of T. Farrell, who suggested to the author that the techniques of [8] might be used to prove the above corollary.

II. CAT CE MAPS.

For any $0 < \varepsilon < 1$, $\mu_\varepsilon : R \to R$ will denote a *CAT* map such that $\mu_\varepsilon(s) = 0$ for $s \leq 1 - \varepsilon$, $\mu_\varepsilon(s) = 1$ for $s \geq 1 - \varepsilon/2$ and $\mu_\varepsilon|(1 - \varepsilon, 1 - \varepsilon/2)$ is a *CAT* homeomorphism onto $(0,1)$. If ε is understood or unimportant we will just write μ.

For any $x = (x_1,\ldots,x_k)$ in $R^k - 0$ there is a unique ray from 0 to x. Let $\bar{x} = (\bar{x}_1,\ldots,\bar{x}_k)$ denote the point where the ray pierces ∂B^k. The value $\left(\sum_{i=1}^{k} x_i^2\right)^{1/2} \cdot \left(\sum_{i=1}^{k} \bar{x}_i^2\right)^{-1/2}$ will be denoted $|x|$. The function $|\ | : R^k - 0 \to R_+$ is then a *CAT* function, corresponding to the standard distance from the origin when $CAT = DIFF$. For $r > 0$ let $rB^n = \{x \in R^n | \ |x| \leq r\}$. An open ε-collar of ∂B^n, $\varepsilon > 0$, will denote $B^n - (1-\varepsilon)B^n$. For M a manifold, let $\mathring{M}$ denote the interior of M.

First we develop some preliminary information about *CAT CE* maps.

LEMMA 2.1: For any open ε-collar C of ∂B^k there is a CAT CE level-preserving concordance $G : B^k \times I \to B^k \times I$, fixed near ∂B^k, from the identity to a map g_1 such that g_1 is a homeomorphism except over $0 \in B^k$ and $g_1^{-1}(0) = B^k - C$.

Proof. Let $\mu = \mu_\varepsilon : R \to R$. Define $\alpha : R^2 \to [0,1]$ by $\alpha(s,t) = (1 - \mu(t)) + \mu(s)\mu(t)$ and $G : B^k \times I \to B^k \times I$ by $G(x,t) = (\alpha(|x|, t)x, t)$. Save for the requirement that G be *CE*, it is clear that G is appropriate. To check that G is *CE*, suppose $G(x,t) = (0,t)$ with $1 - \varepsilon/2 \leq t \leq 1$. Then $\alpha(|x|, t) = \mu(|x|) = 0$ so $|x| \leq 1 - \varepsilon$. Hence $G^{-1}(0,t) = (B^k - C) \times \{t\}$ which is contractible. Now suppose $G(x,t) = (b,t)$ with either $b \neq 0$ or $0 \leq t < 1 - \varepsilon/2$. In either case $\alpha(|x|, t) \neq 0$, so $x = b/\alpha(|x|,t)$, and $G^{-1}(b,t)$ is a single point. Hence G is *CE*. | |

LEMMA 2.2 (CE Alexander trick): Suppose $f_0 : M \to B^k$, $f_1 : M \to B^k$ are CAT CE maps with $f_0 = f_1$ over a neighborhood of ∂B^k. Then there is a CAT CE level-preserving concordance $F : M \times I \to B^k \times I$ from f_0 to f_1, fixed over a neighborhood of ∂B^k.

Proof: Choose an ε-collar C of ∂B^k such that $f_0 = f_1$ over C, and let $G : B^k \times I \to B^k \times I$ be the concordance of 2.1. Then

$g_1f_0 = g_1f_1$. Define $F : M \times I \to B^k \times I$ by

$$f_t(m) = g_{2t}f_0(m) \qquad 0 \leqq t \leqq 1/2$$

$$f_t(m) = g_{2(1-t)}f_1(m) \qquad 1/2 \leqq t \leqq 1. \quad ||$$

Suppose M and N are CAT manifolds, and $f : M \times R^k \to N \times R^k$ is a CAT CE map commuting with projection to R^k. Let D,E be closed subsets of N, U a neighborhood of D, and suppose f is a homeomorphism over a neighborhood of $E \times R^k$.

LEMMA 2.3. (Squeezing lemma) There is a CAT CE level-preserving concordance $F : M \times R^k \times I \to N \times R^k \times I$ such that:

i) The support of F lies in $U \times B^k$.

ii) f_t commutes with projection to R^k.

iii) f_t is a homeomorphism over a neighborhood of $E \times R^k$.

iv) For s near 0 in R^k, $f(m)$ near D, $f_1(m,s) = (f_1(m,0),s)$ in $N \times R^k$.

Proof. *Case 1:* $D = N$. Let $\mu = \mu_\varepsilon$ for some $0 < \varepsilon < 1$ and define $\alpha : R^2 \to [0,1]$ by $\alpha(s,t) = (1-\mu(t)) + \mu(s)\mu(t)$. For m in M, s in R^k, let $f_t(m,s) = (p_1f(m,\alpha(|s|, t)s),s)$. For small t, $\alpha(|s|, t) = 1$, so $f_0 = f$. Since for t near 1 and $|s|$ near 0, $\alpha(|s|, t) = 0$, f_t satisfies *iv)*. Since for $|s| \geqq 1$, $\alpha(|s|, t) = 1$, f_t satisfies *i)*. Condition *ii)* is obvious. Furthermore $f_t(m,s') = (n,s)$ if and only if $s = s'$ and $p_1f(m,\alpha(|s|, t)s) = n$. Thus $f_t^{-1}(n,s) \cong f^{-1}(n,\alpha(|s|, t)s)$, which has the shape of a point, and is a point if n is near E.

Case 2: The general case. Approximate a Urysohn function $N \to [0,1]$ taking D to 0 and $N - U$ to 1 by a CAT function u and replace D by $u^{-1}[0,\gamma]$, where γ is a regular value of $uf : M \times \{0\} \to [0,1]$. We thereby may assume that D is a CAT submanifold of N with ∂D transverse to f. Since f is transverse to ∂D, there are CAT bicollars $c : \partial D \times (-1,1) \to U$ and $\bar{c} : \partial(f^{-1}(D)) \times (-1,1) \to f^{-1}$ (image c) so that $c^{-1}f\bar{c} : \partial(f^{-1}(D)) \times (-1,1) \to \partial D \times (-1,1)$ commutes with projection to $(-1,1)$. Now apply Case 1 to $(\bar{c} \times id_{R^k})$, thereby altering f so that, for some $\delta > 0$, whenever $|(r,s)| < \delta$, $f(\bar{c}(x,r),s) = (c(f(x), r), s)$ for x in $\partial(f^{-1}(D))$. Furthermore, if the collar is chosen small enough, the altered f will still be a homeomorphism over a neighborhood of $E \times R^k$. Without loss of generality, we may assume that if $|s| < 1$, then $|(r,s)| < \delta$.

Let $0 < \varepsilon < 1$ and $\alpha(|s|, t)$ be as in Case 1. Define

$$A_1 = \{(m,s) \mid f(m) \in D \cup \text{image } (c) \text{ and } |s| \leqq 1\}$$

$$A_2 = \{(m,s) \mid f(m) \in N - D \text{ or } |s| \geqq 1 - \varepsilon/2\}.$$

and let

$$f_t(m,s) = \begin{cases} (p_1 f(m,\alpha(|s|, t)s), s) & (m,s) \text{ in } A_1 \\ f(m,s) & (m,s) \text{ in } A_2 \end{cases}$$

For (m,s) in (image $\bar{c}$) $\times B^k$, $p_1 f(m,\alpha(|s|, t)s) = f(m) = p_1 f(m,s)$ so $f_t(m,s) = f(m,s)$. Also, if $1 - \varepsilon/2 \leqq |s| \leqq 1$, then $\alpha(|s|, t) = 1$ so $f_t(m,s) = f(m,s)$. Hence it follows that F is well-defined and CAT.

The only requirements of the Lemma which F does not obviously satisfy is that it be CE, and a homeomorphism near $E \times R^k \times I$. To check this, we first check that $f_t^{-1} f_t(A_i) = A_i$, $i = 1,2$. Indeed suppose that $f_t(m_1,s_1) = f_t(m_2,s_2)$. Then $s_1 = s_2$ and either both (m_1,s_1) and (m_2,s_2) are in A_2 or $|s_1| = |s_2| \leqq 1$. In the latter case, define the arcs $q_i : [0,1] \to N \times B^k$, $i = 1,2$ by $q_i(\beta) = f_t(m_i,\beta s_i)$, connecting $f(m_i,s_i)$ to $f(m_i,0)$. If such an arc intersects (image c) $\times B^k$, then the arc lies totally in (image c) $\times B^k$, by definition of c. Hence either both (m_i,s_i) lie in A_1, or each arc is disjoint from c. In the latter case each uq_i is either less than γ for all β and hence $f(m_i) \in D$ or greater than γ and hence $f(m_i) \in N - D$. Since $f_t(m_1, s_1) = f_t(m_2, s_2)$, $q_1(1) = q_2(1)$ and so $f(m_1)$ is in A_1 if and only if $f(m_2)$ is in D. Thus (m_1, s_1) is in A_1 if and only if (m_2, s_2) is in A_1.

Since $f_t \mid A_2 = f \mid A_2$, f_t is *CE* over $f(A_2)$ and is a homeomorphism when f is. Finally, just as in Case 1, $f_t^{-1}(n,s) \cong f^{-1}(n, \alpha(|s|, t)s)$ for $(n,s) \in f(A_1)$, so f is *CE* over $f(A_1)$ as well, and is a homeomorphism over a neighborhood of $E \times R^k$. ||

Remark. In a *CAT CE* level-preserving concordance $F : M \times I \to N \times I$, the requirement that $f_t = f_0$ for t near 0 and $f_t = f_1$ for t near 1 (which allows us to "glue" *CAT* concordances together) need no longer be checked if F is known to be level-preserving. Indeed, even if F is not level preserving, after an application of the squeezing lemma near $N \times \partial I$ it will be.

LEMMA 2.4 (*Collaring lemma*). *Let $f : M \to N$ be a CAT CE map between CAT n-manifolds. Let $c : (\partial N \times [0,2], \partial N \times 0) \to (N, \partial N)$ and $\bar{c} : (\partial M \times [0,1], \partial M \times 0) \to (M, \partial M)$ be CAT collars of the respective boundaries. Then there is a CAT CE level-preserving concordance $F : M \times I \to N \times I$ from $f_0 = f$ to a map f_1 such that*

i) The support of F lies over $c(\partial N \times (0,2])$

ii) For u in ∂M and r sufficiently small, $f_1\bar{c}(m,r) = c(f_0(m), r)$

Proof. With no loss of generality, assume $f\bar{c}(\partial M \times [0,1]) \subset c(\partial N \times [0,1))$. Pre-compose $f \times id_I$ with a *CAT CE* level-preserving concordance $M \times I \to M \times I$, with support over $f^{-1}c(\partial N \times (0,1])$, from the identity map to a map which, for all m in M, collapses $\bar{c}(m \times I)$ to $\bar{c}(m \times 0)$. After this alteration, f has the new property $f\bar{c}(m,r) = f\bar{c}(m,0)$, for all m in M, $0 \leq r \leq 1$.

Let $\alpha : [0,2] \times I \to [0,2]$ be a *CAT* isotopy of imbeddings such that $\alpha([0,2] \times t) = [t,2]$, α is the projection on $[0,2] \times 0$ and near $2 \times I$, and is the map $(s,t) \to (s + t)$ near $0 \times I$. Let $\mu(t) = \mu_\varepsilon(t)$ some $0 < \varepsilon < 1$.

Define $F : M \times I \to N \times I$ as follows:

1) For $f(m)$ not in image c (and hence m not in image $\bar{c}$) let $f_t(m) = f(m)$.

2) For $f(m) = c(n,s)$, n in ∂N, $0 \leqq s \leqq 2$, but m near $M - \bar{c}(\partial M \times [0,1))$, let $f_t(m) = c(n,\alpha(s,\mu(t)))$

3) For m in ∂M, let $f_t(\bar{c}(m,r)) = c(n,\alpha(0,\mu(t)[1 + r - 1)\mu(1 - r)]))$, where $f(\bar{c}(m,r) = f(\bar{c}(m,0)) = n$ defines n in ∂N.

Since $\alpha(s,\mu(t)) = s$ for s near 2, the map F is well-defined and *CAT* over a neighborhood of $c(\partial N \times 2)$. For r near 1, 3) requires that $f_t(\bar{c}(m,r)) = c(n,\alpha(0,\mu(t)))$, where $f(\bar{c}(m,r)) = n$ in ∂N. Then 2) also requires that $f_t(\bar{c}(m,r)) = c(n,\alpha(0,\mu(t)))$. Thus F is well-defined and *CAT*. Furthermore, for t near 0; $\mu(t) = 0$ so $\alpha(s,\mu(t)) = s$ and hence $f_t = f$. For t near 1, $\alpha(s,\mu(t)) \subset [1,2]$ so whenever $f_t(m) \subset c(\partial N \times [0,1])$, m is in image $\bar{c}$. But for t near 1, $f_t(\bar{c}(m,r)) = c(n,\alpha(0,1 + (r - 1)\mu(r - q)) = c(n,1 + (r - 1)\mu(r - 1)) = c(n,q + (r - 1)\mu(1 - r))$ where $n = f(m)$ in ∂M. Then for small r, $f_t(\bar{c}(m,r)) = c(n,r)$, as required.

To complete the proof, it suffices to show F is *CE*. If n is not in image c, then $f_t^{-1}(n) = f^{-1}(n)$, so f_t is *CE* over N-{image c}. Next consider those points of N of the form $c(n,s)$ with $s < \alpha(0,\mu(t))$, n in ∂N. Then $f_t^{-1}c(n,s) = \{\bar{c}(n,r) \mid c(f(m), \alpha(0,\mu(t)[1 + (r - 1)\mu(1 - r)]) = c(n,s),\ m \text{ in } \partial M,\ 0 \leqq r\} = \{\bar{c}(m,r) \mid f(m) = n,\ \mu(t)[1 + (r - 1)\mu(r - 1)] = s\}$. Since $f \mid \partial M$ is *CE* and $\{r \mid \mu(t)[1 + (r - 1)\mu(r - 1)] = s\}$ is contractible, $f_t^{-1}c(n,s)$ is point-like.

Finally, a point of the form $c(n,s')$ with $s' \geqq \alpha(0,\mu(t))$ may be written as $c(n,\alpha(s,\mu(t)))$, some $0 \leqq s \leqq 2$. First note that $f_t^{-1}(c(n,\alpha(s,\mu(t))) \cap \text{image } \bar{c} = \bar{c}(m,r) \mid \bar{c}(m,0) = n$ and $\alpha(0,\mu(t)[1 + (r - 1)\mu(1 - r)] = \alpha(s,\mu(t))$. But $\alpha(0,\mu(t)[1 + (r - 1)\mu(1 - r)] = \mu(t)[1 + (r - 1)\mu(1 - r)] \leqq \mu(t) \leqq \alpha(s,\mu(t))$

and equality holds if and only if $s = 0$ and $(r - 1)\mu(1 - r) = 0$. But $(r - 1)\mu(1 - r) = 0$ if and only if $\varepsilon \leqq r \leqq 1$. Thus $f_t^{-1}(c(n,\alpha(s,\mu(t)))) \cap \text{image } \overline{c}$ is either empty or, when $s = 0$, equal to $\{\overline{c}(m,r) \mid \overline{c}(m,0) = n,\ \varepsilon \leqq r \leqq 1\}$, Which deformation retracts to $\overline{c}(m,1)$. Hence $f_t^{-1}(c(n,\alpha(s,\mu(t))))$ has the same shape as that part lying off of $\overline{c}(\partial M \times [0,1))$. But there $f_t^{-1}(c(n,\alpha(s,\mu(t)))) = f^{-1}(c(n,s))$ which again is point-like. Hence f_t is everywhere *CE*, and so then is F.

LEMMA 2.5 (Fan lemma). Let $f : M \to N$ be a CAT CE map between CAT n-manifolds. Let $G : \partial M \times I \to \partial N \times I$ be a CAT CE level-preserving concordance with $g_1 = f \mid \partial M$. Then there is a CAT CE level-preserving concordance $F : M \times I \to N \times I$, with $f_1 = f$ and $f_0 \mid \partial M = g_0$, and support over an arbitrarily small collar of ∂N in N.

Proof. Apply the collar Lemma 2.4 to small collars of ∂N and ∂M, altering f so that for *CAT* collars $c : (\partial N \times [0,1], \partial N \times 0) \to (N, \partial N)$ and $\overline{c} : (\partial M \times [0,1], \partial M \times 0) \to (M, \partial M)$, $f\overline{c}(m,r) = c(f(m), r)$ all $0 \leqq r \leqq 1$.

Regard $I \times I$ as lying in R^2 and, for (r,t) in $I \times I$, define $|r,t|$ as appropriate in *CAT* for (r,t) a point in R^2. Define $F : M \times I \to N \times I$ by "fanning out" the map G around $\partial M \times \{0\}$ in $M \times I$. That is, let $F(\overline{c}(m,r), t) = c(p_1G(m, |r,t|), r,t)$ on image $\overline{c} \times I$ when $|r,t| \leqq 1$, and $F(m,t) = (f(m), t)$ otherwise.

III. THE GROUP θ_3^h

Here we develop a different way of viewing the group θ_3^h defined above, a way which will be more useful for our purposes.

LEMMA 3.1. Let M be a CAT manifold, homotopy equivalent to ∂B^{k+1}, and let M_0 denote M with the interior of a CAT B^k removed. Then there is a CAT CE map $M_0 \to B^k$ which is a CAT isomorphism over a neighborhood of ∂B^k.

Proof. We consider ∂B^{k+1} to be the *CAT* union $B_+ \cup_\partial B_-$ of two copies of B^k identified along their boundaries. Let p_+ and p_- be points in $\mathring{B}_+$ and $\mathring{B}_-$ respectively. A homotopy equivalence $g : M \to \partial B^{k+1}$ may be homotoped so that it is *CAT*, p_- is a regular value and $g^{-1}(p_-)$ is a point. Hence we may in fact assume that g is a *CAT* isomorphism over a neighborhood of B_- including an open collar C of ∂B_+. Then $g \mid M - g^{-1}(\mathring{B}_-) \to B_+$ is a *CAT* isomorphism over C. By 1.1, there is a *CAT CE* map $h : B_+ \to B^k$ such that h is a *CAT* isomorphism except over 0, and $h^{-1}(0) = B_+ - C$. Then $hg \mid M - g^{-1}(\mathring{B}_-)$ is the required map. ||

Let ψ_k denote the set of all orientation preserving *CAT CE* maps $f : M \to R^k$ which are homeomorphisms except over a compact set. Two such maps $f_0 : M_0 \to R^k$ and $f_1 : M_1 \to R^k$ are considered equivalent in ψ_k if there is an oriented *CAT* $(k + 1)$-manifold and an orientation preserving *CAT CE* map $F : W \to R^k \times I$ such that

i) F is a *CAT* homeomorphism except over a compact set in $R^k \times I$.

ii) ∂W has two components M_0 and M_1 and $F \mid M_i = f_i : M_i \to R^k \times \{i\}$, $i = 0,1$.

Suppose $f : M \to R^k$ is in ψ_k, and r is chosen so large that f is a homeomorphism over a neighborhood of $R^k - r\mathring{B}^k$. Let $h : r\mathring{B}^k \to R^k$ be a *CAT* isomorphism.

LEMMA 3.2. *The maps f and $hf : f^{-1}(r\mathring{B}^k) \to R^k$ are equivalent in ψ_k.*

Proof. Let $H : R^k \times I \to R^k \times I$ be a *CAT* imbedding which is a concordance from the identity to h^{-1}. Let $W = (f \times id_I)^{-1}H(R^k \times I)$. Then $H^{-1}(f \times id_I) : W \to R^k \times I$ gives the required equivalence.

LEMMA 3.3. *There are natural bijections $\psi_3 \leftrightarrow \theta_3^h$ and $\psi_4 \leftrightarrow 0$.*

Proof. First consider $f : M \to R^4$ in ψ_4. We may assume f is a homeomorphism except over a compact set in $\mathring{B}^4$ and f is transverse to $\partial\mathring{B}^4$. Then $f^{-1}(\partial B^4)$ is a *CAT* manifold homeomorphic to ∂B^4, hence *CAT* isomorphic to ∂B^4 [6]. Let $h : f^{-1}(\partial B^4) \to \partial B^4$ be an orientation reversing *CAT* isomorphism. Regard ∂B^5 as the *CAT* union $B_+ \cup_\partial B_-$ of two copies of B^4, and let $\overline{M}$ denote the manifold $f^{-1}(B^4) \cup_h B^4_-$. Extend $fh^{-1} : \partial B^4_- \to \partial B^4_-$ to a *CAT* map $B^4_- \to B^4_-$ which is a homeomorphism over a neighborhood of ∂B^4_-, and apply Lemma 2.1 as in the proof of 2.2, to alter the extension to a *CAT CE* map. This gives an extension of $f \mid f^{-1}(B^4)$ to a *CAT CE* map $\overline{f} : \overline{M} \to \partial B^5$. It is a well-known theorem of Kervaire that any *CAT* homotopy 4-sphere bounds a *CAT* contractible 5-manifold. Let $\overline{W}$ be such a manifold with $\partial\overline{W} = \overline{M}$. Again apply 2.1 to any extension $\overline{W} \to B^5$ to get a *CAT CE* extension $F : \overline{W} \to B^5$. Let $W = \overline{W} - F^{-1}(\partial B^4_+)$. Then $F|W$ is an equivalence between $f \mid f^{-1}(\mathring{B}^4)$ and $\overline{f} \mid B^4_- \to B^4_-$. The latter is a *CAT* homeomorphism and so, over some smaller ball, is a *CAT* isomorphism. Then, by Lemma 3.2, $f = f \mid f^{-1}(\mathring{B}^4) = \overline{f} \mid B^4 =$ (a *CAT* isomorphism) in ψ_4. But any two *CAT* isomorphisms onto R^4 are *CAT* isotopic, so ψ_4 has only one element.

To any $f : M \to R^3$ representing a class in ψ_3 we may associate a homotopy 3-sphere $\alpha(f)$ constructed as M_+ was above. The manifold $\alpha(f)$ is well-defined, for it is easy to see that different choices of representatives f of the same class in ψ_3 merely alter $f^{-1}(B^3)$ by manifolds *CAT* homeomorphic, hence *CAT* isomorphic to collars $S^2 \times I$ of $f^{-1}(\partial B^3)$. The choice of $h : f^{-1}(\partial B^3) \to \partial B^3$ is unimportant because any orientation preserving *CAT* isomorphism $\partial B^3 \to \partial B^3$ is *CAT* isotopic to the identity.

For any homotopy 3-sphere M, let $M_0 \to B^k$ be the *CE* map of 3.1. The restriction to $\mathring{M}_0 \to \mathring{B}^k \underset{h}{\cong} R^k$ gives a class in ψ_3. This construction induces a map $\beta : \theta_3 \to \psi_3$; it can be shown to be well-defined just as we showed $\psi_4 = 0$. Clearly $\alpha\beta(M) = M$. If $f : M \to R^3$ represents a class in ψ_3, our construction of $\beta\alpha(f)$ gives a map $f' : M' \to R^3$ coinciding with $f : M \to R^3$ where they are not both homeomorphisms. By 3.2, $\beta\alpha(f) = f$ in ψ_3. Thus α and β are bijections.

IV. THE PROOF OF THEOREM A

All theorems are proven by applying an appropriate "handle lemma" to an appropriate triangulation of N.

LEMMA 4.1. Suppose $f : M \to R^k$ is a CAT CE map which is a homeomorphism except over a compact set in R^k. Suppose further either $k \neq 3,4$ or there is a CAT homeomorphism $g : M \to R^k$ which is an isomorphism except over a compact set. Then there is a CAT CE level-preserving concordance $F : M \times I \to R^k \times I$, with compact support, from $f = f_0$ to a CAT homeomorphism $f_1 : M \to B^k$.

Proof. First we construct f_1. Suppose there is a *CAT* homeomorphism $g : M \to R^k$ which is a *CAT* isomorphism outside a compact set. Without loss of generality we may assume f is a homeomorphism and g is a *CAT* isomorphism except over a compact set in $\mathring{B}^k$ and ∂B^k is transverse to $fg^{-1} : R^k \to R^k$. Apply the Squeezing Lemma 2.3 near ∂B^k, altering fg^{-1} so that it coincides with cone $(fg^{-1} \mid \partial B^k)$ near ∂B^k. Now extend $fg^{-1} \mid R^k - B^k$ by cone $(fg^{-1} \mid \partial B^k) : B^k \to B^k$. This extension is a homeomorphism which is *PL* if *CAT* = *PL* and is isotopic, with compact support, to a smooth homeomorphism even if *CAT* = *DIFF* [8]. In either case call the resulting *CAT* homeomorphism $\overline{f} : R^k \to R^k$. Then $f_1 = \overline{f}g : M \to R^k$ is a *CAT* homeomorphism equal to f except over a compact set. Now apply the *CE* Alexander trick (2.2) over a large ball in R^k to obtain the required concordance from f to f_1.

In case g is not given, but $k \neq 3,4$ use the classical result that any *CAT* manifold proper homotopy equivalent to R^k is *CAT* isomorphic to R^k. Let $g : M \to R^k$ be this *CAT* isomorphism.

LEMMA 4.2. Suppose $f : V \to R^k \times R^n$ *is a CAT CE map, transverse to* $R^k \times 0$, *which is a homeomorphism over a neighborhood of* $(R^k - \mathring{B}^k) \times 0$. *Assume either*

i) $3 \neq k \neq 4$

or

ii) for any CAT manifold M homotopy equivalent to ∂B^{k+1} *there is a CAT homeomorphism* $M \to \partial B^{k+1}$

Then there is a CAT CE level-preserving concordance $F : V \times I \to R^k \times R^n \times I$, *with compact support, from* $f_0 = f$ *to a map* f_1 *which is a homeomorphism over a neighborhood of* $R^k \times 0$.

Proof. Case 1. $n = 0$. If $n = 0$ and $3 \neq k \neq 4$, merely apply Lemma 4.1. If $k = 3$ or 4 it suffices, by 4.1, to construct a *CAT* homeomorphism $g : V \to R^k$ which is an isomorphism except over a compact set. With no loss of generality we assume ∂B^k is transverse to f. For U a neighborhood of $R^k - B^k$, there is a *CAT* isomorphism $g_0 : f^{-1}(U) \to U$; this is obvious if *CAT* = *PL* (let $g_0 = f$) and follows from [8 , Theorem 3.1] if *CAT* = *DIFF*. Then $g_0^{-1}(\partial B^k)$ bounds a CAT homotopy k-disk M in V. By hypothesis, there is a *CAT* homeomorphism $M \cup_{g_0} B^k \to \partial B^{k+1}$. By isotoping a regular point into B^k, we may assume the map is an isomorphism near $B^k \subset M \cup_{g_0} B^k$, so there is a *CAT* homeomorphism $g_1 : M \to B^k$ which is an isomorphism near ∂M; after a *CAT* isotopy near ∂M we may take $g_1 = g_0$ near ∂M. Then define $g : V \to R^k$ to be g_0 over $R^k - B^k$ and g_1 over $\mathring{B}^k$.

Case 2. General Case. Since f is transverse to $R^k \times 0$, there is a *CAT* isomorphism h of $f^{-1}(R^k) \times R^n$ onto a neighborhood of $f^{-1}(R^k)$ such that

$$\begin{array}{ccc} f^{-1}(R^k) \times R^n & \xrightarrow{fh} & R^k \times R^n \\ & \searrow \quad \swarrow & \\ & R^n & \end{array}$$

commutes. We may with no loss of generality, assume $V = f^{-1}(R^k) \times R^n$.

Let V_0 denote $f^{-1}(R^k)$. By the squeezing lemma 2.3 there is a *CAT CE* level-preserving concordance $G : V \times I \to R^k \times R^n \times I$, with compact support, from $g_0 = f$ to a map g_1 such that $g_1 \mid V_0 \times \{0\} = f \mid V_0 \times \{0\}$, for v in $f^{-1}(B^k \times 0)$, s in B^n, $g_1(v,s) = (f(v), s) \in B^k \times B^n$, and g_1 is a homeomorphism over a neighborhood of $(R^k - B^k) \times R^n$.

By Case 1 (reversing the roles of 0 and 1) there is a *CAT CE* level-preserving concordance $\overline{F} : V_0 \times I \to R^k \times I$ from $f \mid V_0 = \overline{f}_1$ to a *CAT* homeomorphism $\overline{f}_0$, with compact support over B^k. Then define $F : V_0 \times R^n \times I \to R^k \times R^n \times I$ by

$$f_{1-t}(v,s) = \begin{cases} (p_1\overline{F}(v, |s,t|), s) & \text{for } f(v) \text{ in } B^k \text{ and } |s,t| \leqq 1 \\ g_1(v,s) & \text{otherwise} \end{cases}$$

If $f(v)$ is near ∂B^k in B^k or $|s,t| \leqq 1$ is near 1, $p_1\overline{F}(v, |s,t|) = f(v,0) = g_1(v,s)$, so F is *CAT* and well-defined. It is easy to check that F satisfies all other requirements of the lemma.||

Definitions. Suppose N is *CAT* imbedded in some high-dimensional Euclidean space R^N. We say an imbedding $h : X \to N$ of a simplicial complex X is a *CAT imbedding* if $h^{-1}(\partial N)$ is a subcomplex and, for each k-simplex σ of X there is a neighborhood U of $h(\sigma)$ in R^N and a *CAT* isomorphism $g : R^k \times R^{n-k} \to N$ onto a neighborhood of $h(\sigma)$ such that $g^{-1}h : \sigma \to R^k \times R^{n-k}$ is the linear map of σ onto Δ^k, the standard k-simplex in $R^k \times \{0\}$. (R^N is used solely to avoid a special discussion of ∂N). If X is a combinatorial manifold and h is a homeomorphism, we say h (or X) is a *CAT triangulation* of N. *CAT* triangulations always exist (see e.g. [12]). A *CAT* imbedding $h : X \to N$ is *transverse* to a *CAT* map $f : M \to N$ near X, a closed k-simplex of X, if $g^{-1}f$ is transverse to $R^k \times 0$ over a neighborhood of Δ^k. In particular, $f^{-1}(\overset{\circ}{\Delta})$ is a *CAT* submanifold of M.

It is shown in [7] that, if a *CAT* imbedded $X \subset N$ is transverse to a *CAT* proper map $f : M \to N$ near a subcomplex $K \subset X$, then, if f is transverse to ∂N or $X \cap \partial N \subset K$, there is an arbitrarily small *CAT* isotopy $h_t : N \to N$ from the identity h_0 to a *CAT* isomorphism h_1 such that X is transverse to $h_1 f$.

Proof of Theorem A. *(2 ⟹ 1)* Let $g : M \to \partial B^{k+1}$ be a *CAT* homotopy equivalence and M_0 denote M with a *CAT* B^k removed. By 2.1 there is a *CAT CE* map $M_0 \to B^k$ which is a *CAT* isomorphism over a neighborhood of ∂B^k. The map then extends to a *CAT CE* map $M \to \partial B^{k+1}$. But (2) then gives a *CAT* homeomorphism $M \to \partial B^{k+1}$.

(1 ⟹ 2) The requirement that the concordance be ε-small can be satisfied by requiring that all triangulations, isotopies and collars be sufficiently small. We suppress further details, referring the reader to [9] and his own ingenuity.

Case 1: $f \mid \partial M : \partial M \to \partial N$ is a homeomorphism. By the collaring lemma (1.4), alter f so that it is a homeomorphism over a neighborhood of ∂N and ∂N is transverse to f. Choose a *CAT* triangulation of N, transverse to f, and let K be a subcomplex (near ∂N) over a neighborhood of which f is a homeomorphism. Suppose, inductively, that K contains the $(k-1)$-skeleton of X, $0 \leqq k \leqq n$. Let $\sigma \cong \Delta^k$ be a k-simplex, $g : R^k \times R^n \to N$ a *CAT* isomorphism onto a neighborhood of σ such that $g^{-1}f$ is transverse to $R^k \times \{0\}$ over a neighborhood of $\Delta^k \times \{0\}$. Apply lemma 3.2 to $g^{-1}f$ over $\overset{\circ}{\Delta}{}^k \times R^n$, altering $g^{-1}f$ by a *CAT CE* level-preserving concordance to a map which is a homeomorphism over a neighborhood of $\Delta^k \times \{0\}$. Postcompose the concordance with $g \times id_I$ to similarly alter f to a *CAT CE* map which is a homeomorphism over a neighborhood of $K \cup \sigma$. Now isotope X rel $K \cup \sigma$ so that it is again transverse to X. This completes the inductive step; when $k = n + 1$, f will be a homeomorphism everywhere.

Case 2. The general Case. Apply the fan lemma (2.5) to the *CAT CE* level-preserving concordance $F : \partial M \times I \to \partial N \times I$ given in Case 1 for $f \mid \partial M \to \partial N$. This reduces the general case to Case 1.

V. THE PROOF OF THEOREM B

Suppose $f : M \to N$ is a *CAT CE* map, and $X \subset N$ is a *CAT* imbedded complex which is transverse to f and which is a *CAT* homeomorphism over a neighborhood of X_2, where X_i denotes the i-skeleton of X. For each 3-simplex σ of X_3, $f^{-1}(\mathring{\sigma})$ is a *CAT* manifold *CAT* homeomorphic to σ near $\partial\sigma$. The correspondence of 2.3 associates to $f^{-1}(\mathring{\sigma})$ a unique element $\overline{\alpha}_f(\sigma)$ of θ_3^h. Then $\overline{\alpha}_f$ defines a simplicial cochain in $C^3(X, \theta_3^h)$, that is, a homomorphism from the 3-simplices of X to θ_3^h.

LEMMA 5.1 $\overline{\alpha}_f$ *is a cocycle. That is,* $\delta(\overline{\alpha}_f) = 0$ *in* $C^4(X, \theta_3^h)$.

Proof. We need to show that, for any τ in X_4, $\delta(\overline{\alpha}_f) \mid \tau = 0$. It is shown in [7] that if $f : M^k \to R^k$ is *CAT* transverse to $\partial\Delta^k \subset R^k$, then there is a *CAT* $(k - 1)$-manifold L and homeomorphisms $\overline{c} : B^k \to \Delta^k$, $c : L \times (1 - \varepsilon, 1] \to f^{-1}(\Delta^k)$ onto a neighborhood of $f^{-1}(\partial\Delta^k)$, such that $\overline{c}$ and c are *CAT* isomorphisms on $\mathring{B}^k$ and $L \times (1 - \varepsilon, 1)$ respectively, and such that $|\overline{c}^{-1} fc(\ell, r)| = r$. If we apply this construction near $f : f^{-1}(\tau) \to \tau$, L must be a *CAT* homotopy 3-sphere bounding a *CAT* homotopy 4-disk and homeomorphic to $f^{-1}(\partial\tau)$.

Let σ_i, $i = 1,\ldots,5$, be the faces of τ, oriented so that $\partial\tau = \Sigma\sigma_i$. To each σ_i corresponds an oriented homotopy 3-sphere S_i as in the proof of 3.3; since f is a *CAT* homeomorphism over a neighborhood of the 2-skeleton, $f^{-1}(\partial\tau)$ is homeomorphic to the connected sum $\#_{i=1}^{5} S_i$. By uniqueness of *CAT* structures on 3-manifolds $\#_{i=1}^{5} S_i$ is *CAT* isomorphic to L, which bounds a *CAT* contractible 4-manifold. Hence

$$0 = \sum_{i=1}^{5} \bar{\alpha}_f(\sigma_i) = \bar{\alpha}_f(\partial\tau) = \delta(\bar{\alpha}_f)|\tau \ . \quad ||$$

If X is a simplicial complex, $X \times I$ has a simplicial triangulation extending $X \times \partial I$. If $H : X \times I \to N \times I$ is a concordance which is also a *CAT* imbedding, $H(X \times I)$ will be called a *CAT* imbedded concordance from $h_0(X)$ to $h_1(X)$. Let $f : M \to N$ and $X \subset N$ be as above, $X \times I \subset N \times I$ the product concordance, simplicially triangulated, and $F = f \times id_I : M \times I \to N \times I$.

LEMMA 5.2. Suppose $n \geq 5$ and $\varepsilon : N^n \to (0,\infty)$ is continuous. Then for any β in $C^2(X \times 1, \theta_3^h)$ there is a CAT CE level-preserving concordance $H : M \times I \to M \times I$ such that

i) H is the identity except over an $\varepsilon(b)$ neighborhood of each barycenter b of a 2-simplex in $X \times 1$.

ii) $\bar{\alpha}_{FH} - \bar{\alpha}_F = \delta\beta$ in $C^3(X \times I, \theta_3^h)$. In particular $\bar{\alpha}_{fh_1} - \bar{\alpha}_f = \delta\beta$ in $C^3(X, \theta_3^h)$.

Proof. Without loss of generality, assume $\beta = 0$ except on one 2-simplex τ, with barycenter b. Let $g : L \to B^3$ be a *CAT CE* map which is a *CAT* isomorphism near ∂L and which represents $\beta(\tau)$ in θ_3^h (as defined in 3.1).

CAT imbed $(B^{n-2}, \partial B^{n-2})$ in $(N \times I, \partial(N \times I))$ so that

i) B^{n-2} is within $\varepsilon(b)$ of b

ii) B^{n-2}, ∂B^{n-2} are transverse to F

iii) B^{n-2} intersects transversally and exactly once each 3-simplex σ of $X \times I$ which has τ as a face, and is disjoint from all other 3-simplices and from τ.

Since F is transverse to B^{n-2} and to each 3-simplex σ, there is a tubular neighborhood $\nu : F^{-1}(B^{n-2}) \times B^3 \to M \times I$ of $F^{-1}(B^{n-2})$ such that for each 3-simplex of $X \times \{1\}$ with τ as a face, $(F\nu)^{-1}(\sigma)$ is a fiber $(p \times B^3)$. Remove the interior of image ν from $M \times I$ and attach $F^{-1}(B^{n-2}) \times L$ via the map

$\bar{\nu} \times g : F^{-1}(B^{n-2}) \times \partial L \to \nu(F^{-1}(B^{n-2}) \times \partial B^3)$. The result is a *CAT CE* manifold W and a *CAT CE* map $H : W \to M \times I$ which is the identity off of $F^{-1}(B^{n-2}) \times L$ and is $\bar{\nu} \times g$ on it. Furthermore, for σ any 3-simplex of $X \times I$ which has τ as a face, $(FH)^{-1}(\sigma)$ is obtained from $F^{-1}(\sigma)$ by replacing a copy of B^3 in $F^{-1}(\sigma)$ with L. Thus $\alpha_{FH} = \alpha_F + \delta\beta$.

Now we show there is a *CAT* isomorphism $M \times I \to W$ equal to H^{-1} except over an $\varepsilon(b)$ neighborhood U of $(b,1)$. Take for U a *CAT* $(n + 1)$ disk in $N \times I$, with boundary transverse to FH. Since H and F are *CE*, $(FH)^{-1}(U)$ and $F^{-1}(U)$ are *CAT* homotopy $(n + 1)$-disk. Since $n + 1 \geq 6$, $FH^{-1}(U)$ and $F^{-1}(U)$ are *CAT* $(n + 1)$-disks. Thus the *CAT* isomorphism $H^{-1} : (M \times I) - (F^{-1}(U)) \to W - (FH)^{-1}(U)$ extends to a *CAT* isomorphism $M \times I \to W$. The composition $\bar{H}$ of this *CAT* isomorphism with H fulfills all the requirements of the Lemma except it may not be level-preserving.

However, since $\bar{h}_1$ differs from $\bar{h}_0$ = identity only over an $\varepsilon(b)$ neighborhood of b, we may apply the *CE* Alexander trick (2.2) to find a level preserving *CAT CE* concordance from $\bar{h}_1$ to $\bar{h}_0$, equal to H except over the neighborhood. ||

Let $\dot{\alpha}_f$ in $H^3(X, \theta_3^h)$ denote the cohomology class represented by $\bar{\alpha}_f$.

LEMMA 5.3. Suppose $F : M \times I \to N \times I$ is a CAT CE concordance, $X \times I \subset N \times I$ is a CAT imbedded concordance, F is a CAT homeomorphism over a neighborhood of $(X \times \{i\})_2$ and F is CAT transverse to $X \times \{i\}$, $i = 0,1$. Then $\alpha_{f_0} = \alpha_{f_1}$ in $H^3(X, \theta_3^h)$.

Proof. By the transversality theorem of [7] , $X \times I$ may be isotoped, rel $X \times \partial I$, to an imbedding transverse to F. Now apply 4.2 inductively to the i-skeleton of $(X \times I) - (X \times \partial I)$, $i = 0,1,2$, altering F rel $M \times \partial I$ until it is a *CAT* homeomorphism over $(X \times I)_2$. Then we can define $\bar{\alpha}_F$ and α_F. Let

$j_i : X \times \{i\} \to X \times I$, $i = 0,1$, be the inclusions and $p : X \times I \to X$ be the projection so $p^* j_i^* = (id_{(X \times I)})^* : H^*(X \times I) \to H^*(X \times I)$ and $j_i^* p^* = (id_X)^* : H^*(X) \to H^*(X)$. By definition, $j_i^*(\alpha_F) = \alpha_{f_i}$, so $p^*(\alpha_{f_i}) = p^* j_i^*(\alpha_F) = \alpha_F$. Hence $\alpha_{f_1} - \alpha_{f_0} = j^* p^*(\alpha_{f_1} - \alpha_{f_0}) = h^*(\alpha_F - \alpha_F) = 0$.

Suppose $f : M \to N$ is a *CAT CE* map and $X \subset N$ is any *CAT* imbedded complex. Define $\alpha_f \in H^3(X, \theta_3^h)$ as follows. According to the transversality theorem of [7], there is an isotopy of X in N to a *CAT* imbedded complex *CAT* transverse to f. Now apply 4.2 inductively to X_i, $i = 0,1,2$ altering f to a map f' which is a *CAT* homeomorphism over a neighborhood of X_2. Then $\overline{\alpha}_{f'}$ can be defined, and by 5.3, $\alpha_{f'}$ is well-defined in $H^3(X, \theta_3^h)$. So take α_f to be $\alpha_{f'}$.

If f is a homeomorphism over a neighborhood of a subcomplex K of X then we may fix K and fix f over a neighborhood of K, so that $f' = f$ over a neighborhood of K. Then $\alpha_{f'}(K) = 0$, so $\alpha_{f'}$ is in $C^3(X,K)$ and represents a well-defined class in $H^3(X,K)$ denoted $\alpha_f(X,K)$.

LEMMA 5.4. *Suppose* $f : V \to R^k \times R^n$, $k + n \geq 5$, *is a CAT CE map transverse to* $R^k \times 0$ *and a homeomorphism over a neighborhood of* $(R^k - \overset{\circ}{B}{}^k) \times 0$. *Assume either*

a) $k \neq 3$

b) $k = 3$ *and* $f | f^{-1}(R^3)$ *is trivial in* ψ_3.

Then there is a CAT CE level-preserving concordance $F : V \times I \to R^k \times R^n \times I$, *with compact support, from* $f_0 = f$ *to a map* f_1 *which is a CAT homeomorphism over a neighborhood of* $R^k \times 0$.

Proof. For $k \neq 3,4$ the proof follows from 4.2, so assume $k = 3$ or 4. As in the proof of 4.2, we may assume there is a *CAT* manifold M such that $V = M \times R^n$ and, for m in M and s in R^n,

$f(m,s) = (f(m), s)$ over a neighborhood of $B^k \times R^n$. Since $k = 4$ or by assumption b), there is, according to 3.3, a *CAT CE* map $\overline{f} : W \to R^k \times I$ such that $\overline{f}$ is a *CAT* homeomorphism over $R^k \times 1$, equals $f|M$ over $R^k \times 0$ and is a *CAT* homeomorphism outside a compact set in $\overset{\circ}{B}{}^k \times I$. We may further assume $\partial \dot{B}^k \times I$ is *CAT* transverse to $\overline{f}$.

Let $(R^k \times R^n)_+$ be the $k + n + 1$ manifold, with *CAT* interior, obtained by identifying $B^k \times R^n \times 0$ in one copy of $B^k \times R^n \times I$, with $B^k \times R^n \times 1$ in a disjoint copy of $R^k \times R^n \times I$. Similarly, let V_+ be the manifold with *CAT* interior obtained by identifying $\overline{f}^{-1}(B^k \times 0) \times R^n$ in $W \times R^n$ with $f^{-1}(B^k) \times R^n \times 1$ in $V \times I = M \times R^n \times I$. Then $f \times id_I$ and $\overline{f} \times id_{R^n}$ define a *CE* map $\overline{F} : V_+ \to (R^k \times R^n)_+$ which is *CAT* over the interior and a homeomorphism near $\partial(V_+) - V \times \{0\}$. If $\overline{F}$ were a *CAT CE* map between *CAT* manifolds, it would almost suffice to observe that V_+ is isomorphic to $V \times I$ and $(R^k \times R^n)_+$ is isomorphic to $R^k \times R^n \times I$, and let $F = \overline{F}$. However, as constructed, $(R^k \times R^n)_+$ and V_+ will not be smooth, so we resort to the following artifice:

Using a *CAT* Urysohn function $R^k \times R^n \to I$, with support a ball in $B^k \times B^n$, construct a *CAT* imbedding $h : R^k \times R^n \times I \to (R^k \times R^n_+)$ such that

1) h is the identity except near $B^k \times B^n \times 1$ in $R^k \times R^n \times I$

2) h carries $R^k \times 0 \times 1 \subset R^k \times R^n \times I$ so close to $\partial(R^k \times R^n)_+ - (R^k \times R^n \times 0)$ that $\overline{F}$ is a homeomorphism over a neighborhood of $h(R^k \times 0 \times 1)$.

3) $\overline{F}$ is transverse to $h(R^k \times R^n \times 1)$.

Then by 3), $\overline{F}^{-1} h(R^k \times R^n \times I)$ is a *CAT* submanifold V_h of V_+. Since $\overline{F}$ is *CE* and $h(R^k \times R^n \times I)$ is obtained from $R^k \times R^n \times I$ by attaching a $(k + n + 1)$-ball along $B^k \times B^n \times 1$, it follows that V_h is obtained from $V \times I$ by attaching a homotopy B^{k+n+1} along a homotopy B^{k+n}. Since $k + n \geqq 5$, there is a *CAT* isomorphism $V \times I \cong V_h$, which is the identity except near $f^{-1}(B^k \times B^n) \times 1 \subset V \times 1$. Let F denote the composition

$V \times I \cong V_h \xrightarrow{h^{-1}\overline{F}} R^k \times R^n \times I$. By 2), F is a homeomorphism over a neighborhood of $R^k \times 0 \times 1$.

Thus F satisfies all the requirements of the lemma except that it be level-preserving. But since $f_0, f_1 : V \to R^k \times R^n$ differ only over some compact set in $R^k \times R^n$, it is possible to find a *CAT CE* level-preserving concordance between them, by the *CE* Alexander trick 2.2. ||

THEOREM 5.5. Suppose $X \subset N^n$, $n \geqq 5$, *is a CAT imbedded complex, and* $f : M \to N$ *is a CAT CE map. Then* $\alpha_f = 0$ *in* $H^3(X; \theta_3^h)$ *if and only if there is a CAT CE concordance* $F : M \times I \to N \times I$ *and a CAT imbedded concordance* $H : X \times I \subset N \times I$ *such that*

i) $f_0 = f$, $h_0(X) = X \subset N \times \{0\}$

ii) f_1 *is a homeomorphism over a neighborhood of* $h_1(X)$.

Proof. If the concordances F and H exist, then α_{f_1} is trivial on $h_1(X)$. By 5.3, $\alpha_f = \alpha_{f_0} = 0$ in $H^3(X, \theta_3^h)$.

Suppose, conversely, $\alpha_f = 0$. Then, by definition, there is an isotopy of X and a concordance of f to a map f_1 which is transverse to X, a homeomorphism over X_2, and $\overline{\alpha}_{f_1}(X)$ is a coboundary. Revert to f as notation for f_1. By Lemma 5.2 we may again alter f so that $\overline{\alpha}_f(X) = 0$. Apply Lemma 5.3 to the interior of each 3-simplex of X to alter f so that it is a homeomorphism over a neighborhood of the 3-skeleton. Move X by a small isotopy so it is again transverse to f. Repeat for the 4-skelton. The rest of X is covered by applying 4.2 as in the proof of Theorem A.

Proof of Theorem B. Case 1: $f : \partial M \to \partial N$ is a homeomorphism. By the collaring lemma 2.4, we may assume f is a homeomorphism over a neighborhood of ∂N, and ∂N is transverse of f. Let $h : (X, \partial X) \to (N, \partial N)$ be any *CAT* transverse to $f \mid \partial M$ and hence to f; extend to an isotopy of h to a *CAT* triangulation transverse to f [7].

Since f is a homeomorphism over a neighborhood of ∂N, $\bar{\alpha}_f$ is trivial on ∂X. Using 4.2, alter f by a *CAT CE* level-preserving concordance, fixed near ∂M, so that f is a homeomorphism over X_2. Then α_f defines a cocycle in $C^3(X, X;\theta_3^h)$. As in the proof of 5.3, the corresponding class α_f is well-defined in $H^3(X, X;\theta_3^h)$. Define α in $H^3(N, \partial N; \theta_3^h)$ by $\alpha = (h^{-1})^*(\alpha_f)$; it is independent of the triangulation X because any two *CAT* triangulations of N are *CAT* concordant [12].

Suppose first that f is *CAT CE* concordant to a homeomorphism f_1, fixing $f \mid \partial M$. Since f_1 is a homeomorphism, $\bar{\alpha}_{f_1} = 0$, so $\alpha_f = \alpha_{f_1}$ is trivial. Then $\alpha = (h^{-1})^*(\alpha_f) = 0$ in $H^3(N, \partial N)$.

Suppose, conversely, that α and hence α_f are trivial. Then $\bar{\alpha}_f$ in $C^3(X, \partial X)$ is a coboundary of a 2-chain which is trivial on ∂X. Apply 5.2 to alter f outside a neighborhood of ∂X so that $\bar{\alpha}_f$ is trivial. Repeat the proof of Theorem A, using 5.3 instead of 4.2 near the *3* and *4*-simplices of X.

Case 2. $n \geqq 6$. As in Case 1, choose a triangulation $h : (X, \partial X) \to (N, \partial N)$ and alter f so that f is transverse to X and is a homeomorphism over X_2. Let $\alpha = (h^{-1})^*(\alpha_f)$ in $H^3(N: \theta_3^h)$.

If f is *CAT CE* concordant to a homeomorphism f_1 then α_{f_1} is trivial, and so then are α_f and α.

Conversely, suppose $\alpha = 0$. Then $\bar{\alpha}_f = \delta\beta$ for some cochain β in $C^2(X)$.

Let $i^* : C^*(X) \to C^*(\partial X)$ be induced by inclusion. Construct a *CAT* concordance $H : \partial M \times I \to \partial M \times I$ as given in 5.2 for $-i^*(\beta)$. Apply the fan lemma 2.5 to $(f \mid \partial M \times id_I)H : \partial M \times I \to \partial N \times I$; this alters f near ∂M to a *CAT CE* map f' such that $\bar{\alpha}_{f'} - \bar{\alpha}_f = \delta(-i^*(\beta))$. In particular $\bar{\alpha}_{f'} \mid \partial X = \bar{\alpha}_f \mid \partial X - i^*(\delta\beta) = 0$, so $\bar{\alpha}_{f'}$ is trivial on ∂X. By Case 1, there is a *CAT CE* level-preserving concordance, fixed over $(\partial X)_2$, of $f' \mid \partial M$ to a *CAT* homeomorphism. Again apply the fan lemma to this concordance, altering f' near ∂M so that $f' \mid \partial M \to \partial N$ is a homeomorphism.

Furthermore, $\bar{\alpha}_{f'} = \bar{\alpha}_f + \delta(-i^*(\beta)) = \delta\beta - \delta i^*(\beta) = \delta(\beta - i^*(\beta))$. But $i^*(\beta - i^*(\beta)) = 0$, so $\beta - i^*(\beta)$ is in $C^2(X, \partial X)$. Thus $\bar{\alpha}_{f'}$ is the coboundary of $(\beta - i^*(\beta))$ and $\alpha_{f'}$ is trivial in $H^3(X, \partial X)$. The theorem now follows from Case 1. ||

VI. THE PROOF OF THEOREM C

To any homotopy (indeed homology) 3-sphere M is associated the Rohlin Z_2-invariant, defined as the mod 2 reduction of 1/8 the index of a PL parallelizable 4-manifold with boundary M. This invariant defines a homomomorphism $\rho : \theta_3^h \to Z_2$, and thus homomorphisms $C^3(., \theta_3^h) \to C^3(., Z_2)$ and $H^3(., \theta_3^h) \to H^3(., Z_2)$. We denote the image of $\bar{\alpha}_f$, α_f, and α under these reductions by $\bar{\varphi}_f$, φ_f and φ respectively.

LEMMA 6.1. Suppose $f : V \to R^k \times R^n$, $k + n \geqq 5$, $k \leqq 3$ is a CAT CE map transverse to $R^k \times 0$ and a CAT isomorphism over a neighborhood of $(R^k - \mathring{B}^k) \times 0$. Assume either

a) $k \neq 3$

b) $k = 3$ and $f | f^{-1}(R^3)$ in ψ_3 corresponds to an element γ in θ_3^h with $\rho(\gamma)$ trivial in Z_2.

Then there is a CE level-preserving concordance $F : V \times I \to R^k \times R^n \times I$ with compact support, to a map f_1 which is a CAT isomorphism over a neighborhood of $R^k \times 0$. Moreover, if $k \leqq 2$, F is CAT.

Proof. *Case 1:* $k \leqq 2$. For $CAT = PL$ this is a special case of 3.2. For $CAT = DIFF$ it is an immediate consequence of 4.2 and of Lemma 1.1 of [8].

Case 2: $k = 3$. As in the proof of Theorem A, we may assume there is a CAT 3-manifold M such that $V = M \times R^n$. By the main Theorem 2.1 of [9], there is a CE level-preserving concordance of f, with compact support in $B^3 \times R^n$, to a map f' which is a homeomorphism over a neighborhood of B^3.

First note that the *TOP/CAT* handle-straightening problem represented by f' (see [5]) is equivalent to our problem. Indeed, if f' is isotopic, with compact support, to a map f_1 which is a diffeomorphism over a neighborhood of $R^3 \times 0$ then the concordance from f to f' together with the isotopy from f' to f_1 is the concordance required by 6.1.

On the other hand, if the concordance of 6.1 exists, then, again by 2.1 of [9], the induced concordance $\bar{F} : V \times I \to R^k \times R^n \times I$ from $f' = f_0$ to f_1 may be altered rel $V \times \partial I$ to a homeomorphism. Hence F is a solution to the handle-straightening problem of f'.

It remains only to show that the handle-straightening problem for f' is solvable. The associated torus problem is a homotopy equivalence $M \times T^n \to R^3 \times T^n$, which can be solved if and only if $\rho(\gamma) = 0$ (see [11]).||

Remark. A direct construction of F is relatively easy if one recalls that for M a homotopy 3-sphere, cone$(M) \times R^n$ is a manifold when $3 + n \geq 4$ and is a *CAT* manifold if the Rohlin invariant of $\bar{M}$ is trivial [4], [10].

Proof of C. Alter f as in Theorem A so that f respects boundary collars of ∂M and ∂N. Choose a *CAT* triangulation $h : (X, \partial X) \to (N, \partial N)$ so that X is transverse to f.

First we note that if f is a *CAT* homeomorphism over ∂N we may assume it is a *CAT* isomorphism over a neighborhood of $(\partial X)_3$. This is a tautology if $CAT = PL$. If $CAT = DIFF$, $f | \partial M$ is C^∞ isotopic to a map which is a smooth isomorphism $(\partial X)_3$ [8]. Exploit collars of ∂M and ∂N, then, to alter f over the middle of the collar $(\partial N \times [0,1], \partial N \times 0) \subset (N, \partial N)$ so that f becomes a smooth isomorphism over a neighborhood of $(\partial X)_3 \times \frac{1}{2} \subset \partial N \times \frac{1}{2}$.

Discard $\partial N \times [0, 1/2)$ and $f^{-1}(\partial N \times [0, 1/2))$ from N and M, thereby reducing to the case where f is a *CAT* isomorphism over $(\partial X)_3$.

Case 1. $f : \partial M \to \partial N$ *is a homeomorphism.* As just shown we may assume f is an isomorphism over $(\partial X)_3$. Suppose first that the *CE* concordance F exists, and let $f_1 : M \to N$ be a *CAT* homeomorphism. Even when *CAT* = *DIFF* we may, by 1.1 of [8], assume that f_1 is a *CAT* isomorphism over a neighborhood of X_3. But it was shown in the proof of 6.1 that $\overline{\varphi}_f$ is a cocycle representing the Kirby-Siebenmann obstruction to isotoping (rel ∂) any homeomorphism *CE* concordant (rel ∂) to f to a *CAT* isomorphism over a neighborhood of X_3. Since f_1 exists, this obstruction vanishes, so $\varphi = (h^{-1})^* \varphi_f = 0$ in $H^3(N, \partial N; Z_2)$.

Suppose, coversely, that $\varphi = 0$ so $\varphi_f = 0$. Proceed as in the proof of *B*, using 6.1 instead of 5.3, altering f rel ∂M until f becomes a *CAT* homeomorphism over a neighborhood of $X_3 \cup \partial X$ and an isomorphism over a neighborhood of X_3. Now apply lemma 2.1 of [10] to neighborhoods of all i-simplices of X not in ∂X, $i = 4, 5, \ldots, n$, altering f rel ∂ so that it becomes a (non-*CAT*) homeomorphism over all of N which is still a *CAT* homeomorphism over a neighborhood of $X_3 \cup \partial X$ and a *CAT* isomorphism over X_3. Apply the *TOP/CAT* version of the handle-straightening lemma of [5] near the 4-simplices of X not contained in ∂X. If *CAT* = *PL*, apply the same *TOP/PL* lemma near each i-simplex of X not in ∂X, $i = 5, \ldots, n$ until f becomes a *PL* homeomorphism. If *CAT* = *DIFF* apply instead the TOP/C^∞ handle lemma [8 , preprint 4.2].

Case 2: $n \geqq 6$. When F exists the same proof as in Case 1 shows that φ is trivial in $H^3(N; Z_2)$.

Suppose conversely that φ is trivial. By the main theorem of [9], $f : M \to N$ is *CE* concordant to a homeomorphism f'. In 6.1 we discovered that φ is precisely the obstruction to isotoping f' to a *CAT* isomorphism over X_3. Since $\varphi = 0$, f' is isotopic to a map which is a *CAT* isomorphism over X_3. The proof then proceeds as in Case 1, but we no longer restrict attention only to those simplices outside of ∂X. ||

REFERENCES

1. R. H. Bing, The cartesian product of a certain nonmanifold and a line is E^4, Ann. Math. 70 (1959), 399-412.
2. M. Cohen, Homeomorphisms between homotopy manifolds and their resolutions, Inv. Math. 10 (1970), 239-250.
3. M. Cohen and D. Sullivan, Mappings with contractible point inverses between p.l. manifolds, Notices Amer. Math. Soc. 15 (1968), 186. Abstract #653-363.
4. L. C. Glaser, On the double suspension of certain homotopy 3-spheres, Ann. of Math. 85 (1967), 494-507.
5. R. C. Kirby and L. C. Siebenmann, On the triangulation of manifolds and the Hauptvermutung, Bull. Amer. Math. Soc. 75 (1969), 742-749.
6. E. Moise, Affine structures on 3-manifolds, Ann. Math. 56 (1952), 96-114.
7. M. Scharlemann, Transverse Whitehead triangulations, U.C. Santa Barbara mimeo, 1977.
8. M. Scharlemann and L. C. Siebenmann, The Hauptvermutung for C^∞ homeomorphisms II. A proof valid for open 4-manifolds, Compositio Math. 29 (1974), 253-263.
9. L. Siebenmann, Approximating cellular maps by homeomorphisms, Topology, 11 (1972), 271-294.
10. ________________, Are non-triangulable manifolds triangulable? Topology of manifolds (Proc. Inst. Univ. of Georgia, Athens, Ga., 1969), Markham, Chicago, Ill., 1970, pp. 77-84.
11. ________________, Disruption of low-dimensional handlebody theory by Rohlin's theorem, Topology of manifolds (Proc. Inst. Univ. of Georgia, Athens, Ga., 1969), Markham, Chicago, Ill., 1970, pp. 57-76.
12. J. H. C. Whitehead, On C^1 complexes, Ann. Math. 41 (1940), 804-824.

ON COMPLEXES THAT ARE LIPSCHITZ MANIFOLDS

L. Siebenmann

Department of Mathematics
Universite Paris-Sud
Orsay, France

D. Sullivan

Department of Mathematics
Institut des Hautes Etudes Scientifiques
Bures/Yvette, France

In this paper, we characterize those simplicial complexes that are locally Lipschitz isomorphic to euclidean space. The relevant lemmas concerning the conformal model of hyperbolic geometry as the open unit disc have wider implications and appear as appendices.

The objects of *Lipschitz topology*, cf. [9], are metric spaces X, Y, etc. and the maps $f : X \to Y$ (morphisms) are *(locally) Lipschitz* maps, i.e. each point in X admits a neighborhood U with a constant $K < \infty$ so that for each pair x,y in U, one has

$$d(f(x), f(y)) \leq K\, d(x,y).$$

ISBN 0-12-158860-2

The restriction of f to U is said to be *uniformaly Lipschitz*[1], or more precisely *K-Lipschitz*.

Intuitively speaking, a Lipschitz map is one that obeys locally posted (or temporary) speed limits. Of course, on a compact space X, a Lipschitz map obeys a single speed limit, i.e. it is K-Lipschitz for some $K < \infty$ (depending on f).

THEOREM 1. Suppose that a locally finite simplicial complex X with its barycentric metric is a Lipschitz n-manifold in the sense that each point admits an open neighborhood U and a Lipschitz isomorphism $h : U \to \mathbb{R}^n$. Then the link of every simplex of X is homotopy equivalent to a sphere.

Remark a) : Our proof by Hausdorff measure general position uses only that, for each simplex σ of X, there exist an open $U \subset X$ with $U \cap \sigma \neq \emptyset$ and a homeomorphism $h : U \to \mathbb{R}^n$ such that the restriction $h|(U \cap \sigma)$ is Lipschitz.

Remark b) : R. D. Edwards and J. Cannon [4][1] have shown that, for a closed triangulated (homology) manifold H^k such that $H_*(H^k) = H_*(S^k)$, the double suspension $\Sigma^2 H^k$ is always homeomorphic to the sphere S^{k+2}. It is an amusing consequence of the above that when $\pi_1 H \neq 0$, the homeomorphism *cannot* be Lipschitz on as much as the suspension circle.

[1]*N.B. Many authors differ as to terminology (Sullivan* [17] *included), often preferring to say Lipschitz where this article says uniformly Lipschitz. With the present definition (only !), the Lipschitz property is local so that, for example, every C^1 differentiable map between Riemannian manifolds is Lipschitz, and every piecewise linear map between locally finite simplicial complexes is Lipschitz.*

Proof of theorem 1. An open k-simplex σ of X has an open neighborhood in X, e.g. its open star, that is naturally Lipschitz isomorphic to $\sigma \times c(L)$, where $c(L)$ is the open cone on the link L of σ. Thus local applications of Poincaré duality reveal that L has the integral homology of a $(n - k - 1)$-sphere and is itself an integral homology $(n - k - 1)$-manifold. Hence, when dim L is 0, 1 or 2, the link L is a triangulated sphere.

In view of the Hurewicz theorem, our task is to show that L is simply connected when dim $L \geq 3$, i.e. when σ has codimension ≥ 4. To prove this, consider any loop γ in L, and regard L as the base of a small standard sub-cone $x_0 \times c_\varepsilon(L)$, of radius ε, lying in some U as hypothesized. The cone on γ gives a map $f : B^2 \to U$ into $x_0 \times c_\varepsilon(L)$ contracting γ and meeting σ at $f(0) = x_0$ only.

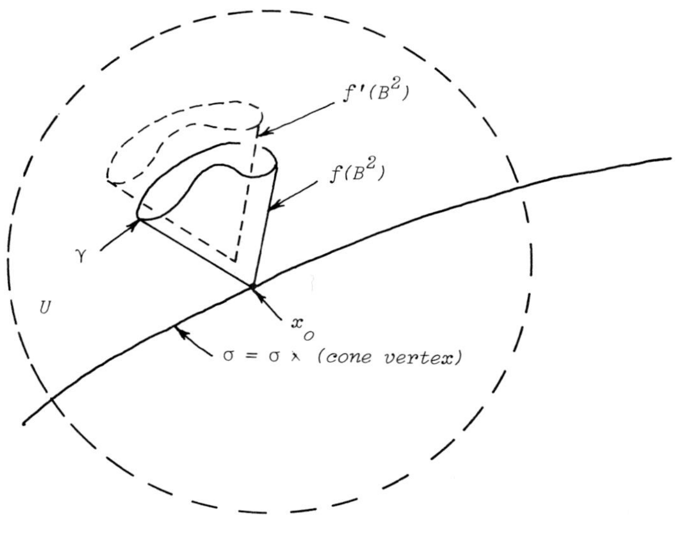

Fig. 1.

The general position lemma below shows that if we identify U with $\mathbb{R}^n$ by h and jiggle f a tiny bit, we get a map $f' : B^2 \to U$ that does not meet σ; thus γ is nul-homotopic in $(\sigma \times cL) - (\sigma \times \text{cone vertex}) = \sigma \times (0,\infty) \times L$ and therefore also in L itself. To apply this lemma, we must observe that $h|(U \cap \sigma)$ Lipschitz implies that the Hausdorff dimension of $h(U \cap \sigma)$ in $\mathbb{R}^n$ is $\leq \dim \sigma$.

GENERAL POSITION LEMMA. Let $A \subset \mathbb{R}^n$ be any subset of Hausdorff dimension $\leq a$. Then any continuous map $f : K \to R^n$ of a finite simplicial k-complex K, with $k < n - a$, can be approximated by a map $f' : K \to (R^n - A)$.

Proof. By simplicial approximation, we can assume (after subdividing K) that f is linear and injective on simplices. Consider a closed t-simplex τ of K and the projection p parallel to $f(\tau)$ to a $(n - t)$-plane P normal to $f(\tau)$; since p is Lipschitz, $p(A) \subset P$ has Hausdorff dimension $\leq a < n-t$ and hence Lebesque measure zero in P. Thus, for almost every t-plane P parallel to $f(\tau)$, the intersection $P \cap A$ is empty. It follows that, for almost every translation T of $\mathbb{R}^n$, one has $Tf(K) \cap A = \emptyset$. We choose a small such T and set $f' = Tf$.

In [9], Luukainen and Vaisala derive stronger general position results from a theorem of Fubini type for Hausdorff measure [7; 2.10.25]. (We thank Vaisala for making the above proof perfectly elementary.)

There is a converse to Theorem 1.

THEOREM 2. (cf. [13]) Let X be a simplicial homotopy n-manifold (without boundary) i.e. a locally finite simplicial complex in which the link of every k-simplex is homotopy equivalent to S^{n-k-1}. Then X is a Lipschitz n-manifold provided that $n \neq 4$.

Remarks: (i) If the 3 and 4-dimensional Poincaré conjectures are true (piecewise linearly), then an easy induction shows something stronger : X^n is a piecewise-linear manifold (for all n). But, they well may be false, see [2]!

(ii) To lift the proviso that $n \neq 4$, the reader will find that it is sufficient (and necessary) to show that for any piece-wise-linear 3-manifold L^3 homotopy equivalent to S^3, the product $L^3 \times S^1$ is Lipschitz isomorphic to $S^3 \times S^1$, cf. proof of Lemma 4 below.

The crux of Theorem 2 is

LEMMA 3. Let L be a triangulated closed 3-manifold that is homotopy equivalent to S^3. Then the simplicial join $L * S^1$ (with barycentric metric) is Lipschitz isomorphic to S^5.

From this point on we shall occasionally rely on the basic results of [17] stating that

(a) a topological manifold X^n, $n \geq 5$, $\partial X = \emptyset$, admits a metric making it a Lipschitz manifold and

(b) any two such Lipschitz structures are Lipschitz isomorphic by a map that (majorant) approximates the identity map.

We prove Lemma 3 by a hyperbolic version of the 'torus-to-suspension' trick in [13]. One can also prove it by the Lipschitz replication technique of [14, §3]; indeed this might lead to a proof of Theorem 2 not using [17]. We leave the reader to explore this alternative.

Proof of Lemma 3. Find a homeomorphism $L^3 \times T^2 \to S^3 \times T^2$ where T^2 is a smooth multiple torus with a metric of constant negative curvature -1. For example, letting $L_0^3 = L^3$ - (open-3-ball), we can identify $L/L_0 = S^3$ topologically; then, the resulting quotient map $L^3 \times T^2 \to S^3 \times T^2$ can be approximated by a homeomorphism [5]; one uniformaly squeezes the sets $L_0^3 \times t$, for $t \in T^2$, by a radial engulfing isotopy of $L^3 \times T^2$ and applies the Bing shrinking criterion.

One can deform $L^3 \times T^2 \to S^3 \times T^2$ to a Lipschitz isomorphism [17], say h_0. The map of fundamental groups is still $id|\pi_1(T^2)$ up to conjugation.

The universal covering $h : L^3 \times \tilde{T}^2 \to S^3 \times \tilde{T}^2$ of h_0 above is a Lipschitz isomorphism that is Γ-equivariant, where $\Gamma = \pi_1(T^2)$. Identifying $\tilde{T}^2$ isometrically to $\mathring{B}^2$ with hyperbolic metric as in [17] and applying Appendix A, we find that h extends uniquely to a Lipschitz isomorphism $H : \Sigma^2 L^3 \to \Sigma^2 S^3$.

Here $\Sigma^2 L^3 = L^3 \times \mathring{B}^3 \cup (\partial B^3)$ has the suspension metric d_s of Appendix A. Now, d_s derives via the natural composed quotient map (of sets) $L \times [0,1] \times S^1 \to L \times B^2 \to \Sigma^2 L$ from the 'join' pseudo-metric on $L \times [0,1] \times S^1$:

$$(x,t,y),(x',t',y') \mapsto (1-\max\{t,t'\})d(x,x')+|t-t'|+\min\{t,t'\}d(y,y').$$

But this pseudo-metric induces a metric on the join $L * S^1$ that one can routinely calculate to be Lipschitz equivalent to the barycentric simplicial metric.

Similarly S^5 and $\Sigma^2 S^3$ are both naturally quotients of $S^3 \times I \times S^1$ by the same relation, the quotient map to S^5 being

$$(x,t,y) \mapsto (\cos\tfrac{\pi}{2}\, tx, \sin\tfrac{\pi}{2}\, ty) \in R^4 \times R^2$$

and one checks that a Lipschitz isomorphism $S^5 \cong \Sigma^2 S^3$ results.

Thus, in all, we have Lipschitz isomorphisms $L * S^1 \cong \Sigma^2 L^3 \overset{H}{\cong} \Sigma^2 S^3 \cong S^5$ as required.

We now can prove

LEMMA 4. Let L^4 be a compact simplicial homotopy 4-manifold that is homotopy equivalent to S^4. Then the simplicial join $L^4 * S^0$ is Lipschitz isomorphic to S^5.

Proof of Lemma 4. $L^4 \times R$ is homeomorphic to $S^4 \times R$, see [13; p. 83, Assertion] or [12, App. I]. Hence, by wrapping up [15], $L^4 \times S^1$ is homeomorphic to $S^4 \times S^1$. Also, Lemma 3 shows that $L^4 \times S^1$ is a Lipschitz manifold (since $L^3 * S^1$ contains cone $(L^3) \times S^1$ Lipschitz embedded). Thus [17] shows that $L^4 \times S^1$ is Lipschitz isomorphic to $S^4 \times S^1$.

There results, on passing to universal coverings, a Lipschitz isomorphism $\Sigma^1 L^4 \cong \Sigma^1 S^4$ for the suspension metrics. (This is a trivial case of Appendix A.)

Thus $L^4 * S^0 \cong \Sigma^1 L^4 \cong \Sigma^1 S^4 \cong S^5$.

Proof of Theorem 2. For $n = 5$, note that every point lies in the open star of some vertex v, which is the open cone on the link L^4 of v, a Lipschitz 5-manifold by Lemma 4.

For $n = 6$, the link L^5 of any vertex v is, by the case $n = 5$ just proved, a Lipschitz 5-manifold. By the topological Poincaré theorem, L^5 is homeomorphic to S^5, so by [17] L^5 is Lipschitz isomorphic to S^5. Thus $v * L^5$ is Lipschitz isomorphic to D^6, which proves X^6 is Lipschitz 6-manifold.

The proof continues in this fashion by induction on n.

This completes the proof of Theorem 2.

Parallel triangulation theories of T. Matumoto and of D. Galewski and R. Stern (see [10][8]) have shown that a necessary and sufficient condition that every topological manifold (without boundary) of a fixed dimension $n \geq 5$ be triangulated by a simplicial homotopy n-manifold is that

(*) *There exist a closed smooth homotopy 3-sphere L^3 with Rohlin invariant 1 in $\mathbb{Z}/2\mathbb{Z}$ such that the connected sum $L^3 \# L^3$ bounds a smooth contractible 4-manifold.*

Combining this with Theorems 1 and 2 and [17], we immediately get

LIPSCHITZ TRIANGULATION THEOREM 5. Fix $n \geq 5$. Every Lipschitz n-manifold (without boundary) of dimension n is Lipschitz isomorphic to some locally finite simplicial complex if and only if (*) is true.

Concluding remark: Theorems 1, 2 and 5 admit fairly obvious generalisations for manifolds with boundary. See [13, p. 84] and [3] for some assistance.

PROBLEMS

1) Is Theorem 1 true replacing Lipschitz embeddings by quasiconformal embeddings ([11]) ?

One can observe, looking back to Theorem 1, that if σ is an open k-simplex, $k < n - 3$, and $h|(U \cap \sigma)$ has a non-singular differential Dh_x at at least one point x, then Link(σ) is simply connected. Indeed a small $(n - k - 1)$- sphere centered at $h(x)$ in any $(n-k)$-plane cutting transversally across the k-plane Image(Dh_x) at $h(x)$ misses $h(U \cap \sigma)$ and maps with degree ± 1 onto Link(σ) proving it to be simply connected.

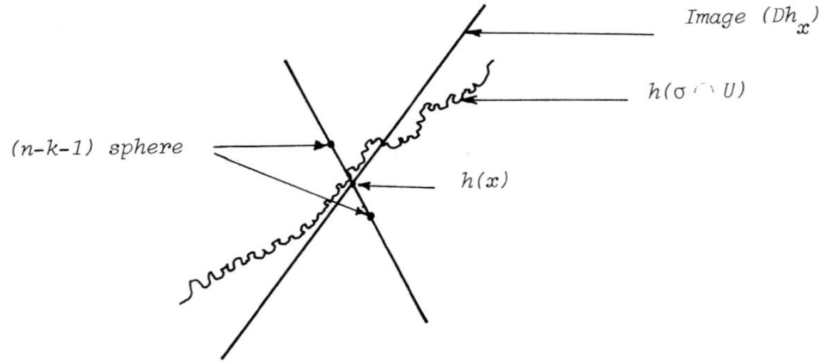

Fig. 2.

However a generic quasi-conformal homeomorphism of the complex plane seems to map the x-axis to a curve that has no tangent line at all!

2) Analytic spaces have (locally defined) Lipschitz structures. For example, the Lipschitz structures of the cusps $x^2 = y^{2k+1}$ in the plane R^2, $k = 1,2,\ldots$ are all non-isomorphic. Up to (local) Lipschitz isomorphism are there only countably many such objects? Is there Lipschitz equisingularity along the strata of a suitable Whitney stratification of analytic spaces?

3) Define the Pontrjagin classes locally in terms of the Lipschitz structure on topological manifolds.

4) Is Rohlin's theorem on signature of 4-manifolds true for Lipschitz 4-manifolds? Perhaps an analytic proof?

5) Can the basic Lipschitz isomorphism extension theorem of [17] (cf. App. A, below) be proved by elementary means? By A. V. Cernavskii's methods? Or can one construct almost parallizable hyperbolic manifolds directly?

APPENDIX ℵ. LIPSCHITZ CONTINUATION

Recall that the open unit ball $\mathring{B}^n \subset R^n$ is a model for hyperbolic n-space when we replace the euclidean Riemannian metric dS_E on $\mathring{B}^n$ by the hyperbolic Riemannian metric $dS_H = dS_E/(1-r^2)$ where $r : R^n \to [0,\infty)$ is euclidean distance from the origin in R^n. We recall that the hyperbolic isometries of $\mathring{B}^n$ are the restrictions to $\mathring{B}^n$ of the Möbius (sphere preserving) transformations of $R^n \cup \infty$ that respect $\mathring{B}^n$ [11, §1].

THEOREM OF LIPSCHITZ CONTINUATION TO ∂B^n. *Let $h : \mathring{B}^n \to \mathring{B}^n$ be a self-map that is K-Lipschitz, $K < \infty$, for the hyperbolic metric, and of hyperbolic distance $\leq \delta$ from the identity map (i.e. for all x in $\mathring{B}^n$ the hyperbolic distance $d(x,h(x))$ is $\leq \delta$).*

Then, for the euclidean metric, h is uniformly Lipschitz; in fact, it is L-Lipschitz where $L < \infty$ is a constant depending *only on* K *and* δ. *Hence,* h *extends uniquely to a Lipschitz self-map* $h : B^n \to B^n$.

Remark: Since the euclidean radius of a ball in $\mathring{B}^n$ of fixed hyperbolic radius r and variable center $x \in \mathring{B}^n$ tends to zero as x tends to ∂B^n, it is clear that *the extension* $h : B^n \to B^n$ *is the identity on* $S^{n-1} = \partial B^n$.

Remark: It is amusing that h need not be a bijective map, although it must be surjective, and that even, if it is bijective, its inverse need not be Lipschitz.

Proof of Theorem. Let D and D_δ in $\mathring{B}^n$ be the compact balls centered at the origin, of hyperbolic radius 1 and $1 + \delta$ respectively. As a map from hyperbolic to euclidean metrics, the identity map $f = id|D_\delta$ is a Lipschitz isomorphism, and as D_δ is compact, both f and f^{-1} are uniformly Lipschitz. Hence $h|D : D \to D_\delta \subset \mathring{B}^n$ is L_0-Lipschitz for euclidean metric where L_0 is a constant depending on K and δ only.

Next, we prove

ASSERTION: If x and x' in $\mathring{B}^n$ lie in <u>*any*</u> *ball D' of hyperbolic radius 1,*

$$(*) \quad |h(x) - h(x')| < L|x - x'| ,$$

where L is a constant depending on K and δ only, i.e. $h|D'$ is L-Lipschitz for the euclidean metric.

From this, note that the theorem follows; indeed (*) holds for any x,x' in $\mathring{B}^n$ as one can see by cutting up the *euclidean* segment x to x' into finitely many segments of *hyperbolic* length ≤ 2.

To prove the assertion, let α be a hyperbolic isometry that carries D onto D'. The conjugate $h' = \alpha^{-1}h\,\alpha$ verifies the same conditions as h, hence is L_0-Lipschitz on D (for the euclidean metric).

Observe that if α were a euclidean similarity, we would have that h' is L_0-Lipschitz on $D \Leftrightarrow h$ is L_0-Lipschitz on $D' = \alpha D$.

In reality, α is only infinitesimally such a similarity, but we prove an *almost similarity lemma* (read it's statement below) that lets us deduce the assertion as follows.

Write y,y' for $\alpha^{-1}(x)$, $\alpha^{-1}(x')$, a pair in D, and write z,z' for $h'(y)$, $h'(y')$, a pair in D_δ.

$$\begin{array}{ccc} x & \xleftarrow{\quad\alpha\quad} & y \\ h \big\downarrow & & \big\downarrow h' \\ h(x) & \xleftarrow{\quad\alpha\quad} & z \end{array} \qquad \text{in } D_\delta$$

When $h(x) \neq h(x')$, we have

$$\frac{|h(x)-h(x')|}{|x-x'|} = \frac{|\alpha(z)-\alpha(z')|}{|z-z'|} \cdot \frac{|h'(y)-h'(y')|}{|y-y'|} \cdot \frac{|y-y'|}{|\alpha(y)-\alpha(y')|}$$

$$\leq L_0 a$$

where a is the approximate similarity constant for the disc D_δ. Thus h is L-Lipschitz on D' where $L = L_0 a$.

We have made essential use of the

APPROXIMATE SIMILARITY LEMMA. Given a compactum $A \subset \overset{\circ}{B}{}^n$, *there exists a constant* $a < \infty$ *such that, for every quadruple* x, x', y, y' *of distinct points in* A *and every hyperbolic isometry* γ *of* $\overset{\circ}{B}{}^n$, *one has*

$$\left|\frac{\gamma(y') - \gamma(y)}{\gamma(x') - \gamma(x)}\right| \leq a \left|\frac{y' - y}{x' - x}\right| .$$

Proof of lemma. Without losing generality, we assume that A is a ball centered at the origin $0 \in \overset{\circ}{B}{}^n$. Each hyperbolic isometry γ is naturally a product $\gamma = \tau\theta$ where τ is the hyperbolic translation along the line 0 to $\gamma(0)$ carrying 0 to $\gamma(0)$, while θ is the rotation $\tau^{-1}\gamma$ fixing 0. Since, θ is an euclidean isometry and respects A, it suffices to prove the lemma for translations as described. Further, by symmetry, it suffices to deal with the translations along a single ray from 0 toward $x_0 \in \partial B^n = S^{n-1}$.

There is a Mobius (sphere preserving) diffeomorphism $f : (B^n - \{-x_0\}) \to\to \mathbb{R}^{n-1} \times [0,\infty)$ with $f(x_0) = (0,0)$, $f(0) = (0,1)$, and sending dS_H to Poincarés Riemannian metric $dSp = dS_E/y$, where y is the projection to $[0,\infty)$. In the Poincaré model (see Figure 4 below), the hyperbolic translations $\overline{\gamma}$ from $f(0)$ towards $f(x_0)$ are euclidean similarities, multiplication by γ for $0 < \gamma \leq 1$. Set $\overline{\gamma} = f\gamma f^{-1}$ and $\overline{x} = f(x)$, etc..., and let K be a Lipschitz constant for both f and f^{-1} on the compacta $f^{-1}\hat{A}$ and $\hat{A}$ respectively, where $\hat{A}$ is the convex hull of $f(A)$ and $f(x_0)$. Then

$$\left|\frac{\overline{\gamma}(\overline{y}') - \overline{\gamma}(\overline{y})}{\overline{\gamma}(\overline{x}') - \overline{\gamma}(\overline{x})}\right| = \left|\frac{\overline{y}' - \overline{y}}{\overline{x}' - \overline{x}}\right| \quad , \text{ by similarity,}$$

while $|\bar{y}' - \bar{y}| \leq k|y' - y|$,
$|\bar{x}' - \bar{x}| \geq \frac{1}{k}|x' - x|$

and $|\bar{\gamma}(\bar{y}') - \bar{\gamma}(\bar{y})| = |f\gamma(y') - f\gamma(y)| \geq \frac{1}{k}|\gamma(y') - \gamma(y)|$,
$|\bar{\gamma}(\bar{x}') - \bar{\gamma}(\bar{x})| = |f\gamma(x') - f\gamma(x)| \leq k|\gamma(x') - \gamma(x)|$.

Thus, the similarily equation gives

$$\frac{k^{-1}|\gamma(y') - \gamma(y)|}{k|\gamma(x') - \gamma(x)|} \qquad \frac{k|y' - y|}{k^{-1}|x' - x|} \quad ;$$

so, the lemma is proved with $a = k^4$.

Quasi-conformal remarks, cf. [11]. The composition of a K-quasi-conformal homeomorphism with a conformal homeomorphism is a K-quasi-conformal homeomorphism. This makes immediate the proof of the above theorem with quasi-conformal in place of Lipschitz[1]. The quasi-conformal version of the continuation theorem of the next appendix is almost as immediate.

As for the last appendix B, beware that if $f : I \times X \to I \times Y$ is an open quasi-conformal embedding respecting projection to I, *i.e. a quasi-conformal isotopy*, then f is necessarily a Lipschitz isomorphism onto its image, i.e. a Lipschitz isotopy.

APPENDIX A. - LIPSCHITZ CONTINUATION TO SUSPENSION

We prove (for Theorem 2) a result whose statement and proof are direct generalizations of the theorem for Lipschitz-continuation from $\mathring{B}^n$ to B^n as proved in Appendix ℵ. It also lets us deform Lipschitz isomorphisms of polyhedra and discuss Lipschitz isotopies (Appendix B).

[1]*At boundary points, the extension is 1-quasi-conformal; elsewhere it is K-quasi conformal.*

Given any metric space V of diameter ≤ 2, there is a *suspension pseudo-metric*[1] d_s on $V \times B^n$ given by

$$(v,x),\ (v',x') \mapsto (1-\max\{|x|,|x'|\})d(v,v') + d_c(x,x')$$

where d_c is the cone metric on B^n given in polar co-ordinates $[0,1] \times \partial B^n \to B^n$, $(\lambda,\hat{x}) \mapsto \lambda\hat{x}$ by

$$d_c : (\lambda\hat{x}, \lambda'\hat{x}') \mapsto |\lambda-\lambda'| + \min\{\lambda,\lambda'\}\ |\hat{x},\hat{x}'|.$$

Remark: One is tempted to simplify d_s, replacing d_c by the (Lipschitz equivalent) euclidean metric. However, as J. Väisälä pointed out to us, the triangle inequality would then not quite hold for the simplified distance function d_σ. By identifying points of $V \times B^n$ whose d_s distance is zero, we obtain a genuine metric d_s on the suspension $\Sigma^n V = V \times B^n/\sim$ where $\sim$ identifies to points the sets $V \times x$, $x \in \partial B^n$; these points $\Sigma^n V - V \times \mathring{B}^n$, naturally and isometrically identified to ∂B^n, form the *suspension sphere*.

Observe that the restriction of d_s to each disc $\{v\} \times B^n$, $v \in V$, is the cone metric of B^n.

Observe that, if $f : V \to W$ is a K-Lipschitz map of metric spaces of diameter ≤ 2, the induced map $\Sigma^n f : \Sigma^n V \to \Sigma^n W$ K-Lipschitz.

Concerning metric spaces of any diameter $\leq \infty$, recall that the modified metric d_λ, $\lambda > 0$,

$$(x,x') \mapsto \min\ \{\lambda, d(x,x')\}$$

is (locally) Lipschitz isomorphic to the old one and has diameter $\leq \lambda$. The metrics d_λ are all mutually uniformly Lipschitz isomorphic. Again, the scaled metrics λd, $\lambda > 0$, are all uniformly Lipschitz isomorphic to d.

[1] *The triangle inequality would fail for diameter* $V > 2$.

We shall consider also the metric on $V \times \mathring{B}^n$ that is the product of the metric on V with the *hyperbolic* metric on $\mathring{B}^n$:

$$(v,x),(v',x') \mapsto d(v,v') + d(x,x').$$

We shall call this, for short, the *product-hyperbolic metric.*

THEOREM OF LIPSCHITZ CONTINUATION TO A SUSPENSION SPHERE. Consider a map $h : V \times \mathring{B}^n \to W \times \mathring{B}^n$ where V, W are metric spaces of diameter ≤ 2. Suppose that, for the product-hyperbolic metric on source and target, h is K-Lipschitz, $K < \infty$, and that $p_2 h : V \times \mathring{B}^n \to \mathring{B}^n$ is of bounded hyperbolic distance $\leq \delta < \infty$ from the projection $p_2 : V \times \mathring{B}^n \to \mathring{B}^n$.

Then, h is L-Lipschitz for the suspension metrics where L is a constant depending on K and δ only. Hence, a L-Lipschitz map $h : \Sigma^n V \to \Sigma^n W$ is induced.

Remark: The extension $h : \Sigma^n V \to \Sigma^n W$ is necessarily the identity on the suspension sphere. This follows from the parallel remark in Appendix ℵ applied to

$$\mathring{B}^n \xrightarrow{v\times} V \times \mathring{B}^n \xrightarrow{h} W \times \mathring{B}^n \xrightarrow{p_2} \mathring{B}^n.$$

The argument proving the theorem is strictly parallel to that in Appendix ℵ and we are content to give an outline discussing the new technical points. V is metric of diameter ≤ 2.

APPROXIMATE SIMILARITY LEMMA. Given a compactum $A \subset \mathring{B}^n$, there exists a constant $a < \infty$ such that, for every quadruple x,x',y,y' of distinct points of $V \times A$ and every hyperbolic isometry α of $\mathring{B}^n$, one has, writing $\gamma = (\mathrm{id}|V) \times \alpha$:

$$\frac{d_s(\gamma(y),\gamma(y'))}{d_s(\gamma(x),\gamma(x'))} \leq a \frac{d_s(y,y')}{d_s(x,x')} \quad ,$$

for the suspension metric d_s on $V \times \mathring{B}^n$.

Proof of Lemma. In imitating the argument in Appendix ℵ, one uses the enhanced comparison with Poincaré's model:

$$(\mathrm{id}|V) \times f : V \times (B^n - \{-x_0\}) \to V \times \{R^{n-1} \times [0,\infty)\}$$

and checks that on each set V x (compactum), it is a uniformly Lipschitz isomorphism, when, on the left, we use the suspension pseudo-metric d_s, and on the right, we use the 'wedge' pseudo-metric

$$(v,x,t),(v',x',t') \mapsto \min\{t,t'\}\, d(v,v') + |x-x'| + |t-t'|$$

where p is projection to $[0,\infty)$. To check this, one needs only to know that f (from Appendix ℵ) is a Lipschitz isomorphism.

The pseudo-metric just mentioned is designed to make $(v,x,t) \mapsto (v,\lambda x,\lambda t)$ a similarity.

Proof of Theorem. Form the metric space X from the disjoint union of the two metric spaces V, W of diameter ≤ 2 by decreeing the distance between v in V and w in W to be 2. Then, X has diameter ≤ 2.

By applying to X this approximate similarity lemma and imitating faithfully the argument of Appendix ℵ, one shows that there exists a constant $L < \infty$ depending on K and δ only, such that, for each ball $D' \subset \mathring{B}^n$ of hyperbolic radius 1, the map h is L-Lipschitz on $V \times D'$ for the suspension metrics.

Finally, one deduces that $h : V \times \mathring{B}^n \to W \times \mathring{B}^n$ is likewise L-Lipschitz by using the fact that, for any pair of points (v,x), (v',x') of $V \times B^n$ with $|x| \leq |x'|$, the d_s distance is exactly the sum of these from (v,x) to (v,x') and from (v,x') to (v',x'). Note this is false for $|x| > |x'|$.

Now, for the pair (v,x'), (v',x'), the map h has just been proved L-Lipschitz, while, for the pair (v,x), (v',x), it is proved K-Lipschitz using the euclidean segment x to x' as in Appendix ℵ.

APPENDIX B. - LIPSCHITZ ISOTOPY EXTENSIONS

Here, exploiting Appendix A, we describe extensions and improvements for the Lipschitz isotopy extension constructed in [17, Theorem 1]. One is that the theory of [16] for deforming open topological embeddings of complexes can be carried out for Lipschitz embeddings. Another is that the isotopy extensions can preserve a Lipschitz condition with respect to parameters.

A *Lipschitz isotopy* of open embeddings of X into Y is an open embedding $F : I \times X \to I \times Y$ that respects projection to $I = [0,1]$ and is a Lipschitz isomorphism onto its image. One sometimes allows I to be a disc or a general metric space.

Note that, with $X = Y = R$, the map $F : (t,x) \mapsto (t,x + \sqrt{t})$ is not a Lipschitz isomorphism, although, for each t, the self map $x \mapsto x + \sqrt{t}$ of R is one.

We shall show, for example, that if we write the Lipschitz isotopy $F(t,x) = (t,F_t(x))$ and suppose $I = [0,1]$, while X, Y are Lipschitz manifolds, or locally finite simplicial complexes, and F_0 is inclusion $X \hookrightarrow Y$, then, for any compactum $C \subset X$, we can find a bijective Lipschitz isotopy $G : I \times Y \to I \times Y$ such that $G = F$ near $I \times C$ and $G_0 = \mathrm{id}|Y$. This is the Lipschitz version of R. Thom's smooth isotopy extension theorem.

DEFORMATION THEOREM FOR LIPSCHITZ EMBEDDINGS FOR COMPLEXES. Let X be a locally finite simplicial complex. We assert the property, cf. [16, §0]:

$\boxed{\mathfrak{L}(X)}$ *For each open set $U \subset X$ and each compactum $B \subset U$, the following holds:* $\boxed{\mathfrak{L}(X\ ;\ B\ ;\ U)}$ *If $h : U \to X$ is an open Lipschitz embedding sufficiently near to the identity $i : U \to X$, there is a rule assigning to h a Lipschitz isomorphism $h' : X \to X$ equal to h on B and the identity outside of U. For h near i, the rule $h \mapsto h'$ can be continuous (for the compact open topology) ; it sends i to* $\mathrm{id}|X$.

The rule $h \mapsto h'$ can be such that, if h respects a subcomplex $Y < X$ and its complement, then h' will too; and if h' fixes $Y < X$, then h' will too.

Remark. An equivalent property $\boxed{\mathcal{L}^*(X\,;\,B\,;\,U)}$ is obtained by assuming $B \subset U$ to be closed (not compact) in X, with closure $(X - B)$ compact, and then relaxing the condition that h' = identity outside U. Still other equivalent statements are sometimes helpful - see [16, §§ 0,2,4].

Remark. Unfortunately this result does *not* apply to analytic spaces with their natural Lipschitz structure, since cusps prevent their triangulations from being Lipschitz (compare Problem 2 in main text).

Proof of Deformation Theorem. The proof, when X is a Lipschitz manifold, is given in [17] (except of course for the statement concerning subcomplexes). We now indicate how to generalize this argument by combining the proof of the topological analog of given in [16], with the technical theorem of this appendix.

Here is the recipe. After reading [17], read [16, §2] to get the plan of proof by induction on *depth* (the depth of an open subset V of a complex is $d = h - h'$ where h is the maximum simplex dimension met in V and h' is the minimum). Then solve the relevant handle problems making two important changes in the argument of [16, §3]. These handle problems amount to proving $\mathcal{L}(R^m \times cL\,;\,B\,;\,U)$ for any $B \subset R^m \times v$ assuming inductively $\mathcal{L}(R^m \times (cL - v))$, where L is any finite complex.

Recall that

$$cL = L \times [0,\infty)/\{L \times 0 = v = \text{vertex}\}$$

is the open cone and that c_1L the radius 1 cone is the quotient of $L \times [0,1]$. The metric can come from the pseudo metric on

$L \times [0,\infty)$ given by

$$(x,t),\ (x',t') \mapsto \min\{t,t'\}d(x,x') + |t-t'| .$$

The barycentric metric for the standard triangulation of c_1L is Lipschitz isomorphic to this one.

CHANGE 1. *Instead of wrapping up to derive a torus problem, use Kirby's punctured torus immersion idea in its hyperbolic form.*

This requires a capping off process as in [17]; beware that $\mathfrak{L}(R^m \times (cL-v))$ is required to make it a problem of type $\mathfrak{L}^*(cK\ ;\ K \times [2,\infty)\ ;\ K \times (1,\infty))$ on an open cone cK on a complex K, which can be solved by wrapping up as in [17]. (The complex K will be a link in $R^m \times cL$ of $0 \times v$, i.e. $K = \Sigma^m L$.)

CHANGE 2. *Adjust the 'horn device' at the unfurling stage of the proof (construction of g_4, g_5 in [16, §3]) so that it functions for Lipschitz.*

One is faced with a Lipschitz isomorphism $g_3 : \mathring{B}^m \times cL \to \mathring{B}^m \times cL$ that is the universal covering of a small Lipschitz isomorphism $g_2 : T^m \times cL \to T^m \times cL$, where T^m is a closed hyperbolic m-manifold. One has arranged that g_2 is the identity outside $T^m \times c_1L$ where $c_1L = L \times [0,1]/\{L \times 0 = v\}$ and that correspondingly g_3 = identity outside $\mathring{B}^m \times c_1L$. The horn modification of g_3

$$g_4 : \mathring{B}^m \times cL \to \mathring{B}^m \times cL,$$

is $g_4 = \Theta g_3 \Theta^{-1}$ where Θ is the Lipschitz automorphism of $\mathring{B}^m \times cL$ defined by

$$\Theta : B^m \times [0,\infty) \times L \ni (x,t,y) \mapsto (x,\varphi(|x|)t,y) \in B^m \times [0,\infty) \times L.$$

Here $\varphi : [0,1] \to [0,1]$ is Lipschitz, decreasing and satisfies $\varphi(t) = 1$ for t near 0 and $\varphi(t) = (1 - t)$ for t near 1.

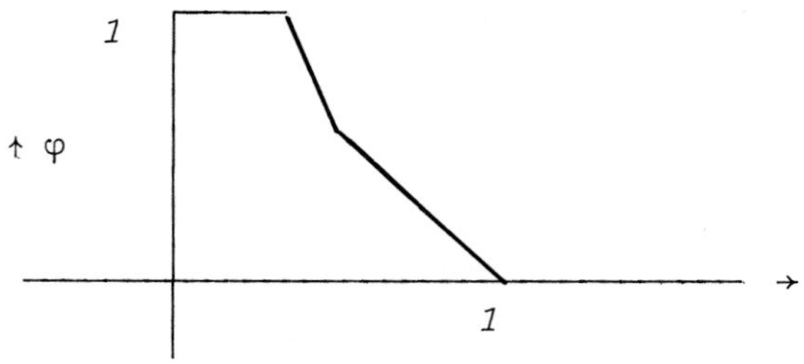

It is easily seen that g_4 extends to a homeomorphism

$$g_5 : R^m \times cL \to R^m \times cL$$

that is the identity outside $\Theta(\mathring{B}^m \times c_1L)$.

ASSERTION. g_5 *is a Lipschitz isomorphism.*

Proof of Assertion. The technical theorem of this appendix shows that

$$g_3 | : \mathring{B}^m \times c_1L \to \mathring{B}^m \times c_1L$$

extends to a Lipschitz isomorphism

$$\bar{g}_3 : \Sigma^m(c_1L) \to \Sigma^m(c_1L).$$

But there is a canonical Lipschitz isomorphism

$$\Theta' : \Sigma^m(c_1L) \to \Theta(\mathring{B}^m \times c_1L) \cup \partial B^m \times v$$

induced by the map Θ. This shows that g_5 gives a Lipschitz automorphism on the compact target of Θ'; since g_5 is the identity elsewhere, the proof is complete.

The verification that the standard map Θ' is indeed a Lipschitz isomorphism is left to the reader.

The isomorphism g_4 essentially solves the handle problem. This completes our outline of the proof of $\mathfrak{L}(X)$.

We now turn to Lipschitz isotopy extensions, and show that the deformation rule $h \mapsto h'$, *as already constructed* in [17] and above, to satisfy $\mathfrak{L}(X\ ;\ B\ ;\ U)$, enjoys a further:

PARAMETER PROPERTY. If $h_t : U \to X$, $t \in I$, gives a Lipschitz isotopy[1] *so near to the inclusion $U \to X$ that the rule $h_t \mapsto h'_t : X \to X$ is defined for all $t \in I$, then (automatically) $h'_t : X \to X$, $t \in I$, likewise gives a Lipschitz isotopy.*

This property is to be checked by looking back over the solutions of the finitely many handle problems that yield the deformation rule $h \mapsto h'$. In each handle problem, one has to check that both *capping off* and *unfurling* preserve Lipschitz isotopies, using the

THEOREM OF LIPSCHITZ PARAMETERS. Let $F : I \times J \times \mathring{B}^n \to I \times J \times \mathring{B}^n$ be a self-map that respects projection to the metric space I, while J is a metric space of diameter ≤ 2. Suppose F is K-Lipschitz, $K < \infty$, for the product metric on $I \times J \times \mathring{B}^n$, where $\mathring{B}^n$ carries the hyperbolic metric, and suppose also that $p_3 \circ F : I \times J \times \mathring{B}^n \to \mathring{B}^n$ is of bounded hyperbolic distance $\leq \delta$ from the projection p_3 to $\mathring{B}^n$. Let $\Sigma^n J = J \times \mathring{B}^n \cup (\partial B^n)$ carry the suspension metric and give $I \times \Sigma^n J$ the product metric. Then F extends to a map $G : I \times \Sigma^n J \to I \times \Sigma^n J$ that is L-Lipschitz where L is a constant depending on K and δ only.

Remark (cf. Appendix A): Necessarily, G is the identity on $I \times (\partial B^n)$ and respects projection to I.

Proof of theorem. This variant of Appendix A is fortunately an easy corollary thereof.

[1]*That is, $H : (t,x) \to (t, h_t(x))$ mapping $I \times U \to I \times X$ is a Lipschitz isomorphism onto its image; here I is any metric space.*

To lighten the notation, we prove only the extreme case where J is a point so that $I \times J \times \mathring{B}^n = I \times \mathring{B}^n$ and $I \times \Sigma^n J = I \times B^n$. Let p_1 and p be the projections from $I \times \mathring{B}^n$ to I and $\mathring{B}^n$ respectively.

Since, it is a local question whether a Lipschitz extension of F exists, we can assume that diameter I is ≤ 2. Appendix A then asserts that F extends uniquely to an L-Lipschitz map

$$H : I \times B^n \to I \times B^n$$

for the *suspension distance function*

$$(t,x),(t',x') \mapsto \{1 - \max(|x|,|x'|)\}d(t,t') + |x - x'|$$

and that $H|(I \times \partial B^n) = \text{id}$. It sufficies to check that H is $2L$-Lipschitz on $I \times B^n$ for the *product metric*

$$d_\pi((t,x),(t',x')) = d(t,t') + |x - x'| .$$

We can assume $1 < L < \infty$, so that

$$d_s(H(t,x),H(t',x') \leq Ld_s((x,t),(x',t')).$$

Hence

$$|pH(t,x) - pH(t',x')| \leq L(d(t,t') + |x-x'|) = Ld_\pi((t,x),(t',x')).$$

But, since H respects projection to I,

$$d(p_1H(t,x),p_1H(t',x')) = d(t,t') \leq Ld_\pi((t,x),(t',x')) .$$

Thus, adding,

$$d_\pi(H(t,x),H(t',x')) \leq 2Ld_\pi((t,x),(t',x')).$$

REFERENCES

1. Cannon, J., Shrinking cell-like decompositions, Annals of Math. (1978), to appear.
2. Cappell, S., and Shaneson, J., Construction of some new four-dimensional manifolds, Bull. Amer. Math. Soc. 82 (1976), pp. 69-70.
3. Cohen, M., Homeomorphisms between homotopy manifolds and their resolutions, Invent. Math. 10 (1970), pp. 239-250.
4. Edwards, R. D., Approximating certain cell-like maps by homeomorphisms, Ann. of Math., to appear.
5. Edwards, R. D., and Glaser, L. C., A method for shrinking decompositions of certain manifolds, Trans. Amer. Math. Soc. 165 (1972), pp. 45-56.
6. Edwards, R. D., and Kirby, R. C., Deformations of spaces of imbeddings, Ann. of Math. 93 (1971), 63-88.
7. Federer, H., Geomet c measure Theory, Grundlehren Series no. 153, Springer, 1969.
8. Galewski, D., and Stern, R., Ann. of Math, to appear and Proc. AMS Stanford Topology Conf. 1976, to appear.
9. Luukkainen, J., and Vaisälä, J., Elements of Lipschitz topology, Ann. Acad. Sci. Fenn. 3 (1977), pp. 85-122
10. Matumoto, T., Variétés simpliciales d'homologie et variétés topologiques métrisables, thèse, Univ. Paris-Sud, 91405 Orsay, 1976. Proceedings of AMS Stanford Topology Conference 1976, to appear.
11. Mostow, G., Rigidity of hyperbolid space forms, Publ. Math. I.H.E.S. 34 (1968), pp. 53-104.
12. Siebenmann, L. C., Approximating cellular maps by homeomorphisms, Notices Amer. Math. Soc. 17 (1970, p. 532 and Topology 11 (1973), pp. 271-294.
13. Siebenmann, L. C., Are non-triangulable manifolds triangulable?, in Topology of manifolds, edited by J. C. Cantrell and C. H. Edwards, (Athens Georgia Conf. 1969), Markham, Chicago, 1970, pp. 77-84.
14. Siebenmann L. C., Topological Manifolds, Proc. I.C.M. Nice (1970), Vol. 2, pp. 143-163, Gauthier-Villars, 1971.
15. Liebenmann, L. C., Pseudo-annuli and invertible cobordisms, Archiv der Math. 19 (1968), pp. 28-35.
16. Siebenmann, L. C., Deformation of homeomorphisms on stratified sets, Comment. Math. Helv. 47 (1972), pp. 123-163. Note errata in 6.25, p. 158: on line 10, N becomes B; on lines 23, 24, 25, $\frac{!}{2}$ becomes $\frac{t}{2}$.
17. Sullivan, D., Hyperbolic geometry and homeomorphisms (these proceedings).

ISOMETRIES OF INNER PRODUCT SPACES AND THEIR GEOMETRIC APPLICATIONS

Neal W. Stoltzfus

Department of Mathematics
Louisiana State University
Baton Rouge, Louisiana

The bordism classification of orientation preserving diffeomorphisms of oriented manifolds and the concordance classification of codimension two knots both depend on a Witt group of isometries of integral inner product spaces. These algebraic obstruction groups are related by localization exact sequences to simpler "building blocks", the Witt classification of Hermitian forms on torsion free modules over an order in an algebraic number field. Using the integral polynomial ring module structure induced by the isometry on the underlying module of the inner product space a new Witt equivalence relation can be defined which, in the knotting situation, is related to the double null concordance relation. Finally, we define a new group of knots under the connected sum operation which is infinitely generated in all dimensions.

I. INTRODUCTION: THE GEOMETRIC PROBLEMS

Two basic problems in the geometry of manifolds have led to algebraic obstructions groups based on isometries of inner product spaces over the rational integers. In this survey we will work in the category of oriented compact smooth manifolds.

ISBN 0-12-158860-2

Problem 1. Classify the diffeomorphisms of a manifold. One of the first secrets of an oriented manifold, unearthed by Poincare, was the discovery of a duality between its homology and cohomology groups in complementary dimensions. In particular, if the manifold V^{2k} is closed, of even dimension $2k$ and $M = H^k(V,Z)/Tor$ is the torsion free part of the middle dimensional cohomology then M admits a $(-1)^k$-symmetric integer valued inner product given $B(x,y) = \varepsilon_*[(x \cup y) \cap [V]]$ where $[V]$ is the orientation class of V, ε_* is the augmentation and $\cup$, $\cap$ are the cup and cap products, respectively. Furthermore, if $T: V \to V$ is an orientation preserving diffeomorphism then we have:

$$B(T^*x, T^*y) = \varepsilon_*[(T^*x \cup T^*y) \cap [V]] = \varepsilon_*[T^*(x \cup y) \cap [V]] =$$

$$\varepsilon_*[(x \cup y) \cap T_*[V]] = B(x,y),$$

that is, T^* preserves the inner product and is therefore an isometry of B.

Diffeomorphisms of a manifold are also important in studying the structure of manifolds of the next higher dimension, by means of the mapping torus construction or the following slight refinement, the open book construction. Let V^n be a manifold with nonempty boundary and let T be a diffeomorphism of V which is the identity on the boundary of V. Then the mapping torus of T, $M_T = V^n \times_T S^1 = V^n \times [0,1]/\{(v,0)\sim(Tv,1)\}$ has boundary $\partial V \times_{Id} S^1 = \partial V \times S^1$. We say that the closed manifold $M^{n+1} = (V^n \times_T S^1) \cup_{V \times S^1} (\partial V \times D^2)$ is an *open book decomposition* with page V^n and binding ∂V. The following theorem has a long history and various cases are due to P. Heegaard, J.W. Alexander [1], H. Winkelnkemper [22], J.P. Alexander [2], and F. Quinn [16].

OPEN BOOK DECOMPOSITION THEOREM. *If M^{2n+1} is a simply connected odd dimensional manifold then M has an open book decomposition.*

A closed simply connected manifold of even dimension greater than six has an open book decomposition if and only if its signature is zero.

In the special case of an open book decomposition of the standard sphere, S^n, with binding a homotopy sphere, the open book is called a fibered knot. In general we have the following definition:

A *knot* is an oriented codimension two submanifold $K^n \subset S^{n+2}$ which is homeomorphic to S^n.

This brings us to our second intractable question:

Problem 2. Classify knots of a given dimension. In this situation, there is a very powerful invariant, discovered and exploited by Seifert [17], which is closely related to the notion of isometry. This invariant is even a complete invariant in a restricted subclass of high dimensional knots [12].

By an easy obstruction theory argument, or an explicit construction in the classical dimension due to Seifert [17], every knot K^n bounds a framed codimension one submanifold V^{n+1} in S^{n+2}, the well known Seifert "surface." Now, if V is even dimensional, the torsion free part of its middle dimensional cohomology, $M = H^k(V^{2k}, Z)/Tor$ supports a $(-1)^k$-symmetric integral inner product (for this conclusion, we need to know that $\partial V = K$ is a homology sphere.) Since W is an oriented codimension one submanifold of the standard sphere with the standard orientation, a positive normal vector field is determined on V. Let i_+ be a diffeomorphism from V to $S^{n+2}\setminus V$ which "pushes" V a small distance along the positive normal direction. Because $k + k = (n+2)-1$, we have an integer linking number $L(x, i_{+*}(y)) = A(x,y)$ defined on M. Consider the endomorphism t of M defined by the relation

$$A(x,y) = B(t(x),y).$$

If $i_-: V \to S^{n+2}\setminus V$ is a small push in the opposite direction, then by relating the cup product pairing to the geometric intersection number, it can be demonstrated that:

$$L(x,i_-^*(y)) = L(i_+^*(x),y) = (-i)^{k+1}A(y,x)$$

and

$$L(x,i_+^*(y)) - L(x,i_-^*(y)) = B(x,y)$$

Hence $B(tx,y) + (-1)^k B(ty,x) = B(x,y)$ yielding the fundamental relation

$$B(tx,y) = B(x,(id-t)y)$$

We call the triple (M,B,t) an isometric structure, following Kervaire [9], because, if t were invertible (and this is true only for fibered knot in high dimensions) then $Id - t^{-1} = s$ is an isometry of B. In other words, the relation requires that the adjoint of t with respect to the inner product B is $Id - t$. Similarly we can restate the fundamental condition for an isometry as an adjointness relation: $B(Tx,y) = B(x,T^{-1}y)$.

This motivates the following definition. Let R be a commutative ring with unit and let I be an R-module.

Definition. An $\varepsilon(=\pm 1)$ symmetric I-valued inner product space over R is a finitely generated R-module M and an ε-symmetric bilinear form $B:M \times M \to I$ such that $Ad\ B:M \to Hom_R(m,I)$-given by $Ad\ B(m)(n) = B(m,n)$ is an R-module isomorphism.

Examples. i) $R = Z$, $M = Z \oplus Z$, $I = Z$ and $Ad\ B$ has matrix $\begin{pmatrix} 0 & 1 \\ 1 & 0 \end{pmatrix}$

ii) $R = Z$, $M = Z/p$, $I = Q/Z$, $B(1,1) = 1/p$.

iii) $R = Z[X]/f(X)$, where $f(X)$ is an irreducible monic polynomial with integral coefficients, $I = R(1/f'(X))$, $M = R$, $B(1,1) = 1/f'(X)$ and $f'(X)$ is the derivative of $f(X)$.

An *isometry* (isometric structure of an inner product space, is an R-module endomorphism $t:M \to M$ satisfying $B(x,t^{-1}y) = B(tx,y)$

(in the parenthetical case $B(tx,y) = B(x,(Id-t)y)$.) One can also study bordism of smooth maps of degree k by positing that the adjoint of t by kt^{-1}. This has been studied by a student of Pierre Conner, Max Warshauer and David Gibbs [6] for the case $k = -1$, the case of orientation reversing diffeomorphisms.

Under the operation of direct sum, the isomorphism classes of triples (M,B,t) form a commutative semigroup. We now introduce into this setting the analog of a relation introduced by E. Witt [23].

Definition. (M,B,t) is *metabolic* if there is a R-submodule H which is invariant under t and such that $H = H^{\perp} = \{m: B(m,H) = 0\}$. Furthermore, two structures, M_0 and M_1 are *metabolic equivalent* if there are metabolic structures H_0 and H_1 such that $M_0 \oplus H_0$ is isomorphic to $M_1 \oplus H_1$. We will denote by $W^{\varepsilon}(I,C^{\infty})$ the group of metabolic equivalence classes of isometries and by $C^{\varepsilon}(I)$, the corresponding group for I-values isometric structures.

As the two geometric problems are exceedingly difficult as originally stated, we now introduce an equivalence relation to ameliorate our chances of finding a solution.

Definition. A diffeomorphism T of V^n is *bordant to zero* if there is an orientation preserving diffeomorphism S of W^{n+1} with $\partial W = V$ and $S|\partial_W = T$. Also, (V_0,T_0) is bordant to (V_1,T_1) if $(V_0 \cup -V_1, T_0 \cup T_1)$ is bordant to zero. Let $\mathcal{W}_n$ denote the commutative group of equivalences classes under the operation of disjoint union.

Two knots (S^{n+2},K_0) and (S^{n+2},K_1) are *concordant* if there is a proper embedding $K_1 \times [0,1] = Z$ in $S^{n+2} \times I$ such that:

$$K_1 \times \{0,1\} = S^{n+2} \times \{0,1\} \cap Z = K_1 \cup -K_0$$

where $-K_0$ denotes K_0 with the opposite orientation. Let C_n be the group of concordance classes of n-dimensional knots in S^{n+2} under the operation of knot connected sum.

We can now state the two fundamental geometric theorems which solve the ameliorated problems in terms of the algebraic obstruction groups just defined. Let Ω_n be the oriented bordism group of n-dimensional manifolds.

THEOREM (Kreck [10]*)* $W_{2n} \xrightarrow{I} W^{\varepsilon}(Z, C_{\infty}) \oplus \Omega_{2n} \oplus \Omega_{2n+1}$ *is injective for* $n > 3$, *where* $I(V,T) = ((H^n(V,Z)/Tor, \cup, T^*), [V], [V_T])$ *and* $\varepsilon = (-1)^n$.

THEOREM (Levine [11]*)* $S: C_{2n-1} \to C^{\varepsilon}(Z)$ *is an isomorphism for* $n > 1$. *However* S *is not injective for* $n = 1$ *as shown by the examples of A. Casson and C. Gordon* [4].

II. LOCALIZATION

We now explore the structure of the Witt groups we have defined. For simplicity, we will state the results only for the isometry case. Analogous statements are true under other adjointness relations.

First, we recognize (by an easy verification) that extension of our coefficient ring R from Z to the rational field Q (by localization at zero or by tensor product) induces an injection: $0 \to W^{\varepsilon}(Z, C_{\infty}) \xrightarrow{i} W^{\varepsilon}(Q, C_{\infty})$. Next, we determine the image by extending the well-known techniques of localization to the isometry case. We first note that a necessary condition for an isometry of a rational inner product (V,B,T) to be the extension of an integral isometry, is that T must satisfy a monic polynomial with integral coefficients, by the Cayley-Hamilton Theorem. Denote by $W_0(Q, C_{\infty})$ the equivalence classes of isometries satisfying this condition.

LOCALIZATION THEOREM. There is a homomorphism $\partial : W_0^{\varepsilon}(Q, C_\infty) \to W^{\varepsilon}(Q/Z, C_\infty)$ *such that the following sequence is exact:*

$$0 \longrightarrow W^{\varepsilon}(Z, C_\infty) \xrightarrow{i} W_0(Q, C_\infty) \xrightarrow{\partial} W^{\varepsilon}(Q/Z, C_\infty)$$

Proof. (*Sketch*) Choose a lattice L in V (i.e. finitely generated Z-submodule of V of maximal rank) with basis $\{\ell_i : 1 \leq i \leq n\}$. By scaling L to $dL = \{v = d\ell : \ell \text{ in } L\}$, for some integer d in Z, we may assume that B restricted to L is integer valued (for example, let $d =$ product of the denominators of $B(\ell_i, \ell_j)$). Furthermore, because T satisfies a monic polynomial of degree with integer coefficients, the lattice spanned by $\{T^i(\ell_j) : 0 \leq i \leq m, 1 \leq j \leq n\}$ is invariant under T. Define the dual lattice $L^\# = \{v : B(v, L) \subset Z\}$. Then $L^\#$ is invariant under T, $L \subset L^\#$ and $L^\#/L$ is a torsion Z-module with an induced endomorphism $\overline{T}$. The composition

$$b : L^\# \times L^\# \xrightarrow{B} Q \to Q/Z$$

induces a Q/Z valued inner product on $L^\#/L$. $\partial\{V, B, T\} = \{L^\#/L, b, \overline{T}\}$.

Example. $V = Q$, $B(1,1) = p^2$, $T = Id$; $L^\#/L = Z/p^2$, $b(1,1) = 1/p^2$, $\overline{T} = id$

As is well-known to any linear algebra student, an endomorphism of an R-module M induces an $R[X]$-module structure on M in which X acts as the endomorphism. When R is a field, we will use this structure to obtain an orthogonal decomposition reminiscent of the classical decomposition of $R[X]$- modules into a direct sum of cyclic modules. Together with the following very useful idea, we can achieve a computation of the second and third groups in the localization exact sequence.

Definition. A structure (M,B,T) is *anisotropic* if the only invariant self-annihilating $(H \subset H^{\perp})$ R-submodule is zero.

Note that, if $H \subset M$ is an invariant R-submodule of an anisotropic structure, then $H \cap H^{\perp} = 0$. If the exact sequence given by the restriction of the adjoint map to H:

$$0 \longrightarrow H^{\perp} \longrightarrow M \xrightarrow{Ad_H} Hom_R(H,I)$$

is short exact (e.g. if R is a field or a Dedekind domain) and if $I = Q/Z$ or $R = Q$ then $M = H \oplus H^{\perp}$ by a dimension or a counting argument. The failure of this direct sum decomposition for $R = Z$ will be featured in a second exact sequence. Since $B(Tx,y) = B(x,T^{-1}y)$, the minimal polynomial of T satisfies the condition $f(X) = a_0 X^{\deg f} f(X^{-1})$. Let $P(R)$ be the set of monic polynomials with coefficients in R satisfying this condition. Also denote by $W_f(R,C_\infty)$ the set of equivalence classes M,B,T satisfying $f(T) = 0$. From the proof of the existence of an anisotropic representative, it can be shown:

THEOREM (Milnor [14]*) If F is a field,* $W^{\varepsilon}(F,C_\infty) \cong \bigoplus_{\substack{f \text{ in } P(F) \\ f \text{ irred.}}} W_f^{\varepsilon}(F,C_\infty)$

THEOREM. $W^{\varepsilon}(Q/Z,C_\infty) \cong \bigoplus_{p \text{ prime no.}} W^{\varepsilon}(F_p, C_\infty)$ *where F_p is the prime field with p elements.*

For a detailed exposition of the above results and the determination of the boundary homomorphism in terms of arithmetic invariants of appropriate algebraic number fields, see the author's monograph [18], Unraveling the Integral Knot Concordance Group.

III. COUPLING EXACT SEQUENCE

The failure of the above orthogonal splitting in the case of integral isometries leads to a family of obstructions which measure this breakdown in methodology. Let f be an irreducible monic integral polynomial in $P(Z)$ and (L,B,T) be an isometry of an integral inner product space. Let $L_f = \{\ell \text{ in } L : f(T)^n(\ell) = 0$ for some $n\}$ and let $L_f^{\#}$ be the dual lattice defined in the boundary construction in the rational vector space V obtained by extending the coefficients of L to the rational field. Then $\partial_f(L,B,T) = \{L_f^{\#}/L_f,\ b, \overline{T|_{L_f^{\#}}}\}$ is well-defined, where b is defined as in the localization theorem.

COUPLING EXACT SEQUENCE

$$0 \longrightarrow \bigoplus_f W_f^{\varepsilon}(Z,C_\infty) \longrightarrow W(Z,C_\infty) \xrightarrow{\oplus\partial_f} \bigoplus_f W_f(Q/Z,C_\infty)$$

is an exact sequence where the sum is taken over all irreducible polynomials f in $P(Z)$.

The following example of an isometric structure of order four with non-trivial coupling invariant ($\partial_f \neq 0$ for some f) in the knot theoretic case may prove instructional. This element is in $C^{+1}(Z)$ which is the concordance obstruction group of $(4n-1)$-knots in S^{4n+1}. The symmetric intersection pairing has matrix:

$$B = \begin{pmatrix} 0 & 0 & 0 & -1 \\ 0 & 0 & 1 & 0 \\ 0 & 1 & 2 & 1 \\ -1 & 0 & 1 & -38 \end{pmatrix}$$

and the associated endomorphism t:

$$t = \begin{pmatrix} -1 & 3 & -1 & 0 \\ 57 & 2 & 0 & -1 \\ 0 & 0 & 0 & 1 \\ 0 & 0 & 19 & 1 \end{pmatrix}$$

The traditional Seifert matrix is:

$$A = tB = \begin{pmatrix} 0 & -1 & 1 & 0 \\ 1 & 0 & 1 & -19 \\ -1 & 0 & 1 & -38 \\ -1 & 19 & 39 & -19 \end{pmatrix}$$

and $A + A^t = B$. The Alexander polynomial is a product $\Delta(Z) = (-19Z^2+39Z-19)(-173Z^2 + 347Z - 173)$. Similar examples can be given in the classical dimension when B is skew-symmetric.

IV. OTHER EQUIVALENCE RELATIONS: METABOLIC, SPLIT AND HYPERBOLIC

The following distinctions are not completely standard and we follow the usage of Pierre Conner [3]. Recall the exact sequence:

$$0 \longrightarrow H^{\perp} \longrightarrow M \longrightarrow Hom_R(H,I) \longrightarrow 0$$

With the appropriate $R[X]$-Module structures, it is also an exact sequence of $R[X]$-modules. (In particular, for g in $Hom_R(H,I)$, $Xg(h) = g(T^*h)$ where T^* is the adjoint of T with respect to B.)

Definition. (M,B,T) is *split* if the above sequence splits as a sequence of $R[X]$-modules for some metabolizer H, that is, there is an invariant R-submodule ($R[X]$-submodule!) H with an invariant complementary summand K of H in M. (M,B,T) is called *hyperbolic* if the complementary summands H and K can both be chosen to be metabolizers.

Examples. i) $R=Z$, $M = Z \oplus Z$, $I = Z$ and $Ad\ B$ has matrix $\begin{pmatrix} 0 & 1 \\ 1 & 1 \end{pmatrix}$ with $T = Id$. This form is split but not hyperbolic. In general the evenness of B is a necessary condition for a structure to split.

ii) $R = Z$, $M = Z/p^2$, $I = Q/Z$, $B(1,1) = 1/p^2$ and $T = Id$ is metabolic but neither hyperbolic nor split.

iii) If S is an automorphism of a finitely generated R-module M then $(M \oplus Hom_R(M,R),\ B((x,y),(u,v)) = y(u) + \varepsilon v(x),\ T(x,y) = (S(x),\ y(S^*(-))))$ is a hyperbolic structure denoted $H(M,S)$.

The localization machinery of the first section breaks down and gives only partial information in the split and hyperbolic cases. However, these algebraic ideas reflect certain known geometric constructions especially in the knot theoretic cases and, in addition, we obtain a new group of knots under the usual operation of knot connected sum. These new groups are non-trivial in even dimensions and in fact are not finitely generated in any dimension.

First, we mention some partial results in the case where is an isometry of finite order.

THEOREM [3] *(P. Conner) If $T^p = Id$, p an odd prime then a metabolic isometry (M,B,T) is split (However, it is necessary to rechoose the metabolizer in certain cases.)*

THEOREM (J.P. Alexander) If $T^{p^n} = Id$, p an odd prime and the induced $Z[Z/p^n]$ module structure on M is projective and B is an even form then metabolic implies hyperbolic.

In the knot concordance case split and hyperbolic are equivalent. Mimicing the metabolic case, stabilization by hyperbolic isometric structures defines an equivalence relation and the resulting set of equivalence classes denoted $CH^{\varepsilon}(I)$ form a group. We now relate this group to the geometric idea of double null concordance studied by D. Sumners [20] and prompted by a question of Fox [5].

Definition. A knot (S^{n+2}, K^n) is doubly null concordant (DNC) if it is a cross section of the trivial knot (S^{n+3}, S^{n+1}), that is, there is a Morse function $f: S^{n+3} \to R$ such that, for some regular value t, $(f^{-1}(t),\ f^{-1}(t) \cap S^{n+1})$ is equivalent to (S^{n+2}, K).

The fundamental result of D. Sumners, ameliorated by C. Kearton [7] is:

THEOREM. If (S^{2n+3}, K^{2n+1}) *is doubly null concordant then the Seifert isometric structure is hyperbolic. The reverse implication holds if* $n \neq 0$ *and the knot is simple (in the sense of Levine* [12].*)*

We now minic our algebraic equivalence relation: Two knots (S^{n+2}, K_0) and (S^{n+2}, K_1) are *DNC equivalent* if there are *DNC* knots H_0 and H_1 such that the knot connected sums $K_0 \# H_0$ and $K_1 \# H_1$ are equivalent knots.

The set of equivalence classes, CH_n, form a group because the one twist-spin of any knot K is trivial by Zeeman [24] and has a cross section $K \# -K$, demonstrating that the inverse of K exists and is given by reversing the orientation.

THEOREM. $CH^{simple}_{2n-1} \quad = \quad CH^{\varepsilon}(Z)$ *for* $n > 1$.

Remark. This group of higher dimensional simple knots lies very close to the classification of simple knots by the S-equivalence relation studied by Murasugi [15], Trotter [21] and Levine [12]. In fact the trivial elements in $CH^{\varepsilon}(Z)$ are the stably hyperbolic structures which have a very simple algebraic description. An interesting question which could be investigated either algebraically or geometrically if the following:

Is a stably hyperbolic structure (stably DNC knot) hyperbolic (DNC)?

We now show that the even dimensional group CH_{2n} is infinitely generated. We define an invariant, which was also studied in a different context by Levine in Knot Modules, I [13]. It is well known that the Seifert surface V^{2n+1} of an even dimensional knot (S^{2n+2}, K^{2n}) admits a linking pairing B ($\varepsilon = (-1)^{n+1}$-symmetric Q/Z-valued inner product) on the torsion subgroup of the integeral homology in dimension n, $M = Tor\ H_n(V,Z)$. We will now construct an isometric structure as follows: First, we note that Alexander duality for V and the universal coefficients theorem give a non-singular pairing p: $Tor\ H_n(V,Z) \times Tor\ H_n(S^{2n+2} \backslash V, Z) \to Q/Z$ because

$$Tor\ H_n(V,Z) \cong Tor\ H^{2n+2-(n)-1}(S^{2n+2} \backslash V, Z) \cong Tor\ H_n(S^{2n+2} \backslash V, Z)$$

This is given geometrically by means of the intersection theory of chains by representing the given homology classes by cycles x on V and y in $S^{2n+2} \backslash V$. Since x is torsion, $mx = \partial c^{n+1}$ for some $(n+1)$-chain c and integer m. Also $y = \partial d^{n+1}$ by the well known computation of the homology of S^{2n+2}. Because ∂c and ∂d are disjoint, the intersection $c \cap d$ is defined and the pairing $p(x, y) = \varepsilon_*(c \cap d)/m$ in Q/Z (where ε_* is the augmentation.) Again using the positive ε-push of V off itself we define $A(x,y) = p(x, i_{+*}(y))$. Mimicing the proof in the odd-dimensional case we have: $B(tx,y) = B(x,(id-t)y)$ where $A(x,y) = B(tx,y)$.

THEOREM. *There is a homomorphism* $S: CH_{2n} \to CH_\varepsilon(Q/Z)$ *which is onto for n greater than one.*

Proof. (*Sketch*) If (S^{2n+2}, K) is *DNC* then there is a Seifert surface V^{2n+1} for K contained in the Seifert surface D^{2n+2} for the trivial knot of which K is a cross section. As V is a co-dimension one orientable submanifold of this disc, it separates the disc into two components W_+ and W_-. The submodules H_0 = kernel i_{+*} : $Tor\ H_n(V) \to H_n(W_+)$ and H_1 = kernel i_{-*} provide a hyperbolic decomposition of the torsion Seifert form.

In the case where $Tor\ H_n(V,Z)$ is of odd order and the knot complement has the $(n-1)$-homotopy type of a circle, the classification results of C. Kearton [8] can be used to demonstrate the sufficiency of the obstruction.

Finally we note that, unlike the even dimensional knot concordance group, which is trivial, CH_{2n} is infinitely generated because of the injections $C^{\varepsilon}(F_p) \subset CH^{\varepsilon}(F_p) \subset CH^{\varepsilon}(Q/Z)$ where the first injection of the infinitely generated group $C(F_p)$ is defined using anistropic representatives.

Further details on the hyperbolic cases will appear in [19] .

REFERENCES

1. Alexander, J.W. A lemma on systems of knotted curves, Proc. Nat. Acad. Sci., 9(1923), p. 93.
2. Alexander, J.P. The Bisection Problem, preprint.
3. Alexander, J.P., Conner, P.E. and Hamrick, G. Odd order Transformation Groups and Inner products, Springer Lecture Notes, 625, 1977.
4. Casson, A. and Gordon, C. Cobordism of classical knots, preprint, Orsay
5. Fox, R.H. Quick Trip Thru Knot Theory in Topology of 3-manifold, ed. M.K. Fort, Jr., Prentice Hall, 1963.
6. Gibbs, D.A. Witt Classes of Integral Representations and Orientation Reversing Maps, Preprint, LSU, 1977
7. Kearton, C. Simple knots which are doubly null concordant. Proc. AMS 52(1975), p. 471.
8. ________, An algebraic classification of some even-dimensional knots, Topology 15(1976), p. 363.
9. Kervaire, M. Knot Cobordism in Codimension Two - "Manifolds - Amsterdam 1970", Springer Lecture Notes, 190.
10. Kreck, M. Bordism of Diffeomorphisms, Bull. AMS 82(1976), p.759.
11. Levine, J. Knot Cobordism in Codimension Two. Comm. Math. Helv. 44 (1968), p. 229.
12. ________, An algebraic classification of some knots of codimension Two. Comm. Math. Helv. 45(1970), p. 185.
13. ________, Knot Modules, I. Trans. AMS (1977).
14. Milnor, J. On Isometries of Inner Product Spaces. Inv. Math. 8(1969), p. 83.
15. Murasugi, K. On a certain numerical invariant of link types, Trans. AMS 117(1965), p. 377.
16. Quinn, F. Open book decomposition and bordism of Automorphisms, preprint, VPI, 1977.
17. Seifert, H. Uber das Geschlecht von Knoten, Math. Ann. 110 (1934), p. 571.
18. Stoltzfus, N.W. Unraveling the Integral Knot concordance Group. Memoir AMS 192 (1977).
19. ________, Algebraic computations of the integral concordance and double null concordance group of knots,(to app.)

20. Sumners; D. Invertible Knot Cobordisms, Comm. Math. Helv. 46(1971), p. 240.
21. Trotter, H. On S-equivalence of Seifert matrices, Inv. Math. 20(1973), p. 173.
22. Winkelnkemper, H. Manifolds as open books, Bull, AMS 79 (1973), p. 45.
23. Witt, E. Theorie der quadratischen Formen in beliebigen Korpern, J. reine u. angew. Math. 176(1937), p. 31.

HYPERBOLIC GEOMETRY AND HOMEOMORPHISMS

Dennis Sullivan

Department of Mathematics
Institut des Hautes Etudes Scientifiques
Bures-sur-Yvette, France

The idea that one can radially identify Euclidean space R^n with the interior of the finite ball B^n has been very useful. In de Rham's book of 1955 a smoothing procedure for currents is based on the fact that the translation group of R^n damps out to the identity in a C^∞ fashion at the boundary of B^n. Thus familiar convolution in R^n can be coordinate wise placed on a general smooth manifold. In 1963 Connell's work on approximating a homeomorphism by a piecewise affine homeomorphism was based on the similar idea that a homeomorphism a bounded distance from the identity on R^n becomes the identity at the boundary at B^n. In 1968 the spectacular results of Kirby, Edwards-Kirby and Kirby-Siebenmann incorporated the Connell idea with the ingenious use of the torus T^n to arrive at periodic and thus bounded homeomorphisms of R^n. (See quotation in "radial engulfing" below.)

We now propose the idea that one may further profit from the idea that the ball B^n with its Euclidean geometry is conformally equivalent to hyperbolic space, the carrier of non-Euclidean geometry.

ISBN 0-12-158860-2

Analogous to de Rham, we note that heat diffusion in hyperbolic space damps out to the identity at the boundary of B^n and we can construct coordinate wise on general manifolds a smoothing procedure with conformal symmetry properties.

Analogous to Connell, we have the simple proposition that a quasi-isometry[1] of hyperbolic space which is a bounded hyperbolic distance from the identity determines a quasi-isometry of B^n (in its Euclidean metric) which is the identity at the boundary. This follows easily since the conformal factor is essentially the Euclidean distance to the boundary, namely,

$$\text{(hyperbolic element)} \sim \text{(Euclidean element)} \cdot \text{(Euclidean distance to boundary)}$$

Similarly, a bounded quasi-conformal[2] homeomorphism of hyperbolic space determines a quasi-conformal homeomorphism of the closed ball which is the identity at the boundary.

Analogous to the Kirby immersion torus devise we have the following

i) there are discrete groups of isometries of hyperbolic space so that the quotient is a compact manifold. (This is classical but non-trivial for $n > 5$, see "hyperbolic space forms" below.)

ii) for each of these there is a finite cover which is parallelizable in the complement of a point. (At the moment the proof of this fact makes use of deep properties of etale homotopy theory in characteristic p, see below.)

[1] *A homeomorphism which distorts the metric by only a bounded amount. Or* φ *and* φ^{-1} *satisfy a uniform Lipschitz condition.*

[2] *A homeomorphism which distorts the conformal geometry by only a bounded amount.*

iii) thus we have desired "hyperbolic manifolds" in each dimension $2, 3, 4, \ldots\ldots, n, \ldots\ldots$ to apply Kirby's immersion device.

Namely, we have M^n so that $M^n - pt$ immerses in R^n, M^n is compact, and the universal cover of M^n is hyperbolic space. We note these manifolds are not related in simple inductive manner as are the torii T^n. For $n = 2$, any surface of higher genus works here.

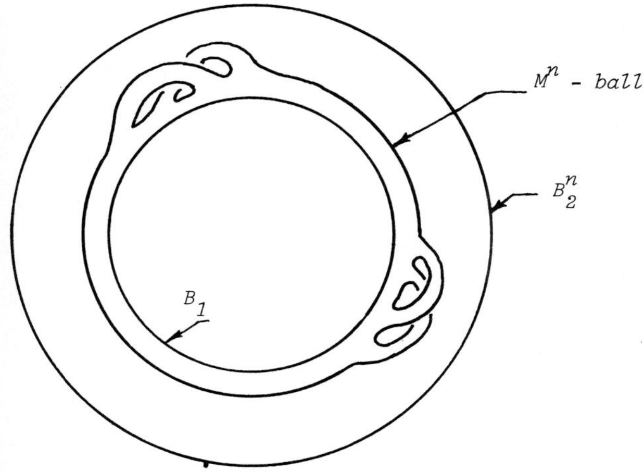

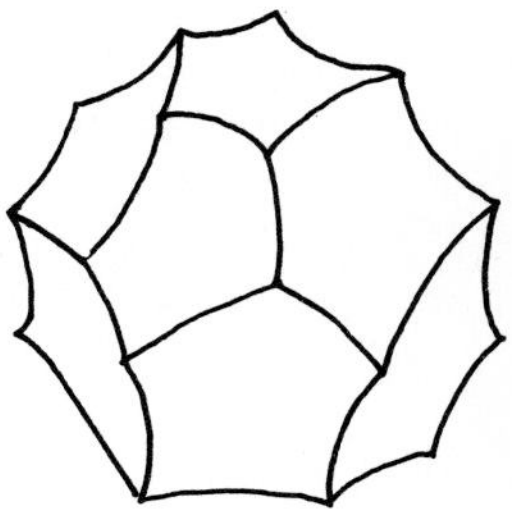

$n = 3$

Fig. 1.

For $n = 3$ one can imagine a discrete group defined by reflecting a symmetrical dodecahedion with dihedral angles equal to $\pi/2$.

iv) Now following Kirby (see figure 1) we can immerse M^n - ball in B_2^n and extend a homeomorphism defined near image (M^n - ball) and sufficiently close to the identity to approximately M^n - ball. Thinking of the deleted ball in polar coordination we can furl (see "furling" below) to obtain commutation with a radial homothety and extend over the ball by infinite repetition. We have extended the quasi-isometry defined near B_1 and close to the identity to a global quasi-isometry of M close to the identity.

We lift to hyperbolic space apply the remark above about bounded quasi-isometries of hyperbolic space to obtain a quasi isometry of B^n which is the identity at the bdry and containing the original on B_1. This is the 0-handle case required for the construction of isotopies as in the Edwards - Kirby, paper (Annals of Math 93 (1971) pp. 63-88.)

v) For the k-handle case we choose a group as in iv) for hyperbolic space of dimension $(n - k)$. We extend this group by geodesic perpendicularity to a group in hyperbolic n-space. Then the fundamental domain has the form $D^k \times D^{n-k}$ ($k = 1$, $n = 3$ is shown in figure 2)

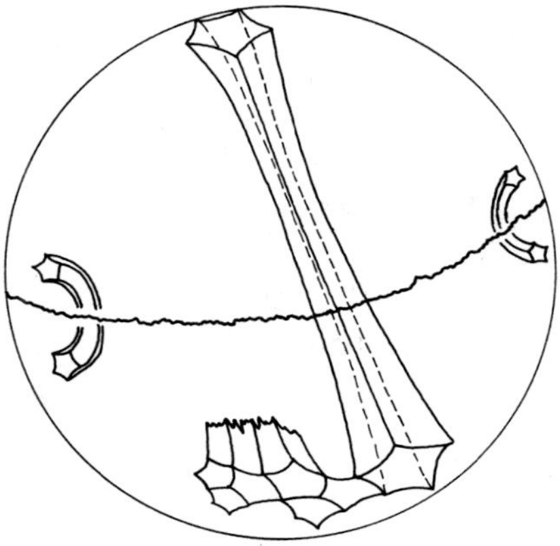

Fig. 2.

Then we treat a homeomorphism defined near the core of a k-handle which is close to the identity and equal to the identity near the bdry. Firstly, we use the Kirby immersion devise as in iv) to obtain an extension to the (fundamental domain - $(n-k)$ ball) x D^k compatible with the group. Then since we have the identity near the bdry we again have an n-dimensional hole which can be filled in by furling and infinite repetition as in iv). We extend by the group and apply the bounded quasi-isometry remark to obtain a quasi-isometry of B^n which is the identity at the bdry and agrees with the original near the core of the k-handle.

Using a fine handle decomposition and iv) and v) inductively we find that in the context of quasi-isometries or quasi-conformal homeomorphisms of any smooth manifold

THEOREM 1. *a) A homeomorphism near a compact domain and pt wise close to the identity can be extended to a global homeomorphism which is the identity outside a slightly larger neighborhood.*

b) Sufficiently close homeomorphisms are connected by a path of homeomorphism, that is they are isotopies.

c) The construction of isotopics is local, relative, compatible with parameters, etc., that is we have the Cernavskii, Kirby-Edwards isotopy theory)

To have the proof for quasi-conformal also we note that everything works the same. We note also there is the stronger extension fact (Gehring) that any quasi-conformal mapping of interior B^n extends to the quasi-conformal on the closed B^n and the boundary maps is also quasi-conformal in dim $n - 1$

Remark. The schoenflies theorem (a collared $(n - 1)$ sphere in S^n bounds a ball), the Annulus theorem (the region between two collared $(n - 1)$ spheres is S^{n-1} x I) and the component problem (an orientation preserving homeomorphism of a ball into R^n is connected to the identity by a path of homeomorphisms) come out nicely in these quasi conformal and quasi isometric contexts.

The isotopy extension of Theorem 1 shows a) connected sum is well defined which allows one to easily give Mazur's proof of Schoenflies with the infinite bad point being homothetic and all right for the quasi contexts, b) the component problem follows since we have derivatives at almost all points. Thus we can use Milnor's isotopy $\{e^{t}\cdot\varphi e^{-t}\}$ near a point to change the map to nearly its derivative and extend this change by theorem 1. Firstly, as usual Schoenflies and the component problem implies the Annulus theorem given isotopy extension (Theorem 1). So we have

COROLLARY 1. In the quasi-conformal and quasi-isometry contexts we have the Schoenflies theorem, the Annulus theorem, and the component problem in all dimensions.

Of course in the topological context the Annulus theorem is unknown in dimension 4 (while Schoenflies is known) and in the *pl* or smooth context Schoenflies is unknown in dimension 4 (while the component problem is known). Only part of the corollary is new (see Väisälä).

Another corollary is the following - by the above quasiconformal homeomorphisms of R^n are stable and thus (using Connell) approximable by piecewise affine homeomorphism in dimensions ≥ 5, (which demonstration we have reached without using Kirby's Annulus theorem).

Now we will change terminology and refer to quasi-isometries as Lipschitz homeomorphisms. We want to discuss approximating arbitrary homeomorphism by Lipschitz homeomorphism. For dimensions less than 4 there is a good classical theory (Moise).

Dimension 4 is unknown and remains so. For dimensions greater than 4 we get full positive results using Connell's radial engulfing and Kirby's Annulus theorem.

As usual it will suffice to study a homeomorphism of a handle $D^k \times D^{n-k}$ into R^n which is Lipschitz near $(\partial D^k \times pt)$. Using Connell's radial engulfing (see below) we spread the Lipschitz property across the k-handle at the cost of an $\in$ discrepancy on nghd $(D^{k-1} \subset \partial D^k)$. This discrepancy is Lipschitz and can by Theorem 1 or v) be extended to a Lipschitz homeomorphism which is the identity outside a small nghd of the discrepancy. The composition of the latter with the former gives the desired relative Lipschitz approximation on the handle.

Applying this result inductively to a fine handle decomposition we have the

THEOREM 2. *A homeomorphism defined near a compact connected region L of R^n, $n \neq 4$, into R^n which is Lipschitz near a non void sub compact K can be approximated pt wise by a Lipschitz homeomorphism which is unchanged near K.*

Remark. We need K to be non void to do the zero handles. If there is no K we use Kirby's annulus theorem for the zero handles to obtain

COROLLARY 2. *Theorem 2 is also true when K is void*

COROLLARY 3. *Topological manifolds of dimension $\neq 4$ have Lipschitz coordinate systems. Such locally Euclidean Lipschitz structures are unique up to homeomorphism close to the identity.*

Corollary 3 follows by standard arguments using Theorem 2 and corollary 2 and fine handle decompositions in coordinate neighborhoods.

CONCLUDING REMARKS

The unique Lipschitz structure on topological manifolds allows one certain geometrical and analytical methods. The discussions in Whitney's book "Geometric Integration" and Federer's book "Geometric measure Theory" are invariant under Lipschitz changes of coordinates.

Thus we have the theory of Hausdorf measure and dimension on k-dimensional subsets $0 \leq k \leq n$. This implies general position, (intersection empty) and complementary dimension transversality (zero dimensional intersection) and almost everywhere transversality (general dimension) on the level of cycles or currents, (federer).

We have also a class of differential forms closed under wedge product, exterior differentiation, and transforming by Lipschitz mappings. These Whitney forms (called flat forms in his book) provided the original motivation[1] for wanting a Lipschitz structure. Locally Whitney forms are defined by the property that they and their exterior derivative (in the sense of distributions) are forms with bounded measurable coefficients. Dualizing Whitney forms give us currents.

On a Lipschitz manifold the L^2-spaces of forms are well defined (as pre-Hilbert spaces) with d as a closed unbounded

[1] *Also very important was Siebenmann's observation (1970) that his counterexample to the pl Hautvermutung was not a counterexample to the analogous Lipschitz statement.*

operator. One can contemplate an index theory which would lead to a new analytic proof of Norvikov's theory about Pontryagin classes. There is a local Gauss Bonnett formula $\chi_M = \int_M U/\text{diagonal}$ where U is a locally defined n form on $M \times M$ defined for example by smoothing the diagonal current (p. 80 de Rham). Furthermore, this equation can be proved directly and analytically taking as χ_M the Euler characteristic for Whitney form cohomology.

There are many other mathematical contexts with a topological component where hyperbolic space and its bdry plays an interesting and important role. One might mention

i) Nielson theory - providing a natural homeomorphism

$$\text{Aut } (\pi_1 \text{ (surfaces)}) \to (\text{quasi-conformal homeo } S^1)$$

ii) Teichmuller space and its boundary

iii) Thurston's canonical surface transformations

iv) Kleinian groups and non-compact hyperbolic 3-manifolds

v) Thurston's canonical hyperbolic structure on many compact 3-manifolds

vi) Anosov flows (geodesic flow on negative curved manifolds) their structural stability, ergodicity, etc.

vii) Mostow's theorem on the rigidity of hyperbolic space forms of finite volume $(n \geq 3)$ and Gehring's quasi-conformal mappings in space

viii) Fefferman's theorem on complex analytic homeomorphisms being smooth up to the boundary.

ix) Margulis and Gromov theorems in discrete groups and Riemannian geometry

x) Furstenburg's work on random walks and Poisson boundary (Corresponding to the beautiful fact that a random continuous path hits a definite point at ∞ with probability 1.)

At the time of the research it was mainly i), vi), and vii) which motivated the ideas here. It is a pleasure to acknowledge the inspiring discussions with Gromov and Thurston on hyperbolic geometry and Edwards and Siebenmann on homeomorphisms.

Further topological references may be found in the body and bibliography of the book by Kirby and Siebenmann "Foundational Essays on Topological manifolds, smoothings, and triangulations." Annals of Math Studies Princeton Univ. Press 1977. For quasi-conformal mappings the body and bibliography of Mostow "Quasi-conformal mappings in n-space and the regidity of hyperbolic space forms" Publications Mathematiques I.H.E.S. No. 34 (1968): (is a good source).

FURLING: M. Brown 1964 unpublished. Also L. Siebenmann, "Pseudo Annuli and invertible cobordisms", Archiv der Math. 19 (1968) 28-35.

THEOREM. If X and Y are compact Hausdorff spaces and h is a almost vertical homeomorphism of X x I onto a nghd of Y x 0 in Y x R, then there is a homeomorphism g so that the following commutes

$$\begin{array}{ccc} X \times I' & \xrightarrow{h} & Y \times R \\ \downarrow i & & \downarrow j \\ X \times S^1 & \xrightarrow{g} & Y \times S^1 \end{array}$$

Remark. The proof is ingenious but only two lines. If X and Y are metric and h is a quasi-isometry so is g. Similar remarks apply to the quasi-conformal context. Here I is an interval, is a smaller interval, I' is determined by an inclusion $I' \subset S^1$ and j by the projection $R \to S^1$.

HYPERBOLIC SPACE FORMS (Compact and almost parallelizable)

i) Let Γ denote matrices preserving the quadratic form $x_1^2 + x_2^2 + \ldots + x_n^2 - \sqrt{2}\, x_{n+1}^2$ with entries of the form $n + m\sqrt{2}$, n, m integers. Then Γ is a group of isometries of hyperbolic space, which is discrete and has compact quotient. All this follows easily using the field automorphism of $Q\sqrt{2}$, and the compactness criterion for lattices in R^n - minimal separation bounded from below and volume bounded from above. (Algebraic group theory is not required as it is for the more general result of Borel's paper in the first volume of Topology.)

A subgroup of finite index (e.g. $(n + m\sqrt{2})_{ij} \equiv \delta_{ij} \bmod 3$) in Γ will have no torsion and define a manifold. This type of construction is the only known way to get compact hyperbolic space forms in n-dimesnions.

ii) It is not clear when such manifolds are almost parallelizable. They are not parallelizable in even dimensions and their analogues in complex hyperbolic space have non-trivial rational Pontryagin classes. However, there is a fortunitous general result which saves us. Namely, "Over a finite polyhedron a vector bundle with *discrete* structure group in a *real orthogonal* group becomes continuously trivial in some finite cover." We can apply this result because a real line bundle over our manifold is the quotient of the positive light cone by the group. Thus its tangent bundle is trivial in some finite cover by the general result. It follows that *some finite cover of a hyperbolic space form is stably parallelizable* (which is even more than almost parallelizable by an elementory obstruction argument).

Now the general result follows because the complexification of a real orthogonal group $O(p,q)$ has the homotopy type of $O(n)$, $n = p + q$. Etale homotopy theory in finite characteristics can be applied (as in Deligne, Sullivan "Fibres complex ..." Compte Rendu t. 281) p. 1081 (1975), to study principal $O(n,\mathbb{C})$ bundles with discrete structure groups (and see they become trivial in finite covers).

The role of the orthogonal group here is crucial not only for the proof but also for the truth of the assertion about bundles. For Millson has recently constructed flat bundles with group $Sl(n,r)$ so that the second Stiefel Whitney class is non-trivial in every finite cover.

CONNELL'S RADIAL ENGULFING: This is lemma 3 of E. H. Connell "Approximating stable homeomorphisms by piecewise linear ones." Annals of Maths 78 (1963) pp. 326-338. It is stated there for $n \geq 7$ where engulfing in codimension 4 follows from straightforward general position and induction. The result is just as true for $n \geq 5$ by the more complicated double induction argument needed for engulfing in codimension 3. This was explained to me by Siebenmann but I don't know a reference.

THEOREM. *On Euclidean space of dimension ≥ 5 there is a homeomorphism which is the identity on one ball and stretches a larger ball over a third larger ball, which almost preserves radii, and which is piecewise linear relative to any given structure.*

It is perhaps historically interesting to add the last para- of this prescient and relevant paper.

"Suppose T_1 and T_2 are two arbitrary pl structures on E^n. It is known (except in dimension 4) that $\exists$ a homeomorphism $h : E \to E$ which is pl from T_1 to T_2 (see Stallings and Moise). If h could be chosen as a bounded homeomorphism, then by Lemma 5 h would be stable and it would follow that all orientation preserving homeomorphisms are stable. Thus *the annulus conjecture in dimension n would be true*. Conversely, if the annulus conjecture were true in all dimensions it would follow from the procedures of this paper that (for $n \geq 7$) h could be chosen to be bounded. Thus *the annulus conjecture is roughly equivalent to this strong form of the Hauptvermutung* for Euclidean space where h is to be chosen as bounded from the identity."

In light of the fact that five years later Kirby (1968) reduced the Annulus conjecture to a "periodic Hauptvermutung" for Euclidean space (and the latter was sufficiently true by surgery (1966) to finish) these remarks of Connell seem almost prophetic.

K_2 AND DIFFEOMORPHISMS OF TWO AND THREE DIMENSIONAL MANIFOLDS

J. B. Wagoner[1]

Department of Mathematics
University of California
Berkeley, California

This article discusses some low dimensional examples of the K_2 invariant for diffeomorphisms defined in [18], [19]. *It is well-known from* [4] *how the geometry of one parameter families of Morse functions fits in with the definition of K_2 in terms of the Steinberg relations. Some of the examples below show the geometric significance of the three relations occurring in Matsumoto's presentation* [8, §11] *of K_2 of a field. There is also an application to the mapping class group.*

I. GENERAL PROPERTIES OF THE Σ INVARIANT

Let M^n be a compact, connected smooth manifold with boundary ∂M the disjoint union $\partial M = N \cup N'$ where N and N' may be empty and are not necessarily connected. Let F be a field with a "conjugation" automorphism $a \to \overline{a}$ of order two. Let $\pi = \pi_1 M$ and suppose we are given a *unitary* representation $\rho : \pi \to F^*$ satisfying $\rho(g)^{-1} = \overline{\rho(g)}$ for g in π. Assume ρ is *acyclic* in the sense that the homology groups $H_*(M,N;\rho)$ of the complex $F \otimes_\rho C_*(\tilde{M},\tilde{N})$ vanish. Here $\tilde{M}$ denotes the universal cover of M and for any subset X of M

[1] *Partially supported by NSF Grant MCS 74-03423.*

ISBN 0-12-158860-2

we let $\tilde{X}$ denote the part of $\tilde{M}$ lying over X. Then as in [9], [15], it is possible to define the Reidemeister-Franz-de Rham torsion

$$\tau_\rho(M,N) \in F^*/_{\pm\rho(\pi)}$$

It is customary when $F = \mathbb{C}$ to consider $\tau_\rho(M,N)$ as lying in $\mathbb{C}^*/S^1 \cong$ positive reals and to call $\tau_\rho(M,N)$ the $\mathbb{R}$- *torsion*. Torsion is a K_1 type invariant associated to an acyclic manifold. The idea is that a diffeomorphism of such an acyclic manifold should have a K_2 type invariant. This is analogous to simple homotopy theory where elements of the Whitehead group $Wh_1(\pi)$ are obstructions to the existence of a product structure on an s-cobordism and elements of $Wh_2(\pi)$ are obstructions to the uniqueness of such product structures. See [4].

According to Matsumoto's theorem [8, §11] the group $K_2(F)$ can be defined as the free abelian group generated by symbols $\{a,b\}$ where a, b lie in F^* subject to the relations

(i) $\{a,bc\} = \{a,b\}\{a,c\}$

(ii) $\{ab,c\} = \{a,c\}\{b,c\}$

(iii) $\{1-a,a\} = 1$ for $a \neq 0,1$.

Here we are writing $K_2(F)$ multiplicatively. Let $K_2^\rho(F)$ denote the quotient of $K_2(F)$ by the subgroup $\{\pm\rho(\pi),\pm\rho(\pi)\}$ generated by symbols $\{\pm\rho(g),\pm\rho(h)\}$ for g and h in π, and let $K_2'(F)$ be the quotient of $K_2(F)$ by the subgroup generated by symbols $\{\lambda,\mu\}$ where λ and μ are in F^* and satisfy $x^{-1} = \bar{x}$. Since ρ is unitary, there is a natural map $K_2^\rho(F) \to K_2'(F)$.

Fix a base point x in M. Let $\mathcal{D}(x)$ be the group of diffeomorphism $f : (M;N,N') \to (M;N,N')$ which fix the base point and preserve the representation ρ. This means $\rho f_* = \rho$ where f_* is the induced map on the fundamental group $\pi = \pi_1(M,x)$ based at x. Give $\mathcal{D}(x)$ the C^∞ topology. Let ρ be acyclic as usual.

THEOREM 1.1. There is a homomorphism

$$\Sigma : \pi_0 \mathcal{D}(x) \to K_2^{\rho}(F)$$

satisfying the product, sum, and duality formulae below.

The construction of Σ uses one parameter Cerf-Morse theory similarly to [4]. As with torsion, Σ can be defined more generally for ρ an r-dimensional acyclic, unitary representation $\rho : \pi \to U(r,F)$ where $U(r,F)$ is the group of r x r *unitary* matrices A over F defined by $AA^* = 1$. A^* is the conjugate transpose of A. In this case Σ takes values in

$$K_2(F) = \lim_{r} H_2(SL(r,F))$$

modulo the image of

$$KU_2(F) = \lim_{r} H_2(SU(r,F)).$$

There is an induced representation formula for lifts of diffeomorphisms of M to a finite sheeted covering space similar to the formula for torsion [13]. Σ can also be defined when N' is empty and M is non-compact with the homotopy type of a finite complex.

When ρ is not acyclic, torsion can still be defined as in [9] provided a choice of basis for $H_*(M,N;\rho)$ is given. Preliminary calculations indicated that Σ can probably be defined for diffeomorphisms inducing unitary transformations of $H_*(M,N;\rho)$ with respect to the basis used to get torsion. When ρ is trivial, it seems likely that a Σ invariant can be defined taking values in $K_2(Q)/K_2(Z)$.

If the diffeomorphisms are allowed to move the base point x there is an indeterminacy. Consider the group $\mathcal{D}$ of diffeomorphisms of $(M;N,N')$ which are isotopic to a diffeomorphism in $\mathcal{D}(x)$. For example, if $x \in \text{int } M$ any diffeomorphism inducing the identity on $H_1(M)$ lies in $\mathcal{D}$. Suppose that in fact $x \in \text{int } M$. Then there is an exact sequence

$$\pi_1(M,x) \xrightarrow{\partial} \pi_0 \mathcal{D}(x) \to \pi_0 \mathcal{D} \to 1.$$

where ∂ is defined as follows: Let g be a smooth loop at x representing an element of $\pi_1(M,x)$ and let $f_t : (M;N,N') \to (M:N,N')$ be an isotopy so that $f_t(x)$ traces out g as $0 \leq t \leq 1$. Then $\partial(g) = f_1$.

PROPOSITION 1.2. $\Sigma_\rho(\partial(g)) = \{\tau_\rho(M,N),\ \rho(g)\}$.

There is a similar result when $x \in \partial M$ and we have a "boundary twist." Let $x \in N'$ and let $N' \times I$ be a collar neighborhood of N' in M so that N' is identified with $N' \times 1$. Let $f \in \mathcal{D}(x)$ have support in $N' \times I$ and assume f is level preserving on $N' \times I$. Let $g \in \pi_1(N',x)$ denote the loop going from $x \times 1$ to $x \times 0$ along $x \times 1$ and then back to $x \times 1$ along $f(x \times I)$.

PROPOSITION 1.3. $\Sigma_\rho(f) = \{\tau_\rho(M,N),\ \rho(g)\}$.

A similar result holds when $x \in N$. The first term in the symbol $\{\ ,\ \}$ is still $\tau_\rho(M,N)$. These two formulae imply

THEOREM 1.4. Let $x \in$ int M. *There is a well-defined homomorphism*

$$\Sigma : \pi_0 \mathcal{D} \to K_2^\rho(F) \text{ mod } \{\tau_\rho(M,N), \rho(G)\}$$

where $G = \pi_1(M,x)$. *If* x *is in* N *or* N', Σ *is well-defined in* $K_2^\rho(F)$ *modulo the subgroup* $\{\tau_\rho(M,N),\ \rho(G)\}$ *where* G *is respectively* $\pi_1(N,x)$ *or* $\pi_1(N',x)$.

Let M and ρ be as usual. Let Q be a closed, connected manifold with base point y. Let $\hat{\rho} : \pi_1(M \times Q,\ x \times y) \to F^*$ be the projection $\pi_1(M \times Q, x \times y) \to \pi_1(M,x)$ followed by ρ. Let $\chi(Q)$ be the Euler characteristic of Q.

PRODUCT FORMULA. $\Sigma_{\hat{\rho}}(f \times 1_Q) = \Sigma_\rho(f)^{\chi(Q)}$.

Assume M is oriented and suppose $f \in \mathcal{D}(x)$ is a diffeomorphism of the triple $(M;N,N')$ which preserves orientation. Let $\overline{f}$ denote f considered as a diffeomorphism of the triple $(M;N',N)$. The conjugation $\overline{} : F \to F$ induces an involution $\overline{} : K_2^\rho(F) \to K_2^\rho(F)$ defined by $\overline{\{x,y\}} = \{\overline{x},\overline{y}\}$. Let $\varepsilon(n) = (-1)^n$.

DUALITY FORMULA. $\Sigma_\rho(\overline{f}) = [\overline{\Sigma}_\rho(f)]^{\varepsilon(n)}$.

For example, suppose F is the complex numbers $\mathbb{C}$. It is known [16] that $K_2'(\mathbb{C})$ is a rational vector space and is isomorphic to the negative eigenspace of $K_2(\mathbb{C})$ under the involution induced by complex conjugation. When M is a closed manifold $\overline{f} = f$ and so $\Sigma_\rho(\overline{f}) = \Sigma_\rho(f)$.

COROLLARY 1.5. If M is closed and dim M *is even, then* $\Sigma_\rho(f) = 1$ *in* $K_2'(\mathbb{C})$.

Let M be a closed manifold which is the union $M = X \cup Y$ with $X \cap Y = \partial X = \partial Y$. Suppose X, Y, and $X \cap Y$ are connected. Let $x \in X \cap Y$. Let $\rho : \pi_1(M,x) \to F^*$ be an acyclic unitary representation for M. Let $\rho_x : \pi_1(X,x) \to F^*$ be the composition of $\pi_1(X,x) \to \pi_1(M,x)$ followed by ρ. Similary for ρ_y. Let ρ_{xy} be $\pi_1(X \cap Y,x) \to \pi_1(M,x)$ followed by ρ. Assume ρ_x, ρ_y, and ρ_{xy} are acyclic for X, Y, and $X \cap Y$ respectively. Let f be a diffeomorphism of M such that $f(X) = X$, $f(Y) = Y$, $f(X \cap Y) = X \cap Y$, and such that f preserves ρ. Let f_X denote f restricted to X. Similarly for f_Y and $f_{X \cap Y}$. Assume $f(x) = x$.

SUM FORMULA. $\Sigma_\rho(f) = \Sigma_{\rho_x}(f_X)\cdot\Sigma_{\rho_y}(f_Y)\cdot\Sigma_{\rho_{xy}}(f_{X\cap Y})^{-1}$.

This version of the sum theorem is not the most general but does give the idea. Methods used in proving the sum formula together with Proposition 1.2 allow one to calculate a number of interesting examples.

Here is how Matsumoto's relations come up geometrically. Bilinearity on the right, relation (i), is needed for Σ to be a homomorphism. Compare Proposition 1.2:

$$\Sigma_\rho(\partial(gh)) = \{\tau_\rho, \rho(gh)\} = \{\tau_\rho, \rho(g)\rho(h)\}$$
$$= \{\tau_\rho, \rho(g)\}\cdot\{\tau_\rho, \rho(g)\} = \Sigma_\rho(\partial(g))\cdot\Sigma_\rho(\partial(h)).$$

Bilinearity on the left has to do with additivity of torsion and the sum formula. Let $g \in \pi_1(X \cap Y, x)$. Then we have

$$\Sigma_\rho(\partial(g)) = \{\tau_\rho(X \cup Y),\ \rho(g)\} = \{\tau_\rho(X)\cdot\tau_\rho(Y)\cdot\tau_\rho(X \cap Y)^{-1}, \rho(g)\}$$
$$= \{\tau_\rho(X), \rho(g)\}\cdot\{\tau_\rho(Y), \rho(g)\}\cdot\{\tau_\rho(X \cap Y), \rho(g)\}^{-1}$$
$$= \Sigma_{\rho_x}(\partial(g))\cdot\Sigma_{\rho_y}(\partial(g))\cdot\Sigma_{\rho_{xy}}(\partial(g))^{-1}.$$

Relation (iii) is connected to the fact that $\Sigma\cdot\partial$ kills anything coming from $\pi_1\mathcal{D}$ in the exact sequence

$$\pi_1\mathcal{D} \xrightarrow{\varepsilon} \pi_1(M,x) \xrightarrow{\partial} \pi_0\mathcal{D}(x) \to \pi_0\mathcal{D}\ .$$

For example, let $M = S^1$ and $x \in S^1$. Let ρ take the generator of $\pi_1(S^1, x)$ to the complex number $\eta \neq 1$ of modulus one. Consider the loop γ in $\pi_1\mathcal{D}$ which rotates the circle through 2π. Then $\partial\circ\varepsilon(\gamma)$ is the identity diffeomorphism so Σ should vanish on it. On the otherhand, Proposition 1.2 shows that $\Sigma\circ\partial\circ\varepsilon(\gamma) = \{1-\eta, \eta\} = 1$.

II. LOW DIMENSIONAL EXAMPLES

Whereas it is not really easy to "visualize" a non-trivial pseudoisotopy and to calculate its Wh_2 invariant, Dehn twists in two and three dimensional manifolds provide nice examples of the Σ invariant.

Let $\gamma \subset M^2$ be an embedded loop in a surface M^2 with a product neighborhood $\gamma \times I \subset M^2$. A *Dehn twist* about γ is a diffeomorphism of M^2 which is the identity outside of $\gamma \times I$ and on $\gamma \times I$ rotates $\gamma \times t$ through the angle $\theta(t)$ going from 0 to 2π as $0 \leq t \leq 1$. Let $T^2 \cong S^1 \times S^1$ be an embedded torus inside a three manifold M^3 with a collar neighborhood $T \times I \subset M^3$. A *Dehn twist* about T is the identity outside $T \times I$ and is $1 \times \tau$: $S^1 \times (S^1 \times I) \to S^1 \times (S^1 \times I)$ where $\tau : S^1 \times I \to S^1 \times I$ is a two dimensional twist. Let $A \simeq I \times S^1$ be an annulus embedded in a three manifold with boundary M^3 so that $\partial A = A \cap \partial M^3$. Let $A \times I$ be a collar neighborhood so that $\partial A \times I \subset \partial M^3$. A *Dehn twist* about A is of the form $1 \times \tau : I \times (S^1 \times I) \to I \times (S^1 \times I)$ where $\tau : S^1 \times I \to S^1 \times I$ is a two dimensional twist.

Example A. Automorphisms of knots. Let $k \subset S^3$ be a smooth knot. Let M denote the closure of the complement of a small tubular neighborhood $S^1 \times D^2$ of k with k identified with $S^1 \times 0$. $H_1(M)$ is infinite cyclic with a generator t represented by $\partial(1 \times D^2) \subset \partial M$. Let F be the field $Q(t)$ of rational functions in t. Let the involution be $\overline{p(t)} = p(t^{-1})$. Let $x \in \partial M$. Let $\rho : \pi_1(M,x) \to Q(t)$ be $\pi_1(M,x) \to H_1(M) \hookrightarrow Q(t)$. Then ρ is acyclic and

$$\tau_\rho(M) = \frac{A_k(t)}{1-t} \bmod\{\pm t^n\}$$

where $A_k(t)$ is the Alexander polynomial of the knot k. See [10].

Let $\mathrm{Diff}^+(S^3,k)$ denote the group of diffeomorphisms of S^3 which preserve k setwise and which preserve the orientations of both S^3 and k. Let $f \in \mathrm{Diff}^+(S^3,k)$. Then f takes the generator t of $H_1(M)$ to itself and hence preserves ρ. Using Theorem 1.4 we get a homomorphism

$$(2.1) \quad \Sigma : \pi_0\mathrm{Diff}^+(S^3,k) \to K_2(Q(t)) \bmod \{A_k(t),\ t^n\}\ .$$

The subgroup of indeterminacy is actually $\{\tau_\rho(M),\ t^n\}$. But $\{\tau_\rho(M), t^n\} = \{\frac{A_k(t)}{1-t}, t^n\} = \{A_k(t), t^n\}$. The group $K_2(Q(t))$ is known: The subgroup $\{\pm t^r,\ \pm t^s\}$ of $K_2(Q(t))$ is just $K_2(Z[t,t^{-1}]) \simeq K_2(Z) \oplus K_1(Z) \simeq Z/2 \oplus Z/2$ generated by $\{-1,-1\}$

and $\{-1, t\}$. See [12]. $K_2(Q(t))$ fits into the localization exact sequence

$$0 \to K_2(Q) \to K_2(Q(t)) \overset{\partial}{\to} \bigoplus_p (Q[t]/p)^* \to 0$$

where p runs through the prime ideals of the ring $Q[t]$ of polynomials in t with rational coefficients. Moreover, given a symbol $\{a(t), b(t)\} \in K_2(Q(t))$, the value of ∂ on it can be easily computed in $(Q[t]/p)^*$ using the tame symbol associated to the ideal p. See [8, §11].

Now let k_1 and k_2 be two knots of S^3 and let $k_1 \# k_2$ denote their connected sum as in the following schematic diagram:

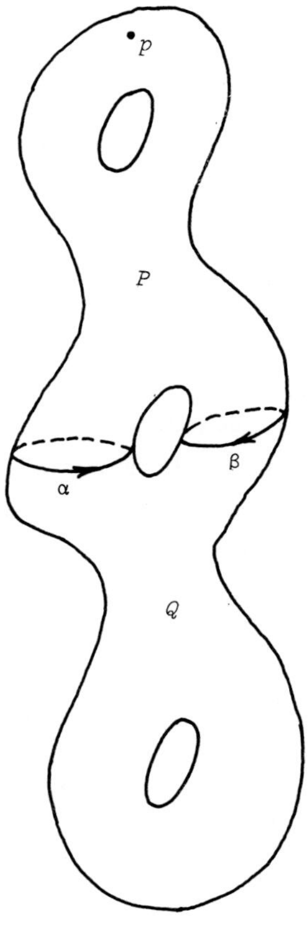

Let M denote as above the closure of the complement of a small tubular neighborhood of $k_1 \# k_2$. Let $A \subset M$ be the annulus obtained by rotating the inner circle about the vertical axis and let $A \times I \subset M$ be a product neighborhood of A. Let $x \in \partial A = (A \times 0) \subset \partial M$ be the base point. Let t be the Dehn twist about A supported in $A \times I$ such that $x \times I$ sweeps around the boundary of the tubular neighborhood counter clockwise (i.e. in the same direction as the generator t of $H_1(M)$). Methods of the sum theorem show that

$$(2.2) \qquad \Sigma_\rho(t) = \{A_2(t),\ t\} \bmod \{A_1(t)A_2(t), t^n\}$$

where for $i = 1,2$ the Alexander polynomial of k_i is $A_i(t)$. The idea is that t is the identity on the complement of k_1 and twists as it goes into the complement of k_2.

PROPOSITION 2.3. Suppose $A_1(t)$ and $A_2(t)$ have prime factors $p_1(t)$ and $p_2(t)$ of degree at least two both of which have no roots of unity. Assume that either $p_1(t)$ does not appear in $A_2(t)$ or that $p_2(t)$ does not appear in $A_1(t)$. Then $\{A_2(t), t\}$ has infinite order in $K_2(Q(t))$ mod $\{A_1(t)A_2(t), t^n\}$.

Proof. We use the tame symbols of [8, §11]. Since $\{A_1(t), t\} = \{A_2(t), t^{-1}\} \bmod \{A_1(t)A_2(t), t^n\}$ it suffices to consider the case when $p_1(t)$ does not appear in $A_2(t)$. Let $a_i = A_i(t)$ for $i = 1,2$. We must show there is no integer $r \neq 0$ so that $\{a_2, t\}^r = \{a_1a_2, t\}^s$ for some integer s. Suppose to the contrary. Then $\{a_2, t\}^{r-s} = \{a_1, t\}^s$. Moreover s must be non-zero: Let ν be the valuation corresponding to the prime ideal $\langle p_2(t)\rangle$. If $s = 0$, then $\{a_2,\ t\}^r = 1$. On the other hand, under the tame symbol for ν the element $\{a_2,\ t\}^r$ goes to

$$\frac{a_2^{\nu(t)r}}{t^{\nu(a_2)r}} \bmod \langle p_2(t)\rangle$$

Since deg $p_2(t) \geq 2$, $\nu(t) = 0$. Hence this element goes to a power of t mod $<p_2(t)>$ and is of infinite order because $p_2(t)$ has no roots of unity. Now let ν denote the valuation corresponding to $p_1(t)$. Applying the tame symbol to $\{a_2, t\}$ gives

$$\frac{a_2^{\nu(t)}}{t^{\nu(a_2)}} \bmod <p_1(t)>$$

which is trivial since $\nu(t) = \nu(a_2) = 0$. But applying it to $\{a_1, t\}^s$ gives a non-zero power of t mod $<p_1(t)>$ which is infinite order as above. q.e.d.

A similar example arises on a closed manifold M^2 obtained by considering two fibrations over the circle which have a torus as boundary, putting them together along that boundary, and then doing a Dehn twist along the torus in the horizontal direction.

Example B. The mapping class group. Let M be a compact, closed surface of genus g. The mapping class group $\mathcal{M}$ is the set of isotopy classes of diffeomorphisms of M. See [1]. Let $\mathcal{M}_0 \subset \mathcal{M}$ be the subgroup of those diffeomorphisms inducing the identity on $H_1(M)$. It has been shown [cf. 7, §2] that $\mathcal{M}_0$ is generated by two types of generators:

(1) a Dehn twist t_α about a simple closed curve α in M which is homologous to zero, and

(2) $t_\alpha \cdot t_\beta^{-1}$ where t_α and t_β are Dehn twists about two disjoint, simple closed curves α and β which are homologous.

Let $t_1, \ldots, t_{2g}$ denote a basis for $H_1(M)$ as well as $2g$ algebraically independent elements generating the transcendental extension $Q(t_1, \ldots, t_{2g})$. There is an obvious choice of a representation $\pi_1(M) \to H_1(M) \hookrightarrow Q(t_1, \ldots, t_{2g})$ which could be used to get a K_2 invariant for elements of $\mathcal{M}_0$, which clearly preserve the representation. But there are two difficulties: first, this representation is not acyclic, and second the Σ invariant would vanish, at least when considered as lying in $K_2'(\mathbb{C})$, because $\dim M = 2$. This is remedied by crossing with the circle. Let s denote the generator of $\pi_1(S^1)$ and also denote an element algebraically independent of $t_1,\ldots,t_{2g}$. Choose a base point p of M and let $1 \in S^1$ be a base point. Let ρ be the unitary representation $\pi_1(M \times S^1, p \times 1) \to H_1(M \times S^1) \hookrightarrow Q(s,t_1,\ldots,t_{2g})$. A direct computation shows that ρ is acyclic for $M \times S^1$ and that

$$\tau_\rho(M \times S^1) = (s-1)^{|\chi(M)|} = (s-1)^{2g-2}$$

First consider the group $M_0(p)$ of isotopy classes of diffeomorphisms of M fixing the base point p. The isotopies are required to fix p also. Any $f \in M_0(p)$ gives rise to a basepoint preserving diffeomorphisms $f \times id$ of $M \times S^1$ which preserves ρ, and the correspondence $f \to \Sigma(f \times id)$ defines a homomorphism

$$(2.4) \qquad \Sigma : M_0(p) \to K_2^\rho(Q(s,t_1,\ldots,t_{2g})).$$

To compute this it suffices to consider the two types of generators. Let $t_\alpha \cdot t_\beta^{-1}$ be a type (2) generator. We can assume $p \in M-(\alpha \cup \beta)$. The set $\alpha \cup \beta$ divides M into two connected surface P, which contains p, and Q, which does not, as in the diagram below.

LEMMA 2.5. For each generator t_β of the first type we have $\Sigma(t_\beta) = 1$. For each generator $t_\alpha \cdot t_\beta^{-1}$ of the second type we have

$$\Sigma(t_\alpha \cdot t_\beta^{-1}) = \{(s-1)^{|\chi(Q)|}, \rho(\alpha)\}.$$

This calculation is made using methods of the sum theorem of §1. $M \times S^1$ is the union of $P \times S^1$ and $Q \times S^1$; $t_\alpha \cdot t_\beta^{-1}$ is the identity on $P \times S^1$ and is a boundary twist on $Q \times S^1$. The result follows as in (1.3).

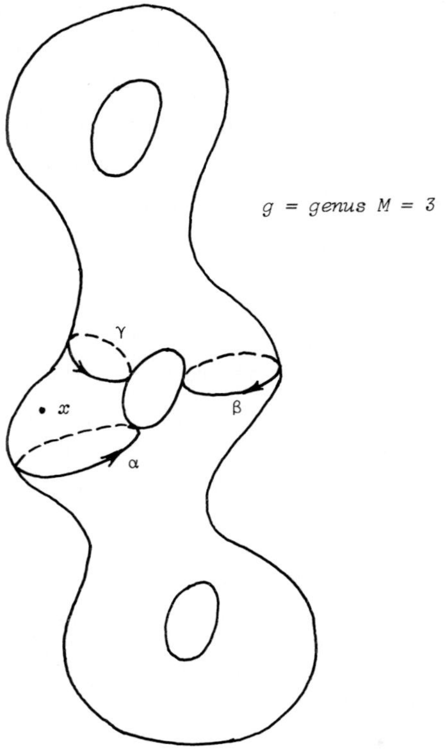

Let $T \subset Q(s,t_1,\ldots,t_{2g})^*$ be the subgroup generated by the t_i.

COROLLARY 2.6. The image of $M_0(p)$ *under* Σ *is the subgroup* $\{(s-1)^2,T\}$ *which is free abelian of rank* $2g$.

That $\{(s-1)^2,T\}$ has rank $2g$ is proved by using tame symbols as in (2.3).

Consider the exact sequence

$$\pi_1(M,p) \underset{\partial}{\to} M_0(p) \to M_0 \to 1.$$

According to (1.2) as well as to (2.5) above the image of $\pi_1(M,p)$ under $\Sigma\circ\partial$ is $\{(s-1)^{2g-2},T\}$. Consequently there is a well-defined homomorphism

$$\Sigma : M_0 \to K_2^\rho(Q(s,t_1,\ldots,t_{2g})) \bmod \{(s-1)^{2g-2},T\}$$

and we have

COROLLARY 2.7. Σ gives rise to a surjective homomorphism

$$M_0 \to [Z/_{(g-1)Z}]^{2g-2}.$$

This bears on the question of Birman [7, §2] as to whether the subgroup $M_0^{(1)}$ of M_0 generated by type (1) generators is of infinite index. Since Σ kills any t_β in $M_0^{(1)}$, this corollary shows the index increases with the genus of the surface. In fact, Dennis Johnson has recently shown $M_0^{(1)}$ does indeed have infinite index. His methods, which use the free differential calculus, how that there is a homomorphism of M_0 onto a certain finitely generated free abelian group plus the same abstract finite abelian group appearing in (2.7).

Computations for non closed surfaces can be worked out as well. For example, let M be a closed surface of genus g and let $\infty \in M$. Consider M, M_0, etc. for the punctured surface $M-\infty$. As usual take the representation ρ to be $\pi_1(M-\infty,p) \to H_1(M-\infty) \hookrightarrow Q(s,t_1,\ldots,t_{2g})$ where p is a point in a product neighborhood of ∞. In this case there is a homomorphism

$$(2.8) \quad \Sigma : M_0 \to K_2^\rho(Q(s,t_1,\ldots,t_{2g}))$$

without any indeterminancy coming from allowing the base point to move. In fact the would be indeterminacy is $\{(s-1)^{2g-2},\rho(\beta)\}$

where β is a loop around ∞ in a small neighborhood. Since β is homologous to zero in $M-\infty$, this error term vanishes. As in (2.5) the subgroup $M_0^{(1)}$ is killed by Σ and has infinite index because the image of M_0 under Σ is the free abelian group $\{(s-1)^2, T\}$ of rank $2g$.

Here is an example to further illustrate how the geometry reflects the bilinearity in Matsumoto's relations. Let α, β and γ be the homologous disjoint curves as follows:

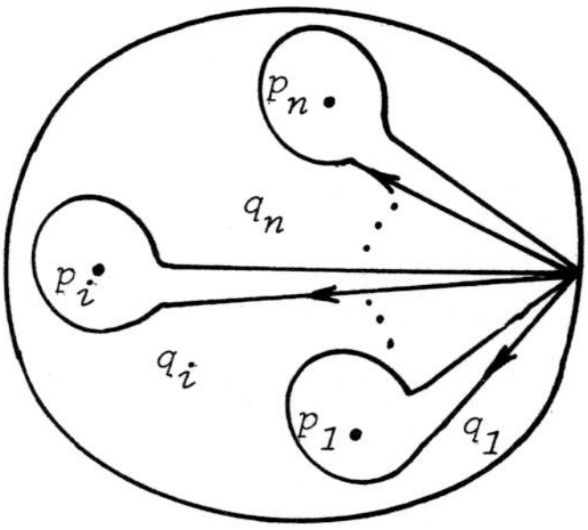

g = *genus* M = 3

Then $t_\alpha t_\gamma^{-1}$ is isotopic to the identity as t_α and t_γ are twists around isotopic oriented curves. In addition $t_\alpha t_\gamma^{-1}$ is isotopic to $(t_\alpha t_\beta^{-1})\cdot(t_\beta t_\gamma^{-1})$. Let $z = \rho(\alpha) = \rho(\beta)$. Applying Σ gives

$$\{(s-1)^2, z\}\cdot\{(s-1)^2, z\} = \{(s-1)^4, z\}$$

which is zero modulo $\{(s-1)^{2g-2}, T\}$ as required by (1.4).

Another example is the pure braid group P_n which can be realized as isotopy classes of diffeomorphisms of the punctured 2-disc $M^2 = D^2 - \{p_1,\ldots,p_n\}$ which keep the boundary fixed and which take each neighborhood of p_i to a neighborhood of the same p_i (i.e. the p_i are left fixed). See [1]. As usual let ρ be the abelianization homomorphism $\pi_1(M^2 \times S^1, x) \to H_1(M \times S^1)$ where $x \in \partial M^2 \times S^1$. According to [1, Cor.1.8.3] each element of P_n induces the identity on $H_1(M^2)$ so the Σ invariant gives a (non-trivial) homomorphism

(2.9) $\Sigma : P_n \to K_2(Q(s,t_1,\ldots,t_n))$.

Here t_i is the homology class corresponding to the loop q_i in the diagram

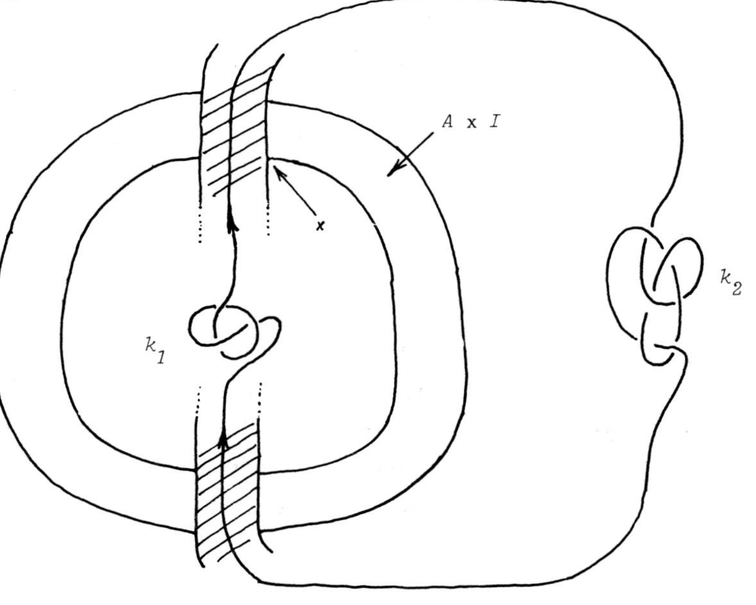

For $n = 2$, P_n is infinite cyclic and generated by A_{12} which corresponds to the boundary twist through 2π. See [1,§1]. Hence $\Sigma(A_{12}) = \{(s-1), t_1 t_2\}$ which is of infinite order.

The Alexander polynomial of a knot is a special case of a general setting where one can associate a K_1 type of invariant to a group G equipped with a unitary representation $\rho : G \to F$. From a presentation of G the free differential calculus can be used to compute this invariant.

In what algebraic generality can a K_2 type invariant be associated to an automorphism of G preserving ρ? How can free calculus methods be used to make computations starting from, say, knowledge of what the automorphism does on generators of a presentation. A nice test case would be the pure braid group P_n. See [1,Cor. 1.8.3]. Johnson's work mentioned above does use free calculus and it would be interesting to work out the direct connection between his methods and the K_2 framework.

Example C. Discrete subgroups. Let G be a three dimensional connected Lie group and let $\Gamma \subset G$ be a discrete subgroup such that G/Γ has the homotopy type of a finite complex. Let the base point x of G/Γ be the identity coset of Γ. Left translation by any element $\gamma \in \Gamma$ is a diffeomorphism of G/Γ fixing the base point and in this section we compute several examples of $\Sigma(\gamma)$.

Let $G = SL_2(R)$ and let $\Gamma \subset G$ be a discrete subgroup. Consider the central extension

$$0 \to Z \to SL_2(R) \to SL_2(R) \to 1$$

where the middle group is the universal cover of $SL_2(R)$. Let $\tilde{\Gamma} \subset \widetilde{SL}_2(R)$ be the inverse image of Γ. Let $M^2 = \widetilde{SL}_2(R)/\tilde{\Gamma} = SL_2(R)/\Gamma$. Left translation by γ on M^3 induces conjugation on $\pi_1(M^3, \Gamma) = \tilde{\Gamma}$ and therefore preserves the abelianization homomorphism $\pi_1(M^3, \Gamma) \to H_1(M^3)$. Assume for the time being that Γ is torsion free. Then there are two cases according to whether $SL_2(R)/\Gamma$ is compact or not.

Case 1. $SL_2(R)/\Gamma$ is compact. We have the unit circle bundle fibration

$$(*) \qquad S^1 \to M^3 \to Q^2$$

where $Q^2 \cong S^1\backslash SL_2(R)/\Gamma$ is a closed compact surface of genus g, Euler class $\chi = 2-2g$, and fundamental group Γ. $H_1(M^3)$ is isomorphic to $Z^{2g} \oplus Z/\chi$ where Z^{2g} is generated by $t_1,\ldots,t_{2g}$ and Z/χ is generated by η. Let t_i also denote $2g$ algebraically independent elements and let η be a primitive χ^{th} root of unity. Let η be the unitary representation $\pi_1(M^3,\Gamma) \to H_1(M^3) \hookrightarrow Q(\eta,t_1,\ldots,t_{2g})$ where the involution on the field is $\bar{t}_i = t_i^{-1}$ and $\bar{\eta} = \eta^{-1}$. A direct calculation shows ρ is acylcic for M^3 and that

$$\tau_\rho(M^3) = (\eta-1)^{-\chi}$$

The homomorphism Σ is defined on the group $\mathcal{D}(x)$ as in §1 taking values in $K_2^\rho(Q(\eta,t_1,\ldots,t_{2g}))$. Since $SL_2(R)$ is connected, left translation by on $SL_2(R)/\Gamma$ is isotopic to the identity (but not keeping the base point $x = \Gamma$ fixed). Therefore γ can be considered as lying in $\mathcal{D}(x)$ and we get a homomorphism

$$(2.10) \qquad \Sigma : \Gamma \to K_2(Q(\eta,t_1,\ldots,t_{2g})).$$

Let $T \subset Q(\eta,t_1,\ldots,t_{2g})$ denote the subgroup generated by the t_i. From (1.2) we have

COROLLARY 2.11. *The image of Σ is the subgroup $\{(\eta-1)^\chi,T\}$ which is free abelian of rank $2g$ for $\chi \neq 6$ and is trivial for $\chi = 6$.*

The value $\chi = 6$ is the only one for which $\eta-1$ is a root of unity and in that case $(\eta-1)^\chi = 1$. When $\chi \neq 6$ one uses tame symbols as in (2.3) to show $\{(n-1)^{\chi}, T\}$ has rank $2g$.

Case 2. $SL_2(R)/\Gamma$ is not compact. As above there is the smooth fibration

$$S^1 \to M^3 \to Q^2$$

but in this case it is trivial because Q^2 is not compact. Let $H_1(M^3) \cong H_1(S^1 \times Q^2)$ be generated by $s \in H_1(S^1)$ and by a basis $t_1, \ldots, t_m$ for $H_1(Q^2)$. Let $s, t_1, \ldots, t_m$ also denote algebraically independent elements. Then we get a homomorphism

$$(2.12) \quad \Sigma : \Gamma \to K_2^{\rho}(Q(s, t_1, \ldots, t_m))$$

and we have

COROLLARY 2.13. *The image of* Σ *is the subgroup* $\{(s-1)^{\chi}, T\}$ *which is free abelian of rank m.*

When Γ has torsion the situation is different. For example, let $\Gamma = SL_2(Z)$. Then $SL_2(R)/\Gamma$ is diffeomorphic to the complement of the trefoil knot in S^3. See [8, §10]. As in Example A the torsion of this manifold is $\frac{t^2-t+1}{1-t}$. Applying (1.2) shows the image of

$$(2.14) \quad \Sigma : SL_2(Z) \to K_2^{\rho}(Q(t))$$

is generated by $\{t^2 - t + 1, t\}$ which has order six.

Another interesting example is the group G constructed as follows: Let $A \in SL_2(Z)$ be hyperbolic (i.e. tr $A > 2$) and let G be the set of elements $(\binom{x}{y}, t)$ where $x, y, t \in R$. Define multiplication in G by

$$(\binom{x}{y}, t) \cdot (\binom{x'}{y'}, t') = (\binom{x}{y} + A^t \binom{x'}{y'}, t + t')$$

Let $\Gamma \subset G$ be the discrete subgroup with $x, y, t \in Z$. G is diffeomorphic to $\mathbb{R}^3$ and there is an exact sequence

$$0 \to R^2 \to G \to R \to 0 .$$

Dividing out by Γ gives a smooth torus bundle over the circle

$$T^2 \to G/\Gamma \to S^1$$

In fact G/Γ is obtained from $T^2 \times I$ by glueing the ends together by $A : R^2/Z^2 \to R^2/Z^2$. See [17]. Let $\rho : \Gamma \to Q(t)$ be the unitary representation taking $\Gamma = \pi_1(G/\Gamma)$ to $\pi_1(S^1)$ generated by t. According to [10] the torsion is given by

$$\tau_\rho = \frac{det(tI - A)}{(t-1)^2}$$

Using (1.2) the image of

$$(2.14) \quad \Sigma : \Gamma \to K_2^\rho(Q(t))$$

is generated by $\{det(tI - A), t\}$ which is of infinite order.

III. ANALYTIC TORSION

We conclude with a remark about "higher analytic torsion." In [13] Ray and Singer defined the analytic torsion of a Riemannian manifold M using the eigenvalues of the Laplacian acting on the deRham complex of forms. J. Cheeger [2] and, independently, W. Muller [11] have proved that this analytic torsion equals R-torsion. Recent work of Waldausen [20], Hsiang-Jahren [6], Hsiang-Farrell [3] and Hsiang [5] shows that π_i Diff$(M) \otimes Q$ is closely related to $K_{i+2}(Z\pi) \otimes Q$ for $i \geq 1$. It is natural to ask in the spirit of the Ray-Singer program whether these higher K-theoretic invariants can be obtained from data arising from the action of diffeomorphisms of M on the deRham complex. As far as

π_0 Diff M is concerned, left translation on G/Γ seems to be a natural test case. One knows from several elementary reasons that for many discrete subgroups Γ of $SL_2(R)$ there is no Riemannian metric which is bi-invariant under Γ. Can a K_2 type invariant for $\gamma \in \Gamma$ be obtained from analyzing how left translation by γ distorts the right invariant metric?

When dim M is even R-torsion vanishes for M compact. But on a complex manifold Ray and Singer [14] used the vertical columns of the Dolbeault complex to define a sequence of $\bar{\partial}$-torsions. Can this program be extended to study how complex automorphisms distort a hermitian metric? An example to consider would again be left translation by elements of Γ on $SL_2(\mathbb{C})/\Gamma$ where the ordinary Σ invariant vanishes for Γ co-compact.

REFERENCES

1. J. S. Birman, Braids, Links, and Mapping Class Groups, Annals Studies No. 82, Princeton University Press (1975).
2. J. Cheeger, Analytic torsion and Reidemeister torsion, Proc. Natl. Acad. Sci. USA Vol. 74, No. 7, pp. 2651-2654, July 1977.
3. F. T. Farrell and W. C. Hsiang, On the rational homotopy groups of diffeomorphisms of discs, spheres, and aspherical manifolds, to appear.
4. A. Hatcher and J. Wagoner, Pseudo-Isotopies of Compact Manifolds, Astérisque No. 6, Société Mathématique de France (1973).
5. W. C. Hsiang, On π_i Diff M^n, preprint, Princeton University.
6. W. C. Hsiang and B. Jahren, On the homotopy groups of the diffeomorphism groups of shperical space forms, to appear.
7. R. Kirby (editor), Problems in low dimensional manifold theory, Proceedings of Symposia in Pure Mathematics (AMS Summer Institute, 1976).
8. J. Milnor, Introduction to Algebraic K-Theory, Annals Studies No. 72, Princeton University (1971).
9. _________, Whitehead torsion, Bull. Amer. Math. Soc. Vol. 72, No. 3 (1966), pp. 358-426.
10. _________, Infinite cyclic coverings, Topology of Manifolds (ed. J. G. Hocking), Prindle, Weber and Schmidt, 1968, pp. 115-133.
11. W. Müller, Analytic torsion and R-torsion of Riemannian Manifolds, preprint, Adademie der Wissenchaften der DDR, Berlin

12. D. Quillen, Higher Algebraic K-Theory: I, Springer Verlag Lecture Notes in Mathematics No. 341, pp. 85-147.
13. D. B. Ray and I. M. Singer, R-torsion and the Laplacian on Riemannian manifolds, Advances in Mathematics Vol. 7 (1971), pp. 145-210.
14. ______________, Analytic torsion for complex manifolds, Annals of Math. Vol. 98, No. 1 (1973), pp. 154-177.
15. G. deRham, M. Kervaire, and S. Maumary, Torsion et type simple d'homotopie, Springer-Verlag Lecture Notes in Mathematics, No. 48.
16. H. Sah and J. Wagoner, Second homology of Lie groups made discrete, Communications in Algebra, 5(6), 1977, pp. 611-642.
17. I. M. Singer, The η-invariant and its relation to real quadratic number fields, Arbeitstage notes by D. Zagier.
18. J. Wagoner, Diffeomorphisms and K_2, preprint, University of California, Berkeley.
19. ______, Diffeomorphisms, K_2, and Analytic Torsion, Proceedings of Symposia in Pure Mathematics (AMS Summer Institute 1976).
20. F. Waldhausen, Algebraic K-theory of topological spaces, I, Preprint, University of Bielefeld.

SHAPE THEORY AND INFINITE DIMENSIONAL TOPOLOGY

CONCORDANCES OF HILBERT CUBE MANIFOLDS AND TUBULAR NEIGHBORHOODS OF FINITE-DIMENSIONAL MANIFOLDS

T. A. Chapman[1]

Department of Mathematics
University of Kentucky
Lexington, Kentucky

In this paper it is shown how some new approximation theorems for maps of manifolds into fiber bundles can be used to prove some well-known results obtained by different methods. These results are (1) the stable equivalence of finite- and infinite-dimensional concordances, and (2) the existence of tublar neighborhoods for codimension 2 locally flat embeddings of finite-dimensional manifolds.

I. INTRODUCTION

If $f : X \to Y$ is a proper map between locally compact spaces and α is an open cover of Y, then f is said to be an α-*equivalence* provided that there is a map $g : Y \to X$ for which $fg : Y \to Y$ is α-homotopic to id and $gf : X \to X$ is $f^{-1}(\alpha)$-homotopic to id. (An α-*homotopy* is one for which the track of each point lies in some element of α; an $f^{-1}(\alpha)$-homotopy is similarly defined, where $f^{-1}(\alpha) = \{f^{-1}(U) \mid U \in \alpha\}$.) The following, which is the main result of [4], serves as our point of departure.

[1]*Supported in part by NSF Grant MCS 76-06929.*

ISBN 0-12-158860-2

APPROXIMATION THEOREM. Let N^n be an n-manifold, $n \geq 5$, and let α be an open cover of N. Then there exists an open cover β of N so that any β-equivalence from an n-manifold M^n to N, which is already a homeomorphism from ∂M to ∂N, is α-close to a homeomorphism.

This result gives homotopy conditions which detech precisely those maps which are "close" to homeomorphisms, where "closeness" is measured in terms of a fine open cover of the target. The purpose of this paper is to show how two well-known results can be deduced from approximation theorems of the above type. These results, which are stated in Theorems 1 and 2 below, concern concordances of Hilbert cube manifolds and tubular neighborhoods of finite-dimensional manifolds. The approximation theorems that we refer to are concerned with the detection of those maps which are "close" to homeomorphisms, where "closeness" is measured not in terms of a fine open cover of the target, but rather in terms of a "nice" open cover of the target. Our first approximation theorem actually generalizes the Approximation Theorem quoted above.

DISC-BUNDLE APPROXIMATION THEOREM. For each open cover α of a manifold B there is an open cover β of B so that if $p : E \to B$ is a disc bundle, $n = \dim E \geq 5$, then any $p^{-1}(\beta)$-equivalence from an n-manifold M^n to E, which is already a homeomorphism from ∂M to ∂E, is $p^{-1}(\alpha)$-homotopic rel ∂M *to a homeomorphism.*

The proof of this result goes exactly like the proof of the Approximation Theorem quoted above. All one has to do is recognize that each handle $B^k \times R^m$ in B becomes a handle in E upon multiplication by $B^{n-(k+m)}$. In order to describe our application of the Disc-Bundle Approximation Theorem we will need some notation.

For any space X let

$$C(X) = \{h : I \times X \to I \times X \mid h \text{ is a homeomorphism, } h = id \text{ on } \{0\} \times X\},$$

where $I = [0,1]$. $C(X)$ is called the space of *concordances* of X and it is equipped with the compact-open topology. Q is used to denote the *Hilbert cube*, which is the countable-infinite product of closed intervals. Note that for any X there is a stability map, $C(X) \xrightarrow{(\times id)_*} C(X \times I)$, obtained by sending each h to $h \times id$. Here is our first application.

THEOREM 1. *If M is a compact topological manifold, then $C(M \times Q)$ is weak homotopy equivalent to the direct limit of the sequence*

$$C(M) \xrightarrow{(\times id)_*} C(M \times I) \xrightarrow{(\times id)_*} C(M \times I^2) \to \cdots$$

This result was established in [1] for M a compact PL manifold. The proof given there mixed a lot of Q-manifold theory and finite-dimensional manifold theory. The proof we give here is somewhat simpler in that only one result from Q-manifold theory is used - namely local contractibility of homeomorphism groups of compact Q-manifolds.

Here is the statement of our second approximation theorem. For a sharper (local) version see Lemma 5.2.

S^1-BUNDLE APPROXIMATION THEOREM. *For each open cover α of a manifold B there is an open cover β of B so that if $p : E \to B$ is an S^1-bundle, $n = \dim E \geq 7$, then any $p^{-1}(\beta)$-equivalence from an n-manifold M^n to E, which is already a homeomorphism from ∂M to ∂E, is $p^{-1}(\alpha)$-homotopic* rel ∂M *to a homeomorphism.*

This result was established in [2]. It was used there to obtain new proofs of Goad's result on approximate fibrations [7] and the Kirby-Siebenmann codimension 2 tubular neighborhood theorem [8]. It is this latter result that we single out as our second application. The proof that we give here is much shorter than the one given in [2]. Instead of working locally through a handle decomposition we give a global proof based on Fadell's notion of the normal fiber space of a locally flat embedding [6].

THEOREM 2. Let M^{n+2} and N^n be manifolds, $\partial N = \phi$ and $n \geq 5$, and assume that N is locally flat in M. Then N has a tubalar neighborhood in M.

Here is how the material in this paper is organized. In Chapt. 2 we present some preliminaries needed for the proof of Theorem 1 which is given in Chapt. 3. Similarly, Chapt. 4 introduces some material needed for the proof of Theorem 2 which is given in Chapt. 5.

II. SOME LEMMAS FOR THEOREM 1

Our main results here are Lemmas 2.2 and 2.4, which are concerned with local contractibility of homeomorphism groups, and Lemma 2.5, which is a result on concordances that is deduced from the Disc-Bundle Approximation Theorem.

Here is the only Q-manifold result needed in this paper. It asserts that for any compact Q-manifold X, $C(X)$ is locally contractible. For an extremely short proof of this result see Step 1 of Theorem 2.2 of [1].

LEMMA 2.1. If X is a compact Q-manifold, then there is a neighborhood U of id in C(X) and a homotopy $H : U \times I \to C(X)$ such that $H_0(h) = h$ and $H_1(h) = id$, for all $h \in U$, and $H_t(id) = id$, for all t.

In Lemma 2.2 below we derive the form of Lemma 2.1 which will be needed in Chapt. 3. For notation let M^m be a compact topological manifold and for any $\varepsilon > 0$ let U_ε denote the subset of $C(M \times Q)$ consisting of all h for which $d(ph,p) < \varepsilon$, where p denotes projection to M and d is a metric on M.

LEMMA 2.2. *There exists an $\varepsilon > 0$ and a homotopy $H : U_\varepsilon \times I \to C(M \times Q)$ such that $H_0(h) = h$ and $H_1(h) = id$, for all $h \in U_\varepsilon$, and $H_t(id) = id$, for all t.*

Proof. Let P denote the subset of $C(M \times Q)$ consisting of all h for which $ph = p$. We will describe below a homotopy $G : C(M \times Q) \times I \to C(M \times Q)$ such that $G_0(h) = h$, for all $h \in C(M \times Q)$, $G_1(P) = \{id\}$, and $G_t(id) = id$, for all t. If $U \subset C(M \times Q)$ is a neighborhood of id as needed in Lemma 2.1, then choose $U_\varepsilon \subset G_1^{-1}(U)$. To get our desired homotopy H we just follow $G|U_\varepsilon \times I$ by the homotopy of Lemma 2.1.

Let $\Gamma(Q)$ denote the cone over Q with cone point v. We may write $\Gamma(Q) = (Q \times [0,1)) \cup \{v\}$. It is easy to construct a homeomorphism of pairs,

$$(I \times Q, \{0\} \times Q) \cong (\Gamma(Q), Q \times \{0\}).$$

This follows immediately from Chapter 1 of [3]. Alternately, it is possible to deduce this result directly without using any Z-set apparatus.

If $h \in C(M \times Q)$, then h may be viewed as a homeomorphism of $M \times \Gamma(Q)$ to itself which is id on $M \times Q \times \{0\}$. Here is how our homotopy $G : C(M \times Q) \times I \to C(M \times Q)$ is constructed. If $h \in P$, then $G|\{h\} \times I$ is just the standard Alexander isotopy of h to id. If $h \in C(M \times Q)-P$, then $G|\{h\} \times I$ is defined by phasing out the Alexander isotopy.

Our next task is to establish an analogue of Lemma 2.2 for finite-dimensional manifolds. We still use M^m and p as above. Let I^n be an n-cell and let $H^n(M)$ be the space of all homeomorphisms of $M \times I^n$ onto itself which are the identity on $\partial(M \times I^n)$. For any $\varepsilon > 0$ let V_ε be the subset of $H^n(M)$ consisting of all h for which $d(ph,p) < \varepsilon$. The following result will be needed in the proof of Lemma 2.4.

LEMMA 2.3. *There exists an $\varepsilon > 0$ and a homotopy $H : V_\varepsilon \times I \to H^n(M)$ such that $H_0(h) = h$ and $H_1(h) = id$, for all $h \in V_\varepsilon$, and $H_t(id) = id$, for all t. Moreover, ε is independent of n.*

Remarks on Proof. Here is an appealing first try. Choose $h \in V_\varepsilon$ and let q denote projection to I^n. By the techniques of Lemma 2.2 we can isotop h to h', where $d(ph',p) < \varepsilon$ and qh' is as close to q as we want. Then the local contractibility result of [5] implies that h' is isotopic to id as required, provided that ε depends on n. To get ε independent of n we need a different approach. What we have in mind is modifying slightly the proof given in [5] to apply to our special situation. This is done by observing that a handle $B^k \times R^{m-k}$ in M yields a handle $B^k \times R^{m-k} \times I^n$. These handles are inserted directly into the proof given in [5].

Now using M, p and I^n as above, let W_ε denote the subset of $C(M \times I^n)$ consisting of all h for which $d(ph,p) < \varepsilon$.

LEMMA 2.4. *There exists an $\varepsilon > 0$ and a homotopy $H : W_\varepsilon \times I \to C(M \times I^n)$ such that $H_0(h) = h$ and $H_1(h) = id$, for all $h \in W_\varepsilon$, and $H_t(id) = id$, for all t. Moreover, ε is independent of n.*

Proof. Choose $h \in C(M \times I^n)$ so that ph is close to p. We will describe an isotopy of h to id which fulfills our requirements. First note that h is isotopic to h', where h' is id on

$$(\{0\} \times M \times I^n) \cup (I \times \partial(M \times I^n)).$$

Moreover, ph' is still close to p, and so $h'|\{1\} \times M \times I^n$ determines an element of $H^n(M)$ which must be isotopic to id by Lemma 2.3. Therefore h' is isotopic to h'', where $h'' = id$ on $\partial(I \times M \times I^n)$. Then one more application of Lemma 2.3 gives an isotopy of h'' to id.

Finally we will need the following result. We still use the notation introduced above.

LEMMA 2.5. For every $\varepsilon > 0$ *there exists a* $\delta > 0$ *so that if* $f : I \times M \times I^n \to I \times M \times I^n$ *is a* $p^{-1}(\delta)$*-equivalence and* $f = id$ *on* $\{0\} \times M \times I^n$, *then* f *is* $p^{-1}(\varepsilon)$*-close to an element* g *of* $C(M \times I^n)$. *Moreover,* δ *is independent of* n.

Remark. When we say that f is a $p^{-1}(\delta)$-equivalence, we mean that f is a $p^{-1}(\alpha)$-equivalence, where α is the open cover of M consisting of all open sets having diameter $< \delta$.

Proof. It is easy to construct a map $f' : I \times M \times I^n \to I \times M \times I^n$ such that $f' = id$ on

$$(\{0\} \times M \times I^n) \cup (I \times \partial(M \times I^n)),$$

and pf' is as close to pf as we want. Then f' is still a $p^{-1}(\delta)$-equivalence. By the homotopy extension theorem we may assume that $f'(\{1\} \times M \times I^n) \subset \{1\} \times M \times I^n$. Using the natural retraction of $I \times M \times I^n$ to $\{1\} \times M \times I^n$ it is easy to see that $f'|\{1\} \times M \times I^n$ is a $p^{-1}(\delta)$-equivalence to $\{1\} \times M \times I^n$. Then by the Disc-Bundle Approximation Theorem there is a homeomorphism h of $\{1\} \times M \times I^n$ onto itself which is $p^{-1}(\varepsilon_1)$-homotopic to $f'|\{1\} \times M \times I^n$ rel $\partial(\{1\} \times M \times I^n)$, where ε_1 is small. By the homotopy extension theorem we may then construct a map $f'' : I \times M \times I^n \to I \times M \times I^n$ so that $f'' = f'$ on $(\{0\} \times M \times I^n) \cup (I \times \partial(M \times I^n))$, $f'' = h$ on $\{1\} \times M \times I^n$, and pf'' is close to

pf'. Then we again use the Disc-Bundle Approximation Theorem to get a homeomorphism $g : I \times M \times I^n \to I \times M \times I^n$ which agrees with f'' on $\partial(I \times M \times I^n)$ and for which pg is close to pf''.

III. PROOF OF THEOREM 1

Before getting into proof of Theorem 1 we will need a lemma, and for this we will first need some more notation. Let $Q = [0,1]^\infty$ and let $I^n = [0,1]^n$, thus giving a canonical decomposition $Q = I^n \times Q_{n+1}$. We identify I^n with $I^n \times \{(0,0,\ldots)\}$ in Q. As usual, p denotes projection to the manifold M.

LEMMA 3.1. If $h \in C(M \times Q)$ and $\varepsilon > 0$ is given, then there exists an n and $g \in C(M \times I^n)$ so that $g \times id_{Q_{n+1}}$ is $p^{-1}(\varepsilon)$-close to h.

Proof. Choose n large and consider the composition,

$$f : I \times M \times I^n \hookrightarrow I \times M \times Q \overset{h}{\to} I \times M \times Q \overset{\text{proj}}{\to} I \times M \times I^n.$$

For any $\delta > 0$ we can choose n large enough so that f is a $p^{-1}(\delta)$-equivalence. By Lemma 2.5, f is $p^{-1}(\varepsilon/2)$-close to an element g of $C(M \times I^n)$. Then $g \times id$ is $p^{-1}(\varepsilon)$-close to h for n large.

Proof of Theorem 3.1. Let $C^s(M)$ denote the direct limit of

$$C(M) \overset{(\times\, id)_*}{\to} C(M \times I) \overset{(\times\, id)_*}{\to} C(M \times I^2) \to \cdots \quad .$$

There is a natural map $i : C^s(M) \to C(M \times Q)$ defined by sending $h \in C(M \times I^n)$ to $h \times id_{Q_{n+1}}$ in $C(M \times Q)$. We must prove that i is a weak homotopy equivalence.

Let Δ^k be a k-simplex and let $h : \Delta \times I \times M \times Q \to \Delta \times I \times M \times Q$ be a fiber preserving homeomorphism such that $h = id$ on $\Delta \times \{0\} \times M \times Q$ and $h = g \times id$ on $\partial\Delta \times I \times M \times Q$, where $g : \partial\Delta \times I \times M \times I^n \to \partial\Delta \times I \times M \times I^n$ is a fiber preserving homeomorphism. (The fibers in $\Delta \times I \times M \times Q$ are

$\{t\} \times I \times M \times Q$.) We want a fiber preserving isotopy of h to $\bar{g} \times id$ rel

$$(\partial\Delta \times I \times M \times Q) \cup (\Delta \times \{0\} \times M \times Q),$$

where $\bar{g} : \Delta \times I \times M \times I^{n+q} \to \Delta \times I \times M \times I^{n+q}$ is a fiber preserving homeomorphism. This will prove that i is a weak homotopy equivalence. It will suffice to construct $\bar{g}$ so that $\bar{g} = g \times id$ on

$$[(\partial\Delta \times I \times M \times I^n) \cup (\Delta \times \{0\} \times M \times I^n)] \times I^q,$$

and so that h is $p^{-1}(\varepsilon)$-close to $\bar{g} \times id$, for some small ε. This follows because the local contractibility result of Lemma 2.2 will then provide our isotopy of h to $\bar{g} \times id$.

We will construct $\bar{g}$ inductively over the skeleta in a fine subdivision of Δ. Choose such a subdivision, and for each vertex v of Δ consider

$$f_v : I \times M \times I^n \hookrightarrow I \times M \times Q \xrightarrow{h_v} I \times M \times Q \xrightarrow{\text{proj}} I \times M \times I^n.$$

Note that $f_v = g_v$ for $v \in \partial\Delta$. By Lemma 3.1 we can choose q large enough so that for each v there is a $\bar{g}_v \in C(M \times I^{n+q})$ such that $p(\bar{g}_v \times id)$ is close to ph_v and $\bar{g}_v = g_v \times id$ if $v \in \partial\Delta$. This defines $\bar{g}$ over the 0-skeleton.

To define $\bar{g}$ over the 1-skeleton note that if v_1, v_2 are vertices which span a 1-simplex σ, then Lemma 2.4 enables us to extend $\bar{g}$ over this 1-simplex. We can do this simultaneously to extend $\bar{g}$ over the entire 1-skeleton. Continuing this process inductively over the skeleta we will eventually arrive at our $\bar{g}$.

IV. SOME LEMMAS FOR THEOREM 2

The purpose of this section is to define and establish some properties of Fadell's notion of a normal fiber space of a locally flat embedding [6]. Let (M^{n+k}, N^n) be a locally flat manifold pair, $\partial N = \phi$, and let

$$E = \{\omega | \omega \text{ is a path in } M, \ \omega(t) \in N \text{ iff } t = 0\}.$$

Then E is a subset of the path space M^I and we give it the compact-open topology. Define $p : E \to N$ by $p(\omega) = \omega(0)$. Here is the only result that we will need from [6].

LEMMA 4.1. *$p : E \to N$ is a Hurewicz fibration with fibers that are homotopy equivalent to S^{k-1}.*

Now define

$$\overline{E} = E \cup (\cup\{\omega \in M^I | \omega \text{ is a constant path in } N\}).$$

We will need the following result in Lemma 4.3 below.

LEMMA 4.2. *There exists a neighborhood U of N in M and a map $\alpha : U \to \overline{E}$ so that $\alpha(U - N) \subset E$, $\alpha(x)(1) = x$, for all $x \in U$, and $\alpha(N) \subset \overline{E} - E$.*

Proof. Locally N has product neighborhoods. This implies that for each $x \in N$ there is a neighborhood G of x and a pseudo-isotopy $h_t : G \to M$, $t \in I$, so that $h_0 = id$, $h_1(G) \subset N$, $h_t | N \cap G = id$, and h_t is an open embedding, for $t < 1$. It is easy to see that we can piece together these local pseudo-isotopies to obtain a neighborhood U of N and a pseudo-isotopy $H_t : U \to M$, $t \in I$, so that $H_0 = id$, $H_1(U) \subset N$, $H_t | N = id$, and H_t is an open embedding, for $t < 1$.

We now define $\alpha : U \to \overline{E}$. For any $x \in U$ consider the path $H_t(x)$, which starts at x and ends at $H_1(x) \in N$. Then define $\alpha(x) = \omega$, where $\omega(t) = H_{1-t}(x)$.

We will need one more definition before Lemma 4.3. Let $M(p)$ denote the mapping cylinder of $p : E \to N$. Set-wise it is just

$$M(p) = (E \times (0,1]) \cup N.$$

A topology on $M(p)$ is defined by requiring $W \subset M(p)$ to be a subbasis element if either (1) W is open in the product topology of $E \times (0,1]$, or (2) $W = (p^{-1}(G) \times (0,\varepsilon)) \cup G$, where G is open in N and $0 < \varepsilon < 1$. $M(p)$ is nothing but a decomposition $E \times [0,1] \coprod N/\sim$, where $\sim$ is the equivalence relation generated by $(\omega,0) \sim \omega(0)$, but it does not have the quotient topology.

LEMMA 4.3. There is a neighborhood U of N in M, a neighborhood G of N in $M(p)$, and maps $f : U \to M(p)$, $g : M(p) \to M$ so that

(1) $f = id$ and $g = id$ on N,

(2) $f(U - N) \subset E \times (0,1]$ and $g(E \times (0,1]) \subset M - N$,

(3) $gf : U \to M$ is homotopic to id via a homotopy which is id on N and which takes $U - N$ into $M - N$ at each level,

(4) $fg : G \to M(p)$ is homotopic to id via a homotopy which is id on N and which takes $G - N$ into $E \times (0,1]$ at each level.

Proof. We have already constructed U in Lemma 4.2. Let d be a metric on M which is normalized so that $d(x,y) \leq 1$, for all x and y. Define $f : U \to M(p)$ by $f(x) = (\alpha(x), d(x,N))$, for $x \in U - N$, and $f(x) = x$, for $x \in N$. It is easy to see that f is continuous. Define $g : M(p) \to M$ by $g(\omega,t) = \omega(t)$, for all $(\omega,t) \in E \times (0,1]$, and $g(x) = x$, for $x \in N$. Again g is continuous. Finally $G \subset g^{-1}(U)$ will be a neighborhood of N to be specified below. Clearly (1) and (2) of our required conditions are satisfied.

For (3) choose any $x \in U - N$ and note that

$$gf(x) = g(\alpha(x), d(x,N)) = \alpha(x)(d(x,N)).$$

So we have an obvious path from $gf(x)$ to x provided by the restriction of $\alpha(x)$ to $[d(x,N),1]$. All of these paths piece together to define our required homotopy $gf \simeq id$.

Before establishing condition (4) it will be convenient to examine the following special situation. Let $W \subset M$ be an open set for which there is a homeomorphism of pairs,

$$(W, W \cap N) \cong (R^n \times R^k, R^n),$$

where $R^n \equiv R^n \times \{0\}$. We will identify W with $R^n \times R^k$. Let ω_1, ω_2 be elements of E which lie in $R^n \times R^k$ and for which $\omega_1(1) = \omega_2(1)$. We want to describe a path in $E \cap W^I$ from ω_1 to ω_2 which depends continuously on ω_1 and ω_2.

Let ω_3 be the "vertical" straight-line path in $R^n \times R^k$ from x to (x,y), where $\omega_1(1) = (x,y) \in R^n \times R^k$. It is easy to find a path in $E \cap W^I$ from ω_1 to an element ω_1', where $\omega_1'(0) = \omega_3(0)$ and $\omega_1'(1) = \omega_3(1)$. This just requires composing ω_1 with an isotopy of $R^n \times R^k$ that moves only the R^n-coordinates. Then obtaining a path in $E \cap W^I$ from ω_1' to ω_3 is a simple matter of using a process like the Alexander trick. This gets us from ω_1 to ω_3. Similiarly, we can go from ω_2 to ω_3.

Now returning to condition (4) choose any $(\omega, t) \in G \cap (E \times (0,1])$ and note that

$$fg(\omega,t) = f(\omega(t)) = (\alpha(\omega(t)), d(\omega(t),N)).$$

Let ω' be the element of E which is a reparametrization of $\omega|[0,\omega(t)]$. Then define $h : G \to M(p)$ by $h(\omega,t) = (\omega', d(\omega(t),N))$, for $(\omega,t) \in G \cap (E \times (0,1])$, and $h(x) = x$, for $x \in N$. It is easy to get a homotopy of id to h, via a homotopy which is id on N and which takes $G \cap (E \times (0,1])$ into $E \times (0,1]$ at each level. So there remains the problem of getting a homotopy of h to $fg|G$.

For any $(\omega,t) \in G \cap (E \times (0,1])$ we have

$$fg(\omega,t) = (\alpha(\omega(t)), d(\omega(t),N)),$$
$$h(\omega,t) = (\omega', d(\omega(t),N)).$$

Also we have $\omega'(1) = \alpha(\omega(t))(1)$. Note that if G is chosen close enough to N in $M(p)$, then $\alpha(\omega(t))$ and ω' have small diameters. The special situation described above gives us a path in E from ω' to $\alpha(\omega(t))$. Thus by applying the special situation above we can choose G close enough to N and successively homotop h to h_1, h_1 to h_2, h_2 to h_3, ..., so that $\lim_{i\to\infty} h_i = fg$. Note that each homotopy, $h_i \simeq h_{i+1}$, is obtained by working locally in a product neighborhood of the form $R^n \times R^k$. Also there are only a finite number of the h_i if N is compact.

V. PROOF OF THEOREM 2

We adopt the notation of Chapt. 4. Our first task will be to show how to construct a candidate for our tubular neighborhood. In Chapt. 4 we studied a Hurewicz fibration $p : E \to N$ with fiber homotopy equivalent to S^1. It is well known that the inclusion of the homeomorphism group of S^1 into the space of self-equivalences of S^1 is a homotopy equivalence. Thus E is fiber homotopy equivalent to a fiber bundle $q : E \to N$ with fiber S^1. We may therefore form the mapping cylinder $M(q)$ containing N. For our proof of Theorem 2 we will prove that there is a neighborhood of N in M which is homeomorphic to a neighborhood of N in $M(q)$. This will suffice since N obviously has a tubular neighborhood in $M(q)$.

The following is an analogue of Lemma 4.3 for $M(q)$.

LEMMA 5.1. There is a neighborhood U_1 of N in M, a neighborhood G_1 of N in $M(q)$, and maps $f_1 : U_1 \to M(q)$, $g_1 : M(q) \to M$ so that

(1) $f_1 = id$ and $g_1 = id$ on N,

(2) $f_1(U_1 - N) \subset E \times (0,1])$ and $g_1(E \times (0,1]) \subset M - N$,

(3) $g_1 f_1 : U_1 \to M$ is homotopic to id via a homotopy which is id on N and which takes $U_1 - N$ into $M - N$ at each level,

(4) $fg : G_1 \to M(q)$ is homotopic to id via a homotopy which is id on N and which takes $G_1 - N$ into $E \times (0,1]$ at each level.

Proof. Recall the notation of Lemma 4.3,

$$f : U \to M(p), \quad g : M(p) \to M.$$

Let $u : E \to E$ be a fiber homotopy equivalence, with fiber homotopy inverse $v : E \to E$. Define $\bar{u} : M(p) \to M(q)$ by $\bar{u}(\omega, t) = (u(\omega), t)$, for $(\omega, t) \in E \times (0,1]$, and $\bar{u}|N = id$. Let $\bar{v} : (M(q) \to M(p)$ be similarly defined. Then there are obvious homotopies $\bar{u}\,\bar{v} \simeq id$ rel N and $\bar{v}\,\bar{u} \simeq id$ rel N.

Now define $U_1 = U$, $f_1 = \bar{u}f$, $g_1 = g\bar{v}$, and let $G_1 = (\bar{v})^{-1}(G)$. It is easy to check that these choices meet our requirements.

Here is a sharper form of the S^1-Bundle Approximation Theorem which we will need in the proof of Theorem 2. The proof is exactly the same.

LEMMA 5.2. *Let B be a manifold, let $C \subset U \subset B$, where C is closed and U is open, and let α be an open cover of B. Then there is an open cover β of B so that if $p : E \to B$ is an S^1-bundle, $n =$* dim $E \geq 7$*, then any proper map $f : M^n \to E$ which is a $p^{-1}(\beta)$-equivalence over $p^{-1}(U)$ and which is already a homeomorphism from $\partial(f^{-1}p^{-1}(U))$ to $\partial p^{-1}(U)$, is $p^{-1}(\alpha)$-homotopic to a map $g : M \to E$ which is a homeomorphism over $p^{-1}(C)$. Moreover, g may be chosen to agree with f over $\partial(p^{-1}(U)) \cup (E - p^{-1}(U))$, and to be homotopic to f* rel *$f^{-1}(\partial(p^{-1}(U)) \cup (E - p^{-1}(U))$.*

Remark. When we say that $f : M \to E$ is a $p^{-1}(\beta)$-equivalence over $p^{-1}(U)$, we mean that there is a map $f_1 : p^{-1}(U) \to M$ so that $ff_1 : p^{-1}(U) \to E$ is $p^{-1}(\beta)$-homotopic to the inclusion and $f_1 f|f^{-1}p^{-1}(U)$ is $f^{-1}p^{-1}(\beta)$-homotopic to the inclusion. Generally "over" means "when restricted to the inverse image of."

Proof of Theorem 2. We will prove that there is a neighborhood A of N in M and a neighborhood B of N in $M(q)$ so that $A - N$ is homeomorphic to $B - N$. This homeomorphism will enjoy the property that it extends via the identity on N to a homeomorphism of A onto B.

Choose maps $\varphi_i : E \to (0,1]$ so that $\varphi_{i+1} \leq \varphi_i$. Then define

$$B_i = \{(x,t) \in E \times (0,1] \,|\, \varphi_{i+1}(x) \leq t \leq \varphi_i(x)\}.$$

It is easy to see that φ_1 may be chosen so that $f_1 : U_1 - N \to E \times (0,1]$ is proper over $\cup B_i$. (This requires that the U of Lemma 4.2 be cut down.) By Lemma 5.1 we can choose the φ_i far enough apart so that f_1 is a $(q \times id)^{-1}(\alpha)$-equivalence over each B_i, where α is a fine open cover of $N \times (0,1]$ and $q \times id$ is the map

$$q \times id : E \times (0,1] \to N \times (0,1] .$$

Then Lemma 5.2 implies that f_1 can be homotoped to $f_2 : U_1 \to M(q)$ so that $f_2 = id$ on N, $f_2(U_1 - N) \subset E \times (0,1]$, f_2 is a homeomorphism over each B_{2i+1}, and f_2 is still a $(q \times id)^{-1}(\beta)$-equivalence over each B_{2i}, where β is a fine open cover of $N \times (0,1]$. Now applying Lemma 5.2 once again to the maps $f_2|f_2^{-1}(B_{2i})$ we can homotop f_2 to f_3, which is a homeomorphism over $\cup B_i$.

REFERENCES

1. T. A. Chapman, Concordances of Hilbert cube manifolds, Trans. Amer. Math. Soc. 219 (1976), 253-268.
2. ______________, Approximating maps into fiber bundles by homeomorphisms, preprint.
3. ______________, Lectures on Hilbert cube manifolds, C.B.M.S. Regional Conference Series in Math., number 28, 1976.
4. T. A. Chapman and Steve Ferry, Approximating homotopy equivalences by homeomorphisms, American J. of Math., to appear.
5. R. D. Edwards and R. C. Kirby, Deformations of spaces of embeddings, Ann. of Math. 93 (1971), 63-88.
6. Edward Fadell, Generalized normal bundles for locally-flat imbeddings, Trans. A.M.S. 114 (1965), 488-513.
7. Robert E. Goad, Local homotopy properties of maps and approximation by fiber bundle projections, Dissertation, University of Georgia, Athens, Georgia, 1976.
8. R. C. Kirby and L. C. Siebenmann, Normal bundles for codimension 2 locally flat imbeddings, Lecture notes in Math. 438 (1974), 310-324.

UNIVERSAL OPEN PRINCIPAL FIBRATIONS[1]

David A. Edwards

Department of Mathematics
University of Georgia
Athens, Georgia

Harold M. Hastings[2]

Department of Mathematics
Hofstra University
Hempstead, N.Y.

Let G be a compact metric group. We shall construct classifying spaces for open principal G-fibrations over compact metric spaces. All G-bundles are open Hurewicz fibrations; however the converse requires additional hypotheses, for example, that G is a Lie group and the total space is completely regular (A. Gleason [5]*). Our theorem uses R. Milgram's universal bundle construction* [7]*, related results of N. E. Steenrod* [9]*, J. Cohen's approximation theorem for open principal G-fibrations* [2]*, and our Steenrod homotopy theory* [3]*. A detailed proof will appear elsewhere* [4].

I. INTRODUCTION

Assume that a compact metric group G acts principally *(on the right)* on a compact metric space E. If the projection $\pi : E \to E/G$ is open and is a Hurewicz fibration, p is called an *open principal G-fibration.*

[1] *Based upon a talk by the second-named author.*

[2] *Partially supported by NSF grant number MCS 77-01628.*

ISBN 0-12-158860-2

Such maps arise naturally as limits of towers of bundle maps. There are natural open principal fibrations $S \to S^1$ where S is any *solenoid*, that is, the limit of a sequence

$$\{S^1 \xleftarrow{n_1} S^1 \xleftarrow{n_2} S^1 \xleftarrow{n_1} \ldots\}$$

where $S^1 = \{z \,|\, |z| = 1\}$ and "n_k" denotes the n_k-*th* power map. Another example, but with G not compact, is motivated by sequences of bifurcations: the projection $R^\infty \to T^\infty$ induced from the covering maps $R^n \to T^n$, is an open principal fibration.

J. P. May obtained a coarser classification of open principal fibrations [6]. A. K. Bousfield and D. M. Kan obtained an earlier, but more complicated description of homotopy limits than ours [1], §4.

II. J. COHEN'S APPROXIMATION

Let $\{G_n\}$ be a Lie series for G (L. Pontryagin [8]). Cohen associated to an open principal fibration $G \to E \to X$ a tower

$$\begin{array}{ccccc}
\vdots & & \vdots & & \vdots \\
\downarrow & & \downarrow & & \| \\
G_{n+1} & \longrightarrow & E_{n+1} & \longrightarrow & X \\
\downarrow{\scriptstyle \pi_{n+1}} & & \downarrow{\scriptstyle \pi'_{n+1}} & & \| \\
G_n & \longrightarrow & E_n & \longrightarrow & X \\
\downarrow & & \downarrow & & \| \\
\vdots & & \vdots & &
\end{array}$$

of G_n-*bundles* over X where $E_n = Ex_G G_n$, and π'_{n+1} is a principal bundle map with fibre $\ker(\pi_n)$. We shall call such a tower a *principal* $\{G_n\}$-*fibration* over X. Let $K_{\{G_n\}}(X)$ denote the *class of isomorphism classes of principal* $\{G_n\}$-*fibrations* over X, and $K_G(X)$ the *class of isomorphism classes of open principal*

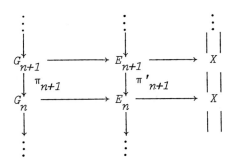

G-fibrations over X. J. Cohen proved the existence of natural isomorphisms

$$k_G(X) \underset{\lim}{\overset{\{\cdot \times G_n\}}{\rightleftarrows}} k_{\{G_n\}}(X)$$

III. CLASSIFICATION

For a topological group H, let $H \to EH \to BH$ denote Milgram's classifying space [7] applied to H. Let

$$\xi_{\{G_n\}} \equiv \{G_n \to EG_n \to BG_n\} \text{ , and}$$

$$\hat{\xi}_G\colon \; G \to \hat{E}G_n \to \hat{B}G_n$$

be the limit of $\xi_{\{G_n\}}$. There is a natural weak homotopy equivalence [4] $BG \to \hat{B}G$.

CLASSIFICATION THEOREM [4]. *There are natural isomorphisms*

$$k_{\{G_n\}}(X) \underset{\text{pullback}}{\rightleftarrows} [X, \{BG_n\}] \text{ , and}$$

$$k_G(X) \underset{\text{pullback}}{\rightleftarrows} [X, BG]$$

The second isomorphism follows from the first isomorphism by [2] and [3], however $[X, \{BG_n\}]$ requires some interpretation. We proved in [4] that $\{BG_n\}$ is a *tower of fibrations*.

IV. HOMOTOPY LIMITS

Let $\{Y_n\}$ be a tower of fibrations. There are two natural definitions of homotopy classes of maps from a *space* X to $\{Y_n\}$:

a "weak" definition

$$\lim\{[X,Y_n]\}$$

and a "strong" definition which requires coherence. The weak and strong definitions correspond to Čech and Steenrod homotopy theory - see [3] for precise definitions, and the following.

THEOREM [3]. *With X and* $\{Y_n\}$ *as above, Steenrod homotopy classes of maps from X to* $\{Y_n\}$ *are precisely the rigid maps from X to* $\{Y_n\}$, $\lim\{Top(X,Y_n)\}$ *modulo a homotopy relation induced by rigid maps from* $X \times I$ *to* $\{Y_n\}$.

Corollary [3]. For $\{Y_n\}$ as above, the topological limit $\lim\{Y\}$ is *also* the homotopy limit $\operatorname{holim}\{Y_n\}$.

Thus strong (Steenrod) homotopy theory is natural whenever homotopy limits are required, for example, for the isomorphism $[X,\hat{B}G]\cong[X,\{BG_n\}]$ in the classification theorem.

REFERENCES

1. A. K. Bousfield and E. M. Kan, Homotopy limits, completions, and localizations, Lecture Notes in Math. 304, Springer, Berlin-Heidelberg-New York, 1973.
2. J. Cohen, Inverse limits of principal fibrations, Proc. London Math. Soc. (3) 27 (1973), 178-192.
3. D. A. Edwards and H. M. Hastings, Čech and Steenrod homotopy theory with applications to geometric topology, Lecture Notes in Math., 542, Springer, Berlin-Heidelberg-New York, 1976.
4. _________, and _________, Classifying open principal fibrations, Trans. Amer. Math. Soc., (to appear).
5. A. Gleason, Spaces with a compact Lie group of transformations, Proc. Amer. Math. Soc. 1 (1950), 39-43.
6. J. P. May, Classifying spaces and fibrations, Mem. Amer. Math. Soc. 155 (1975).
7. R. J. Milgram, The bar construction and abelian H-spaces, III. J. of Math. 11 (1967), 242-250.

8. L. S. Pontryagin, Topological groups, Trans. A. Brown, 2nd ed., Gordon and Breach, New York, 1966.
9. N. E. Steenrod, The topology of fibre bundles, Princeton Univ. Press, Princeton, N. J., 1951.

COMPACTA WITH THE HOMOTOPY TYPE OF FINITE COMPLEXES

Ross Geoghegan[1]

Department of Mathematics
State University of New York
Binghamton, N.Y.

It is unknown[2] whether there exists a compact topological space X which is homotopy equivalent to a complex but not to any finite complex. The status of this problem is reviewed, and two new theorems are announced. Together, they show that if such a space X exists, it cannot be constructed as the inverse limit of a sequence of compacta of finite homotopy type and fibrations; because such an inverse limit is either not shape equivalent to any complex, or is homotopy equivalent to a finite complex.

This talk is about the following well-known

Problem: Let the compact topological space X be homotopy equivalent to a (CW) complex. Must X be homotopy equivalent to a finite complex?

I will describe some of the interesting known results, and will then state some results of mine which seem to have a bearing on the problem. Unfortunately, the problem as stated is still open[2].

J.H.C. Whitehead proved [17] that *a compact space which is dominated by a complex is dominated by a finite complex.* Thus

[1]*Supported in part by NSF Grants MCS 75-10377 and MCS 77-00104*
[2]*See note added in proof at the end of the paper.*

ISBN 0-12-158860-2

the Problem is unchanged if we assume instead that X is dominated by a finite complex. (X is *dominated* by L if there are maps $X \overset{u}{\to} L$ and $L \overset{d}{\to} X$ such that $d \circ u$ is homotopic to 1_X.)

The algebraic topology of the Problem was worked out in two steps. The first was the easy case: X simply-connected. Then the answer is "yes". Using the relative Hurewicz theorem, one attaches cells to make a map $K \overset{f}{\to} X$ where K is a finite complex and f gives an isomorphism on homology: the process terminates at a finite dimension, because subgroups of free abelian groups are free. This case is sometimes attributed to de Lyra [8] though it was probably known to many. It is given as an exercise in [14].

The second step was the attempt to extend this method to the non-simply-connected case: then the process of attaching cells as above need not terminate, because instead of a subgroup of a free abelian group, one is dealing (using universal covers) with a finitely generated projective submodule of a free module, over the integral group ring $Z[\pi_1(X)]$. This may or may not be free ("stably free" is the correct concept) and *the equivalence class of that module in* $\tilde{K}_0(Z[\pi_1(X)])$ *is the obstruction to* X *being homotopy equivalent to a finite complex.* This "Wall obstruction" was described definitively by Wall in [15], though de Lyra [9] discovered it independently for the case of finite fundamental group. A very elegant exposition is in [2].

Wall showed that every possible obstruction is realized: there are infinite complexes dominated by finite complexes but not homotopy equivalent to finite complexes. However, he did not construct compact spaces with this property. See also [9].

Shortly before, M. Mather proved in an elementary page [10] that *if* X *is dominated a finite complex then* $X \times$ *(circle) is homotopy equivalent to a finite complex (* X *need not be compact).* An immediate corollary is: *if* X *is dominated by a finite complex then* X *is homotopy equivalent to a finite-dimensional complex.* In [17; Appendix], J.H.C. Whitehead had asked explicitly if this were true for X compact. It is reasonable, after Mather's answer, to think of our Problem as the surviving version of Whitehead's question.

In a more algebraic setting, Mather's theorem was generalized by Gersten [6]: *if X is dominated by a finite complex, and K is a finite complex of zero Euler characteristic, then $X \times K$ is homotopy equivalent to a finite complex.* The point is that, under those hypotheses, the Wall obstruction can be calculated to be zero.

Returning to the Problem, we next consider the special cases: (i) X a compact topological n-manifold, (ii) X a compact Hilbert Cube manifold and (iii) X a compact (metric) ANR. (Compact ANR's are obviously retracts of subsets of the Hilbert Cube, Q, of the form (compact polyhedron) $\times$ Q, so compact ANR's are certainly dominated by finite complexes; manifolds are well-known to be ANR's.)

The answer to our Problem is "yes" in each case: (i) is due to Kirby-Siebenmann [7], (ii) to Chapman [1], and (iii) to West [16]. In each case much more is proved: a compact n-manifold has a PL normal disk bundle; a compact Q-manifold is homeomorphic to (compact polyhedron) x Q; a compact ANR is the image of a Q-manifold under a hereditary and fine homotopy equivalence (CE map). Thus the fact that these spaces are homotopy equivalent to finite complexes is, as it were, a minor corollary of theorems in geometric topology whose proofs involve handle straightening and other "local" techniques at the level of "homeomorphism theory" rather than homotopy theory. The passage from Case (iii) (compact retracts) to the general Problem (compact homotopy retracts) appears to lead from a local problem to an essentially global one. *If this is correct* (it is only a guess) then one should not expect to parley the ideas of [7], [1] and [16] into a positive solution to the Problem; and one should look for a negative solution, a compactum (≡ compact metric space) homotopy equivalent to a complex, but not to a finite complex. I will end by describing some of the difficulties I have met in trying to construct such an example. It will be a theorem that the (to me) most natural method can never yield the required example!

There is a weaker notion than "homotopy equivalent to a complex" called "shape equivalent to a complex". No shape theory will be needed for what follows; only the definition that a compactum X *is shape equivalent to a (possibly infinite) complex* L if there is a homotopy commutative diagram of maps

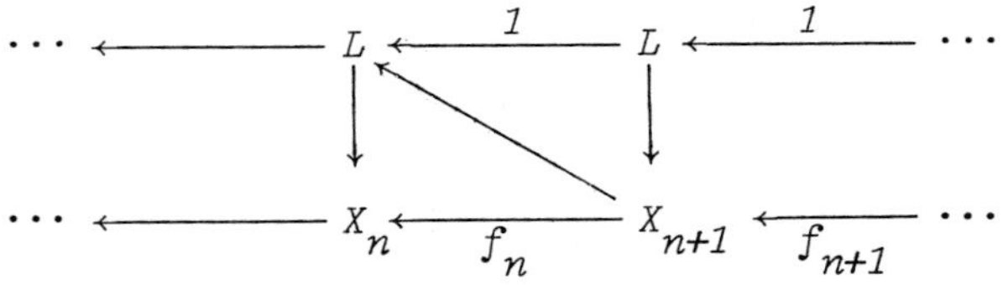

(†)

where each X_n is a compactum homotopy equivalent to a finite complex, and X is homeomorphic to $\varprojlim_n \{X_n; f_n\}$. This definition is equivalent to the usual one; see [5; Remark 2.1], and up to cofinal subsequences it does not depend on the choice of sequence whose inverse limit is X.

There are compacta shape equivalent to complexes, but not shape equivalent to finite complexes [3]. A fortiori, these compacta are not homotopy equivalent to finite complexes. Unfortunately, the examples in [3] are not homotopy equivalent to infinite complexes either, so, as they stand, they do not provide a negative solution to the Problem. Let X be one of those compacta: X is given as a certain inverse limit of finite complexes, whose exact description may be omitted here. Suppose X is shape equivalent to the complex L. I would like to change X into a compactum homotopy equivalent to L; if I could, the Problem would be solved.

There are very few theorems characterizing spaces which are homotopy equivalent to complexes. Apart from J.H.C. Whitehead's theorem quoted at the start of this talk, the ones that come to mind are Milnor's theorem on function spaces [11], and G. Segal's theorem on spaces nicely covered by pieces homotopy equivalent to complexes [13]. None of these is suitable. Here is a new one: *if* $X \equiv \varprojlim_n \{X_n; f_n\}$ *is as in (†) above, and if each* f_n *is a (Hurewicz) fibration, then X is homotopy equivalent to L.* This theorem can be deduced fairly easily from the abstract machinery in Ch 3 of [4] and Ch 1 of [12], but in [18] I eliminate that machinery and give an elementary proof. I call the inverse limit of a sequence of fibrations between compacta a fibered compactum; the theorem can be restated: *a fibered compactum which is shape equivalent to a complex is homotopy equivalent to that complex.*

The program, then, is to take one of the compacta $X \equiv \varprojlim_n \{X_n; f_n\}$ in [3], shape equivalent to a complex L, and, by altering the X_n*'s* and the f_n*'s*, to construct a fibered compactum shape equivalent (hence homotopy equivalent) to L. But the program cannot be carried out, because my second theorem in [5] reads: *a fibered compactum which is homotopy equivalent to a complex is homotopy equivalent to a finite complex.* The proof is a Serre spectral sequence argument.

The lesson is that in order to turn the inverse-limit-shape-examples into compacta homotopy equivalent to complexes, but not to finite complexes, one must find some other way of recognizing an inverse limit as having the homotopy type of a complex. At present, I know of none.

Added in Proof: S. Ferry has announced a negative solution to the Problem.

REFERENCES

1. T. A. Chapman, Lectures on Hillbert Cube manifolds, C.B.M.S. Lecture Notes, Vol. 28, American Mathematical Society, Providence, 1976.
2. T. A. Chapman and S. Ferry, Obstruction to finiteness in the proper category, preprint.

3. D. A. Edwards and R. Geoghegan, Sahpes of complexes, ends of Manifolds, homotopy limits and the Wall obstruction, Ann. of Math. 101 (1975), 521-535, with a correction 104 (1976), 389.
4. D. A. Edwards and H. M. Hastings, Cech and Steenrod homotopy theories with Applications to Geometric Topology, Lecture Notes in Mathematics, vol. 542, Springer-Verlag, Berlin and New York, 1976.
5. R. Geoghegan, Fibered stable compacta have finite homotopy type, submitted.
6. S. Gersten, A product formula for Wall's obstruction, Amer. J. Math. 88 (1966), 337-346.
7. R. Kirby and L. Siebenmann, Foundational Essays on topological manifolds, smoothings and triangulations, Annals of Mathematics Studies, vol. 88, Princeton University Press, Princeton 1977.
8. C. B. de Lyra, On a conjecture in homotopy type as polyhedra, Bol. Soc. Math. São Paulo 12 (1957), 43-62.
9. ____________, On a conjecture in homotopy, Anais Acad, Bras. Ciênc, 37 (1965), 167-184.
10. M. Mather, Counting homotopy types of manifolds, Topology 4 (1965), 93-94.
11. J. Milnor, On spaces having the homotopy type of CW complexes, Trans. Amer. Math. Soc. 90 (1959), 272-280.
12. D. G. Quillen, Homotopical Algebra, Lecture Notes in Mathematics, vol. 43, Springer-Verlag, Berlin and New York, 1967.
13. G. Segal, Classifying spaces and spectral sequences, Publ. math. I.H.E.S. 34 (1968), 105-112.
14. E. H. Spanier, Algebraic Topology, McGraw-Hill, New York, 1966.
15. C.T.C. Wall, Finiteness conditions for CW-complexes, Ann. of Math. 81 (1965), 55-69.
16. J. E. West, Mapping Hilbert Cube manifolds to ANR's, Ann. of Math. 106 (1977) (to appear).
17. J.H.C. Whitehead, A certail exact sequence, Ann. of Math. 52 (1950), 51-110.
18. R. Geoghegan, The inverse limit of homotopy equivalences between towers of fibrations in a homotopy equivalence--a simple proof. (preprint).

MANIFOLDS MODELLED ON THE DIRECT LIMIT OF HILBERT CUBES

Richard E. Heisey[1]

Department of Mathematics
Vanderbilt University
Nashville, Tennessee

The main result of this paper is that if M is a Q^∞-manifold, then the projection map $p : M \times Q^\infty \to M$ is a near homeomorphism. A corollary of this result is that any homotopy equivalence between Q^∞-manifolds is homotopic to a homeomorphism. It is also shown that Q^∞-manifolds can be triangulated. A corollary of this result is that any Q^∞-manifold is homeomorphic to $U \times R^\infty$, where U is an open subset of the Hilbert cube and $R^\infty = \text{dir lim } R^n$, R the reals.

If $\mathcal{U}$ is an open cover of a space Y then two maps $f, g: X \to Y$ are *$\mathcal{U}$-close* if for each $x \in X$ there is a $U \in \mathcal{U}$ such that $\{f(x), g(x)\} \subset U$. A map $f : X \to Y$ is a *near homeomorphism* if for each open cover $\mathcal{U}$ of Y there is a homeomorphism of X onto Y which is *$\mathcal{U}$-close* to f.

Let Q denote the Hilbert cube, which we regard as the product $\Pi_{i=1}^{\infty} I_i$, where $I_i = I = [-1, 1]$. Let Q^n denote the cartesian product of n copies of Q. Let $0 = (0, 0, 0, \ldots) \in Q$, and let $i_n : Q^n \to Q^{n+1}$ be the inclusion $i_n(x_1, \ldots, x_n) = (x_1, \ldots, x_n, 0)$. Then by Q^∞ we mean the dir lim $\{Q^n; i_n\}$. By

[1] *Supported in part by a Summer Research Fellowship at Vanderbilt University, 1976.*

ISBN 0-12-158860-2

M and N we denote paracompact, connected Q^∞-manifolds. (As shown in [7, Theorem II-1] any separable, reflexive, infinite-dimensional Banach space with its bounded weak topology is homeomorphic to Q^∞.)

The following three results are proven in [9].

THEOREM 1. (Stability, [9]). $M \times Q^\infty \cong M$. Here $\cong$ denotes "is homeomorphic to."

THEOREM 2. (Open embedding, [9]). M embeds as an open subset of Q^∞.

THEOREM 3. (Classification by homotopy type, [9]). If M and N have the same homotopy type, then $M \cong N$.

The main result of this paper is the following improvement of Theorem 1.

THEOREM 4. The projection map $p : M \times Q^\infty \rightarrow M$ is near homeomorphism.

The proof of Theorem 4 is a direct limit argument refining the techniques of [8]. (Proofs of this result and the results below are given in Lemma 1-5.) A consequence of Theorem 4 and [7, Theorem II.9] is the following.

THEOREM 5. If $f : M \rightarrow N$ is a homotopy equivalence, then f is homotopic to a homeomorphism $h : M \rightarrow N$.

At the conference[2] Chapman, Ferry, Liem and West observed that the techniques and results presented here should now make

[2]*The Conference on Geometric Topology at the University of Georgia, August 1977.*

triangulation of Q^∞-manifolds possible. This is indeed the case, and we take the liberty of presenting the details here.

THEOREM 6. (Triangulation). $M \cong |K| \times Q^\infty$ *where* K *is a countable, locally-finite, simplicial complex.*

The proof of Theorem 6 uses a relative triangulation theorem for Q-manifolds established by Chapman in [1] and the theorem of West that $|K| \times Q$ is a Q-manifold [13, Corollary 5.2]. A direct consequence of Theorem 6 is that Q^∞-manifolds factor as the product of a Q-manifold and $R^\infty = \text{dir lim } R^n$, R the reals. In fact, we obtain an even stronger result.

THEOREM 7. (Factorization). $M \cong U \times R^\infty$ *where* U *is an open subset of* Q.

This last theorem generalizes the result of the author in [6] that $Q^\infty \cong Q \times R^\infty$.

Proofs. We begin with a few elementary lemmas. The proofs of the first two are straightforward and are omitted.

LEMMA 1. Let C *be a compact subset of the locally compact metric space* (X,d). *Let* $\mathcal{U}$ *be a collection of open subsets of* X *whose union contains* C. *Then there is an* $\varepsilon > 0$ *such that for each* $x \in C$, $B(x,\varepsilon) \subset U$ *some* $U \in \mathcal{V}$. *(Here* $B(x,\varepsilon) = \{y \in X \mid d(x,y)<\varepsilon\}$.*)*

LEMMA 2. Let $f : X \to Y$ *be continuous where* (X,d) *is a compact metric space and* (Y,d') *is a metric space. Let* $y \in Y$ *be such that* $f^{-1}(y) \subset W$, W *an open subset of* X. *Then there is an* $\varepsilon > 0$ *such that* $f^{-1}\{B(y,\varepsilon)\} \subset W$. *(Here* $B(y,\varepsilon) = \{x \in Y \mid d'(x,y)< \varepsilon \}$*).*

LEMMA 3. Let $f : X \to Y$ *be continuous where* (Y,d) *is a locally compact metric space and* X *is a compact subspace. Let* $\mathcal{V}$ *be an* *open cover of Y such that f is V-close to the identy. Then there is an $\varepsilon > 0$ such that if $g : Y \to Y$ is any map $\{B(y,\varepsilon) \mid y \in Y\}$-close to the identity, then $g \circ f$ is V-close to the identity.*

Proof. Let $y \in f(X)$. For each $x \in f^{-1}(y)$ choose $V_{x,y} \in V$ such that $\{x,y\} \subset V_{x,y}$. Finitely many of the $V_{x,y}$, say $V_{y_1}, \ldots, V_{y_{k(y)}}$, cover compact $f^{-1}(y)$. Choose $\delta_y > 0$ such that $B(y,\delta_y) \subset V_{y_1} \cap \ldots \cap V_{y_{k(y)}}$. By Lemma 2 we may choose ε_y such that $0 < \varepsilon_y < \delta_y$ and such that $f^{-1}\{B(y,\varepsilon_y)\} \subset V_{y_1} \cup \ldots \cup V_{y_{k(y)}}$. Let $U = \{B(y,\varepsilon_y) \mid y \in f(X)\}$. By Lemma 1 there is an $\varepsilon > 0$ such that if $y \in f(X)$, then $B(y,\varepsilon) \subset U$ some $U \in U$. Now let $g : Y \to Y$ be $\{B(y,\varepsilon) \mid y \in Y\}$ *-close to the identity.* Given $x \in X$, let $y = f(x)$, $z = gf(x)$. Then $d(y,z) < \varepsilon$ implies $\{y,z\} \subset B(\zeta,\varepsilon_\zeta)$ some $\zeta \in f(X)$. Therefore, $x \in f^{-1}\{B(\zeta,\varepsilon_\zeta)\} \subset V_{\zeta_1} \cup \ldots \cup V_{\zeta_{k(\zeta)}}$. Also, by choice of $\varepsilon_\zeta < \delta_\zeta$, $z \in V_{\zeta_1} \cap \ldots \cap V_{\zeta_{k(\zeta)}}$. Thus, $\{x,z\} \subset V_{\zeta_j}$, some j, as required.

If $H : X \times I \to Y$ is a homotopy we denote $H(\{x\} \times I)$, the track of x, by $T(x)$. Also, we say that H is *limited* by a collection V of subsets of Y if for each $x \in X$, $T(x) \subset V$, some $V \in V$.

LEMMA 4. *Let $H : X \times I \to Y$ be a homotopy where (X,d) is a compact metric space and (Y,d') is a metric space. Let V be an open cover of Y such that H is limited by V. Then there is an $\varepsilon > 0$ such that for every $x \in X$ there is a $V \in V$ such that $B(T(x),\varepsilon) = \{y \in Y \mid d'(y,T(x)) < \varepsilon\} \subset V$.*

Proof. For $x \in X$ choose $V_x \in V$ such that $T(x) \subset V_x$ and then choose $\varepsilon_x > 0$ such that $B(T(x), 2\varepsilon_x) \subset V_x$. Let ρ be the sup metric on $X \times I$, and choose $\delta_x > 0$, $x \in X$, such that if $\rho((x,t), (y,s)) < \delta_x$ then $d'(H(x,t), H(y,s)) < \varepsilon_x$.

Let $B(x_1, \delta_{x_1})$, ..., $B(x_k, \delta_{x_k})$ be a finite subcover of the open cover $\{B(x,\delta_x^1) \mid x \in X\}$ of X, and set $\varepsilon = \min\{\varepsilon_{x_1}, \ldots, \varepsilon_{x_k}\}$.

Given $x \in X$ choose $i \in \{1, \ldots, k\}$ such that $x \in B(x_i, \delta_{x_i})$. Then $T(x) \subset B(T(x_i), \varepsilon_{x_i})$ so that $B(T(x), \varepsilon) \subset B(T(x_i), \varepsilon_{x_i} + \varepsilon) \subset V_{x_i}$ as required.

LEMMA 5. *Let U be an open subset of Q^∞. Then $U = \text{dir lim}\, M_n$ where M_n is a Q-manifold in $U^n = Q^n \cap U$ and $M_n \subset \text{Int}_{U^{n+1}} M_{n+1}$.*

Proof. By elementary reasoning (cf. the proofs of Proposition III.1, Proposition III.2 of [7]) $U = \text{dir lim}\, C_n$ where C_n is a compact metric space, $C_n \subset U^n$, $C_n \subset C_{n+1}$, $n=1, 2, \ldots$ By [9, Lemma 2] if $K \subset W \subset G$ where K is closed in Q and W and G are open in Q, then there is an open subset V of Q such that if $\bar{V}$ is the closure of V in G, then $K \subset V \subset \bar{V} \subset W$ and $\bar{V}$ is a Q-manifold. Thus, we may choose an open subset V_1 of U^1 such that $\bar{V}_1$, the closure of V_1 in U^1, is a compact Q-manifold and $C_1 \subset V_1 \subset \bar{V}_1 \subset U^1$. Inductively, choose open subsets V_n, $n \geq 1$, of U^n such that if $\bar{V}_n$ is the closure of V_n in U^n then $\bar{V}_n$ is a compact Q-manifold and $C_n \cup \bar{V}_{n-1} \subset V_n \subset \bar{V}_n \subset U^n$. Since $\bar{V}_n \supset C_n$, it follows that $U = \text{dir lim}\, \bar{V}_n$. Setting $M_n = \bar{V}_n$ we obtain our desired sequence of Q-manifolds.

Proof of Theorem 4. Let M be a paracompact, connected Q^∞-manifold. By Theorem 2 we may assume that $M = U$, an open subset of Q^∞. By Lemma 5 $U = \text{dir lim}\, M_n$, M_n a compact Q-manifold in U^{2^n}, $M_n \subset \text{Int}_{U^{2^{n+1}}} M_{n+1}$.

Let $\mathcal{V}$ be any open cover of U. We must show there is a homeomorphism $h : U \times Q^\infty \to U$ such that h and the projection map p are $\mathcal{V}$-close. For $n \geq 1$, let $\mathcal{V}^n = \{V^n = V \cap Q^n \mid V \in \mathcal{V}\}$.

Let $d(\{x_i\}, \{y_i\}) = \sum(1/2^i)\ |x_i - y_i|$ be the metric on Q (and hence on U), and for finite products $X = X_1 \times \ldots \times X_n$ let the metric on X be $\sup\{d_i(a_i, b_i) \mid i=1, \ldots, n\}$ where d_i

is the metric on X_i. If (X,d) is a metric space and $A \subset X$ we use the notation $B(A,\varepsilon; X)$ for $\{x \in X \mid d(A,x) < \varepsilon\}$.

To begin, using Lemma 1, choose ε_1 such that $0 < \varepsilon_1 < 1$ and such that (a) $B(M, 2\varepsilon_1; U^4) \subset V^4$ some $V \in \mathcal{V}$ for every $m \in M_1$ and (b) $B(M_1, 2\varepsilon_1; U^4) \subset \text{Int}_{U^4} M_2$. Let $Q(\varepsilon_1) = \prod_1^\infty [-\varepsilon_1, \varepsilon_1] \times \prod_1^\infty [-\varepsilon_1, \varepsilon_1] \subset Q^2$. Note $Q(\varepsilon_1) \subset B(0, 2\varepsilon_1; Q^2)$. Let $g_1 : Q(\varepsilon_1) \to Q^2$ be the homeomorphism that stretches each $[-\varepsilon_1, \varepsilon_1]$ linearly onto $[-1,1]$. Let $h_1 : M_1 \times Q(\varepsilon_1) \to M_1 \times Q^2$ be the homeomorphism $h_1(m,q) = (m, g_1(q))$. Consider

$$
\begin{array}{ccc}
M_1 \times Q(\varepsilon_1) & & \\
\downarrow h_1 & \searrow^{i_1} & \\
M_1 \times Q^2 & \xrightarrow{\beta_1} & M_2 \times Q^4
\end{array}
$$

where $i_1(m,0) = ((m,q), 0)$ and $\beta_1(m,q) = ((m,0), (q,0))$. Define $H_1 : M \times Q(\varepsilon_1) \times I \to M_2 \times Q$ by $H_1(m,q,t) = t\beta_1 h_1(m,q) + (1-t)i_1(m,q)$. Then H_1 is a homotopy between the embeddings $\beta_1 h_1$ and i_1 of $M_1 \times Q(\varepsilon_1)$ into $M_2 \times (Q^2 \times 0) \subset M_2 \times Q^4$. Also, $H_1(m,q,t) = ((m, (1-t)q), (tg_1(q), 0)) \in B(m, 2\varepsilon_1; U^4) \times Q^4 \subset V^4 \times Q^4$ some $V \in \mathcal{V}$ so that H_1 is limited by $\mathcal{V}^4 \times Q^4$. Note, too, that $B(m, 2\varepsilon_1; U^4) \subset \text{Int}_U M_2$ by (b) so that H_1 maps into $M_2 \times Q^4$ as indicated. By [2, Lemma 19.1] there is an ambient, invertible isotopy $G_2 : M_2 \times Q^4 \times I \to M_2 \times Q^4$ such that $(G_2)_1 i_1 = \beta_1 h_1$ and such that G_2 is limited by $\mathcal{V}^4 \times Q^4$.

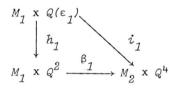

Suppose, inductively, that we have defined ε_k, $Q(\varepsilon_k)$, h_k, i_k, β_k, G_{k+1}, $k=1, \ldots, n$, and α_k, $k=1, \ldots, n-1$; (we assume $n \geq 2$; the step from $n=1$ to $n=2$ is like the following general inductive step) such that (i) $\beta_k : M_k \times Q^{2k} \to M_{k+1} \times Q^{2k+1}$, $i_k : M_k \times Q(\varepsilon_k) \to M_{k+1} \times Q^{2k+1}$ and $\alpha_{k-1} : M_{k-1} \times Q(\varepsilon_{k-1}) \to M_k \times Q(\varepsilon_k)$ are given by $\beta_k(m,q) = ((m,0), (q,0))$, $i_k(m,q) = ((m,q),0)$ and $\alpha_{k-1}(m,q) = ((m,q),0)$, (ii) $h_k : M_k \times Q(\varepsilon_k) \to M_k \times Q^{2k}$ is a homeomorphism such that for $k \geq 2$ $\beta_{k-1} h_{k-1} = h_k \alpha_{k-1}$, and (iii) G_{k+1} is an ambient invertible isotopy on $M_{k+1} \times Q^{2^{k+1}}$ limited by $V^{2^{k+1}} \times Q^{2^{k+1}}$ such that $(G_{k+1})_1 i_k = \beta_k h_k$.

Consider

$$\begin{array}{cccc} \xrightarrow{\alpha_{n-1}} & M_n \times Q(\varepsilon_n) & & \\ & \downarrow h_n & \searrow i_n & \\ \xrightarrow{\beta_{n-1}} & M_n \times Q^{2^n} & \xrightarrow{\beta_n} & M_{n+1} \times Q^{2^{n+1}} \end{array}$$

Let $\beta_{n+1}: M_{n+1} \times Q^{2^{n+1}} \to U^{2^{n+2}} \times Q^{2^{n+2}}$ be the inclusion $\beta_{n+1}(m,q) = ((m,0),(q,0))$. Then $\beta_{n+1}G_{n+1}$ is limited by $V^{2^{n+2}} \times Q^{2^{n+2}}$. By Lemma 4 we may choose $0 < \varepsilon_{n+1} < 1$ such that for each $(m,q) \in M_{n+1} \times Q^{2^{n+1}}$, $B(\beta_{n+1}G_{n+1}(\{(m,q)\} \times I), 2\varepsilon_{n+1}; U^{2^{n+2}} \times Q^{2^{n+2}}) \subset V^{2^{n+2}} \times Q^{2^{n+2}}$ some $V \in \mathcal{V}$. We may certainly also require that $B(M_{n+1}, 2\varepsilon_{n+1}; U^{2^{n+2}}) \subset \operatorname{Int}_{U^{2^{n+2}}}(M_{n+2})$. Set $Q(\varepsilon_{n+1}) = \prod_1^\infty [-\varepsilon_{n+1}, \varepsilon_{n+1}] \times \ldots \times \prod_1^\infty [-\varepsilon_{n+1}, \varepsilon_{n+1}] \subset Q^{2^{n+1}}$. Let $g_{n+1}: Q(\varepsilon_{n+1}) \to Q^{2^{n+1}}$ be the homeomorphism which stretches each factor $[-\varepsilon_{n+1}, \varepsilon_{n+1}]$ linearly onto $[-1,1]$. Let $f_{n+1}: M_{n+1} \times Q(\varepsilon_{n+1}) \to M_{n+1} \times Q^{2^{n+1}}$ be the homeomorphism $f_{n+1}(m,q) = (m, g_{n+1}(q))$. Let $h_{n+1} = (G_{n+1})_1 f_{n+1}$. Noting that by choice of ε_n, $M_n \times Q(\varepsilon_n) \subset M_{n+1}$, we define $\alpha_n: M_n \times Q(\varepsilon_n) \to M_{n+1} \times Q(\varepsilon_{n+1})$ by $\alpha_n(m,q) = ((m,q),0)$. Then $f_{n+1}\alpha_n = i_n$ and $h_{n+1}\alpha_n = (G_{n+1})_1 f_{n+1}\alpha_n = (G_{n+1})_1 i_n = \beta_n h_n$. Define $i_{n+1}: M_{n+1} \times Q(\varepsilon_{n+1}) \to M_{n+2} \times Q^{2^{n+2}}$ by $i_{n+1}(m,q) = ((m,q), 0)$. Define $H_{n+1}: M_{n+1} \times Q(\varepsilon_{n+1}) \times I \to M_{n+2} \times Q^{2^{n+2}}$ by $H_{n+1}(m,q,t) = (1-t)i_{n+1}(m,q) + t\beta_{n+1}f_{n+1}(m,q) = (1-t)((m,q),0) + t((m,0),(g_{n+1}(q),0))$. Define $T_{n+1}: M_{n+1} \times Q(\varepsilon_{n+1}) \times I \to M_{n+2} \times Q^{2^{n+2}}$ by

$$T_{n+1}(m,q,t) = \begin{cases} H_{n+1}(m,q,2t), & t \in [0,1/2] \\ \beta_{n+1}G_{n+1}(f_{n+1}(m,q), 2t-1), & t \in [1/2,1]. \end{cases}$$

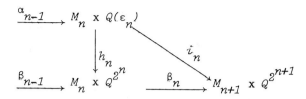

Then T_{n+1} is a homotopy between i_{n+1} and $\beta_{n+1}h_{n+1}$. Also, $T_{n+1}(\{(m,q)\} \times I) \subset B(\beta_{n+1}G_{n+1}(\{f_{n+1}(m,q)\} \times I), 2\varepsilon_{n+1}; U^{2n+2} \times Q^{2n+2}) \subset V^{2n+2} \times Q^{2n+2}$ some $V \in \mathcal{V}$ by choice of ε_{n+1}. Thus, T_{n+1} is limited by $\mathcal{V}^{2n+2} \times Q^{2n+2}$. Also, i_{n+1} and $\beta_{n+1}h_{n+1}$ are Z-embeddings into the Q-manifold $M_{n+2} \times Q^{2n+2}$. By [2, Lemma 19.1] there is an ambient invertible isotopy G_{n+2} on $M_{n+2} \times Q^{2n+2}$ limited by $\mathcal{V}^{2n+2} \times Q^{2n+2}$ and such that $(G_{n+2})_1 i_{n+1} = \beta_{n+1}h_{n+1}$. Then ε_{n+1}, $Q(\varepsilon_{n+1})$, h_{n+1}, i_{n+1}, β_{n+1}, G_{n+2} and α_n satisfy the conditions of the inductive step. By induction we have the communtative diagram

$$\begin{array}{ccc} M_k \times Q(\varepsilon_k) & \xrightarrow{\alpha_k} & M_{k+1} \times Q(\varepsilon_{k+1}) \\ \downarrow{\scriptstyle h_k} & & \downarrow{\scriptstyle h_{k+1}} \\ M_k \times Q^{2^k} & \xrightarrow{\beta_k} & M_{k+1} \times Q^{2^{k+1}} \end{array}$$

for each $k \geq 1$. The sequence $\{h_k\}$ induces a homeomorphism h_∞:dir lim$\{M_k \times Q(\varepsilon_k); \alpha_k\} \to$ dir lim$\{M_k \times Q^k; \beta_k\}$. Each α_k is just inclusion, so the first direct limit is U. Using [5, Corollary III.2] it follows that the second direct limit is $U \times Q^\infty$. Thus, setting $h = (h_\infty)^{-1}$, we have a homeomorphism of $U \times Q^\infty$ onto U.

To see that h is $\mathcal{V}$-close to the projection, let $(m,q) \in U \times Q^\infty$, $x = h(m,q) \in U$. Then $x \in M_k$ some k, and $(m,q) = h_\infty(x) = h_k(x,0)$. From (iii) of our induction hypothesis it follows that $\beta_k h_k(x,0)$ and $i_k(x,0)$ are in some $V^{2k+1} \times Q^{2k+1}$. Thus $\{(m,0),(x,0)\} \subset V^{2k+1}$, so that $\{p(m,q), h(m,q)\} \subset V^{2k+1}$ as required.

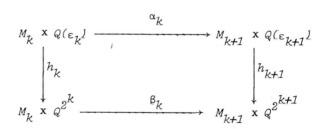

Proof of Theorem 5. Let $f : M \to N$ be a homotopy equivalence. Because of Theorem 2 we may assume that M and N are open subsets of Q^∞. By [5, Corollary III.3] sets of the form $(U_1 \times U_2 \times \dots) \cap Q^\infty$, U_i open in Q, are a basis for Q^∞. It follows that there are open covers $\mathcal{U}, \mathcal{V}$ of M, N, respectively, consisting of convex sets. By Theorem 4, there are homeomorphisms $h_1 : M \times Q^\infty \to M$, $h_2 : N \times Q^\infty \to N$ such that h_1 is $\mathcal{U}$-close to the projection $p_1 : M \times Q^\infty \to M$ and h_2 is $\mathcal{V}$-close to $p_2 : N \times Q^\infty \to N$. Define $g : M \to M \times Q^\infty$ by $g(m) = (m,0)$. Then $h_1^{-1} \sim g$ and $h_2 \sim p_2$. (By "$\sim$" we denote "is homotopic to.") By [7, Theorem II.1 and Theorem II.9] there is a homeomorphism $k : M \times Q^\infty \to N \times Q^\infty$ such that $k \sim (f \times id)$. Then $h = h_2 k h_1^{-1} : M \to N$ is a homeomorphism, and $h \sim p_2(f \times id)g = f$ as required.

We will use the following in the proof of Theorem 6.

LEMMA 6. Let $\{X_i \mid i = 1, 2, \dots\}$ be a sequence of compact polyhedra such that each X_i is a subpolyhedron of X_{i+1}. Then $X = \text{dir lim } X_i$ has the homotopy type of a countable, locally-finite, simplicial complex (c.l.f.s.c.).

Proof. In the terminology of [4] the array $X_1 \hookrightarrow X_2 \hookrightarrow X_3 \hookrightarrow \dots$ is a closed expanding sequence. (Here $\hookrightarrow$ denotes inclusion.) Since each X_i is an ANR (metric) by [4, Corollary 6.4] (see also [11, Appendix]) $X = \text{dir lim } X_i$ has the homotopy type of an ANR and hence of a CW-complex [4, Lemma 6.6] But X is Lindelof and thus has the homotopy type of a countable CW-complex [12, Proposition 2]. Hence, by [12, Theorem 1], X has the homotopy type of a c.l.f.s.c. (see also [14, Theorem 24]).

Proof of Theorem 6. Let M be a paracompact, connected Q^∞-manifold. By Theorem 2 we may assume M is an open subset of Q^∞, and by Lemma 5 we may write $M = \text{dir lim } M_n$ where M_n is a compact Q-manifold in U^n and M_n is a Z-set in M_{n+1}. By the Relative Triangulation Theorem of [1] (see also [2, Lemma 37.1]), there are homeomorphisms $h_i : M_i \to X_i \times Q$, $i = 1, 2, \ldots,$ where (a) each X_i is a (necessarily finite) polyhedron, (b) X_i is a subpolyhedron of X_{i+1}, and (c) h_{i+1} extends h_i. Let

$$g_i = h_i \times id : M_i \times Q^i \to X_i \times Q \times Q^i = X \times Q^{i+1}.$$

For $n = 1, 2, \ldots,$ consider

$$\begin{array}{ccc} M_n \times Q^n & \xrightarrow{\alpha_n} & M_{n+1} \times Q^{n+1} \\ \downarrow g_n & & \downarrow g_{n+1} \\ X_n \times Q^{n+1} & \xrightarrow{\beta_n} & X_{n+1} \times Q^{n+2} \end{array}$$

where $\alpha_n(m,q) = (m, (q,0))$ and $\beta_n(x,q) = (x, (q,0))$. Since each such diagram commutes, $\{g_n \mid n = 1, 2, \ldots\}$ induces a homeomorphism $g : M \times Q^\infty \cong \text{dir lim } \{M_n \times Q^n; \alpha_n\} \to \text{dir lim } \{X_n \times Q^{n+1}; \beta_n\} \cong X \times Q^\infty$ where $X = \text{dir lim } X_i$. By the preceding lemma X has the homotopy type of a c.l.f.s.c. $|K|$. By [13, Corollary 5.2] $|K| \times Q$ is a Q-manifold. By [6, Theorem 4] $Q^\infty \cong Q \times R^\infty$. It follows that $|K| \times Q^\infty \cong (|K| \times Q) \times R^\infty$ is a Q^∞-manifold having the same homotopy type as $X \times Q^\infty$. By Theorem 3 $X \times Q^\infty \cong |K| \times Q^\infty$ Thus, $M \cong M \times Q^\infty \cong X \times Q^\infty \cong |K| \times Q^\infty$ as required.

Proof of Theorem 7. Let M be a paracompact, connected Q^∞-manifold. By Theorem 6 $M \cong |K| \times Q^\infty$ where K is a c.l.f.s.c. By [6, Theorem 4] $Q^\infty \cong Q \times R^\infty$ and by [10, Theorem 1] $[0,\infty) \times R^\infty \cong R^\infty$. Thus, $M \cong (|K| \times Q \times [0,\infty)) \times R^\infty$. By [13, Corollary 5.2] $|K| \times Q$ is a Q-manifold, and by [3, Theorem 4] (see also [2, Corollary 16.3]) $|K| \times Q \times [0,\infty)$ embeds as an open subset of Q. This completes the proof.

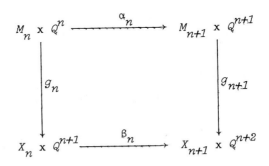

REFERENCES

1. T. A. Chapman, "All Hilbert Cube Manifolds are Triangulable," preprint.
2. ______, "Lectures on Hilbert Cube Manifolds," CBMS series of the AMS, 28, 1976.
3. ______, "On the Structure of Hilbert Cube Manifolds," Compositio Math. 24 (1972), 329-353.
4. V. L. Hansen, "Some theorems on Direct Limits of Expanding Sequences of Manifolds," Mathematica Scandinavica 29 (1971), 5-36.
5. R. E. Heisey, "Contracting Spaces of Maps on the Countable Direct Limit of a Space," Trans. Amer. Math. Soc. 193 (1974), 389-412.
6. ______, "A Factorization of the Direct Limit of Hilbert Cubes," Proc. Amer. Math. Soc. 54 (1976), 255-260.
7. ______, "Manifolds Modelled on R^∞ or Bounded Weak-* Topologies," Trans. Amer. Math. Soc. 260 (1975), 295-312.
8. ______, "Open Subsets of R^∞ are Stable," Proc. Amer. Math. Soc. 59 (1976), 377-380.
9. ______, "Stability, Open Embeddings, and Classification of Bounded weak-* Manifolds," General Topology and its Applications 6 (1976), 327-338.
10. D. W. Henderson, "A Simplicial Complex Whose Product with any ANR is a Simplicial Complex," General Topology and its Applications 3 (1973), 81-83.
11. J. Milnor, "Morse Theory," (Annals of Math. Studies 51), Princeton University Press, 1963.
12. J. Milnor, "On Spaces Having the Homotopy Type of a *CW* - Complex," Trans. Amer. Math. Soc. 90 (1959), 272-280.
13. J. E. West, "Infinite factors which are Hilbert Cubes," Trans. Amer. Math. Soc. 150 (1970), 1-25.
14. J. H. C. Whitehead, "A Certain Exact Sequence," Annals of Mathematics 52 (1950), 51-110.

CONCORDANCE CLASSES OF FREE ACTIONS OF COMPACT LIE GROUPS ON INFINITE DIMENSIONIAL MANIFOLDS

Vo Thanh Liem

Department of Mathematics
Louisiana State University
Baton Rouge, La.

Every free action of a compact Lie group G on a $Q(\ell_2)$-manifold is G-concordant to a tame free action. Also, if the free action is tame on a $Q(\ell_2)$-manifold, invariant Z-set N, then the concordance can be chosen to be relative to N. Consequently, every free n-toral action on $T^n \times Q$ is T^n-concordant to the trivial one.

Given a compact topological group G, two G-actions φ and η on a topological space M is *G-concordant* if there exists a G-action Ψ on $M \times I$ preserving $M \times 0$ and $M \times 1$ such that $\Psi | M \times 0 = \varphi$ and $\Psi | M \times 1 = \eta$.

A G-action φ on $Q(\ell_2)$-manifold M is said to be *tame* if the orbit space M/G is a $Q(\ell_2)$-manifold, and *wild* otherwise. We always denote $p: M \to M/G$ the orbit map.

In this note, extending the main result of [4] : "*If M is a closed topological manifold of dimension n > 5, then any free circle action on M is S^1-concordant to a tame free action*", we will present the following theorem.

ISBN 0-12-158860-2

MAIN THEOREM. Given a compact Lie group G, every free G-action on a $Q(\ell_2)$-manifold is G-concordant to a tame free action.

As in [4], there are both relative version and uniqueness statement for free actions on compact Q-manifolds.

Moreover, from a result of Torunczýk (Thm. 5′, [6]), there are many wild free T^n-actions on $T^n \times Q$; however, as an application of Main Theorem, all are T^n-concordant to the trivial free T^n-action.

Definition. A retraction $r : X \to Y$ is a *straight deformation retraction* if there exists a deformation $\{\Psi_t\}$, $0 \leq t \leq 1$, of X such that $\Psi_0 = id$, $\Psi_1 = r$ and $r \circ \Psi_t = r$ for each t.

Any map $f : X \to Y$ is a *near homeomorphism* if for each open cover U of Y, there is a homeomorphism $h : X \to Y$ such that f and h are U-close.

The following Lemma is a straight application of

LEMMA A. [7] *Let $f : X \to Y$ be a straigth deformation retraction from an ℓ_2-manifold X onto an ℓ_2-manifold Y. Then f is a near homeomorphism.*

In this note, we only concern with separable metric spaces. Q denotes $\prod_1^\infty[-1,1]$ and ℓ_2 for $\prod_1^\infty(-1,1)$. Usual notion in infinite dimensional manifold theory as Z-set, infinite simple homotopy equivalence (s.h.e.), cellular map, etc... is referred to [2].

A surjective map $f : X \to Y$ is *proper* if the preimage of a compact set is compact.

From a Gleason's result, it follows that the orbit map $p : X \to X/G$ is a principle fiber bundle if X is a metric space (Thm.5.8,II,[1]). The following Lemma can be obtained by an observation.

LEMMA C [6]. *Let X be a locally compact ANR. If there is a Z-set A in X such that X-A is a Q-manifold, then X is a Q-manifold itself.*

LEMMA D [2]. *Let X be an ANR and Y a Z-set in X. If X-Y is an ℓ_2-manifold, then X is an ℓ_2-manifold itself.*

Definition. Let X be a locally compact space. $C_0(X)$ denotes the space of positive continuous functions f defined on X such that for every $\varepsilon > 0$, there is a compact $K \subset X$ such that $f(X-K) \subset (0,\varepsilon)$. If $f \in C_0(X)$,

$$||f|| = sup \{|f(x)| \ : x \in X\}$$

LEMMA 1. Let S^1 act freely on a Q-manifold M and N a Q-manifold, invariant Z-set in M. If S^1 acts tamely on N, then there is a proper map $\varphi : (M/S^1)\text{x}Q \to M/S^1$ which is near the projection map and $\varphi|(N/S^1)\text{x}Q$ is a homeomorphism onto N/S^1.

Proof. Let $r : (M/S^1)\text{x}Q \to M/S^1$ be the projection. Then r is obviously a CE-map. Given $\varepsilon \in C_0(M/S^1)$, then $r|(N/S^1)\text{x}Q$ has an ε-approximation, say σ, that is a homeomorphism. Since M/S^1 is an ANR and $(N/S^1)\text{x}Q$ is a closed subset of $(M/S^1)\text{x}Q$, if $||\varepsilon||$ is small, we may extend σ to a proper map φ such that

(i) φ and r are ε-close,

(ii) $\varphi \simeq_p r$.

LEMMA 2. Let S^1 act freely on a Q-manifold M and N a Z-set in M as in Lemma 1. Then, there is a homeomorphism $\Psi : (M/S^1)\text{x}Q\text{x}R \to (M/S^1)\text{x}R$ such that

(i) Ψ *is an extension of* $\varphi|[(N/S^1)\text{x}Q] \times \text{id}$,

(ii) Ψ *is near the projection.*

Proof. The map $r \times id$ is a near homemorphism, since it is a CE-map (Thm 43.2, [2]). Now, since $r \times id$ is properly homotopic to $\varphi \times id$ by the construction of φ in Lemma 1, we may approximate $\varphi \times id$ by a homeomorphism γ which is as close to the projection as we desire, if we choose $||\varepsilon||$ small enough in Lemma 1. Now, since $\gamma|(N/S^1)\times Q\times R$ is very close to $\varphi|[(N/S^1)\times Q]\times id$ and since N/S^1 is a Z-set in M/S^1 by Lemma 4 below, there is a small isotopy $H_t : (M/S^1)\times R \to (M/S^1)\times R$ such that

(i) $H_0 = id$,

(ii) $H_1\gamma|(N/S^1)\times Q\times R = \varphi|[(N/S^1)\times Q] \times id$.

Therefore, $\Psi = H_1\gamma$ *is a desired homeomorphism.*

LEMMA 3. Let S^1 act freely on an ℓ_2-manifold M and N be an ℓ_2-manifold that is invariant Z-set in M. If S^1 acts tamely on N, then there is a homeomorphism $\Psi : (M/S^1)\times \ell_2\times R \to (M/S^1)\times R$ such that

(i) $\Psi|(N/S^1)\times \ell_2\times R = \mu\times id$, where μ is a homeomorphism from $(N/S^1)\times \ell_2$ onto N/S^1,

(ii) $p_R \circ \Psi$ is near the projection $(M/S^1)\times \ell_2\times R \to R$, where $p_R : (M/S^1)\times R \to R$ is the projection.

Proof. By homotopy extension theorem, there is a map $\gamma : (M/S^1)\times \ell_2 \to M/S^1$ homotopic to the projection $r : (M/S^1)\times \ell_2$ M/S^1 such that $\gamma|(N/S^1)\times \ell_2$ is a homeomorphism onto N/S^1.

Now, similar to the proof of Lemma 2, since $r\times id$ is a straight deformation retraction, it is a near homeomorphism. Therefore, $\lambda\times id$ can be homotoped to a homeomorphism that does not move R-factor too far.

Finally, the construction of Ψ is similar to that of Lemma 2.

THEOREM 1. Every free S^1-action on a Q-manifold M is S^1-concordant to a tame free S^1-action on M.

Proof. Let B denote the orbit space M/S^1 and $\Psi : B \times Q \times R \to B \times R$ be a homeomorphism with $\|\varepsilon\|$ very small and $p_R \circ \Psi$ near the projection $B \times Q \times R \to R$. So, we may assume that $B \times 0 \subset \Psi(B \times Q \times (0,1))$.

Let U denote $\Psi(B \times Q \times 1)$ and V the subset of $B \times R$ bounded by $B \times 0$ and U.

If $\|\varepsilon\|$ is small enough, we can show that the following inclusions are infinite s.h.e.'s

(i) $B \times 0 \subset V$,

(ii) $U \subset V$.

Now consider $\tilde{V} = p^{-1}(V) \subset M \times R$. By use of the collar structure along U in $B \times R$, it follows that $\tilde{V}$ is a Q-manifold whose topological frontier in $M \times R$ is $M \times 0 \cup p^{-1}(U)$. Moreover, by use of Q-manifold theory and fiber bundle theory, we can show that the following inclusions are infinite s.h.e.'s

(i) $M \times 0 \subset \tilde{V}$,

(ii) $p^{-1}(U) \subset \tilde{V}$.

Hence, we can show that

$(\tilde{V}, M \times 0, p^{-1}(U)) \approx (M \times [0,1], M \times 0, M \times 1)$.

Therefore, the free action on $\tilde{V}$ induces a free action on $M \times [0,1]$ which provides a desired concordance. The proof of Theorem 1 is complete.

Before proving a relative form of Theorem 1, we need the following Lemma (all spaces are ANR's).

LEMMA 4. *Let $p : E \to B$ be a locally trivial F-fiber bundle and Z a subset of B. Then, Z is a Z-set in B if $p^{-1}(Z)$ is a Z-set in E.*

Proof. Let U be an open subset of B. It suffices to show that the inclusion $U - Z \subset U$ is a homotopy equivalence. However, it follows directly from the five Lemma and the following commutative diagrams

$$\begin{array}{ccccccccc}
\to & \Pi_{q+1}(U-Z) & \to & \Pi_q(F) & \to & \Pi_q(p^{-1}(U-Z)) & \to \\
 & \downarrow & & \| & & \downarrow\approx & \\
\to & \Pi_{q+1}(U) & \to & \Pi_q(F) & \to & \Pi_q(p^{-1}(U)) & \to
\end{array}$$

in which $\Pi_q(p^{-1}(U-Z)) \to \Pi_q(p^{-1}(U))$ is an isomorphism for every q.

THEOREM 2. Let N be a Q-manifold, Z-set in M. If N is invariant and the free S^1-action is tame on N, then the S^1-concordance in Theorem 1 can be chosen to be relative to N.

Proof. By Lemma 4, $Z = N/S^1$ is a Z-set in B. By Lemma 2, we have a homeomorphism

$$\Psi : B\times Q\times E \to B\times R$$

such that $\Psi | Z\times Q\times R = \varphi \times id$, where φ is a homeomorphism from $Z\times Q$ onto Z. The proof is similar that for the absolute case.

THEOREM 3. Every free S^1-action on ℓ_2-manifold M is S^1-concordant to a free tame S^1-action on M. Given a Z-set N in M, if N is an ℓ_2-manifold and the action is tame on N. then the concordance can be chosen to be relative to N.

Proof. The proof is similar to that of Theorem 1 and 2 by use of Lemma 3 and 4. However, we do not have to pay attention on Whitehead's torsion, since ℓ_2-manifolds are classified by their homotopy types [5].

Let a compact Lie group G act freely on a $Q(\ell_2)$-manifold whose orbit space $B = M/G$. Then, $B\times R^n$ is an $Q(\ell_2)$-manifold, where n is the dimension of G.

By use of Lemma C and D, it follows that $BxI^{n-1}xR$ is an $Q(\ell_2)$- manifold. Now, it is clear that the product of the given G-action and the identity on I^{n-1} is a free G-action on MxI^{n-1} whose orbit space is BxI^{n-1}.

Parallel to the proof of Theorem 1, 2 and 3, we can show the following theorem.

MAIN THEOREM. Every free G-action on a $Q(\ell_2)$-manifold M is G-concordant to a tame free G-action on M. If a $Q(\ell_2)$-manifold N is an invariant Z-set in M, then the concordance can be chosen to be relative to N.

Proof. Actually the proof of Theorem 1 and 3 shows that the product free action above is G-concordant to a free tame G-action. Moreover, $p^{-1}(BxI^{n-1}x0) \approx MxI^{n-1}x0$, $p^{-1}(U') \approx M$ are Z-sets in $\tilde{V} = p^{-1}(V) = MxI^{n-1}x[0,1]$ ($\approx Mx[0,1]$).

By use of unknotting property of Z-set in $Q(\ell_2)$-manifolds, there exists a homeomorphism

$$\eta \ : \ MxI^{n-1}x[0,1] \ \rightarrow \ Mx[0,1]$$

such that $\eta(Mx0x0) = Mx0$ and $\eta(p^{-1}(U')) = Mx1$. So, we obtain the absolute form of the theorem.

For the proof of the relative form, η must be chosen with care so that $\eta(x,0,t) = (x,t)$ for every x in N. This can be done since N is a Z-set in M.

Corollary 1. Let M be a compact Q-manifold. Then, T^n-concordant free tame actions on M are tamely T^n-concordant.

Proof. Apply the relative form of Main Theorem to the pair $Mx([0,1], \{0,1\})$, since we may assume that the action is free on $Mx[0,1]$ by III.7 of [1].

COROLLARY 2. For every compact Q-manifold M, there are countably many free T^n-actions on M up to T^n-concordance.

Proof. Since there are only countably many finite complexes, up to simple homotopy equivalence, it follows from Triangulation Theorem and Classification Theorem of Chapman for compact Q-manifolds [2] that there are countably many compact Q-manifolds.

On the other hand, for every compact Q-manifold B, the set of concordance classes of principle T^n-fiber bundle over B is one-to-one correspondant to the set

$$[B, BT^n] = [B, K(Z^n, 2)] = H^2(B, Z^n)$$

which is countable.

So, Corollary 2 is proved by use of Corollary 1.

THEOREM 5. Every free T^r-action on $T^n \times Q$, $r \leq n$, is T^r-concordant to the product of the trivial free T^r-action on $T_1^r \times Q$ and the identity on T_1^{n-r}, where $T_1^r \times Q \times T_1^{n-r}$ is homeomorphic to $T^n \times Q$.

Proof. By Theorem 4 and Classification Theorem [2], it suffices to show that the orbit space $(T^n \times Q)/T^r$ has homotopy type of T^{n-r}.

Since $(T^n \times Q)/T^r$ is a compact ANR (Lemma B), $[(T^n \times Q)/T^r] \times Q$ is a compact Q-manifold. So, there is a compact finite-dimensional manifold X such that

$$[(T^n \times Q)/T^r] \times Q \approx X \times Q$$

Let $\tilde{X}$ be the total space of the induced T^r-fiber bundle from the inclusion map $X \xrightarrow{\times 0} X \times Q$. Then, $\tilde{X}$ is a finite-dimensional manifold and $\tilde{X} \simeq T^n$. Now it follows from Theorem 5.2 in [3] that $\tilde{X} \simeq T^{n-r}$. Hence, so does $(T^n \times Q)/T^r$.

COROLLARY 3. Every free T^n-action on T^nxQ is T^n-concordant to the trivial free action.

By use of the homotopy sequence of fiber bundle, we can prove the following Lemma.

LEMMA 5. Let $p : E \to B$ be a S^1-fiber bundle between ANR's. If E has homotopy type of S^1, then

either (i) B is contractible,

or (ii) $B \simeq S^1 \mathrm{x} K(Z,2)$,

or (iii) $B \simeq K(Z_p,1)$ for some integer p.

COROLLARY 4. A free S^1-action on $S^1 \mathrm{x} \ell_2$ is S^1-concordant to a tame free S^1-action whose orbit space has one of the following three types:

(i) a point,

(ii) $S^1 \mathrm{x} K(Z,2)$,

(iii) $K(Z_p,1)$ for some integer p.

REFERENCES

1. G. E. Bredon, Introduction to compact transformation groups, Pure and Applied Math. Series #6, Academic Press (1972).
2. T. A. Chapman, Lectures on Hilbert cube manifolds, C.B.M.S. Regional Conference Series in Math. #28, A.M.S. (1976)
3. P.E. Conner and D. Montgomery, Transformation groups on a *K(π,1) I*, Michigan Mathematical Journal, Vol. 6 (1959) 4, 405-412.
4. A. L. Edmonds, Taming free circle actions, Proc. Amer. Math. Soc. 62 (1977) 2, 337-343.
5. D.W. Henderson, Infinite dimensional manifolds are open subset of Hilbert space, Topology 9 (1970), 25-34.
6. H. Torunczýk, On CE-images of the Hilbert cube and characterization of *Q*-manifolds, mimeographed.
7. D. W. Curtis and Vo Th Liem, Infinite product which are homeomorphic to Hilbert space, to appear.

CELL-LIKE MAPS, APPROXIMATE FIBRATIONS AND SHAPE FIBRATIONS

T. B. Rushing[1]

Department of Mathematics
University of Utah
Salt Lake City, Utah

The purpose of this expository paper is to outline and motivate results on shape fibrations contained in two papers by Sibe Mardešić and the author, [16], [17], *and in the dissertations of Thomas McMillan,* [19], *and Allen Matsumoto,* [14]. *The first two papers,* [16], [17], *define the notion of a shape fibration and establish basic properties. The dissertation of Thomas McMillan investigates the relationship between the notions of shape fibration, cell-like map and herditary shape equivalence. The dissertation of Allen Matsumoto develops an obstruction theory for shape fibrations with applications to cell-like maps.*

In order to place shape fibrations in their proper perspective, it seem desirable to first briefly discuss the related concepts of cell-like map and approximate fibration.

I. CELL-LIKE MAPS

The original definition of cell-like space applied only to finite dimensional compacta. A space was said to be *cell-like*

[1]*Supported by NSF grant MCS77-03978.*

ISBN 0-12-158860-2

if it is homeomorphic to a cellular subspace of some manifold. A *cell-like map* is a map whose point-inverses are cell-like spaces. The following proposition made it natural to extend the notion of cell-like to infinite dimensions by defining a *cell-like space* to be a space having the shape of a point.

PROPOSITION (Lacher [11], [12]. *For a finite-dimensional compactum X, the following statements are equivalent:*

(a) X admits an embedding as a cellular subset of some manifold;

(b) X has property UV^{∞}*; and*

(c) X has the shape of a point.

In order to indicate the importance that cell-like maps have assumed, we now mention three results. We shall have a particular interest in the first of these in Section 5. (See Corollary 10 and Question 4.)

THEOREM (Siebenmann [20]*). The set of cell-like maps of a closed n-manifold M onto a closed n-manifold N,* $n \geq 5$*, is precisely the closure of the set of homeomorphisms of M onto N in the space of maps of M into N. (For* $n = 3$*, it is required that "cell-like" be replaced by "cellular" and this case was also done by Armentrout, e.g.* [1]*.)*

THEOREM (Chapman [1]*). If X,Y are Q-manifolds and* $f : X \to Y$ *is a cell-like map, then f is a proper homotopy equivalence.*

Work of J. West [22], Chapman [5] or R. Edwards [10], and R. Edwards [9] implies the following theorem.

THEOREM. If $f : X \to Y$ *is a cell-like map between compact ANR's, then f is a simple homotopy equivalence.*

For an excellent expository article on the theory of cell-like maps, the reader is referred to [12] by C. Lacher, who was one of the pioneers in the subject.

We now briefly relate shape fibrations to cell-like maps. More discussion will appear later in the paper. Every cell-like map $p : E \to B$ between finite-dimensional metric compacta is a shape fibration ([16], Theorem 5). Also, every cell-like map $p : E \to B$ between compact ANR's is an approximate fibration, or equivalently, a shape fibration (e.g. [16], Theorem 4). Hence, shape fibrations pretty much constitute a generalization of cell-like maps. (We shall discuss the relationship between cell-like maps and shape fibrations for the infinite dimensional case in Section 4.) Now, a cell-like map always has fibers (point inverses) of the same shape--in particular, the shape of a point. Similarly, shape fibrations always have fibers of the same shape whenever the range is path connected ([16], Theorem 3). Many examples of shape fibrations exist where the fibers have non-trivial shape. In general, a map which has all fibers of the same shape will not be a shape fibration because the definition of shape fibration will necessitate that the fibers be put together somehow "continuously". It just happens to be the case that it is difficult to pack fibers of trivial shape together "discontinuously".

Early in the study of cell-like maps between locally compact ANR's, it was observed by Lacher [13], and Armentrout and Price [2] that maps of polyhedra into the range can be ε-lifted to the domain. Thus, cell-like maps enjoy a property very close to the defining property for fibrations. This led Coram and Duvall [6] to abstract the ε-lifting property and to study the class of maps between locally compact ANR's which possess it. Their theory is the subject of the next section.

II. APPROXIMATE FIBRATIONS

The following two definitions are due to Coram and Duvall, [7]. A map $p : E \to B$ between metric spaces has the *approximate homotopy lifting property* (AHLP) with respect to a space X provided that given an open covering ε of B, given a map $h : X \to E$, and given a homotopy $H : X \times I \to B$ such that $ph = H_0$, then there exists a homotopy $\tilde{H} : X \times I \to E$ such that $\tilde{H}_0 = h$ and the maps $p\tilde{H}$ and H are ε-close. A proper map $p : E \to B$ between locally compact metric ANR's is an *approximate fibration* if p has the approximate homotopy lifting property for all metric spaces.

Perhaps the simplest example of an approximate fibration which fails to be a fibration is given by any monotone map between two intervals with at least one non-degenerate point inverse, e.g., the map $p : [-1,1] \to [0,1]$ which is the identity on $[0,1]$ and maps $[-1,0]$ to 0. Coram and Duvall show that maps between locally compact ANR's which can be approximated by fibrations are approximate fibrations ([6], Proposition 1.1). By using this fact, they observe that a map $p : S^1 \times S^1 \to S^1$ which has one point inverse a "meridional" Warsaw circle and other point inverses "meridional" simple closed curves provides an example of an approximate fibration.

Coram and Duvall establish for approximate fibrations some analogues to theorems about fibrations. In particular, they show that fibers are FANR's ([6], Corollary 2.5), that fibers have the same shape whenever the base is path connected ([6], Theorem 2.12), and that there is an exact sequence of shape groups for an approximate fibration ([6], Corollary 3.5). The proofs of these results strongly use the requirement that the base space and total space are ANR's.

When reduced to ANR's, the notion of shape fibration agrees with the notion of approximate fibration ([16], Corollary 1); consequently, shape fibrations are a true generalization of approximate fibrations. Although approximate fibrations constitute a significant generalization of fibrations between ANR's, Mardešić and the author felt that the requirement that base space and total space be ANR's is too restrictive. (We shall illustrate this restriction with some examples in the next paragraph.) The concept of shape fibration applies to locally compact metric spaces. Another disadvantage of approximate fibrations is that they do not include the notion of a pull-back. In particular, the restriction of an approximate fibration need not be an approximate fibration. (For restrictions, not only may total space and base space fail to be ANR's, but the AHLP may also fail ([14], Example 3.2)). However, in Section 5 we shall see that shape fibrations do include pull-backs. Thus, for instance, one may do obstruction theory. As a consequence, the theory of shape fibrations yields, in a natural fashion, interesting consequences even for the special case of approximate fibrations (e.g., [14], Theorem 5.12 and Theorem 6.7) and cell-like maps (e.g. [14], Corollaries 5.13, 5.14, 5.15 and 5.16). We shall elaborate more on this in Section 5.

We conclude this section by mentioning some examples of shape fibrations which fail to be approximate fibrations. As a first example notice that such a nice bundle map as the canonical projection of a solenoid to S^1 fails to be an approximate fibration since a solenoid is not an ANR. As a second example, we give a map similar to the Coram-Duvall example mentioned above, which fails to have the AHLP. Let A be the limit arc of the Warsaw circle W and fix $b_0 \in S^1$. Let $E = (W \times S^1)/(A \times b_0)$ and let $p : E \to S^1$ be the map induced by the projection $W \times S^1 \to S^1$. Then $p^{-1}(b_0)$ is a 1-sphere and all other point inverses $p^{-1}(b)$, $b \neq b_0$, are copies of W. As a third even simpler example, let $p : W \to W/A \approx S^1$ be the quotient map. Then $p^{-1}(b_0) = A$ and all other point inverses $p^{-1}(b)$, $b \neq b_0$, are singletons. More generally, every cell-like map $p : E \to B$ between finite dimensional metric compacta is a shape fibration ([16], Theorem 5), however if either E or B is not an ANR, then p fails to be an approximate fibration.

III. SHAPE FIBRATIONS

The second and third examples in the preceeding paragraph illustrate that in non-ANR cases it is inappropriate to simply require that the map p have the AHLP. The basic idea in defining shape fibrations is to proceed in a similar manner to the development of shape theory [18] by replacing E and B by ANR-sequences $\underline{E} = (E_i, q_{ij})$, $\underline{B} = (B_i, r_{ij})$ with inverse limits E and B respectively, and by replacing p by a level map $\underline{p} = (p_i) : \underline{E} \to \underline{B}$, where the maps $p_i : E_i \to B_i$ form a commutative diagram which induces the map p. Finally, the AHLP for the map p is replaced by a new property called the AHLP for $\underline{p}$. From the point-of-view of Brosuk shape theory [3], this essentially amounts to allowing the approximate liftings of homotopies to take place in neighborhoods of E in the Hilbert cube $Q \supset E$ rather than in just E itself.

We are now ready to give precise definitions of some fundamental concepts. In particular, we shall define shape fibration, and the closely related notion *-fibration (see Definitions 4 and 5 below). We consider inverse sequences $\underline{E} = (E_i, q_{ij})$, $\underline{B} = (B_i, r_{ij})$ of metric compacta. If all E_i and B_i are ANR's, we speak of ANR-sequences. A level map of sequences $\underline{p} : \underline{E} \to \underline{B}$ is a sequence of maps $\underline{p} = (p_i)$ where $p_i : E_i \to B_i$ and for every i and $j \geq i$ we have $p_i q_{ij} = r_{ij} p_j$.

Definition 1. A level map $\underline{p} : \underline{E} \to \underline{B}$ has the *homotopy lifting property* (HLP) w.r.t. a space X provided that each i admits a $j \geq i$ such that for any maps $h_j : X \to E_j$, $H_j : X \times I \to B_j$ with $p_j h_j = H_{j0}$, there exists a homotopy $\tilde{H}_i : X \times I \to E_i$ with $\tilde{H}_{i0} = q_{ij} h_j$ and $p_i \tilde{H}_i = r_{ij} H_j$. (Every such j is called a lifting index for i.)

Definition 2. A level map $\underline{p} : \underline{E} \to \underline{B}$ has the *approximate homotopy lifting property* (AHLP) w.r.t. a space X provided each i and each $\varepsilon > 0$ admit a $j \geq i$ and a $\delta \geq 0$ with the following property: whenever maps $h_j : X \to E_j$ and $H_j : X \times I \to B_j$ are such that $d(p_j h_j, H_{j0}) \leq \delta$ then there is a homotopy $\tilde{H}_i : X \times I \to E_i$ with $d(\tilde{H}_{i0}, q_{ij} h_j) \leq \varepsilon$ and $d(p_i \tilde{H}_i, r_{ij} H_j) < \varepsilon$. (Every such j is called a lifting index for (i, ε) and δ is called a lifting mesh for (i, ε).)

Definition 3. Let $\underline{p} : \underline{E} \to \underline{B}$ be a level map of compact sequences. Let $\lim \underline{E} = (E, q_i)$ and $\lim \underline{B} = (B, r_i)$, where $q_i : E \to E_i$ and $r_i : B \to B_i$ are the natural projections. The unique map $p : E \to B$ such that for every i, $p_i q_i = r_i p$, is said to be *induced* by $\underline{p}$ or to be the *limit* of $\underline{p}$.

Definition 4. A map between metric compacta $p : E \to B$ is called a *shape fibration* (respectively, *$*$-fibration*) provided it is induced by a level map $\underline{p} : \underline{E} \to \underline{B}$ of ANR-sequences (respectively, compact sequences) satisfying the AHLP (respectively, the HLP) with respect to any metric space X and the lifting index and lifting mesh do not depend on X.

Definition 5. Let E and B be locally compact metric spaces. A proper map $p : E \to B$ is a *shape fibration* ($*$-fibration) provided for every compact subset $C \subset B$ the restriction $p|p^{-1}(C) : p^{-1}(C) \to C$ is a shape fibration ($*$-fibration) between compact metric spaces.

We now enumerate some basic results concerning shape fibrations which appear in [16] and [17].

THEOREM 1. Let $\underline{p} : \underline{E} \to \underline{B}$ and $\underline{p}' : \underline{E}' \to \underline{B}'$ be level maps of ANR-*sequences which induce the same map $p : E \to B$. If $\underline{p}$ has the* AHLP, *then so does $\underline{p}'$.*

COROLLARY 1. If E and B are compact ANR'*s, then a map* $p : E \to B$ *is a shape fibration if and only if it is an approximate fibration.*

Corollary 1 can in fact be established in the case E and B are only locally compact (Proposition 5, [16]). The HLP is not invariant even for level maps of ANR-sequences inducing the same map, e.g. Example 3, [16].

THEOREM 2. Let $\underline{p} : \underline{E} \to \underline{B}$ *be a level map of* ANR-*sequences which induces a map* $p : E \to B$. *If* $\underline{p}$ *has the* AHLP, *then there is a level map* $\underline{p}' : \underline{E}' \to \underline{B}$ *of* ANR-*sequences which induces the same map* p *and which has the* HLP.

COROLLARY 2. Every shape fibration $p : E \to B$ *between metric compacta is a* $*$-*fibration.*

The converse to Corollary 2 is not true. In fact, there are fibrations which fail to be shape fibrations (see Example 4, [16], and Example 1, [17]). At the end of this section we shall give a bundle map which fails to be a shape fibration!

THEOREM 3. If the map $p : E \to B$ *between metric compacta is a* $*$-*fibration (a shape fibration) and* $x,y \in B$ *can be joined by a path in* B, *then the fibers* $X = p^{-1}(x)$ *and* $Y = p^{-1}(y)$ *have the same shape.*

Note that Theorem 3 can be extended trivially to locally compact metric spaces.

Question 1. Let $p : E \to B$ be a shape fibration and let $x,y \in B$ be points belonging to the same component. Do the fibers $p^{-1}(x)$ and $p^{-1}(y)$ have the same shape?

Of course Theorem 3 answers Question 1 affirmatively whenever B is path connected. More generally, Corollary 1 of [17] answers Question 1 affirmatively whenever each two points of B are contained in a continuum of trivial shape (see Corollary 3 below). Also Mardešić [15] has obtained a positive answer to Question 1 in the special case that one fiber has the shape of a compact ANR. However, the general question remains. The analogous question for $*$-fibrations has a negative answer, e.g. Example 4 of [16] or Example 1 of [17]. The converse to Question 1 is false, that is, shape equivalence of fibers does not imply shape fibration (see Examples 5 and 6 of [16]).

THEOREM 4. Every cell-like map $p : E \to B$ between compact ANR*'s is an approximate fibration and equivalently a shape fibration.*

By [13] and [7], Theorem 4 also holds for locally compact ANR's.

THEOREM 5. Let E and B be finite-dimensional locally compact metric spaces and $p : E \to B$ a cell-like map. Then, p is a shape fibration.

Although many cell-like maps are shape fibrations by the above two theorems, not all are such. For instance, the Taylor map is not one (see Example 6, [16]).

THEOREM 6. Let $p : E \to B$ be a shape fibration between metric compacta and let B have the shape of a point. If $e \in E$, $b = p(e)$, $F = p^{-1}(b)$, then the inclusion $u: (F,e) \to (E,e)$ is a pointed shape equivalence.

COROLLARY 3. Let $p : E \to B$ be a shape fibration between metric compacta. If the points $b_0, b_1 \in B$ are contained in a subcontinuum

$B' \subseteq B$ of trivial shape, then the fibers $F_0 = p^{-1}(b_0)$ and $F_1 = p^{-1}(b_1)$ have the same shape.

COROLLARY 4. *Let $p : E \to B$ be an approximate fibration between compact* ANR*'s. For every point $e \in E$ the fiber (F,e), $F = p^{-1}(p(e))$, is a pointed* FANR.

THEOREM 7. *Let $p : E \to B$ be a shape fibration between metric compacta and let $e \in E$, $b = p(e)$, $F = p^{-1}(b)$. Then p induces an isomorphism of the homotopy pro-groups*

$$\underline{p}_* : \text{pro-}\pi_n(E,F,e) \to \text{pro-}\pi_n(B,b) .$$

COROLLARY 5. *Let $p : E \to B$ be a shape fibration between metric compacta and let $e \in E$, $b = p(e)$, $F = p^{-1}(b)$. Then, p induces an isomorphism of the shape groups*

$$p_* : \check{\pi}_n(E,F,e) \to \check{\pi}_n(B,b) .$$

THEOREM 8. *Let $p : E \to B$ be a shape fibration between metric compacta, $e \in E$, $b = p(e)$, $F = p^{-1}(b)$. Then there is an exact sequence of homotopy pro-groups as follows:*

$$\cdots \to \text{pro-}\pi_n(F,e) \to \text{pro-}\pi_n(E,e) \to \text{pro-}\pi_n(B,b) \to \text{pro-}\pi_{n-1}(F,e) \to \cdots .$$

COROLLARY 6. *(Coram and Duvall) Let $p : E \to B$ be an approximate fibration between compact* ANR*'s, and let $e \in E$, $b = p(e)$, $F = p^{-1}(b)$. Then there is an exact sequence of shape groups as follows:*

$$\cdots \to \check{\pi}_n(F,e) \to \check{\pi}_n(E,e) \to \check{\pi}_n(B,b) \to \check{\pi}_{n-1}(F,e) \to \cdots .$$

In Section 9 of [16] the notions of HLP and AHLP are extended from level maps of inverse sequences to arbitrary maps of arbitrary inverse systems. However, we shall not repeat those definitions here.

We conclude this section with a question and a discussion of that question.

Question 2. Is there a reasonable characterization of those bundle maps (and/or those fibrations) which are shape fibrations?

Certainly any fibration (consequently, any bundle map) between ANR's is a shape fibration. It is easy to see that the bundle map which canonically projects a solenoid to S^1 is also a shape fibration, although the solenoid is not an ANR. However, the following example, which was pointed out to the author by David Edwards, illustrates that there are bundle maps (with base space an ANR) which fail to be shape fibrations. For examples of fibrations which fail to be shape fibrations see Example 4, [16] and Example 1, [17].

Example 1. Let $p : E \to S^1$ be the projection from the origin of the plane continuum E (indicated below) to S^1. Similarly, let $p_i : E_i \to S^1$ be the projection of E_i to S^1. Then p is induced by $\underline{p} = (p_i)$ which is a level map of the indicated ANR-sequences. However, it is easily seen that $\underline{p}$ does not have the AHLP for the continuum X indicated; hence by Theorem 1 above, p is not a shape fibration. However, p is clearly a bundle map.

IV. CELL-LIKE MAPS WHICH ARE SHAPE FIBRATIONS

Since a cell-like map has all point-inverses of the same shape (and an outstandingly simple shape at that), in view of Theorem 3, it is a priori reasonable to expect that all cell-like maps are shape fibrations. However, we have already mentioned

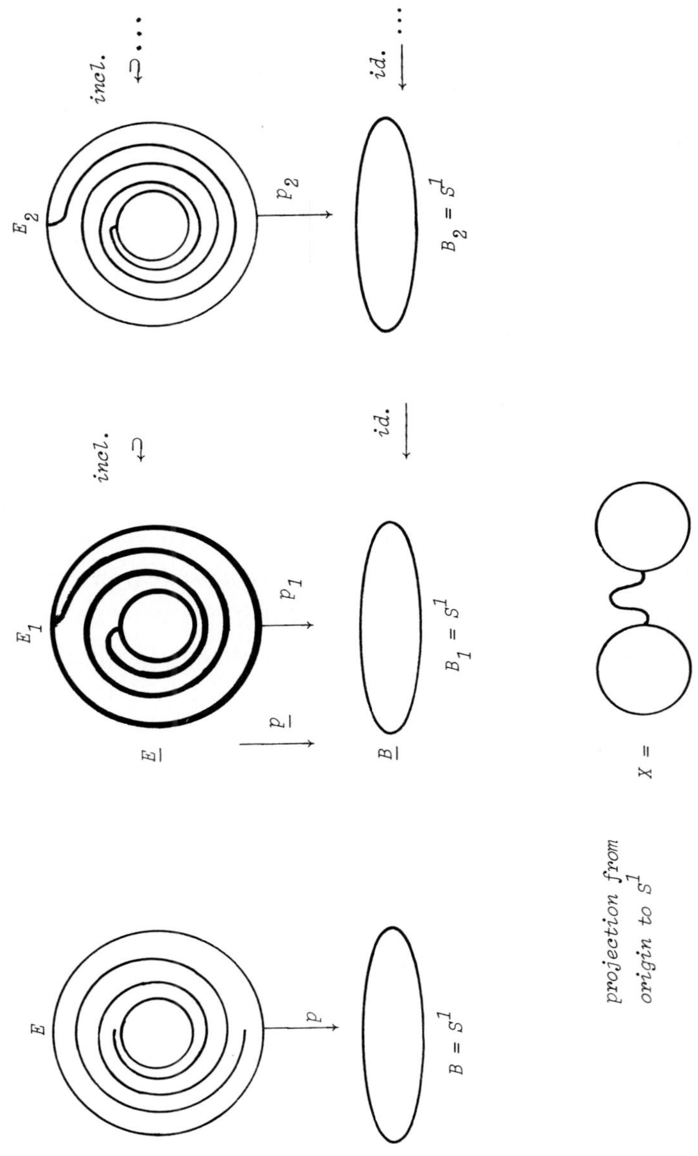
E
p
B = S^1
E
p
B
E_1
p_1
B_1 = S^1
incl.
id.
E_2
p_2
B_2 = S^1
incl.
id.
projection from origin to S^1
X =

that such is not the case as evidenced by the Taylor map [21] (see Example 6, [16]). Theorem 5 above does assure us that in finite dimensions every cell-like map is a shape fibration. In fact, it is not difficult to prove that in finite dimensions the notions of cell-like map, cell-like shape fibration, and hereditary shape equivalence are all equivalent (see Chapter III, [19]). Hence the interest lies in determining the relationships between cell-like maps and shape fibrations in infinite dimensions.

The following theorem isolates a class of cell-like maps which are shape fibrations--the hereditary shape equivalences.

THEOREM 9. (T. McMillan [19]) If $p : E \to B$ is an hereditary shape equivalence between compact metric spaces, then p is a shape fibration.

The converse of Theorem 9 is not true. Indeed, Edwards and Hastings ([8], pp. 196-200) give an example of a cell-like shape fibration which fails to be a shape equivalence.

Question 3. Can one give a reasonable sufficient condition for a shape fibration to be an hereditary shape equivalence? (For instance, let $p : E \to B$ be a shape fibration between compact metric spaces such that for all $b \in B$ there is an $e \in p^{-1}(b)$ such that $p : (E,e) \to (B,b)$ is a shape equivalence. Then is p an hereditary shape equivalence?)

V. OBSTRUCTION THEORY FOR SHAPE FIBRATIONS WITH APPLICATIONS TO CELL-LIKE MAPS

In this section we shall discuss an obstruction theory for shape fibrations developed by Allen Matsumoto, [14].

In Section 2 we remarked that approximate fibrations do not include the notion of a pull-back, and that, in particular, the restriction of an approximate fibration need not be an approximate fibration. There are obvious examples of approximate fibrations $p : E \to B$ and subsets $B' \subset B$ such that $p|p^{-1}(B') : p^{-1}(B') \to B'$ fails to be an approximate fibration because neither $p^{-1}(B')$ or B' are ANR's. The next example shows that restrictions may also fail to satisfy the AHLP.

Example 2. (A. Matsumoto) Let A be the limit arc of the Warsaw circle W. Embed W in $S^1 \times S^1$ so that there is a homeomorphism $h : (S^1 \times S^1)/W \to S^1 \times (S^1/x_0)$ for some point $x_0 \in S^1$. Then the map $p : S^1 \times S^1 \to S^1 \times S^1$ obtained by collapsing A to a point is an approximate fibration. However, the restriction $p|W : W \to p(W) \approx S^1$ does not satisfy the AHLP.

A restriction of a shape fibration is always a shape fibration (see Prop. 4, [16]), hence one has hope that the pull-back of a shape fibration will again be a shape fibration. (Note that the above fact and Corollary 1 imply that a restriction of an approximate fibration to ANR's is again a shape fibration.)

Let $p : E \to B$ and $f : B' \to B$ be maps. Recall that the map $\pi : f^*(E) \to B'$, $f^*(E) = \{(x,y) \in B' \times E : f(x) = p(y)\}$, $\pi(b,e) = b$, is called the *pull-back* of f. The next theorem asserts that the pull-back of a shape fibration is a shape fibration.

THEOREM 10. (A. Matsumoto) Let E,B,B' be compact metric spaces, $p : E \to B$ a shape fibration, and $f : B' \to B$ a map. Then there is a level map $\underline{\pi} : \underline{E}' \to \underline{B}'$ of ANR sequences which induces $\pi : f^*(E) \to B'$ and which has the HLP. Hence, $\pi : f^*(E) \to B$ is a shape fibration.

The classical lifting problem in homotopy theory is to determine when a map $f : X \to B$ into the base space of a fibration $p : E \to B$ has a lift into E, i.e., a map $F : X \to E$ such that $pF = f$. The cross-section problem is to determine when a fibration $p : E \to B$ has a cross-section, i.e., a map $s : B \to E$ such that $ps = 1_B$. These two problems are easily seen to be equivalent

The lifting problem includes the cross-section problem since a lift of the identity map of B is a cross-section. Conversely, if $s : X \to f^*(E)$ is a cross-section then $gs : X \to E$, where $g(x,e) = e$, is a lift of f. Obstruction theory gives a reasonable solution to these problems. In particular, when X (for liftings) or (for cross-sections) is a CW-complex then the obstruction to obtaining the desired lift over the $(n + 1)$-skeleton, once it is defined on the n-skeleton, lies in the $(n+1)$-singular cohomology group with coefficients in the n-th homotopy group of the fiber.

In the case of shape fibrations (in particular, for approximate fibrations), the examples in Chapter 2 readily show that one cannot hope to obtain exact liftings and cross-sections. The appropriate analogue in the shape category is the notion of an approximate lift. Matsumoto solves the approximate lifting problem in the shape category by using inverse sequences of ANR's, because the basic methods used for fibrations are more adaptable to shape fibrations than to approximate fibrations. In particular, the pull-back assured by Theorem 10 is of fundamental importance in the development. Once lifts are obtained in terms of inducing maps of ANR-sequences, then approximate lifts result for the limit map by means of Theorem 11 (stated below). An obstruction cohomology class is obtained by using Čech cohomology with coefficients in the shape groups of the fiber which yields a shape theoretic result entirely analogous to the homotopy result mentioned in the preceeding paragraph.

The theory requires that X (for liftings) of B (for cross-sections) be a *shape complex*. This notion is due to T. McMillan. We shall not give the precise definition of a shape complex here. However loosely speaking a shape complex $K = \lim (K_i, h_{ij})$ is an inverse limit of CW-complexes K_i and is defined inductively to be the union of "n-skeleta" K^n where $K^n = \left(K_i^n, h_{ij}\, K_j^n\right)$ is the inverse limit of a certain type sequence of CW-complexes $K_i^n \subset K_i$.

It turns out trivially that every *CW*-complex is a shape complex and there are many other examples of shape complexes as well (for instance the p-adic solenoids).

Let us now define the appropriate lifting properties.

Definition 6. Let $p : E \to B$ be a shape fibration and $f : K \to B$ be a shape fibration and $f : K \to B$ be a map of a shape complex K. Suppose $\underline{f} : \underline{K} \to \underline{B}$ and $\underline{p} : \underline{E} \to \underline{B}$ are level maps of ANR sequences inducing f and p, respectively, where $\underline{K} = (K_i, h_{ij})$ is a "defining" sequence for K, $\underline{B} = (B_i, r_{ij})$, $\underline{E} = (E_i, q_{ij})$. We say that $\underline{f}$ has an n-lift if there exist maps $f_i^n : K_i^n \to E_i$ satisfying $p_i f_i^n \simeq f_i | K_i^n$ and $f_i^n h_{ij} | K_j^n \simeq q_{ij} f_j^n$. The level map $\underline{f}^n = (f_i^n)$ is called an n-lift of $\underline{f}$.

Definition 7. Suppose that $p : E \to B$ is an approximate fibration. For $\varepsilon > 0$, we say that $f : K \to B$ has an (n, ε)-lift if there exists a map $f^{n,\varepsilon} : K^n \to E$ such that $pf^{n,\varepsilon}$ and $f|K^n$ are ε-close. We call $f^{n,\varepsilon}$ an (n, ε)-lift of f. If f has an (n, ε)-lift for every $\varepsilon > 0$, we say that f *has approximate* n-lifts.

THEOREM 11. (A. Matsumoto) Suppose that $p : E \to B$ is an approximate fibration induced by $\underline{p} : \underline{E} \to \underline{B}$ and that $f : K \to B$ is induced by $\underline{f} : \underline{K} \to \underline{B}$. If $\underline{f}$ has an n-lift, then f has approximate n-lifts.

THEOREM 12. (A. Matsumoto) Suppose that $p : E \to B$ is a shape fibration between compact metric spaces and $f : K \to B$ is a map. Then, given level maps $\underline{p} : \underline{E} \to \underline{B}$ and $\underline{f} : \underline{K} \to \underline{B}$ inducing p and f and given an n-lift $\underline{f}^n$ of $\underline{f}$, there exists a cohomology class $\Omega^{n+1} = \Omega^{n+1}(\underline{f}^n) \in \check{H}^{n+1}(K^{n+1}; \check{\pi}_n(F))$ satisfying:

(1) If $\check{\pi}_{n+1}(F) = 0$ and $\Omega^{n+1} = 0$, then there is an $(n + 1)$-lift $\underline{f}^{n+1}$ of $\underline{f}$ extending $\underline{f}^n | \underline{K}^{n-1}$ up to homotopy, and

(2) If $p : E \to B$ is an approximate fibration and $\Omega^{n+1} = 0$, then f has approximate $(n + 1)$-lifts.

Definition 8. An *ε-cross-section* of a map $p : E \to B$ is a map $s : B \to E$ such that ps is ε-close to 1_B. We say that p has *approximate cross-sections* if p has an ε-cross section for each $\varepsilon > 0$.

COROLLARY 7. *If $p : E \to B$ is an approximate fibration and $\check{\pi}_m(F) = 0$ for $0 \leq m \leq n$, then f has approximate $(n + 1)$-lifts for every $f : K \to B$.*

COROLLARY 8. *If $p : E \to B$ is a cell-like map between compact* ANR'*s, then every map $f : K \to B$ of a shape complex has n-lifts for $0 \leq n \leq$* dim *K*.

COROLLARY 9. *If $p : E \to B$ is a cell-like map between compact* ANR'*s, then p has approximate cross-sections.*

COROLLARY 10. *If $f : M^m \to N^n$, $2n + 1 \leq m$, is a cell-like map from an m-manifold M to a n-manifold N and $\varepsilon > 0$ is given, then there exists an embedding $g : N \to M$ such that $d(fg, 1_N) < \varepsilon$.*

Note that Corollary 10 is a trivial range analogue of Siebenmann's *CE*-approximation theorem stated in Section 1. We wonder if such a result holds between the trivial range and co-dimension *0*. (Observe that Corollary 9 does imply such a result when 'embedding' is replaced by 'map' even in dimension 4 and in the case (if possible) that the dimension of *M* is greater than the dimension of *N*.)

Question 4. If $f : M^m \to N^n$, $m \leq n$ is a cell-like map between manifolds (possibly with boundary), then for every $\varepsilon > 0$ is there an embedding $g : N \to M$ such that $d(fg, 1_N) < \varepsilon$?

REFERENCES

1. S. Armentrout, Cellular decompositions of 3-manifolds that yield 3-manifolds, Mem. Amer. Math. Soc., 107 (1971)
2. S. Armentrout and T. M. Price, Decompositions into compact sets with UV properties, Trans. Amer. Math. Soc., 141 (1969), 433-442.
3. K. Borsuk, Concerning homotopy properties of compacts, Fund. Math. 62 (1968), 223-254.
4. T. A. Chapman, Cell-like mappings of Hilbert cube manifolds. Solution of a handle problem, General Topology Appl. 5 (1975), 123-145.
5. ______________, Cell-like mappings, Lecture Notes Vol. 428, Springer-Verlag, Berlin and New York (1974), 230-240.
6. D. Coram and P. Duvall, Approximate fibrations, Rocky Mountain J. Math., 7 (No. 2) (1977), 275-288.
7. ______________, Approximate fibrations and a movability condition for maps, Pacific J. Math. 72 (1977), 41-56.
8. D. A. Edwards and H. M. Hastings, Čech and Steenrod Homotopy Theories with Applications to Geometric Topology, Springer Lecture Notes, Vol. 542, Berlin, 1976.
9. R. D. Edwards, Compact ANR's are Q-manifold factors, (to appear).
10. R. D. Edwards, The topological invariance of simple homotopy type, (to appear).
11. R. C. Lacher, Cell-like spaces, Proc. Amer. Math. Soc., 20 (1969), 598-602.
12. ______________, Cell-like mappings and their generalizations, Bull. Amer. Math. Soc., 83 (1977) 495-552.
13. ______________, Cell-like mappings, I, Pacific J. Math., 30 (1969), 717-731.
14. Allan Matsumoto, Obstruction theory for shape fibrations with applications to cell-like maps, Ph.D. dissertation (1977), University of Utah.
15. Sibe Mardešić, Comparing fibres in a shape fibration, (to appear).
16. Sibe Mardešić and T. B. Rushing, Shape fibrations, I, General Topology and Appl., (to appear).
17. ____________________________, Shape fibrations, II, Rocky Mountain J. Math., (to appear).
18. S. Mardešić and J. Segal, Shapes of compacta and ANR-systems, Fund. Math., 72 (1971), 41-59.
19. T. C. McMillan, Cell-like maps which are shape fibrations, Ph.D. dissertation (1977), University of Utah.
20. L. C. Siebenmann, Approximating cellular maps by homeomorphisms, Topology 11 (1972), 271-294.
21. J. L. Taylor, A counter-example in shape theory, Bull. Amer. Math. Soc., 81 (1975), 629-632.
22. J. E. West, Compact ANR's have finite type, Bull. Amer. Math. Soc., 81 (1975), 163-165.

A WEAK FLATTENING CRITERION FOR COMPACTA IN 4-SPACE

T. B. Rushing[1]

Department of Mathematics
University of Utah
Salt Lake City, Utah

Gerard A. Venema

Department of Mathematics
Institute for Advanced Study
Princeton, New Jersey

Suppose that $X, Y \subset E^4$ are compacta of shape dimension ≤ 1. The main result of this paper gives an embedding condition on X and Y which insures that $Sh(X) = Sh(Y)$ is equivalent to $E^4 - X \approx E^4 - Y$. As a consequence we obtain: (1) a cellularity criterion for cell-like subsets of E^4; (2) a class of 1-dimensional ANR's in E^4 whose homotopy types are characterized by the homeomorphism types of their complements; and (3) a condition which insures that certain homeomorphic open subsets of E^4 are PL homeomorphic.

I. INTRODUCTION

In this note we use the following property to study the complements of compacta embedded in Euclidean 4-space E^4.

DEFINITION. A compactum $X \subset E^4$ had the *DISK PUSHING PROPERTY* if for every PL disk $\Delta^2 \subset E^4$ with $\partial\Delta^2 \cap X = \emptyset$ and for every $\varepsilon > 0$ there exists a PL homeomorphism $h : E^4 \to E^4$ such that $h(\Delta^2) \cap X = \emptyset$ and $h = id$ outside the ε-neighborhood of X.

[1] *The first author was supported by NSF grant MCS 7703978.*

ISBN 0-12-158860-2

Our main result is a weak flattening theorem for compacta in 4-space. As usual, the notation $Sh(X) = Sh(Y)$ means that X and Y have the same shape (in the sense of [1, Chapter VII]). See Chapter II for a definition of the shape dimension of $X (Sd(X))$.

THEOREM 1. Suppose $X, Y \subset E^4$ are compacta such that $Sd(X) \leq 1$, $Sd(Y) \leq 1$ and each has the disk pushing property. Then $Sh(X) = Sh(Y)$ if and only if $E^4 - X$ is homeomorphic with $E^4 - Y$.

Our theorem can be viewed as a generalization to dimension 4 of the finite dimensional complement theorems of [5], [2], [3], [4] and [8]. However the hypothesis we use is stronger so that the 2-dimensional engulfing in those proofs can be avoided. In Chapter II we show that if a compactum $X \subset E^4$ has shape dimension ≤ 2, then 1-dimensional polyhedra can always be pushed off X. Thus it is only necessary to assume the condition about 2-disks. It would be of substantial interest to prove or disprove Theorem 1 with a weaker condition in the hypothesis (e.g. the Inessential Loops Condition [8]).

One corollary of Theorem 1 is that the disk pushing property is a criterion for the cellularity of cell-like continua in 4-space. (A continuum X is *CELL-LIKE* if $Sh(X) = Sh(\text{cell})$.)

COROLLARY 1. If $X \subset E^4$ is a cell-like continuum, then X is cellular if and only if X has the disk pushing property.

The simplest way to prove Corollary 1 would be to use Chapter II of this paper to find small PL neighborhoods of X with 1-spines (thus avoiding the 2-dimensional engulfing of [5]) and then to use the codimension 3 engulfing arguments of [5] to complete the proof.

Every compact 1-dimensional ANR can be embedded in E^4 in such a way that it has the disk pushing property [2, Proposition 3.4]. A second corollary of Theorem 1 is that the homeomorphism type of the complement of the ANR embedded in this way determines the homotopy type of the ANR.

COROLLARY 2. Suppose $X, Y \subset E^4$ *are compact 1-dimensional ANR's having the disk pushing property. Then* X *and* Y *have the same homotopy type if and only if* $E^4 - X \approx E^4 - Y$.

The final corollary which we mention is a version of the Hauptvermutung for certain open PL 4-manifolds. If $Sh(X) = Sh(Y)$, then the homeomorphism from $E^4 - X$ to $E^4 - Y$ constructed in the proof of Theorem 1 is a PL homeomorphism. On the other hand, in proving the converse we need only to assume the existence of a topological homeomorphism from $E^4 - X$ to $E^4 - Y$ in order to conclude that $Sh(X) = Sh(Y)$. Thus the following result is a corollary of the proof of Theorem 1.

COROLLARY 3. Suppose $X, Y \subset E^4$ *are compacta having shape dimension* ≤ 1 *and having the disk pushing property. Let* $M = E^4 - X$ *and* $N = E^4 - Y$. *Then* M *and* N *are topologically homeomorphic if and only if* M *and* N *are PL homeomorphic.*

II. FINDING NEIGHBORHOODS WITH 1-SPINES

We begin by making some definitions. If M is a PL n-manifold, a polyhedron $K \subset M$ is called a *SPINE* of M if $M \searrow K$. In particular, if M is a regular neighborhood of K, then K is a spine of M. If $\varepsilon > 0$ and $X \subset E^4$, then $N_\varepsilon(X) = \{p \in E^4 | \text{dist}(p,X) < \varepsilon\}$. For any compact metric space, the *SHAPE DIMENSION* of X is defined by $Sd(X) = \min\{\dim Y | Y \text{ is a compact metric space with } Sh(X) = Sh(Y)\}$. The key property of shape dimension which we need is this: If $X \subset E^n$ is a compactum with $Sd(X) = k$, then for every neighborhood U of X there exists a neighborhood V of X in U and a k-dimensional polyhedron $K \subset U$ such that the inclusion $V \hookrightarrow U$ is homotopic in U to a map of V into K (see the proof of Lemma 1 in [8]).

LEMMA. Suppose $X \subset E^4$ *is a compactum with shape dimension* ≤ 2 *and* $L^1 \subset E^4$ *is a compact, 1-dimensional polyhedron. They for every* $\varepsilon > 0$ *there exists a PL homeomorphism* $h : E^4 \to E^4$ *such that* $h(L^1) \cap X = \emptyset$ *and* $h|E^4 - N_\varepsilon(X) = id.$

Proof. The existence of the homeomorphism h will follow from Stalling's engulfing theorem [7] provided that $\pi_0(U, U - X) = 0$ and $\pi_1(U, U - X) = 0$ for every neighborhood U of X. The fact that $\pi_0(U, U - X) = 0$ is obvious, so we concentrate on $\pi_1(U, U - X)$. Let $\Delta^1 = [0,1]$.

Let $U \supset X$ and $f : (\Delta^1, \partial\Delta^1) \to (U, U - X)$ be given. Since $Sd(X) \leq 2$, there exists a PL manifold neighborhood V of X and a polyhedron $K \subset U$ with $\dim K \leq 2$ such that the inclusion $V \hookrightarrow U - f(\partial\Delta^1)$ is homotopic in $U - f(\partial\Delta^1)$ to a map of V into K. By general position we may assume that f is transverse to ∂V and that $f(\Delta^1) \cap K = \emptyset$. We also assume that U is connected by concentrating only on the component of U containing $f(\Delta^1)$ if necessary.

Let $\tilde{U}$ denote the universal covering space of U and let $p : \tilde{U} \to U$ denote the projection. Let $\overline{V} = p^{-1}(V)$ and let $\overline{f} : (\Delta^1, \partial\Delta^1) \to (\tilde{U}, \tilde{U} - p^{-1}(X))$ denote a lift of f. Since the diagram

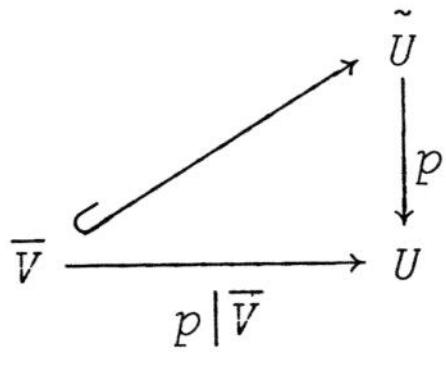

commutes, the homotopy lifting property can be used to lift the homotopy of V downstairs to a homotopy $H : \overline{V} \times I \to \tilde{U}$ such that H_0 = inclusion and $H_1(V) \subset p^{-1}(K)$. Throw H and $\overline{f}$ into general position.

We claim that $\partial\overline{V}$ does not separate $\overline{f}(0)$ from $\overline{f}(1)$ in $\tilde{U}$. If it did, then there would exist a component B of $\partial\overline{V}$ such that

$\overline{f}(\Delta^1)$ crosses B an odd number of times. But then $(H|B \times I)^{-1}(\overline{f}(\Delta^1))$ is a compact 1-manifold with an odd number of endpoints. Since no such 1-manifold exists, we can join $\overline{f}(0)$ and $\overline{f}(1)$ with an arc α which lies in $\tilde{U} - \partial\tilde{\overline{V}}$ and hence in $\tilde{U} - p^{-1}(X)$.

Since $\tilde{U}$ is simply connected, $\overline{f}(\Delta^1) \cup \alpha$ is null homotopic in $\tilde{U}$. Projecting that homotopy down, we see that f is homotopic (rel $\partial\Delta^1$) to a map $f' : \Delta^1 \to U - X$. Thus $\pi_1(U, U - X) = 0$.

THEOREM 2. Suppose $X \subset E^4$ is compact. Then the following are equivalent.

A. X has arbitrarily small PL neighborhoods with 1-spines.

B. $Sd(X) \leq 1$ and X has the disk pushing property.

Proof. ($A \Rightarrow B$). The fact that X has shape dimension ≤ 1 follows from [6, Theorem 3.1] and the fact that X has the disk pushing property follows from general position.

($B \Rightarrow A$). Given $\varepsilon > 0$, find a compact PL manifold neighborhood M of X such that $M \subset N_\varepsilon(X)$. Let M^1 and M^2 denote the 1- and 2-skeleta of M respectively and let M^1_* denote the dual 1-skeleton. We find a PL homeomorphism $h : E^4 \to E^4$ such that $h(M^2) \cap X = \emptyset$ and $h|E^4 - \text{int } M = id$. Then $X \subset N$ for some regular neighborhood N of M^1_* in $N_\varepsilon(X)$, so the proof will be complete.

By the Lemma above, there exists a PL homeomorphism $h_1 : E^4 \to E^4$ such that $h_1(M^1) \cap X = \emptyset$ and $h_1|E^4 - \text{int } M = id$. We next use the disk pushing property to push the 2-simplices of $M^2 - M^1$ off X one at a time. After one of these 2-simplices has been pushed off X we move points only in such a small neighborhood of X that the 2-simplex is never moved again. The homeomorphism h is the composition of h_1 followed by these pushes.

III. PROOF OF THEOREM 1.

Let $X, Y \subset E^4$ be compacta such that $Sd(X) \leq 1$, $Sd(Y) \leq 1$ and each has the disk pushing property. By Theorem 2, X and Y have arbitrarily small PL neighborhoods with 1-spines and thus, in the terminology of [2], they are in standard position.

Suppose $Sh(X) = Sh(Y)$. Since X and Y are in standard position, the proof of part (a) of Theorem 1 in [2] (see [2] pp. 273-275) shows that $E^4 - X \approx E^4 - Y$. The property of X and Y needed there is that X and Y each have small neighborhoods with spines having dimension in the trivial range with respect to the dimension of the ambient space. This means that any homotopy of a neighborhood of X or Y can be approximated with an ambient isotopy.

Now Suppose that $E^4 - X \approx E^4 - Y$. There exist metric compacta X' and Y' such that $Sh(X) = Sh(X')$, $Sh(Y) = Sh(Y')$, $\dim X' \leq 1$ and $\dim Y' \leq 1$. Embed X' and Y' in E^4 as pseudo-polyhedral sets [3, Theorem 3.3]. By the first part of Theorem 1, $E^4 - X \approx E^4 - X'$ and $E^4 - Y \approx E^4 - Y'$ and thus $E^4 - X' \approx E^4 - Y'$. But now [3, Theorem 1.1] implies that $Sh(X') = Sh(Y')$.

REFERENCES

1. K. Borsuk, Theory of Shape, Monografie Matematyczne, Tom 59, Polish Scientific Publishers, Warsaw, 1975.
2. T. A. Chapman, Shapes of finite-dimensional compacta, Fund. Math. 76 (1972), 261-276.
3. R. Geoghegan and R. R. Summerhill, Concerning the shapes of finite-dimensional compacta, Trans. Amer. Math. Soc. 179, (1973), 281-292.
4. J. G. Hollingsworth and T. B. Rushing, Embeddings of shape classes of compacta in the trivial range, Pacific J. Math. 60 (2), (1975), 103-110.
5. D. R. McMillan, Jr., A criterion for cellularity in a manifold, Ann. of Math. 79 (1964), 327-337.
6. S. Nowak, Some properties of fundamental dimension, Fund. Math. 85 (1974), 211-227.
7. J. R. Stallings, The piecewise linear structure of Euclidean space, Proc. Cambridge Philos. Soc. 58 (1962), 481-488.
8. G. A. Venema, Embeddings of compacta with shape dimension in the trivial range, Proc. Amer. Math. Soc. 55 (1976), 443-448.

BASED-FREE ACTIONS OF FINITE GROUPS ON HILBERT CUBES WITH ABSOLUTE RETRACT ORBIT SPACES ARE CONJUGATE[1]

J. E. West

Department of Mathematics
Cornell University
Ithaca, New York

R. Y. -T. Wong

Department of Mathematics
University of California at Santa Barbara
Santa Barbara, California

I. INTRODUCTION

Every finite group, G, can act on the Hilbert Cube, Q, semi-freely with a single fixed point. Since Q has the fixed point property, this must be the simplest sort of action, which we term *based-free*. A particularly pretty construction of a based-free G-action, σ, on a Hilbert cube is the following, and we term any topologically conjugate one *standard*. (See [2].) Let G act on

[1]*Partially supported by the N.S.F.*

ISBN 0-12-158860-2

itself by left translation and extend this in the natural way to the cone, $C(G)$, on G. Let Q_G be the product of (countably) infinitely many copies of $C(G)$ and take for σ the diagonal action. R. D. Anderson proved that Q_G is a Hilbert cube [1] (see also [7], [10], and [4]) and posed in the mid-sixties the problem of classifying based-free G-actions on Q. Note that if we identify $C(\mathbb{Z}_2)$ with $[-1,1]$, then $Q_{\mathbb{Z}_2}$ is a standard model for Q and σ is the antipodal map. The above construction works for arbitrary compact Lie groups, providing a standard based-free tame G-action on Q for each compact Lie group, G[2]. (We say that a based-free action is *tame* if the space of free orbits is a Q-manifold.) One of the sources of our persistent interest in these actions is that because Q -{point} is contractible, they are one-point compactifications of universal G-bundles and the classification of these G-actions revolves around the proper homotopy types of the derived Q-manifolds classifying spaces, with that of the standard actions being the simplest possible. (See [2] for a more complete discussion.)

Previously established, for based-free tame G-actions, ρ, on Q, G compact, Lie, are

A. (Wong [12]) *If $G = \mathbb{Z}_p$, p prime, then ρ is standard if and only if the fixed point has a basis of contractible, invariant neighborhoods.*

B. (West [11]) *If $G = \mathbb{Z}_2$, τ is a G-action on a Peano continuum X with nowhere dense fixed point set, and S is the collection of τ-invariant closed subsets of X, then $2^X/S$ is a Hilbert cube, and the induced action is (based-free and) standard.*

(B is motivated by D below.)

C. ((1) Edwards-Hastings [6], (2) Berstein-West [2]).

If $M(\rho)$ is the space of free orbits of ρ, then

(1) for finite G, ρ is standard if (and only if) the end $E(\rho)$,

of $M(\rho)$ is homotopy equivalent to $E(\sigma)$ in pro-Ho(Top)

(2) ρ *is standard if (and only if)*

(i) $E(\rho)$ *is movable, or*

(ii) $M(\rho)$ *is proper homotopy equivalent to* $M(\sigma)$, *or*

(iii) $M(\rho)$ *admits a proper deformation to infinity, i.e., a proper map* $f : M(\rho)\times[0,1) \to M(\rho)$ *with* $f_0 = id$.

(In C, the problem is precisely that a sufficiently general "Whitehead Theorem" is not available in the "Proper" Category. See [6].)

D. (Berstein-West [2]). ρ *is standard if (and only if) it splits as the diagonal action on a countably infinite product of finite-dimensional absolute retracts, each equipped with a based-free G-action.*

E. (Hastings, Berstein-West [2]). *There exists a non-standard based-free tame G-action on Q if and only if there exists an inverse system*

$$\begin{array}{ccccc} \cdots \to (E_i,E) & \xrightarrow{\tilde{f}_i} & (E_{i-1},E) \to \cdots \\ \big\downarrow /G & & \big\downarrow /G \\ \cdots \to (B_i,B) & \xrightarrow{f_i} & (B_{i-1},B) \to \cdots \end{array}$$

of free G-actions on CW-pairs such that

(i) $E \to B$ *is a universal G-bundle,*

(ii) (B_i,B) *is a relatively finite CW-complex,*

(iii) $f_i|B = id$,

(iv) $\tilde{f}_i$ *is null-homotopic, and*

(v) $f_1 \circ \cdots \circ f_i$ *is always essential as a map of pairs.*

F. *Nobody has an example of a non-standard based-free tame action.*

Our main results are

1. *If G is finite, then ρ is standard if and only if the orbit space Q/ρ is an absolute retract.*

2. *If G is an n-torus, then ρ is standard if and only if Q/ρ is an absolute retract.*

(Our methods do not extend to groups with infinitely many non-zero homotopy groups.) The difficulty in constructing non-standard actions is illustrated by D and E above and by the following, which was remarked to us by R. Geoghegan in a discussion of 1. and 2.

3. *(Corollary) If G is finite or an n-torus, then ρ is non-standard if and only if Q/ρ fails to have the homotopy type of a CW-complex.*

II. NOTATIONAL APPARATUS

We set up our notation for finite G; the proof in the toral case is virtually identical. Basic Q-manifold results are in Chapman's excellent monograph [4].

Let ρ be a based-free G-action on Q with fixed point b and absolute retract orbit space Q/ρ. Because Q is homogeneous, we are at liberty to choose our favorite model for (Q,b), which is $(\prod_{i=1}^{\infty}[0,1]_i, (0,0,\ldots))$. Let $p : Q \to Q/\rho$ be the orbit map with $Q_0 = Q-\{b\}$ and $M = p(Q_0)$. We use b to denote $p(b)$. As $p \mid Q_0$ is a covering projection (indeed, a universal G-bundle, for Q_0 is convex), M is a Q-manifold, hence $M=K\times Q$ for some locally finite simplicial complex K, which we identify with $K\times\{0\} \subset K\times Q = M = p(Q_0)$. Hence, $Q_0 = \tilde{K}\times Q$. We denote by K^b the set $K \cup \{b\}$ and generally write L^b for $L \cup \{b\}$ whenever L is a subcomplex of K. For L infinite, L^b is the one-point compactification. To describe the end of K we choose a sequence $K = K_0 \supset K_1 \supset K_2 \supset \ldots$ of connected subcomplexes of K such that $K_{i+1} \subset \text{int } K_i$, $K - K_i$ is always finite, and $\cap K_i = \emptyset$. Clearly we may also arrange a sequence of convex neighborhoods C_i of b in Q such that $p^{-1}(K_i \text{x} Q) \subset C_i \subset p^{-1}(K_{i-1} \text{x} Q)$. From this it is immediate that the boundary of $p^{-1}(K_i)$ in $p^{-1}(K_{i-1})$ may always be assumed connected.

If $\varphi : X \to K^b$ is a mapping, we denote $\max\{n | \varphi(X) \subset K_n^b\}$, if it exists, by $n(\varphi, X)$.

We adopt more or less standard notation from p.ℓ. topology, but do not distinguish $|L|$ from L. Our simplices are always closed unless otherwise indicated. We use $st(X,L)$ for $\cup\{\sigma \in L | \sigma \cap X \neq \emptyset\}$ and $\mathring{st}(X,L)$ for $\cup\{\mathring{\sigma} | \sigma \in L,\ \sigma \cap X \neq \emptyset\}$ for subsets X of L. We also need explicity parameterized "regular" neighborhoods of the interiors of simplices and of open cells of the form $\mathring{\Delta}\text{x}\mathring{I}$, where $\Delta \in K$. We call these *"pinched" neighborhoods* and require them to satisfy the following: If $\sigma \in L$, a pinched neighborhood of σ in L, $N(\sigma,L)$, is a copy of $st(\mathring{\sigma},L'')$, where L'' is the second barycentric subdivision of L, together with the strong deformation retraction, c, to σ given by projection along the join lines from the link of σ in L to σ. This gives $N(\sigma,L)$ the structure of the reduced mapping cylinder of $c' = c|bd\ N(\sigma,L)$, reduced "under" $\partial\sigma$, i.e., of $bd\ N \times [0,1] \cup \sigma/\sim$ where $(x,1) \sim c(x)$ and $(y,t) \sim y$ for each $x \in bd\ N$ and $y \in \partial\sigma$. We denote points of N by $[x,t]$. The deformation retraction of N given by this mapping cylinder structure, being just projection along the join lines, preserves $N \cap \tau$ for each $\tau \in L$. By "smaller" or "larger" pinched neighborhoods, we shall mean just expansions or (slight) contractions of N along these join lines.

We shall also be dealing with $\Delta\text{x}I$ in $K\text{x}I$ for $\Delta \in K$ and a triangulation J of $K\text{x}[0,1)$ which triangulates each $\Delta\text{x}[0,1)$. In this context, by a *pinched neighborhood of* $\Delta\text{x}I$ *in* $K\text{x}I$ we shall mean $\Delta\text{x}I \cup \mathring{st}(\mathring{\Delta}\text{x}\mathring{I},J'')$ with the analogous mapping cylinder parameterization, which also preserves intersections with simplices of J.

III. LEMMAS

Unfortunately, we use nine, but most are trivial or easy.

LEMMA 1. ρ *is standard if and only if there is a strong deformation retraction*

$$f : K^b \times I \to K^b \quad \text{with} \quad f^{-1}(b) = K\times\{1\} \cup \{b\}\times I .$$

(This is clearly equivalent to (C2 iii) above.)

LEMMA 2. Denoting by j *the inclusions* $K_i \to K_{i-1}$*, we have*

(a) *a commutative diagram*

$$\begin{array}{ccccccccc} \cdots \to & \pi_1 K_{i+1} & \xrightarrow{j_*} & & \pi_1(K_i) & \longrightarrow & & \pi_1(K_{i-1}) & \to \cdots \\ & & \searrow & \nearrow & & \searrow & \nearrow & & \\ \cdots & = & = & G & = & = & G & = & \cdots \end{array}$$

(b) $j_* : \pi_n K_i \to \pi_n K_{i-1}$ *the zero homomorphism for* $n > 1$*, and*

(c) *each* $\pi_1 K_i$ *finitely generated.*

(This follows immediately from the fact that $p|Q_0$ is a covering map, the convexity of the C_i-$\{b\}$'s and the definition of the K_i's.)

LEMMA 3. We may choose K *and the* K_i*'s so that each* $j : K_i \to K_{i-1}$ *is an isomorphism on* π_1 *and so that* $\pi_2 K_i = 0$.

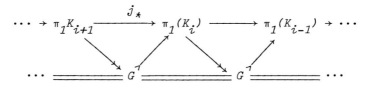

Proof. This is essentially standard. Let $Q_0 = \tilde{K}\times Q$ be induced by K and p, where $\tilde{K}$ is the universal cover of K, and denote by $\tilde{K}_i$ $p^{-1}(K_i)$. From the convexity of the sets $C_i - \{b\}$ we discover that the kernel of $j_* : \pi_1 K_i \to \pi_1 K_{i-1}$ is the normal subgroup $p_*(\pi_1 \tilde{K}_i)$, which is finitely generated. We attach a finite collection of discs to each $\tilde{K}_i$ to kill this kernel, obtaining complexes L_i and extensions $k_i : L_i \to K_{i-1}$ of the j's which are injective on π_1. Borrowing a few factors of $[0,1]$ from Q if need be we arrange that k_i is a $p.\ell.$ embedding with image missing the sites of attachment of L_{i-1}. Setting N_i to be the relative mapping cylinder of $k_i : (L_i, K_i) \to (K_{i-1}, K_i)$, i.e., the true mapping cylinder with $K_i \times I$ collapsed to the target K_i, we obtain $P = \bigcup_{i \geq 1} N_i$, all as simplicial compexes if sufficient care is taken. Now P collapses to K by a collection of mutually separated finite collapses and is thus (infinite) simple homotopy equivalent to K, so $P \times Q$ is homeomorphic to $K \times Q$ and we may replace K by P with a suitable subsequence of $\{P_i = \bigcup_{j \geq 1} N_j\}_i$ replacing $\{K_i\}_i$. This shows we may assume j a π_1 isomorphism. Now $\pi_2 K_i$ is finitely generated, so we may repeat the process to produce K_i's with trivial π_2.

LEMMA 4. Each map $\varphi : \partial\Delta^k \to K_i^b$, $k = 1,2,3$, *extends to a map* $\psi : \Delta^k \to K_i^b$.

Proof. Let T be a triangulation of $\partial\Delta^k - \varphi^{-1}(b)$, and choose $v \in \text{int}(\Delta^k)$. Let $S = v * t$ be the join triangulation of $(\Delta^k - v * \varphi^{-1}(b)) \cup \{v\}$. Use the path-connectivity of the K_j^b's to map $v * w$ to a path from $\varphi(w)$ to $b = \psi(v)$ in $K^b_{n(\varphi,w)}$ for each vertex w of T. Since the pairs (K_m, K_{m+n}) are 2-connected, this map extends to one of S into K_i^b sending $v * \sigma$ into $K^b_{n(\varphi,\sigma)}$ for each $\sigma \in T$. Let ψ extend this to Δ^k by setting $\psi(v * \varphi^{-1}(b)) = b$.

LEMMA 5. For each map $\varphi : (\Delta^k, \partial\Delta^k) \to (K_i^b, K_i)$, $k \geq 3$, *there is a map* $\psi : \Delta^k \to K_{i-1}$ *extending* $\varphi|\partial\Delta^k$ *and a homotopy* Ψ *from* φ *to* ψ *in* K^b_{i-k} *which is stationary on* $\partial\Delta^k$.

Proof. ψ comes from Lemma 2(b); Ψ comes from the same place as Lemma 4 with the difference being that in the induction on the skeleta of T, we only have that $\pi_j K_m \to 0 \in \pi_j K_{m-1}$ for $j \geq 3$.

LEMMA 6. K^b *is an absolute retract.*

Proof. $r : Q/\rho = K\times Q \cup \{b\} \to K^b$ induced by projection is a retraction.

LEMMA 7. *There is a strong deformation retraction* $g : K^b\times I \to K^b$ *of* K^b *to* b *with* $g^{-1}(b) \cap K^3\times I = K^3\times\{1\}$.

Proof. Let $n(X) = n(i,X)$ for $i : X \to K$ the inclusion whenever $X \subset K$. From the path-connectivity of the K_i^b's there is a map $g_v : v\times I \to K^b_{n(v)}$ with $g_v^{-1}(b) = \{(v,1)\}$ for each $v \in K^0$. Now Lemma 2(b) guarantees that the map $g_0 : K^b\times\{0,1\} \cup K^0\times I$ so defined extends by an induction to $g_3 : K^b\times\{0,1\} \cup K^3\times I$ with $g_3^{-1}(b) = K^b\times\{1\}$. Lemma 6 now guarantees an extension g to all of $K^b\times I$.

Now let $\varphi : (\Delta^k\times I, \Delta^k\times\{0\}, \Delta^k\times\{1\}) \to (K^b,K,b)$ be a map, $k \geq 2$; let J be a relatively fine triangulation of $\Delta^k\times[0,1)$; $A = st(\Delta^k\times\{0\} \cup \partial\Delta^k\times[0,1),J)$ or $st(\Delta^k\times\{0\},J) \cup \partial\Delta^k\times[0,1)$, and suppose that $J^2 \cap A \cap \varphi^{-1}(b) = \emptyset$.

LEMMA 8. *There are a map* $\psi : (\Delta^k\times I, \Delta^k\times\{1\}) \to (K^b,b)$ *extending* $\varphi|A$ *with* $J^2 \cap \psi^{-1}(b) = \emptyset$ *and a homotopy* Ψ *from* φ *to* ψ *such that*

(a) Ψ *is stationary on* $A \cup \Delta^k\times\{1\}$ *and on the complement of* $\cup\{N(\sigma^i,J)\,|\,\sigma^i \in J,\ i \leq 2\}$ *and*

(b) $n(\Psi,\sigma\times I) \geq n(\varphi,\sigma)$ *for each* $\sigma \in J$ *(for which* $n(\varphi,\sigma)$ *is defined.)*

Proof. Set $S = J^2 \cap A$. As S is simply-connected, $\varphi|S$ lifts to $\tilde{K}$. Note that each $\tilde{K}_i$ is 2-connected, so we may extend this to a map $\tilde{\lambda} : J^2 \to K$ for which, with $\lambda = p\circ\tilde{\lambda}$, $n(\lambda,\sigma) \geq n(\varphi,\sigma)$ whenever $n(\varphi,\sigma)$ is defined. The map λ, which will be $\psi|J^2$, provides us with sufficient information that when we unsnarl J^1 from b we do not create embarrassing loops in K.

Denote by T^i the set of i-simplices of J not lying in A. For each $v \in T^0$, let ω_v be a path in $K^b_{n(\varphi,v)}$ from $\varphi(v)$ to $\lambda(v)$ (in $K^b_{n(\lambda,v)}$, if $\varphi(v) = b$) and define ψ_0 by

$$\psi_0(x) = \begin{cases} \varphi(x), & \text{if } x \notin \cup\{N(v,J) \mid v \in T^0\} \\ \varphi([y,2t]), & \text{if } x = [y,t] \in N(v,J),\ 0 \le t \le 1/2 \\ \omega_v(2t-1), & \text{if } x = [y,t] \in N(v,J),\ 1/2 \le t \le 1. \end{cases}$$

For each $\tau = \tau^1 \in T^1$, apply Lemma 4 to the mapping $\psi_0 | \tau x\{0\} \cup \lambda o(pjn) | (\partial\tau xI \cup \tau x\{1\})$ to obtain a map $\zeta_\tau : \tau xI \to K^b_{n(\varphi,\tau)}$, and define $\psi_1 : \Delta^k xI \to K^b$ by

$$\psi_1(x) = \begin{cases} \psi_0(x), & \text{if } x \notin \cup\{ N(\tau,J) \mid \tau \in T^1\} \\ \psi_0([y,2t]), & \text{if } x = [y,t] \in N(\tau,J),\ 0 \le t \le 1/2 \\ \zeta_\tau([y,1],2t-1), & \text{if } x = [y,t] \in N(\tau,J), \\ \quad 1/2 \le t \le 1. \end{cases}$$

For each $\sigma = \sigma^2 \in T^2$, use Lemma 4 to procure an extension $\eta_\sigma : \sigma xI \to K^b_{n(\varphi,\sigma)}$ of the map $\psi_1 | \sigma x\{0\} \cup \lambda o(pjn) | (\partial\sigma\ xI \cup \sigma x\{1\})$, and define $\psi : \Delta^k xI \to K^b$ by

$$\psi(x) = \begin{cases} \psi_1(x), & \text{if } x \notin \cup\{ N(\sigma,J) \mid \sigma \in T^2\} \\ \psi_1([y,2t]), & \text{if } x = [y,t] \in N(\sigma,J),\ 0 \le t \le 1/2 \\ \eta_\sigma([y,1],2t-1), & \text{if } x = [y,t] \in N(\sigma,J), \\ & \quad 1/2 \le t \le 1. \end{cases}$$

The required homotopy Ψ has been implicitly demonstrated in the construction of ψ.

Now suppose that J triangulates $\tau x[0,1)$ for each $\tau \in \Delta = \Delta^k$, and that $k \ge 4$. Let $B = st(\Delta x\{0\} \cup (\Delta)^3 x[0,1),J)$, where $(\Delta)^3$ is the 3-skeleton of Δ; let $C = J^2 - \text{int } B$; let $N = \cup\{N(\sigma,J) \mid \sigma \in C\}$, and let $D = (\Delta x[0,1) - \text{int } N) \cup J^2$.

LEMMA 9. D is simply connected.

Proof. Because $\Delta \times [0,1)$ deforms to $\Delta \times [0,1) - \text{int } B$, the latter is simply connected and C, being its 2-skeleton, is, too. N is a regular neighborhood of C, so it, too, is simply connected. From general position and the mapping cylinder structure of regular neighborhoods, $bd\ N$ and $\Delta \times [0,1) - \text{int } N$ are also simply connected. Because N deforms to C preserving intersection with J^2, $N \cap J^2$ is simply connected. Therefore, from van Kampen's Theorem we have $bd\ N \cup (N \cap J^2)$ and thence $D = (\Delta \times [0,1) - \text{int } N) \cup (bd\ N \cup (J^2 \cap N))$ are simply connected.

IV. PROOF OF THE MAIN THEOREM

THEOREM 1. Based-free actions of finite groups on Hilbert cubes are standard if and only if their orbit spaces are absolute retracts.

Proof. Let ρ be a based-free action on Q of the finite group G. We adopt our standing notation assuming the conclusion of Lemma 3. By definition, ρ is standard if and only if conjugate to σ. As Q_G is clearly equivariantly contractible to the infinite vertex, Q_G/σ is locally contractible at b. As Q_G/σ is the one-point compactification of the locally compact absolute neighborhood retract $Q_G/\sigma - \{b\}$, this ensures that it is an absolute neighborhood retract, and as it is contractible, it must be an absolute retract. Therefore the condition is necessary.

To show that it is sufficient, we shall show that if Q/ρ is an absolute retract, then there is a strong deformation retraction of Q/ρ to b moving no point to b until the last instant, that is, we shall show that the condition of Lemma 1 is satisfied, i.e., that Q_0/ρ satisfies (C 2 iii). By hypothesizing that Q/ρ is an absolute retract, we are giving ourselves a strong deformation retraction to b, and we proceed to "unsnarl" it by making three successive inductive adjustments.

By Lemma 1, we need only produce a strong deformation retraction $f : K^b \times I \to K^b$ of K^b to b with $f^{-1}(b) = K^b \times \{1\} \cup \{b\} \times I$. By Lemma 7, there is a strong deformation retraction g of K^b to b with $g^{-1}(b) \cap K^3 \times [0,1) = \emptyset$.

Let J be a triangulation of $K \times [0,1)$ satisfying

(1_J) for each simplex Δ of K, $\Delta \times [0,1)$ is triangulated by some subcomplex of J,

(2_J) $st\ (K \times \{0\} \cup K^3 \times [0,1), J) \cap g^{-1}(b) = \emptyset$, and

(3_J) if $\sigma \in J$ and $\sigma \cap g^{-1}(b) \neq \emptyset$, then $n(g,\sigma) \geq d(\sigma)(d(\sigma)+1)^2/2 + m(\sigma)$,

where $d(\sigma) = \dim st(\sigma, J)$ and $m(\sigma) = \max\{m \mid \sigma \subset K_m \times [0,1)\}$.

Step 1: Preliminary Deformation. Our first induction will free J^2 from b in order that the new deformation, β, will map J^2 in the appropriate way so that we can continue to eliminate the "singularities" skeleton by skeleton. However, we have no control *a priori* over the location in $K \times [0,1)$ of those inessential loops which are at present sent into essential loops of K and will thus have to be deformed out through b and back into K as inessential loops in the construction of β. In particular, as we are going to make an induction on skeleta which results in lowering $m(\sigma)$ at each stage, we are in danger of losing continuity at b (in the last induction, which will remove the singularities introduced by our freeing of J^2). Therefore, we first deform the zone in which these singularities will occur sufficiently far toward b that we do not run into trouble:

By using Lemma 3 inductively on the skeleta of J^2 and employing pinched neighborhoods for extension purposes as in the proof of Lemma 8, we may deform g relative to $K \times \{0,1\}$ to a new deformation α such that

(1_α) $\alpha^{-1}(b) = g^{-1}(b)$

(2_α) if $\sigma \in J^2$ and $\sigma \cap g^{-1}(b) = \emptyset = \sigma \cap K \times \{0\} \cup K^3 \times I$, then

$n(\alpha, N(\sigma, J)) \geq n(g, \sigma) + d(\sigma)(d(\sigma)+1)^2/2$, and

(3_α) for each $\tau \in J$, $n(\alpha, \tau) \geq n(g, \tau)$. (The pinched neighborhoods exployed in the construction of α will of course have to be larger than the $N(\sigma, J)$'s of (2_α). Henceforth, we shall use $N(\sigma, J)$ to refer only to those pinched neighborhoods fixed in (2_α).)

Step 2: Positioning J^2. By applying Lemma 8 judiciously to the successive "skeleta" $K^i \times [0,1]$ of $K \times [0,1]$ and extending to $K^b \times I$ via pinched neighborhoods of $\Delta^i \times I$ and the homotopies Ψ given by Lemma 8, we obtain a sequence $\alpha = \alpha_3, \alpha_4, \ldots$ of strong deformation retractions of K^b to b satisfying the following, with $A = st(K \times \{0\} \cup K^3 \times [0,1), J)$:

(i_α) $\alpha_{i+j} | A \cup K^i \times I = \alpha_i | A \cup K^i \times I$,

(ii_α) $\alpha_i^{-1}(b) \cap (A \cup (J^2 \cap K^i \times [0,1))) = \emptyset$,

(iii_α) for every $\sigma \in J$, $n(\alpha_i, \sigma) \geq n(\alpha, \sigma)$,

(iv_α) for every $\sigma \in J^2$ with $\sigma \cap g^{-1}(b) = \emptyset = \sigma \cap K \times \{0\}$, $n(\alpha_i, N(\sigma, J)) \geq n(g, \sigma) + d(\sigma)(d(\sigma)+1)^2/2$, and

(v_α) $\alpha_i^{-1}(b) - \{b\} \times I - K \times \{1\} - \mathring{s}t(g^{-1}(b), J) \subset \cup\{\mathring{N}(\sigma, J) \mid \sigma \in B\}$, where $B = J^2 - \mathrm{int}(A)$.

(In order to obtain (iv) and (v) from Lemma 8 in this manner, we must in the construction of α_i take the pinched neighborhoods appearing in Lemma 8, which will be neighborhoods in $\Delta^i \times [0,1)$ for $\Delta^i \in K$, to be smaller than the $N(\sigma, J)$'s of (2_α) above and use pinched neighborhoods M of the $\Delta^i \times I$'s in $K \times I$ which are so close to $\Delta^i \times I$ that their parameterized retractions $r : M \to \Delta^i \times I$ have the property that if L is one of the first sort of neighborhoods, then $r^{-1}(L) \subset \cup\{\mathring{N}(\sigma, J)\}$.) We set $\beta = \lim \alpha_i$ and note that

(1_β) $\beta^{-1}(b) \subset g^{-1}(b) \cup \cup\{N(\sigma, J) \mid \sigma \in B\}$,

(2_β) $\beta^{-1}(b) \cap J^2 = \emptyset$,

(3_β) $n(\beta, \sigma) \geq n(\alpha, \sigma)$ for each $\sigma \in J$,

(4_β) $n(\beta, N(\sigma, J)) \geq d(\sigma)(d(\sigma)+1)^2/2 + n(g, \sigma)$ for each $\sigma \in B - \mathring{s}t(g^{-1}(b), J)$.

Step 3: Freeing $\overset{\circ}{st}(g^{-1}(b),J)$. Our second process eliminates the singularities in $\overset{\circ}{st}(g^{-1}(b),J)$ as follows: For each $\sigma = \sigma^3 \in J$ such that $\sigma \cap g^{-1}(b) \neq \emptyset$, we obtain from Lemma 5 a map $\psi_\sigma : \sigma \to K^b_{n(\beta,\sigma)-1}$ extending $\beta|\partial\sigma$ and a homotopy Ψ_σ from $\beta|\sigma$ to ψ_σ in $K^b_{n(\beta,\sigma)-3}$ which is stationary on $\partial\sigma$. Using pinched neighborhoods $M(\sigma,J)$ and Ψ_σ as in the proof of Lemma 8 we obtain an extension β_3 of $\beta|K^b \times I - \cup\{M(\sigma,J)|\sigma \in J^3,\ \sigma \cap g^{-1}(b) \neq \emptyset\}$ with

(i_β) $\beta_3^{-1}(b) \cap J^3 \cap \overset{\circ}{st}(g^{-1}(b),J) = \emptyset$, and

(ii_β) $n(\beta_3,\tau) \geq n(\beta,\tau)-3$ for each $\tau \in J$.

(We have created new singularities, but only in $\overset{\circ}{st}(g^{-1}(b),J)-J^3$.) This process may be carried out inductively over the skeleta of J, and we obtain thus a sequence $\beta = \beta_2, \beta_3, \ldots$ of strong deformation retractions of K^b to b such that

(a_β) $\beta_i|K^b \times I - \overset{\circ}{st}(g^{-1}(b),J) = \beta|K^b \times I - \overset{\circ}{st}(g^{-1}(b),J)$,

(b_β) $\beta_{i+j}|J^i = \beta_i|J^i$, and

(c_β) $n(\beta_i,\tau) \geq n(\beta,\tau) - i(i+1)/2$ for each $\tau \in J$.

We let $\gamma = \lim \beta_i$ and note that

(1_γ) $\gamma^{-1}(b) = \beta^{-1}(b) - \overset{\circ}{st}(g^{-1}(b),J) \subset K \times \{1\} \cup \{b\} \times I \cup (\cup\{\overset{\circ}{N}(\sigma,J)|\sigma \in B\}-J^2)$,

(2_γ) for each $\tau = \tau^i \in J$ with $\tau \cap g^{-1}(b) \neq \emptyset$, $n(\gamma,\tau) \geq d(\tau)(d(\tau)+1)^2/2+m(\tau) - i(i+1)/2 \geq d(\tau)^2(d(\tau)+1)/2 + m(\tau)$, and

(3_γ) for each $\sigma \in B - \overset{\circ}{st}(g^{-1}(b),J)$, $n(\gamma,N(\sigma,\tau)) \geq d(\sigma)^2 (d(\sigma)+1)/2 + n(g,\sigma)$.

((1_γ) follows from (a_β), (b_β), (1_β), (2_β), (i_α), (ii_α), and (v_α);

(2_γ) follows from (b_β), (c_β), (3_β), (3_α), and the choice of J;

(3_γ) follows from (a_β) and (4_β).)

Step 4: Eliminating singularities of γ. This is our last induction. Let $\Delta = \Delta^4 \in K$ and set $T_\Delta = B \cap \Delta \times [0,1)$. Now $N_\Delta = \cup\{N(\sigma,J) \cap \Delta\times[0,1) \mid \sigma \in T\}$ is a regular neighborhood of T_Δ in $J \cap \Delta\times[0,1)$ which misses $\partial\Delta\times[0,1) \cup \Delta\times\{0\}$. Moreover, $\gamma^{-1}(b) \cap \Delta\times[0,1) \subset \text{int } N_\Delta - J^2$. Let $C_\Delta = (\Delta\times[0,1) - \text{int } N_\Delta) \cup (J^2 \cap \Delta\times[0,1))$. By Lemma 9, C_Δ is simply connected; therefore $\gamma|C_\Delta$ lifts to $\tilde{K}$. In fact, we may lift γ on a small neighborhood C'_Δ of C_Δ in $\Delta\times[0,1)$. Using the convexity of the $C_i-\{b\}$'s, we choose an extension $\tilde{\lambda}_\Delta$ of such a lifting to all of $\Delta\times[0,1)$ such that, with $\lambda_\Delta = p\circ\tilde{\lambda}_\Delta$, $n(\lambda_\Delta,\tau) \geq n(\gamma,\tau)-1$ for all $\tau \in J \cap \Delta\times[0,1)$. By triangulating $D_\Delta = \Delta\times[0,1) - \text{int } C'_\Delta$ and using Lemma 5 inductively, together with pinched neighborhoods in D_Δ, we obtain a homotopy ψ_Δ from $\gamma|\Delta\times I$ to the natural extension of λ_Δ to $\Delta\times I$ which is stationary on $C'_\Delta \cup \Delta\times\{1\}$ and such that for each $\tau = \tau^i \in J \cap \Delta\times[0,1)$, $n(\psi_\Delta,\tau\times I) \geq n(\gamma,\tau) - i(i+1)/2$.

Now by using the ψ_Δ's and small pinched neighborhoods of the $\Delta\times I$'s in $K\times I$, we obtain $\gamma_4 : K^b\times I \to K^b$ agreeing with λ_Δ on each $\Delta\times I$ for $\Delta = \Delta^4 \in K$, agreeing with γ off $\cup\{N(\sigma,J) \mid \sigma \in B\}$, and with the property that for $\sigma \in B$, $n(\gamma_4,N(\sigma,J)) \geq n(\gamma,N(\sigma,J))-d(\sigma)(d(\sigma)+1)/2$. Observe that because the deformations of the pinched neighborhoods of $\Delta\times I$ in $K\times I$ preserve intersections with J^2, $\gamma_4^{-1}(b) \subset \{b\}\times I \cup K\times\{1\} \cup(\cup\{\mathring{N}(\sigma,J) \mid \sigma \in B\}) - (J^2 \cup K\times\{0\} \cup K^4\times[0,1))$.

If we now examine the situation in and near $\Delta\times I$ for Δ a 5-simplex of K, we find it slightly different than we had before for 4-simplices in that the $N(\sigma,J)$'s meet $\partial\Delta\times[0,1)$, although $\gamma_4^{-1}(b)$ does not. We may continue our process of freeing the skeleta of $K\times[0,1)$ from b with the next step being the generic inductive construction.

Again we set $T_\Delta = B \cap \Delta\times[0,1)$ and $N_\Delta = \cup\{N(\sigma,J) \cap \Delta\times[0,1) \mid \sigma \in T_\Delta\}$. This time, we let $C_\Delta = (\Delta\times[0,1) - \text{int } N_\Delta) \cup (J^2 \cap \Delta\times[0,1)) \cup (\partial\Delta\times[0,1))$. Now again by Lemma 9, C_Δ is simply connected.

We proceed exactly as in the construction of γ_4, producing a succession $\gamma = \gamma_3, \gamma_4, \gamma_5 \ldots$ of maps, each a strong deformation retraction of K^b to b, satisfying the following:

(a_γ) $\gamma_i | K^b \times I - \cup\{N(\sigma,J) | \sigma \in B\} = \gamma | K^b \times I - \cup\{N(\sigma,J) | \sigma \in B\}$,

(b_γ) $\gamma_{i+j} | K^i \times I = \gamma_i | K^i \times I$,

(c_γ) $\gamma_i^{-1}(b) \cap K^i \times [0,1) = \emptyset$, and

(d_γ) for each $\sigma \in B$, $n(\gamma_i, N(\sigma,J)) \geq n(\gamma_{i-1}, N(\sigma,J)) - d(\sigma)(d(\sigma)+1)/2$.

We let $f = \lim \gamma_i$. Now $f : K^b \times I \to K^b$, and $f^{-1}(b) = K \times \{1\} \cup \{b\} \times I$. To check that f is continuous at b, we note that for $\sigma \in B$, $f | N(\sigma,J) = \gamma_{d(\sigma)} | N(\sigma,J)$, so that by (d_γ) we have $n(f, N(\sigma,J)) \geq n(\gamma, N(\sigma,J)) - d(\sigma)^2 (d(\sigma)+1)/2 \geq \min\{n(g,\sigma), m(\sigma)\}$ (by (2_γ) and (3_γ)). Thus, the proof is complete.

THEOREM 2. If G is the n-torus $(S^1)^n$, based-free tame G-actions on Hilbert cubes are standard if and only if their orbit spaces are absolute retracts.

Proof. This is virtually identical with the proof of Theorem 1 except that as G is a $K(\mathbb{Z}^n, 1)$, $M = Q_0/\rho$ is a $K(\mathbb{Z}^n, 2)$ and we must increase the dimension of everything in the argumentation in order to get the liftings we need, e.g., instead of $k>3$ in Lemma 9, we need to use $k>5$ for the general position and replace van Kampen's Theorem by the Mayer-Vietoris Theorem, using the Hurewicz isomorphism.

V. CONCLUSION

Here is the Corollary remarked by R. Geoghegan.

COROLLARY. Let G be finite or an n-torus. A based-free tame G-action ρ on Q is standard if and only if Q/ρ has the homotopy type of a CW-complex.

Proof. By Theorems 1 and 2, ρ is standard if and only if Q/ρ is an absolute retract. By Lemma 2, $\pi_n Q/\rho = 0$ for each n. Thus, Q/ρ is an absolute retract if and only if it is an absolute neighborhood retract. Being the one-point compactification of the locally compact absolute neighborhood retract Q_0/ρ, Q/ρ is an absolute neighborhood retract if and only if it is locally contractible at the singular orbit (b), for then it will satisfy the criterion of Lefschetz [3; Theorem 8.1, p. 112]. This follows immediately if there is a deformation of Q/ρ which moves the singular orbit (and in our situation is equivalent to the existence of such a deformation). If Q/ρ is homotopy equivalent to a locally contractible space, we may produce from the local contractions of that space either a contraction of a neighborhood of the singular orbit to that orbit or a deformation of Q/ρ which moves it.

VI. RELATED QUESTIONS AND PROBLEMS

1. *Is there a non-standard, based-free, tame G-action on Q for some compact Lie group G?* (This is obviously a fundamental question. Relieving any of the conditions of Theorem E makes examples relatively easy to come by. See, for example [2] and [6; Theorem 5.5.15]. The latter points out that the compactness of the geometric situation introduces significant difficulties.)

2. *Clarify the situation resulting from the failure of "Whitehead-Type Theorem" for infinite-dimensional objects in Shape Theory, Pro-Homotopy Theory, and Proper Homotopy Theory.* (This was a primary motivation of [6]. See particularly Chapter

5, Section 5 which contains a discussion and some highly relevant examples. This is a difficult but centrally located problem.)

3. *Does Theorem 2 hold for more complicated groups?*

4. *Do Theorems 1 and 2 hold for semi-free tame G-actions on Q with a Z-set arc or cell or more complicated fixed point set?*

5. *Classify the based-free tame $\mathbb{Z}$-actions on Q.* (The standard *G*-action, for *G* locally compact, Lie, ought to be the analogue of our diagonal action on the infinite product of the cone of *G*, where we use in this case the cone of the one-point compactification reduced at "infinity". Note the implications of T. W. Tucker's wild translations of R^3 [9].)

6. *Classify principal G-actions on Q* x *[0,1).* (Note Chapman's result [4] that for *M* a *Q*-manifold, homeomorphisms and homotopy equivalences of *M* x *[0,1)* are virtually the same concept.)

7. *Study the effect of stabilization by application of the hyperspace functor(s).* (D. W. Curtis and R. M. Schori have shown that 2^X is *Q* whenever *X* is a Peano continuum [5]. All considerations of this sort are greatly facilitated by H. Torunczyk's proof [8] that a compact metric absolute neighborhood retract *A* is a *Q*-manifold if and only if for each *n* any two maps into *A* of the *n*-cell can be approximated by maps with disjoint images. See [11] for further questions.)

REFERENCES

1. R. D. Anderson, The Hilbert cube as a product of dendrons, Notices Amer. Math. Soc. 11 (1964). Abstract # 614-649.
2. I. Berstein and J. E. West, Based-free compact Lie group actions on Hilbert cubes, Proceedings of the 1976 Stanford Summer Institute on Topology (to appear).

3. K. Borsuk, Theory of Retracts. P.A.N. Mathematical Monographs, v. 44 (1966) Warsaw.
4. T. A. Chapman, Lectures on Hilbert cube manifolds. C.B.M.S. Regional Conference Series in Mathematics #28 (1976). American Mathematical Society (Providence, R.I.).
5. D. W. Curtis and R. M. Schori, Hyperspaces of Peano continua are Hilbert cubes, Fund. Math. (to appear).
6. D. A. Edwards and H. M. Hastings, Čech and Steenrod Homotopy Theories with Applications to Geometric Topology. Lecture Notes in Mathematics K. 542 (1976). Springer-Verlag.
7. A. Szankowski, On factors of the Hilbert cube, Bull. Acad. Polon. Sci. 17 (1969), 703-709.
8. H. Toruńczyk, On CE-images of the Hilbert cube and characterization of Q-manifolds, Fund. Math., (to appear).
9. T. W. Tucker, Some non-compact 3-manifold examples giving wild translations of R^3, (mimeographed).
10. J. E. West, Infinite products which are Hilbert cubes, Trans Amer. Math. Soc. 150 (1970), 1-25.
11. ______, Induced Involutions on Hilbert cube hyperspaces, Topology Proceedings 1 (1976), 281-293. (Proceedings of the 1976 Auburn University Topology Conference.)
12. R. Y.-T. Wong, Periodic actions on the Hilbert cube, Fund. Math. LXXXV (1974), 203-210.

MISCELLANEOUS PROBLEMS

HOW TO BUILD A FLEXIBLE POLYHEDRAL SURFACE

Robert Connelly[1]

Department of Mathematics
Cornell University
Ithaca, N. Y.

I. INTRODUCTION

The conjecture that every embedded polyhedral surface is rigid is very natural and probably existed even in the time of Euler (see Euler [4] and Gluck [5]). Here we say a polyhedron flexes if, when we regard its faces as made of some rigid material that does not bend but are joined with a flexible tape, then the whole polyhedron moves in three-space in such a way tnat the polyhedron changes shape (i.e., the distance between some pair of points changes). If a polyhedron (or more precisely a complex) has no flexes, then we say it is rigid. In 1813, Cauchy [2] showed that a convex polyhedron is rigid (if its "natural" faces are kept rigid). Since then there does not seem to be any drastic extensions of his result.

Here we show how to build a counterexample to the above rigidity conjecture. The proof that this is an honest example will appear in [3], and we will give here only instructions on how to build your own example out of cardboard and tape. This uses two of the flexible octahedra of Bricard [1]. The builder

[1]*Supported by a N.S.F. Grant.*

ISBN 0-12-158860-2

should be careful, however, since the flex keeps the model embedded only for a short interval, and unfortunately this model was invented for ease of explanation and not for ease of construction or demonstration. Also, if thick material is used for the faces it is not a bad idea to tape adjacent faces together with the tape on the inside of the dihedral angle.

II. THE CONSTRUCTION

Part I: For the first stage of the construction we draw a rectangle in the plane as in figure 1.

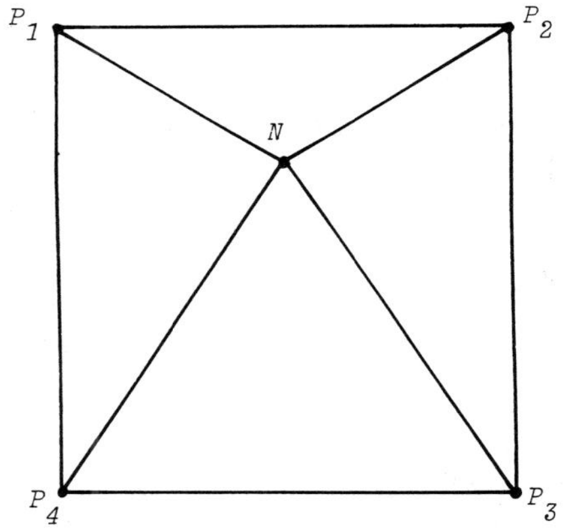

Fig. 1.

We next choose a point N on the perpendicular bisector of P_1, P_2 near P_1, P_2 in the rectangle $\langle P_1, P_2, P_3, P_4\rangle$ as in figure 1. This is used only as a guide for the next construction.

Construct out of paper or cardboard 4 tents or pyramids with an *open triangular base* over each of the 4 triangles $\langle P_1, N, P_2\rangle$; $\langle P_2, N, P_3\rangle$; $\langle P_3, N, P_4\rangle$; $\langle P_4, N, P_1\rangle$. Be sure to choose the apex of each pyramid so that no two pyramids meet except along their bases. Tape adjacent pyramids together (along the edges $\langle N, P_1\rangle$, $\langle N, P_2\rangle$, $\langle N, P_3\rangle$, $\langle N, P_4\rangle$.)

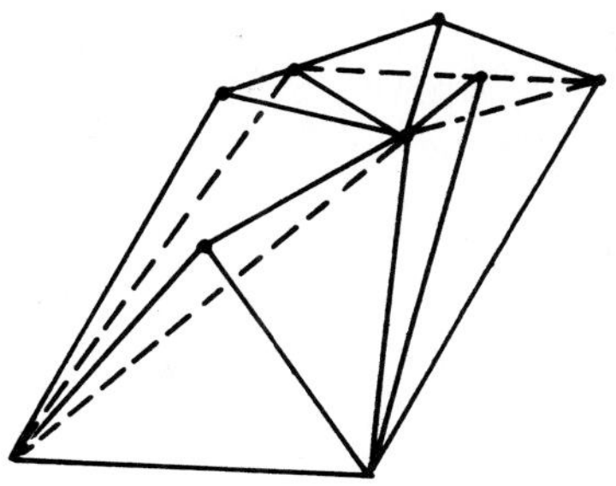

Fig. 2.

The surface now obtained has the rectangle $<P_1, P_2, P_3, P_4>$ as its boundary and looks like 4 mountains with 4 level rivers meeting at N as in figure 2.

Now make another exact copy of the same surface as in figure 2, but before it is taped together cut out two relatively small bites or notches which we describe below.

In figure 1 we find another point S near $<P_3, P_4>$ inside the rectangle $<P_1, P_2, P_3, P_4>$ so that S is the reflection of N through the center point of the rectangle. (so $<N, P_2>$ and $<S, P_4>$ have the same length, and $<N, P_1>$ and $<S, P_3>$ have the same length etc.)

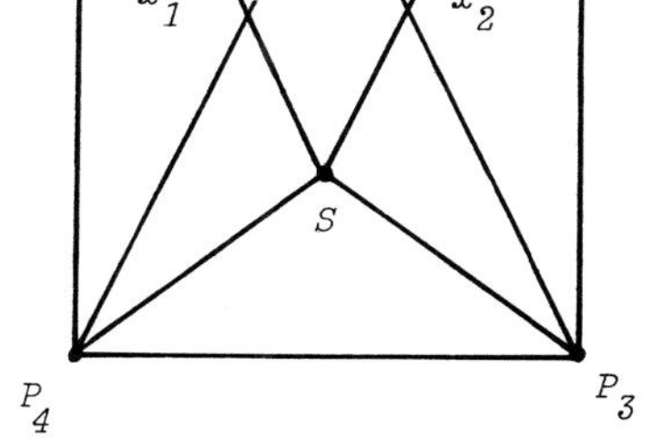

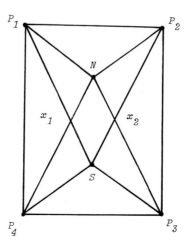

Fig. 3.

Let x_1 be the intersection of the edges $\langle N, P_4\rangle$ and $\langle S, P_1\rangle$ and x_2 the intersection of the edges $\langle N, P_3\rangle$ and $\langle S, P_2\rangle$ in figure 3. Let A_1 be apex of the pyramid or tent corresponding to the triangle $\langle P_1, N, P_4\rangle$, A_2 the apex corresponding to $\langle P_4, N, P_3\rangle$, and A_3 the apex corresponding to $\langle P_2, N, P_3\rangle$. First lay out the two triangles $\langle A_1, P_4, N\rangle$ and $\langle A_2, P_4, N\rangle$ in the plane as in figure 4.

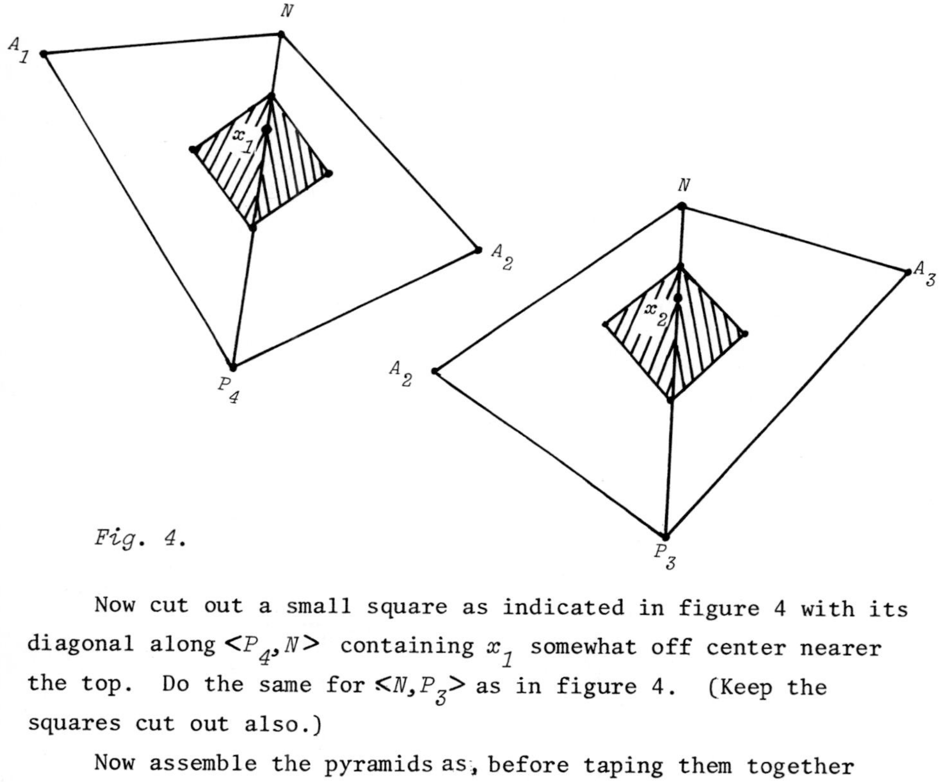

Fig. 4.

Now cut out a small square as indicated in figure 4 with its diagonal along $\langle P_4, N\rangle$ containing x_1 somewhat off center nearer the top. Do the same for $\langle N, P_3\rangle$ as in figure 4. (Keep the squares cut out also.)

Now assemble the pyramids as before taping them together along their bases to get the surface as in figure 2 except with 2 holes over part of 2 rivers, as in figure 5.

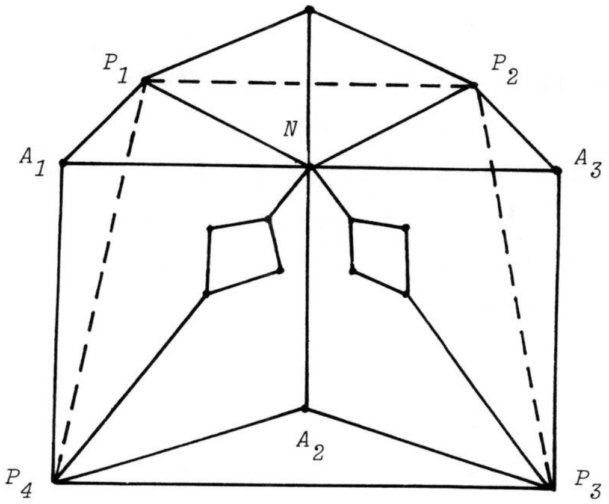

Fig. 5.

Part II: The Crinkle

We now concentrate on how to fill in the holes that we cut out of one of the surfaces in part I.

Let $\langle q_1, q_2, q_3, q_4 \rangle$ be one of the squares that were cut out in part I. Regard it as a diamond with two (right) triangles $\langle q_1, q_2, q_4 \rangle$ and $\langle q_4, q_2, q_3 \rangle$ taped together.

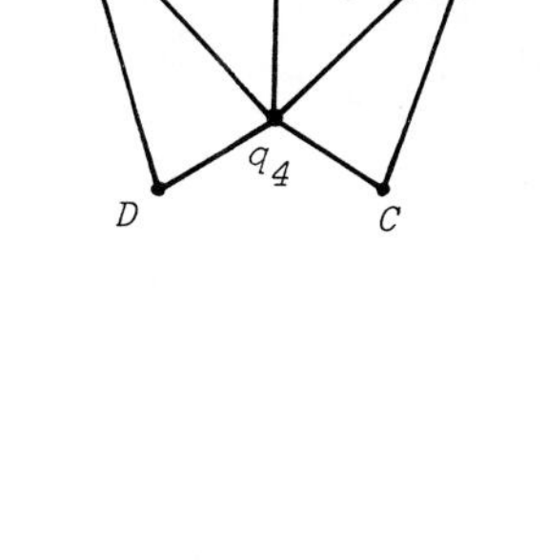

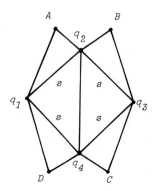

Fig. 6.

Note $\langle q_2 q_4 \rangle$ used to be part of one of the "rivers". Let the side of the square have length s. Choose some length ε about a quarter the length of s, and make 4 isosceles triangles with sides of length s, s, ε. Tape the 4 isosceles triangles to the square as in figure 6 with the ε edges touching q_2 or q_4. Now with the square held fixed push the vertices A and B down until they meet and tape $\langle A, q_2 \rangle$ and $\langle B, q_2 \rangle$ together. Similarly pull C and D up and tape $\langle C, q_4 \rangle$ and $\langle D, q_4 \rangle$ together.

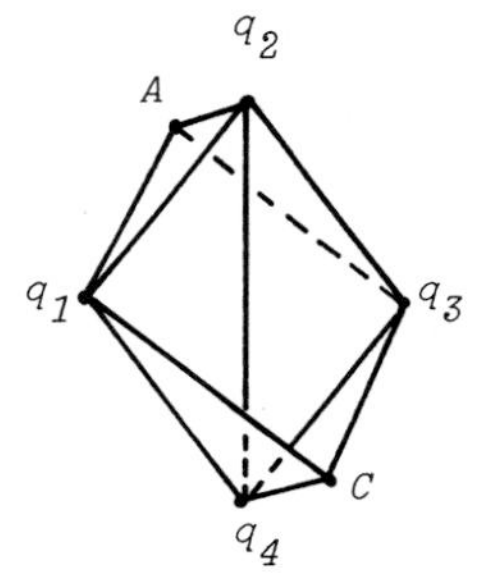

Fig. 7.

This is the crinkle. It flexes as the dihedral angle along $\langle q_2, q_4 \rangle$ changes. Note it is a surface again with boundary $\langle A, q_1, C, q_3 \rangle$. Figure 8 is another view after the crinkle is flexed a bit.

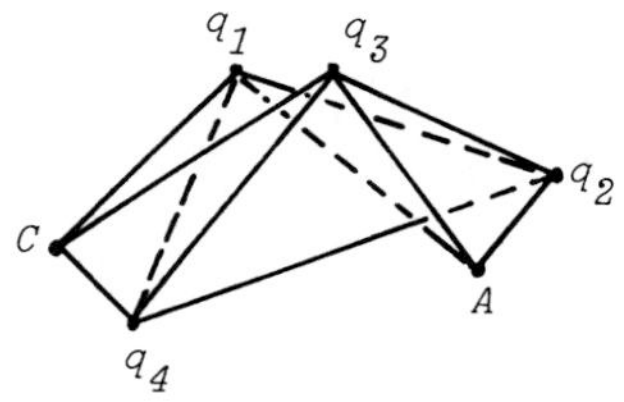

Fig. 8.

The point of this is that even as the crinkle is flexed the distance between A and C is constant and equal to the distance between q_2 and q_4 (i.e., $2s$). Thus if the crinkle is flexed to the proper angle the boundary of the crinkle will exactly fit into the boundary of a hole in the surface of figure 5.

So tape a crinkle of figure 8 with A the point nearest N into the one hole of the surface of figure 5. Make another crinkle and tape it onto the other whole in exactly the same way with the vertex nearest N.

You should now have two surfaces with the rectangle as boundary, one as in figure 2 and one surface with the crinkles in the holes as in figure 9.

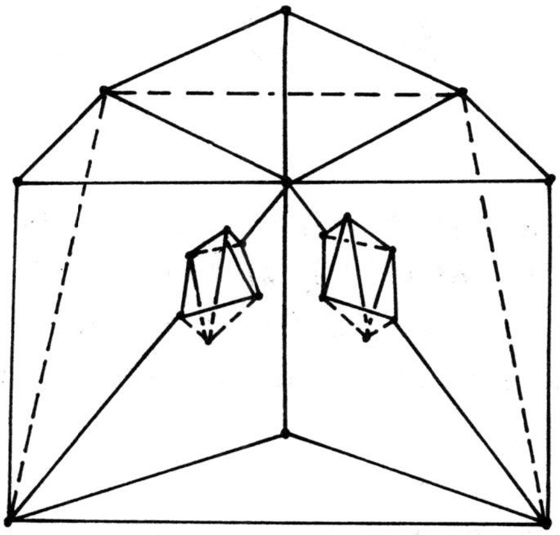

Fig. 9.

Now take the uncrinkled surface and rotate it about a line parallel to $\langle P_1, P_2 \rangle$ so that the mountains hang down below the plane of the rectangle. Make sure the smallest pyramid is in the back on the top and in the front on the bottom surface. Then tape the rectangular boundaries of the crinkled and uncrinkled surfaces together. If the crinkles are positioned properly along their edges there will be no self-intersections, and it will flex at least for a small time interval until it hits itself.

The final surface should look something like figure 10, where some of the hidden lines are left out.

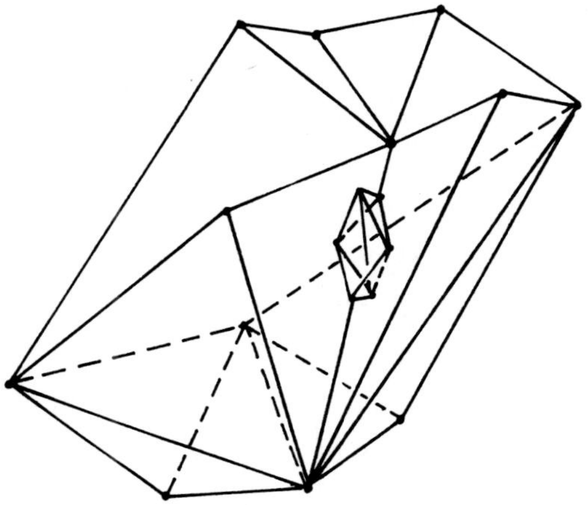

Fig. 10.

Note that if neither of the two final surfaces were crinkled then the intersection of the top with the bottom would contain the two points $<x_1, x_2>$ of figure 3. The effect of the crinkles is to move the top surface away from these points thus creating an embedding. (See fig. 8)

If one wishes, it is possible to save some labor by making two of the mountains on each side flat and making only one crinkle. This will eliminate only one half of the flex. If the mountains flattened are those over $<P_1, P_2, N>$ and $<P_1, N, P_4>$ and the left crinkle is left out, it will still flex.

REFERENCES

1. R. Bricard, "Mémoire sur la theorie de l'octaedre articlé", J. Math. Pures Appl. (5) 3 (1897), 113-148.
2. A. L. Cauchy, "Sur les polygones et polyedres, Second Memoire", J. École Polytechnique 9, Tom 19 (1813), 87-98.
3. R. Connelly, "A Counterexample to the Rigidity Conjecture for Polyhedra", to appear in the Journal of I.H.E.S. no. 67, pp. 333-338.
4. L. Euler, Opera postuma, I, Petropoli (1867) 494-496.
5. H. Gluck, "Almost all simply connected closed surfaces are rigid" Lecture Notes in Math, 438, Geometric Topology, (1975), 225-239.

A SHORT LIST OF PROBLEMS

John G. Hollingsworth

Department of Mathematics
University of Georgia
Athens, Georgia

One of the sessions at the conference was devoted to a discussion of problems that are, in the collective opinion of the conference participants, likely to be of central interest or importance for the next few years. Some of these problems are listed below, with very brief comments, and some references. No attempt was made to order the problems listed according to significance, interest, or difficulty.

I. The Poincaré Conjecture in dimensions 3 and 4

II. The Smith Conjecture

III. The Novikov Conjecture

IV. The h-Cobordism Conjecture in dimension 4.

V. Compute θ_3^H and the homomorphism $\mu : \theta_3^H \to \mathbb{Z}/2$.

VI. Can a p-adic group act effectively on a manifold?

VII. The Cell-like Mapping Problem

VIII. The Group Action Problem on Q.

IX. The Homeomorphism Group Problem.

X. Does there exist a closed, almost parallelizable 4-manifold of index 8?

ISBN 0-12-158860-2

I. The Generalized Poincare Conjecture in dimension n: If M is a compact n-dimensional manifold without boundary and if M has the homotopy type of S^n, then M is homeomorphic to S^n.

For $n \neq 3,4$ the conjecture is true. (See Smale [31], Stallings [33], Newman [26], Connell [7], Brahana [1]). If M is a smooth homotopy n-sphere, $n \geq 5$, M is homeomorphic to S^n, but for $n \geq 7$ need not be diffeomorphic to S^n. (Milnor [23], and Kervaire-Milnor [15].) If M is a pl homotopy n-sphere, $n \geq 5$, then M is pl homeomorphic to S^n.

For $n = 3$ this is the classical Poincaré Conjecture, and has perhaps received as much attention as any problem in topology.

II. The Generalized Smith Conjecture in dimension n: If $\mathbb{Z}/p$ acts effectively the n-sphere S^n with fixed point set F, and if F is a tame (n-2)-sphere, then F is unknotted.

Griffen shows [12] that the conjecture is false for $n > 4$ and for $n = 4$ provided p is even. Summers [34] used different techniques and produced more counterexamples, including the cases $n = 4$ and p odd. If $n = 3$, we have the classical Smith Conjecture, which has been verified if $p = 2$ by Waldhausen [36].

III. The Novikov Conjecture: For closed, oriented manifolds the "higher signatures" $\sigma_\pi(M)$ are homotopy invariants, where $\sigma_\pi(M)$ is defined as follows: For $f : M \to K(\pi,1)$, $\sigma_\pi(M) = f_*([M] \cap \ell(M)) \in H_*(\pi, \mathbb{Q})$. In this formula, $[M]$ is the orientation class of M and $\ell(M)$ is the Hirzebruch class of M.

If M has dimension $4k$ and π is the trivial group, then $f_*([M] \cap \ell[M])$ is just the index, or signature of M (by the Hirzebruch index theorem); hence the term "higher signatures".

In the case that π is a free abelian group, the conjecture has been established by Farrell-Hsing and by Kasparov using the splitting theorem of Farrell-Hsiang [8] Lusztig has also given a different proof. Cappell [4] has established the Novikov conjecture for a large class of groups, using his codimension one splitting theorem. It has been shown that this class does not contain all finitely presented groups.

Wall [37] constructs a map $\ell_\pi : H_*(\pi, \mathbb{Q}) \to L_*(\pi) \otimes \mathbb{Q}$, and shows that it suffices to show that ℓ_π is injective. Other formulations and generalizations of the conjecture are given in [24], [25].

IV. The h-cobordism Conjecture in dimension n: If W is a 1-connected compact n-manifold with two boundary components M_0 and M_1, and if each of the inclusions $M_i \to W$ is a homotopy equivalence, then W is isomorphic (diffeomorphic, pl homeomorphisms, or homeomorphic, depending on the structure possessed by W) to the product $M_0 \times [0,1]$.

For $n \geq 6$ Smale proved this conjecture in the smooth category, Mazur, Stallings and Zeeman in the pl category and Kirby-Siebenmann in the topological category. The generalization to s-cobordism is known to fail in dimension 4 or dimension 5, but it is not known which.

An affirmative solution in dimension 4 yields both the conjectures of *I*, but it is not clear that a counterexample would shed significant light on either.

V. Computation of $\theta_3^H \xrightarrow{\mu} \mathbb{Z}/2$. If Σ_1 and Σ_2 are closed oriented pl 3-manifolds with the homology of S^3, we define $\Sigma_1 \sim \Sigma_2$ if there is a compact 4-dimensional pl manifold W such that ∂W is the disjoint union of Σ_1 and Σ_2, and such that the inclusions $\Sigma_i \to W$ induce isomorphisms on homology. θ_3^H is the set of equivalence classes of oriented homology 3-spheres. The connected sum operation induces on θ_3^H an abelian group structure.

If Σ is a homology 3-sphere, let W be a compact parallelizalbe pl 4-manifold with boundary Σ. The index of W is a multiple of 8. Define $\mu(\Sigma) = 1/8$ (Index W mod 16). According to Rohlin [30] this is well defined. Then μ induces a homomorphism $\theta_3^H \to \mathbb{Z}/2$; we denote it μ also.

It is known that μ is not the zero homomorphism; in fact, if H is the dodecahedral space, then $\mu(H) = 1$. It is not known if there is an element M of finite order in θ_3^H with $\mu(M) = 1$. Suppose there is such an element, and let

$$\ell = \min\{n \mid \exists \text{ element } M \text{ of order } n \text{ with } \mu(M) = 1\}.$$

If $\ell = 2$, then each topological manifold M of dimension greater than 4 (or 5 if $\partial M \neq \emptyset$) admits a simplicial triangulation [10], [21].

If $\ell > 2$, there is a topological manifold that is not homeomorphical to a simplicial complex.

Let G be the kernel of μ. If M is an n-dimensional topological manifold with $n > 4$ (or $n > 5$ if $\partial M \neq \emptyset$), then there is a well defined obstruction $\eta \in H^5(M;G)$ to the existence of a simplicial triangulation of M. If $\eta = 0$, so that M has a simplicial triangulation, then triangulations are classified (up to simplicial concordance) by elements of $H^4(M;G)$.

Finally, $\eta = \beta_* k_M$ where β_* is the Bockstein associated with the exact sequence

$$0 \to G \to \theta_3^H \xrightarrow{\mu} \mathbb{Z}/2 \to 0$$

of coefficient groups and k_M is the Kirby-Siebenmann obstruction to a pl triangulation on M.

VI. *Can a p-adic group act effectively on a manifold?*

A conjecture very closely related to Hilberts Fifth Problem is the conjecture that every compact group that acts effectively on a manifold in a Lie group. If G is such a group, then either G is a Lie group or G contains an p-adic group A_p and A_p acts effectively on the manifold. Hence the conjecture has some historical interests.

If the p-adic group A_p acts effectively on the n-dimensional manifold M, then the dimension of the orbit space M/A_p is either infinite or $N + 2$. (Yang, [39]).

VII. The Cell-Like Mapping Problem: Does there exist a cell-like mapping from the n-cell onto a space of dimension greater than n?

If $f : I^n \to Y$ is such a mapping, then

1. The dimension of Y is infinite
2. For $n \geq 3$, Y is not an ANR.
3. If $F \subset Y$ and dim $F < \infty$, then dim $F \leq n$.
4. Cohomological Dimension $Y \leq n$.

VIII. The Group Actions Problem: Are all semi-free $\mathbb{Z}/2$ actions on the Hilbert cube Q equivalent?

A restatement of the problem is: If $h : Q \to Q$ is an involution with only one fixed point, is h congugate to the "standard" involution that reverses orientation of each coordinate? If $O(h)$ is the orbit space of h with the fixed point deleted, then $O(h)$ is a Q-manifold with a well defined simple homotopy type. West has shown that the conjugacy class of h is completely determined by the proper homotopy type of $O(h)$. Hence the problem is to find a suitable Whitehead theorem for infinite dimensional spaces.

IX. The Homeomorphism Group Problem: If M is a compact manifold, let $H_\partial(M)$ be the group of homeomorphisms $h : M \to M$ for which $f|\partial M : \partial M \to \partial M$ is the identity. Conjecture: $H_\partial(I)^n$ is an ℓ_2-manifold, where ℓ_2 is separable Hilbert space.

The conjecture has been established for $n = 1$ by Anderson and $n = 2$ (Luke-Mason [19]). For $n \geq 3$ the result is unknown, but equivalent (by Geoghegan and Torunczyk [11] and [35]) to showing that $H_\partial(I^n)$ is an absolute retract. Haver has observed that $H_\partial(M^n)$ is an ℓ_2-manifold provided $H_\partial(I^n)$ is.

X. Does there exist a closed, almost parallelizable 4-manifold of index 8?

If M^4 is a closed, almost parallelizable 4-manifold, then the intersection form is even, so the index of M is a multiple of 8. If M is pl, then according to Rohlin, [30], the index of M is a multiple of 16. Hence such a manifold would not admit a pl structure.

Since the transversality theorems have been so important in the topology of manifolds, perhaps the question should be stated: Is the topological transversality theorem true for codimension 4. Such a theorem implies an affirmative answer to question X.

REFERENCES

1. Brahana, H. R. Systems of Circuits on two manifolds, Ann. of Math. 23 (1922) 144-168.
2. Bredon, G. E., Compact Transformation Groups, Academic Press, New York 1972.
3. G. E. Bredon, Frank Raymond and R. F. Williams, p-adic transformations. Trans. Amer. Math. Soc. 99 (1961) 488-498.
4. Cappell, S. E., On homotopy invariance of higher signatures.
5. Cappell, S. E. and Shaneson, J. L., A note on the Smith Conjecture, Topology, 17 (1978) 105-107.
6. Chapman, T. A., Lectures on Hilbert cube manifolds, CBMS regional conference series in math. No. 28. Providence R. I. 1976.
7. Cornell, E. H., A topological h-cobordism theorem for $n \geq 5$, Ill. J. Math., 11 (1967) 300-309.
8. Farrell, F. T., and W. C. Hsiang, Manifolds with $\pi_1 = \mathbb{Z} \times_\alpha G$, Amer. J.

9. Ferry, S., The homeomorphism group of a compact Hilbert cube manifold is an ANR, Ann. of Math. 106 (1977) 101-119.
10. Galewski, D. E. and Stern, R. J., Simplicial Triangulations, to appear.
11. Geoghegan, R., On spaces of homeomorphisms, embeddings, and functions I, Topology 11 (1972) 159-177.
12. Giffen, C. H., The Generalized Smith Conjecture, Amer. J. Math. 88 (1966) 187-198.
13. Gordon, C. M., On the higher dimensional Smith Conjecture, Proc. London Math. Soc. (3) 29 (1974) 98-110.
14. Hsiang, W. C., A splitting theorem and the Kunneth formula in algebraic K-theory, in "Algebraic K-theory and its geometric applications", Lecture Notes no. 108, Springer-Verlag (1969) 72-77.
15. Kervaire, M. A. and Milnor, J., Groups of homotopy spheres I., Ann. Math. 77 (1963) 504-537.
16. Kirby, R. C. and Siebenman, L. C., Foundational Essays on Topological Manifolds, Smoothings, and Triangulations. Annals of Math Studies No. 88, Princeton University Press (1977).
17. Kozlowski, G., Images of ANR's, to appear in Trans AMS.
18. Lacher, R. C., Cell-like mappings and their generatlizations, Bull. AMS 83 (1977) 495-552.
19. Luke, R., and W. K. Mason, The space of homeomorphisms of a compact two-manifold is an absolute neighbothood retract. Trans. AMS 164 (1972) 275-285.
20. Mason, W. K., The space of all homeomorphism of a two-cell which fix the cell's boundary is a absolute retract. Trans. AMS 161 (1971) 185-205.
21. Matsumoto, Takao, Homology Manfiolds, mimeographed notes.
22. Matsumoto, Takao, Variétiés simpliciales d'homologie et variétiés topologiques metrisables, Thesis, University d-Paris-Scud 91405 Orsay (1976)
23. Milnor, J., On manifolds homeomorphic to the 7-sphere. Ann. Math. 64 (1956) 399-405.
24. Mishchenko, A. S., Hermitian K-theory. The theory of characteristic classes and methods of functional analysis. Russian Math. Surveys 31 (1976) 71-138.
25. Mishchenko, A. S., Homotopy invariants of multiply connected manifolds III, Higher Signatures. IZV. Akad. Nauk SSSR Ser. Mat. 3 (1971) 1316-1355 = Math. USSR - IZV. 5 (1971) 668-679.
26. Newman, M.H.A. The engulfing theorem for topological manifolds, Ann. of Math. 84 (1966) 555-571.
27. Novikov, S. P., On manifolds with free abelian fundamental group and their applications. IZV Akad Nauk SSSR Ser. Mat. 30 (1966) 207-246 (Russian)
28. Novikov, S. P., Pontrjagin classes, the fundamental group and some problems in stable algebra, in "Essays on topology and related topics, Memoirs dédiés a G. de Raham" Springer-Verlag (1967) 147-155.

29. Pao, P. S., Non-linear circle actions on the 4-sphere. to appear.
30. Rohlin, V. A., New results in the theory of 4-dimensional manifolds, (Russian) Dokl. Akad. Nauk. SSSR 84 (1952) 221-224.
31. Smale, S., Generalized Poincaré conjecture in dimensions greater than four, Ann. of Math. 74 (1961) 391-406.
32. Smith, P. A., Transformation groups of finite period II. Annals. of Math. (2) 40 (1939) 690-711.
33. Stallings, J., Polyhedral Homotopy Spheres, Bull. AMS 66 (1960) 485-488.
34. Summers, D. W., Smooth $\mathbb{Z}_p$-actions on spheres which leave knots pointwise fixed, Trans. AMS 205 (1975) 193-203.
35. Torunczyk, H., Absolute retracts as factors of normed linear spaces, Fund. Math. (1974) 51-67.
36. Waldhausen, F., Über involutionen der 3-sphere, Topology 8 (1969) 81-91.
37. Wall, C.T.C., Surgery or compact manifolds, Academic Press, 1970.
38. West, J. E., Mapping Hilbert cube manifolds to ANR's, Ann. of Math. 106 (1977) 1-18.
39. Yang, C. T., p-adic transformation groups, Mich. Math. J. 7 (1960) 201-218.
40. Yang, C. T., Hilberts Fifth Problem and related problems on transformation groups, Mathematical Developments arising from Hilbert Problems, Symposium in Pure Mathematics (1976) 142-146.
41. Zeeman, C., The generalized Poincaré conjecture, Bull. AMS 67 (1971) 270.

SOME BORSUK-ULAM TYPE THEOREMS

Jean E. Roberts

Department of Mathematics
Oakland University
Rochester, Michigan

This note is concerned with generalizations of the classical Borsuk-Ulam theorem, in particular those in which the antipodal map is replaced by a map $f : S^n \to S^n$ of prime period p. A brief survey of known generalizations is given. We obtain under a less restrictive hypothesis a result of Munkholm's and thereby, for the case p = 2, give a positive answer to a question of Conner and Floyd's.

The classical Borsuk-Ulam theorem states:

THEOREM (Borsuk-Ulam). For any map $f : S^n \to \mathbb{R}^n$ there is an $x \in S^n$ with $f(x) = f(-x)$.

This was proved in 1933 by Borsuk [1]. Then Bourgin took up the question of the dimension of the set of points x with $f(x) = f(-x)$, and in 1955 Bourgin [2], and independently Yang [13] proved:

THEOREM (Bourgin-Yang). For any map $f : S^n \to \mathbb{R}^k$ the covering dimension of $\{x \in S^n | f(x) = f(-x)\}$ is at least $n - k$.

ISBN 0-12-158860-2

In 1960 Conner and Floyd [5] and [6] proved the following generalization.

THEOREM (Conner-Floyd). For any fixed-point free, differentiable involution $t : S^n \to S^n$ *and map* $f : S^n \to M$ *into a differentiable k-manifold M satisfying*

$$f_n = 0 : H_n(S^n, Z_2) \to H_n(M, Z_2) \quad ,$$

the covering dimension of $\{x \in S^n | f(x) = f(-x)\}$ *is at least* $n - k$.

At that time Conner and Floyd [6] asked if any of the following generalizations could be obtained:

1. Can all differentiability hypotheses be eliminated?
2. Can S^n be replaced by a closed n-manifold which is a mod 2 homology n-sphere?
3. Can M be replaced by a space that is not a manifold?

In 1960 Munkholm [8] gave a positive answer to the first question for M a compact manifold.

The question also arises of what generalizations can be obtained if the fixed-point free involution is replaced by a fixed-point free map t of prime period p. Here there are two possibilities. In one the set being considered is $\{x | f(x) = f(t^i(x))$ for some i, $i = 1,\ldots,p-1\}$. In the other the set of interest is $\{x | f(x) = f(t^i(x))$ for all i, $i = 1,\ldots,p-1\}$.

The first possibility has been taken up by Connett and Cohen in [3] and by Cohen and Lusk in [4]. The second possibility was considered by Munkholm [9], and in 1969 he published:

THEOREM (Munkholm). Let S^n *be a closed n-manifold which is a mod p homology n-sphere, and let* $t : S^n \to S^n$ *be a fixed-point free map of prime period p. Let M be a compact,* Z_p*-orientable k-manifold and* $f : S^n \to M$ *a map satisfying:*

1) If $p = 2$, $f_n = 0 : H_n(S^n, Z_2) \to H_n(M, Z_2)$

2) There is a point Y_0 *in* M *and a map* f_0 *homotopic to* f *so that* f_0 *maps at least* $p - 1$ *points of each orbit to* Y_0.

Then the covering dimension of $\{x \in S^n | f(x) = f(t(x)) = \ldots = f(t^{p-1}(x))\}$ *is at least* $n - k(p - 1)$.

Notice that condition 2) is satisfied if M is contractible, in particular for $M = \mathbb{R}^k$, or if $S^n = S^n$. Hence Munkholm has a positive answer to the second question of Conner and Floyd for the case $M = \mathbb{R}^k$. Still, condition 2) seems an unnatural restriction, and Bourgin suggested that the result might be obtained without assuming condition 2) by utilizing a certain cohomology class defined by Haefliger. Using this idea we were able to prove the following theorem and thereby to give a positive answer to Conner and Floyd's second question for M any compact manifold.

THEOREM. Let S^n *be a closed n-manifold which is a mod-p homology n-sphere, and let* $t : S^n \to S^n$ *be a fixed-point free map of prime period p. Let M be a compact,* Z_p *- orientable k-manifold and* $f : S^n \to M$ *a map satisfying,*

if $p = 2$, $f_n = 0 : H_n(S^n, Z_2) \to H_n(M, Z_2)$.

Then the covering dimension of $\{x \in S^n | f(x) = f(t(x)) = \ldots = f(t^{p-1}(x))\}$ *is at least* $n - k(p - 1)$.

Proof. Let A denote $\{x \in S^n | f(x) = f(t(x)) = \ldots = f(t^{p-1}(x))\}$, and let π denote Z_p. Then t generates an action of π on S^n. Note that cov dim A = cov dim A/π and hence that it suffices to show that $\bar{H}_c^{n-k(p-1)}(A/\pi) \neq 0$. The proof consists largely of showing that a certain cohomology class $u_\pi \in H^{k(p-1)}(E_\pi \times_\pi M^p)$, defined by Haefliger in [7], does not vanish under $f_\pi{}^* : H^*(E_\pi \times_\pi M^p) \to H^*(S^n/\pi)$ where $f_\pi : S^n/\pi \to E_\pi \times_\pi M^p$ is a map induced by f. All coefficients are Z_p.

Let π act on M^p by cyclic permutation of the factors, and consider M to be a subspace of M^p via the diagonal embedding. The class u_π is defined to be the dual under Poincare duality of the orientation class of $E_\pi \times_\pi M$. Haefliger has shown in [7] that the image of u_π under the map induced by the inclusion $j : E_\pi \times_\pi (M^p - M) \to E_\pi \times_\pi M^p$ is not 0.

The map f induces $f'_\pi : S^n/\pi \to S^n \dot{\times}_\pi M^p$ defined by $[[\, x \,]]_\pi \mapsto [[\, x, (f(x), f(t(x)),\dots,f(t^{p-1}(x)))]]_\pi$. Let h be a representation of S^n in the universal space E_π, and define $f_\pi : S^n/\pi \to E_\pi \times_\pi M^p$ to be the composite $(h \times_\pi 1) \circ f'_\pi$. Let $\tilde{f}_\pi$ denote the restriction of f_π to $(S^n - A)/\pi$. Since $\tilde{f}_\pi$ factors through $E_\pi \times_\pi (M^p - M)$, we have $\tilde{f}_\pi{}^*(u_\pi) = 0$.

To see that $f_\pi{}^*(u_\pi) \neq 0$, we use a result of Haefliger [7] giving an explicit formulation of the class u_π in terms of the characteristic classes of Wu. Results of Steenrod [11] yield that $H^*(E_\pi \times_\pi M^p)$ may be canonically identified as an $H^*(\pi)$-algebra with $H^*(\pi) \otimes D + N$, where D denotes the diagonal elements of $H^*(M^p)$ and N denotes the image of the norm homomorphism.

For each prime p and positive integer j, Wu has defined in [12] a class $U^j_{(p)}$ belonging to $H^{2j(p-1)}(M, Z_p)$ if $p > 2$ or to $H^j(M)$ if $p = 2$, such that if $\omega \in H_n(M)$ is the orientation class and $\alpha \in H^*(M)$,

$$< S_q^{\,j}(\alpha), \omega > \; = \; < U^j_{(2)} \cup \alpha, \omega > \quad \text{if} \;\; p = 2$$

$$< P^j(\alpha), \omega > \; = \; < U^j_{(p)} \cup \alpha, \omega > \quad \text{if} \;\; p > 2 \quad .$$

Then Haeflinger has shown that

$$u_p = \begin{cases} \displaystyle\sum_{j=0}^{[[\frac{k}{2}]]} \mu^{k-2j} (U^j_{(2)})^2 + \delta_2 , & p = 2 \\[2ex] \displaystyle\lambda \sum_{j=0}^{[[\frac{k}{2p}]]} (-1)^j \, \mu^{\frac{p-1}{2}(k - 2jp)} (U^j_{(p)})^p + \delta_p, & p > 2, \end{cases}$$

where μ is the generator of the polynomial algebra in $H^*(\pi)$, δ_p is the element in $H^*(M^p)$ dual to the diagonal, and λ is $(-1)^{m/2}$ or $(-1)^{(m-1)/2}\frac{p-1}{2}!$, depending on whether p is even or is odd.

Recall that there is a canonical identification of $H^*(S^n/\pi)$ with $H^*(E_\pi \times_\pi S^n)$. Then it is not difficult to see that $f^*: H^*(E_\pi \times_\pi M^p) \to H^*(S^n/\pi)$ may be identified with the composite map $H^*(E_\pi \times_\pi M^p) \xrightarrow{(1 \times_\pi f^p)^*} H^*(E_\pi \times_\pi (S^n)^p) \xrightarrow{(1 \times_\pi T)^*} H^*(E_\pi \times_\pi S^n)$, where $T : S^n \to (S^n)^p$ is defined by $x \to (x, t(x), \ldots, t^{p-1}(x))$.

By dimension arguments it can be shown that

$$(1 \times_\pi f^P)^*(u_\pi) = \begin{cases} \mu^k (f^*(U^0_{(2)}))^2, & p = 2 \\ \lambda\mu^k (f^*(U^0_{(p)}))^p, & p > 2 \end{cases}$$

Then a spectral sequence argument shows that this does not map to 0 under $(1 \times_\pi T)^*$

Thus we have $f_\pi^*(u_\pi) \neq 0$, but in the cohomology exact sequence for the pair $(S^n/\pi, (S^n - A)/\pi)$, $f^*(u_\pi)$ maps to $f_\pi^*(u_\pi) = 0$. Hence there is a nonzero antecedent in $H^{k(p-1)}(S^n/\pi, (S^n - A)/\pi)$. So, $H_{k(p-1)}(S^n/\pi, (S^n - A)/\pi) \neq 0$, and by Alexander duality $\overline{H}_c^{m-k(p-1)}(A/\pi) \neq 0$. The theorem follows.

REFERENCES

1. K. Borsuk, "Drei satze uber die n-dimensionale Euclidische sphare", Fund. Math. 20 (1933), 170-190.
2. D. G. Bourgin, "On some separation and mapping theorems", Comment. Math. Helv. 29 (1955), 199-214.
3. F. Cohen and J. E. Connett, "A coincidence theorem related to the Borsuk-Ulam theorem", Proc. Amer. Math. Soc. 44 (1974), 218-220.
4. F. Cohen and E. L. Lusk, "Coincidence point results for spheres with free Z_p-actions", Proc. Amer. Math. Soc. 49 (1975), 245-252.

5. P. E. Conner and E. F. Floyd, Differentiable Periodic Maps, Springer-Verlag, Berlin, 1974.
6. ______________, "Fixed point free involutions and equivariant maps", Bull. Amer. Math. Soc. 66 (1960), 416-441.
7. A. Haefliger, "Points multiples d'une application et produit cyclique reduit", Am. J. of Math. 83 (1961), 57-70.
8. H. J. Munkholm, "A Borsuk-Ulam theorem for maps from a sphere to a compact topological manifold", Ill. J. Math. 13 (1969), 116-124.
9. ______________, "Borsuk-Ulam type theorems for proper Z_p-actions on (mod p homology) n-spheres", Math. Scand. 24 (1969), 167-185.
10. J. E. Roberts, "A Stronger Borsuk-Ulam type theorem for proper Z_p-actions on mod p homology n-spheres", Proc. Amer. Math. Soc. (to appear)
11. N. E. Steenrod and D.B.A. Epstein, Cohomology Operations, Annals of Mathematics Studies No. 50, Princeton University Press, 1962.
12. Wen-Tsun Wu, "Classes caracteristiques et i-carres d'une variete", Comptes Rendus, Paris, 230 (1950), 508-511.
13. C. T. Yang, "Continuous functions from spheres to Euclidean spaces", Annals of Math. 62 (1955), 284-292.